# Experimental Techniques of GLASS SCIENCE

*Edited by*


Catherine J. Simmons
Department of Materials
Science and Engineering
University of Florida
Gainesville, Florida

*and*

Osama H. El-Bayoumi
U.S. Air Force
Rome Laboratory
Hanscom AFB, Massachusetts



The American Ceramic Society
Westerville, Ohio

Library of Congress Cataloging-in-Publication Data

Experiemental techniques of glass science / edited by Catherine J. Simmons and Osama H. El-Bayoumi.
p. cm.
Includes index.
ISBN 0-944904-58-0
1. Glass--Congresses. I. Simmons, Catherine J. II. El-Bayoumi, Osama H.
TP848.E97 1993
620.1'44--dc20 93-40
CIP



Printed in the United States of America.

2 3 4–97 96 95

ISBN 0-944904-58-0

# PREFACE

There has been a growing interest in applications of glass for advanced systems, ranging from large optical imaging systems and lasers to lightweight, high-strength composite materials, that has focused attention on glass research and development. This has emphasized the need to utilize the powerful analytical techniques currently available for materials characterization in order to gain a better understanding of the structure and properties of glass. It is also clear that there are many scientists and engineers in industry and in academia who require a coherent and comprehensive collection of information on how to perform experiments to characterize glass from all aspects–optical, electrical, mechanical, theoretical, etc.

The monographs in this volume were largely inspired by five tutorial sessions organized for the Glass and Optical Materials Division at the 89th and 90th Annual Meetings of the American Ceramic Society. Those tutorials gave in-depth coverage to a broad range of analytical techniques deemed useful in determining the structure of amorphous materials. The sessions also included tutorials on glass properties as they relate to structure.

Several excellent books exist that treat various aspects of glass science in detail, of which the following are a few examples: *Properties of Glass* by Morey (Reinhold, New York, 1954), *Physics of Amorphous Materials* by Elliott (Longman Scientific, 1984), *Properties and Applications of Glass* by Rawson (Elsevier Scientific, Amsterdam, 1980), *Glass Structure* by Spectroscopy by Wong and Angell (Dekker, New York, 1976; out of print), and *Glasses and the Vitreous State* by Zarzychi (Cambridge University Press, Cambridge, 1991). However, many advances in glass science and technology have occurred in recent years, and new developments in several important subject areas have not been covered in the depth and detail they warrant. As a result, 14 topics, most from the tutorial sessions, were selected to be reviewed in detail for this book. The authors of these chapters are currently recognized experts in their respective fields and the material contained here is up-to-date at the time of publication.

This volume should provide the reader with an introduction to selected subjects relating to glass science, as well as with detailed information on methods for conducting experiments to characterize glass properties. Examples of methods for analysis and interpretation of data are included. The authors also address inherent limitations and artifacts of the analytical techniques.

These chapters are written as a reference and a manual to facilitate orientation of newcomers, including advanced undergraduate and graduate students, to the study of glass science and structure. It is hoped that academics and scientists involved in teaching, research, manufacturing, and use of glass in industry will find this volume informative and useful to their problem solving.

Catherine J. Simmons
Osama H. El-Bayoumi

# Table of CONTENTS

# CHAPTER 1

## OPTICAL PROPERTIES AND CHARACTERIZATION

Dr. James W. Fleming
AT&T Bell Laboratories
600 Mountain Ave.
Murray Hill, NJ 07974

### INTRODUCTION

Characterization or measurement of the optical properties of glass such as transparency, refractive index, and reflectivity is the most important analysis generally applied to glasses in research and industry. Glasses have unique optical properties including isotropy, transparency, continuously variable dispersive properties as a function of composition, and ease of fabrication into a wide range of shapes. As a result, glass is the material of choice for a wide range of applications.[1–5] In this chapter, methods of determination of glass optical properties will be discussed.

When electromagnetic radiation of any wavelength impinges on a glass, there is an interaction between the material structure and the radiation that varies in nature depending on the energy level of the radiation, the characteristic energy levels associated with the material, and the presence of defects or inclusions in the structure.

A certain portion of the light is reflected from the surface. If the glass being examined is a homogeneous material, the amount of light reflected increases with increasing interaction between the glass structure and the radiation. High reflectivity at normal incidence generally means that the energy level of the radiation is close to a characteristic energy ($E=h\nu$) in the glass. At energies significantly removed from the characteristic energy levels of the material, the radiation is mostly transmitted.

Transmission can also be modified by structural defects in the glass as will be addressed in more detail below. This is not the same type of transmission as one would observe in free space. In a glass or other material, radiation continuously interacts with the structure and the resultant transmitted energy is a superposition of the impinging electromagnetic energy and electromagnetic components of the scattered light generated through the interaction of structure and radiation. The result is that the velocity of propagation in the optical region of the spectrum is slowed which gives rise to the material characteristic of refractive index.

Additionally, the scattering is not completely efficient and the propagating light loses some of its transmitted intensity. Defects and inclusions can enhance the scattering loss and also attenuate the propagating light through its conversion to heat. This scattering and absorption is typically wavelength dependent.

While the physical phenomena described above pertains to all electromagnetic energy, in this chapter discussion will be limited to characterization methods principally employed in the optical portion of the spectrum. That is the region in wavelength from about 759 nm to 393 nm as defined by the detection capability of the human eye. In terms of energy this is about 1.6 to 3.1 eV. The intrinsic optical characteristics in this energy range are a result of interactions with fundamental structural oscillators in the material that exhibit intense electromagnetic activity in the infrared and ultraviolet.[2,3,6–9] This range is where the characteristic structural energies of most glasses are located - at least those glasses that exhibit transparency in the optical range. Other glasses, such as many chalcogenides, don't exhibit optical transparency because the fundamental oscillators resulting from electronic polarizability that are on the high energy side of the visible range are very close in energy to that of visible light. In figure 1, the approximate ranges for transmission of various glasses are indicated.[10]

The optical characteristics of transmission, absorption, and reflection fundamentally related to the interaction between the atomic structure of glass and impinging radiation in the optical range will be the basis for discussion of the various methods used to measure the optical properties of glass such as refractive index dispersion, attenuation or transmission, scattering, and reflection. The objective of this chapter is to bring the reader in touch with some of the most valuable tools used for optical characterization and aid in the choice of measurement methods.

## REFRACTIVE INDEX DISPERSION MEASUREMENT

The characterization of the refractive index of glass is fundamental to the design and fabrication of optical systems in which it is used. Design of optical devices depends on the proper choice of glass compositions exhibiting

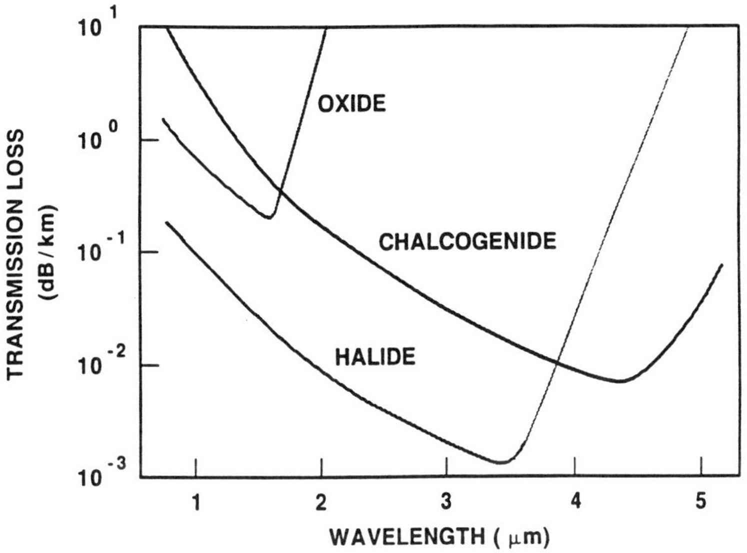


Figure 1. Regions of transparency for three general glass types.

the appropriate optical behavior. For example, in the case of lens design for visible wavelength systems, knowledge of the refractive index dispersion, $dn/d\lambda$, for each lens glass composition is essential. The standard quantities employed are the partial dispersions of the materials. These are characteristic quantities that indicate the variability in refraction in different wavelength ranges. Such calculations require precise measurement of index at several wavelengths in the visible range. Optical communications and other infrared applications for glass require an extension of the index determination into the infrared - currently out to about 1.6 μm. Certain properties, such as chromatic dispersion, important to the optimum design of achromatic lens systems, have taken on new importance in fiber optics. In optical devices using lenses for image transmission achromatic lens designs improve image quality by dispersing all light wavelengths in close to the same way. In fiber optic systems it is not image but bandwidth that is strongly dependent on this chromatic dispersion.[11–14] To take advantage of the high potential information carrying capacity offered by fiber, it is important to very accurately measure refractive index dispersion of glasses comprising lightguides.

## Measurement Methods

### Single Surface Methods

In the development, fabrication, or quality control of optical glasses, perhaps the most important and useful response measured is refractive index at the wavelength emitted by sodium vapor, 589.3 nm. This measurement can aid in rapid determination of compositional variations and other physical changes in the specimen.

As in most measurements of optical properties, some sample surface preparation is required. In the case of optical measurements, this preparation is generally the most time consuming step. The quality of optical polish will largely determine the precision of the measurement. Typically the maximum instrument precision can only be achieved by polishing the sample to better than $\lambda/4$ flatness. If bulk glass samples are available, measurement of refractive index at a specific wavelength is accomplished most easily using a method that requires only one polished surface of the sample to be analyzed. The following are such measurement techniques.

#### *Abbe' Refractometer*

The Abbe' refractometer consists of a calibrated hemispherical prism, as shown schematically in figure 2, which is mounted in a specific location in relation to an observing telescope. The measurement is based on total internal reflection from the sample surface. This phenomenon will only occur appropriately for measurement if the prism has a higher refractive index than the sample to be measured. The telescope can be rotated about the prism and along a scale that is used to access the degree of movement. This movement is confined to a plane perpendicular to the sample surface to be measured. When light from a monochromatic source of a specific visible wavelength for which the instrument has been calibrated (typically $Na_D$ - 0.5893 μm) is incident on the prism, such that its flat surface is illuminated from below or at grazing incidence from above, the phenomenon of total internal reflection causes the illumination to be visible in the observing telescope.

Total internal reflection occurs when the incident light approaches the interface at an angle greater than or equal to the critical angle, $\theta_c$. Now, if a flat piece of glass is placed on this surface such that it interacts with the illumination, then it will alter the reflection in a precise way, depending upon the refractive index of the sample. The interaction is assured by placing a thin layer of high index liquid at the sample-prism interface. The optical response seen in the observing telescope at a specific location with respect to the prism is an interface between a light and dark field. The position of this

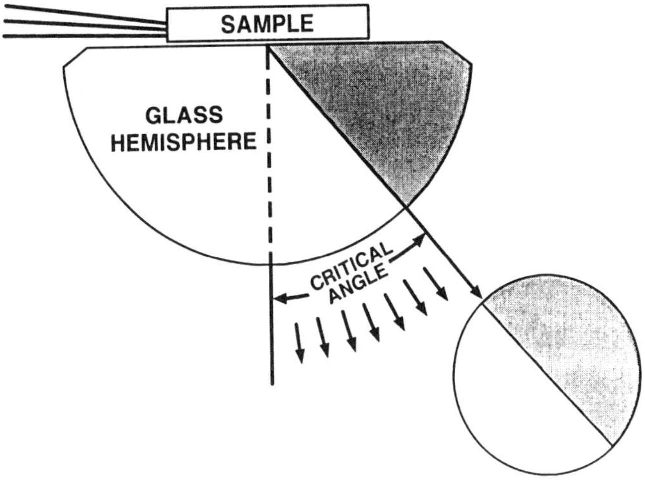


Figure 2. Principle elements of an Abbe refractometer.

interface is a measure of the critical angle of the specimen. From the relation

$$n' = n \sin \theta_{c,} \qquad (1)$$

where n is the refractive index of the prism, it is possible to determine the refractive index of the specimen, n′, for the specific wavelength. Commercially available refractometers are calibrated for directly reading the refractive index rather than the critical angle.[15,16]

This boundary between two fields of dissimilar illumination is sharp for monochromatic light but diffuse for white light. A sharp interface is necessary for high precision measurement. Measurement accuracy is also dependent on the degree of homogeneity of the prism, the precision of prism-observing telescope placement, the homogeneity of the sample, and the quality of the contacting surfaces. Typically, accuracy is about $\pm 1 \times 10^{-3}$. The simplicity of the technique and strong dependence of refractive index on composition make this one of the most attractive methods of sample evaluation. While the apparatus could be adapted to dispersion measurement, it is not commonly done. Dispersion is generally determined using the methods described below.

The Abbe' refractometer is useful for characterizing refractive index at a specific wavelength for variable temperatures. Commercially available equipment will measure the index of materials in the range of 1.3 to 1.7. Special prisms are available for some apparatus to extend the range. Because the instrument is dependent on a significant positive difference between the index of the prism and sample, the index of the prism material usually sets the limits for index determination.

*Refractive Index Measurement Using Brewster's Angle*

Light is partially polarized when reflected off a polished glass surface as shown schematically in figure 3.

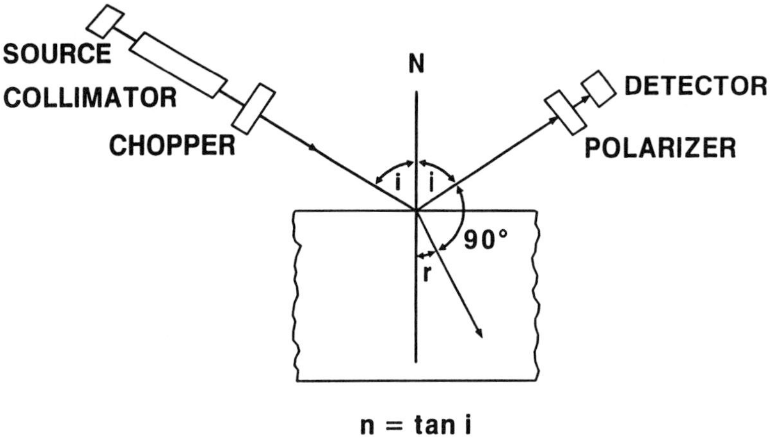


Figure 3. Representation of an apparatus for measuring refractive index using Brewster's angle.

The degree of polarization is dependent upon the quality of the reflecting surface, the refractive index of the reflecting material, and the angle of incidence of the light. Light incident on a glass surface at an angle i with respect to a normal to the glass surface is partially reflected at an angle i with respect to the normal and partially refracted at an angle r with respect to the normal. According to Brewster's law, the degree of polarization of the reflected light is maximized when the quantity (i+r) = 90. Using this phenomenon, it is possible to measure refractive index of the reflecting surface. The angle at which the polarization is most efficient can be determined using a collimated light source, a detector on a goniometer, and a polarizer in the path of the reflected beam, as shown in figure 3. The polarizer is located perpendicular to the plane of polarization of the reflecting surface. In this way the detector is made specifically sensitive to the effectiveness of polarization via reflection. From Snells law and the above relation between i and r, it is clear that

$$n = \sin i/\sin(90-i) = \tan i. \tag{2}$$

For this measurement it is important that the measured surfaces be optically flat and that the detector be very stable and sensitive. The minimum in reflectivity detected is very broad which limits the accuracy with which the index can be determined. However, this method is simpler to implement than prism specimen methods such as those discussed below and provides for measurement of index at any wavelength.

### Prism Methods

If one needs to measure refractive index accurately over a broad wavelength range, one can use a measurement apparatus that is essentially a spectrometer in which the glass specimen (a prism) is placed where the calibrated analysis prism or grating would typically be. A beam of light consisting of one or more well defined monochromatic wavelengths incident on a face of a prism sample is refracted in a precise, well understood way. From information about the prism geometry and measurement of the refracted angle for any given wavelength, the dispersion of the prism material can be determined.

Sample preparation becomes very important for proper implementation and measurement accuracy when using this method. Prism faces should be finished to better than $\lambda/10$ flatness.

Prism techniques use some easily identifiable characteristic optical phenomena such as angle of minimum deviation of the incident radiation, its autocollimation through reflection off the back face of the prism or a precise prism geometry to calculate the index at a given wavelength. These three methods are discussed below.

#### *Minimum Deviation*

Of the prism methods, the minimum deviation technique has the potential for greatest sample flexibility in terms of dealing with sample geometry composition, imperfections, atmosphere, temperature, etc. It also has the potential for the highest measurement accuracy. However, it is probably the most time consuming of the dispersion methods. The apparatus consists of a light source, a light detector, reflecting optics, a rotatable prism support table, and a precision angle measurement device. Assembly of the equipment is shown schematically in figure 4.[17] Light may originate from any spectrally narrow, well calibrated source such as a monochrometer, spectral emission lamp, selective interference filter, or absorption source. Selected emission sources and the associated applicable wavelengths emitted are listed in Table 1.

Table 1. List of potential wavelengths and sources that might be used in index measurement.

| | | | | | |
|---|---|---|---|---|---|
| Hg | 0.313169 | Cd | 0.643847 | Hg | 1.6932 |
| Hg | 0.334148 | Cs | 0.794411 | Hg | 1.8130 |
| Hg | 0.365015 | Cs | 0.852110 | Hg | 1.9700 |
| Hg | 0.404656 | Cs | 0.876138 | Hg | 2.3253 |
| Hg | 0.435835 | Cs | 0.894350 | $H_2O$ | 2.6767 |
| Cd | 0.479991 | Cs | 1.0074 | $H_2O$ | 2.7397 |
| Cd | 0.508582 | Hg | 1.01398 | PS | 3.30350 |
| Hg | 0.546074 | He | 1.08297 | PS | 3.41997 |
| Hg | 0.576959 | Hg | 1.12866 | PS | 3.50791 |
| Hg | 0.578012 | Cs | 1.3589 | $CO_2$ | 4.23227 |
| Hg | 0.579065 | Hg | 1.3622 | $CO_2$ | 4.28743 |
| He | 0.587561 | Cs | 1.4695 | PS | 5.14403 |
| Na | 0.589262 | Hg | 1.52952 | PS | 5.3447 |

Minimum deviation occurs at light propagation angle, D, at which a parallel beam of light refracted at the front face of a prism is normal to the prism angle bisector. Such propagation occurs when the prism is adjusted with respect to the source and detector such that the angle between the incident light and the refracted light is at a minimum. The minimum angle is found by moving the detector in the direction of minimum deviation while simultaneously adjusting the prism. If the angle of minimum deviation is measured for a specific wavelength along with prism angle, A, the refractive index can be determined from the relation

$$n = \sin \tfrac{1}{2}(D+A)/\sin\tfrac{1}{2}A. \quad (3)$$

To ensure accurate measurement of these angles, the prism must be precisely oriented with respect to the axis of rotation of the angle measurement apparatus. For this purpose an autocollimator is attached to the refractometer such that the parallel beam of light from the autocollimator is centered on

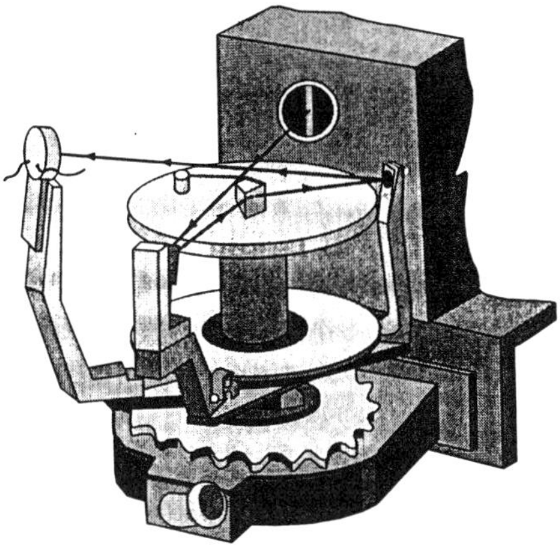

Figure 4. Schematic of a minimum deviation refractometer.

and normal to the axis of rotation of the angle measurement apparatus. By reflecting this light off each prism face such that it returns from both faces to the same point in the autocollimator, one can assure that the faces are mutually perpendicular to the autocollimator beam and so mutually parallel to the axis of the goniometer. The objective is to measure the angle of minimum deviation for a number of source wavelengths in the range of interest. Since it is difficult to determine the path of the unrefracted light, the angle of minimum deviation is determined by measuring the angle at which minimum deviation occurs on both sides of the prism and dividing the determined angle between them by two. A number of measurements at each face are necessary to ensure accuracy. It is advisable to control temperature while making measurements, since the magnitude of dn/dT is comparable to the standard deviation of the technique.

Other sources of measurement error are placement of the prism, which should be at the center of the axis of rotation of the goniometer, and measurement of the prism and deflection angles.[17] By differentiating equation 2, it is evident that measurement of n to an accuracy dn will be possible if

$$dD < 2(dn/n)\tan(A+D)/2. \quad (4)$$

If (A+D)/2 is 45 and n is 2.0, then D must be measured to better than 20" for .0001 accuracy in n. Also, from differentiating equation 1

$$dA < 2(\sin A/2/\sin D/2)dn. \quad (5)$$

For A = 40, D = 46, and n = 2.0, prism angle, A must be determined to better than 12" for the above accuracy. There is a maximum prism angle limit which is established by that angle at which total internal reflection of the incident light occurs. This is determined from the relation

$$A<2 \sin^{-1}(1/n). \tag{6}$$

*Autocollimation*

The dispersion measurement task can be simplified substantially by using the autocollimation method while maintaining the same measurement precision that is available using minimum deviation. In this technique a prism, which is prepared to the same specifications as outlined above, is mounted as the dispersing element in a Littrow spectrometer. A schematic of such an apparatus is shown in figure 5.[18,19] Basically, it consists of the same components as in the minimum deviation apparatus, but the key differences are that the light is made to traverse a set optical path to the back face of the sample prism, is reflected off this face and returned along the same path to a fixed detector. For this simplification there is an additional task in sample preparation. One face of the prism must be made highly reflective after polishing by depositing a reflective metal coating such that both surfaces of the coating are reflective.

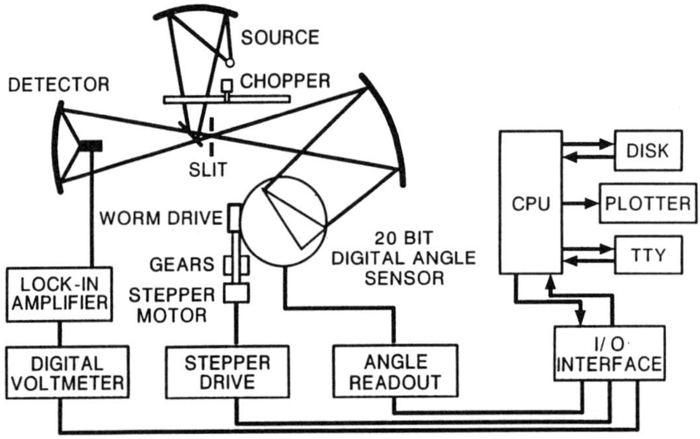


Figure 5. A schematic diagram of an autocollimation refractometer.

Mounting the prism so that the faces are mutually perpendicular to the axis of rotation of the angle sensor and assuring proper prism location are as important here as in the minimum deviation measurement method. The autocollimation scope is used here as described above for minimum deviation. For determining the refractive index, three measurements of angular position are made as shown schematically in figure 6: $A_1$, the deviated ray, which returns upon itself; $A_2$, the reflected ray from the front surface of the prism; and $A_3$, the ray reflected from the rear face. The latter two provide for angular orientation of the prism in the apparatus and measurement of the prism angle.

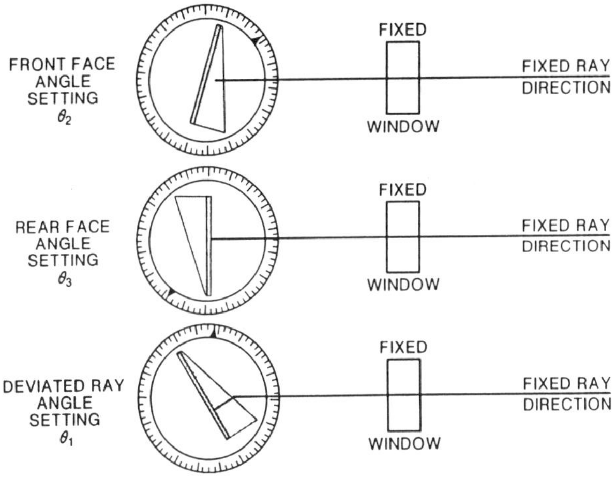


Figure 6. The three angle measurements required for index measurement using autocollimation.

The refractive index for any given wavelength is then determined from the relation:

$$n = \sin(A_2 - A_1)/\sin(\pi - (A_3 - A_2)). \qquad (7)$$

Unlike the minimum deviation method, in this method only prism rotation is necessary for measurement of the necessary angles. Thus, the apparatus is relatively simple to operate and automation is very tractable. The accuracy of the index measurements is dependent on the accuracy and precision of the angle measurement. The principal methods employed to achieve high precision are the use of Moire fringe gratings[(19)] and a digital angle sensor.[(18)] Both give continuous output of accurate measurement angle information. This can be recorded simultaneously with the spectroscopic dispersion signal obtained from a lock-in amplifier detector system. Using multiple monochromatic wavelength sources such as a mercury lamp, many indices can be determined in a single rotation of the prism.

In certain situations it is desirable to use a modification of the above technique wherein an external mirror is fixed to the prism table such that its face is behind the prism with respect to the incident light and perpendicular to light refracted by the prism. This mirror replaces the one formed on the prism face. For this configuration, a fourth angle measurement, $A_4$, is required - the angle of this back mirror. The relation from which the index can be obtained is as follows:

$$n = \left[ \sin^2 i_2 + \left( \frac{\sin^2 i_1 + \cos A \sin i_2}{\sin A} \right)^2 \right]^{1/2}, \tag{8}$$

$$\text{where } i_1 = A_2 - A_1, \tag{9}$$

$$i_2 = \pi - (A_4 - A_3), \tag{10}$$

$$A = \pi - (A_3 - A_2). \tag{11}$$

*V-Block Measurement Technique*

The V-Block method of refractive index measurement is a dispersion method also requiring a prism shaped specimen. In terms of accuracy and simplicity it ranks between the Abbe' refractometer and autocollimation methods. The technique requires a prism sample that is machined to a precise geometry. The prism sits into a glass block, as indicated schematically in figure 7.[16,20] The block has been machined to have a V shaped groove with optically polished faces. The angle of the V groove and the prism angle should be identical. Typically this angle is 60°.

A collimated light beam passing through this solid block, the external faces of which are optically polished and oriented normal to the incident light, is subject to dispersion at the interfaces between the standard V-block and the sample. The dispersion of the beam occurs in a plane containing the axis of the collimator that is perpendicular to the prism edge. A detector can be oriented in this plane and calibrated with respect to the angle of beam displacement or linear displacement from the point of no deflection at a specific distance from the prism sample. From this deflection measurement and knowledge of the refractive index of the V-block glass, the refractive index of the sample can be determined using the following equation:

$$n = \left\{ 1/3 \left[ 2 \sin^2 i + \sqrt{3}\, n' \sin i + 2n'^2 + (2\sqrt{3} \sin i + n')\sqrt{n'^2 - \sin^2 i} \right] \right\}^{1/2}, \tag{12}$$

where n′ is the refractive index of the V block.

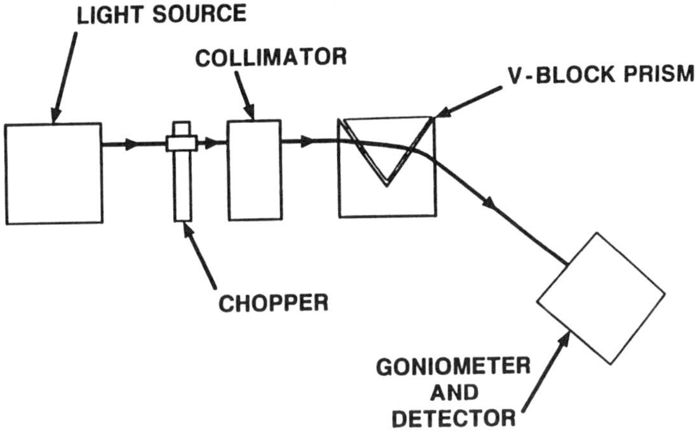


Figure 7. Schematic diagram of a V-block refractometer.

The measurement accuracy using the V-Block apparatus is dependent on the precision of calibration and sample preparation. One must measure the prism angle very accurately and the sample surfaces must be polished to 1/10 wave flatness as in the other techniques. Obtaining such precise measurements of angle will require an apparatus similar to the minimum deviation spectrometer or autocollimation spectrometer. With these factors optimized, accuracy of 0.00004 in index can be achieved.[(16)] The V-block can be made of any glass, but silica is preferred. Because of the support which can be provided by the V-block, it is possible to use this method to measure the refractive index of glasses through their softening points. Silica has a very high softening point compared to most glasses.

Index Matching Methods

Situations often occur in which one cannot use one of the above techniques. For example, the available sample might be of insufficient size for preparation of a polished specimen having adequate surface area for measurement. In such a case it is possible to measure the specimen through a comparison technique where the unknown is matched against a standard. The standard may be an index matching oil or another solid. Matching methods are suitable for determining the refractive index at a single wavelength, but are much too time consuming and imprecise to be used for a dispersion analysis. The measurement accuracy is generally 0.001.

*Immersion Methods*

The method of central illumination or the Becke line method is an index measurement technique commonly employed in mineralogy for determination of the refractive index of small grains. This determination is preferably done with a microscope having better than 50X magnification. For optimum accuracy, monochromatic light should be used for illumination. Small fragments of the glass are immersed in a droplet of liquid of known refractive index on a microscope slide. Light from a microscope condenser impinging on the interface between a fragment and the liquid is reflected and refracted. Depending on the relative indices of the two media, as a result of dispersion, a line or border of higher intensity than the average illumination of the field will appear near the interface of the two when viewed through the eyepiece of the microscope, as shown in figure 8. Its location will be in the material at this interface with the higher index.

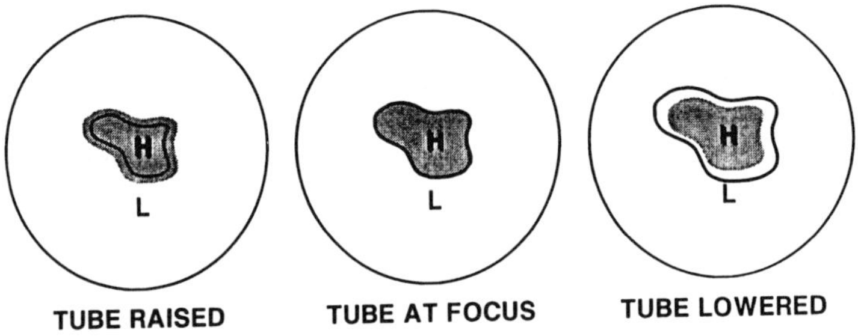


Figure 8. Drawings showing the expected pattern of the refracted illumination when employing the Becke line method.

If there is an index match, no Becke line will be observed. The measurement process involves an iterative procedure in which liquids of known refractive indices are used for fragment immersion until there is a match.

A helpful way to identify the Becke line is by the moving the microscope tube in and out of focus. When the tube is moved up, the Becke line moves into the medium of higher index and conversely, when the tube is lowered, the line moves into the medium of lower index as indicated in figure 8.

Two factors can contribute to errors in this measurement. Immersion media have a strong dependence of index on temperature, and they tend to be hydroscopic with absorbed water changing their index.

*Interferometric Index Measurement*

Interferometric index measurement is also very applicable to small samples.[21] It was used extensively in the early period of the development of optical waveguides to accurately determine the refractive index profile of lightguide preform and fiber specimens. A thin section of glass polished to a high degree of flatness and parallelism is required. Additionally, the index of a portion of the sample must be known because this is a comparative technique. The simplest and most accurate form of this method requires an interference microscope. These are very expensive, so I will describe a version which uses an ordinary microscope. The sample is mounted in a Fabry-Perot etalon as shown in figure 9.[12] The bottom plate of the etalon is a completely reflecting mirror and the top is a partially reflecting mirror.

Monochromatic light from top illumination passing through the sample interferes with the incident beam or the beam reflected off the top mirror, causing fringes to be seen in the visual field of the microscope. Both mirrors must be perfectly flat to avoid fringe distortion. Location of the interference fringes is dependent on the refractive index of the glass specimen through which the light is propagating. Specific fringes can be tracked along the sample and correlated with a specific region of the sample surface. Thus a fringe can be tracked from the standard to the unknown. The shift in fringe placement is related to the difference in index between sample and unknown. To determine the refractive index, one needs to measure the distance, D, between the fringes and the fringe displacement in the sample region, S. The refractive index is then calculated from the relation

$$n = n_s + \lambda S/Dd, \tag{13}$$

where $n_s$ is the refractive index of the standard and d is the sample thickness.

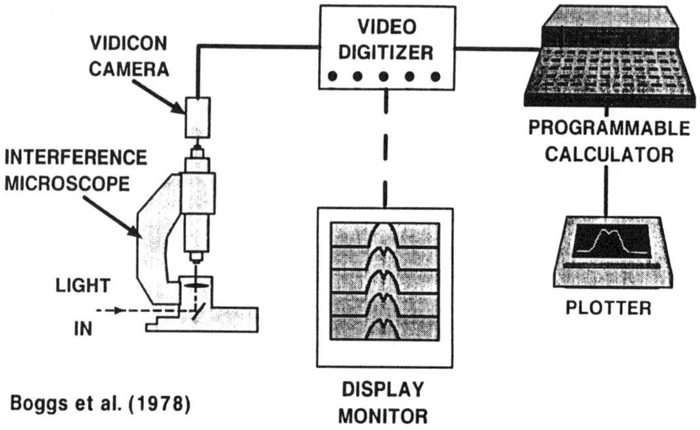


Figure 9. Schematic of an optical microscope set up for measurement of refractive index using interferometry.

This method is very useful for specimens for which there is a continuous change in index, but it is difficult to apply to samples in which the index changes significantly and abruptly. A step index change between standard and unknown causes a discontinuous shift in the fringes, and S is not easily determined.

Sample preparation is very complex. The very thin specimen required must be finely polished. Uniformity of polish when the glasses comprising the specimen vary significantly in hardness is often difficult.

**Data Analysis**

Having measured the refractive index of a glass at several wavelengths, it is often very useful to model the refractive index dispersion behavior so that the data can be employed in calculations such as those necessary for optical equipment design or for index estimation. The important information to be conveyed is the value of the index at a given wavelength and the change in index with wavelength, $dn/d\lambda$, at that wavelength. Physically, the relative magnitude of the dispersion at a given wavelength indicates semi-quantitatively the position of the fundamental absorptions with respect to that wavelength.

Figure 10 indicates schematically the typical variation of refractive index with wavelength of a material which is transparent in the optical region. Note that the slope and value of refractive index practically alludes to the fundamental material absorptions.

The most often employed dispersion model is that of the constringence or Abbe' number, ν, given by the following relation: [5,13]

$$\nu = \frac{n_D - 1}{n_F - n_C}, \tag{14}$$

where $n_D$ is the refractive index @ .5893 μm,
$n_F$ is the refractive index @ .4861 μm,
$n_C$ is the refractive index @ .6563 μm.

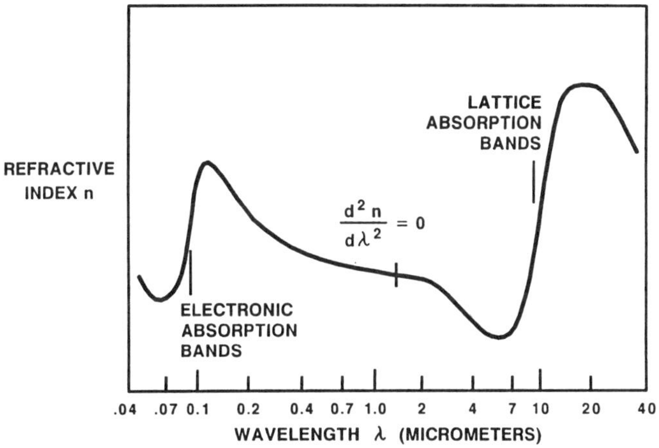


Figure 10. Typical variation of refractive index with wavelength of a glass transparent in the visible range.

Many commercially available glasses have been characterized using this number, which permits extensive comparisons of various glasses and can often help with physical understanding of the optical characteristics of a newly measured glass. A graph showing the behavior of a number of glasses is given in figure 11. The constringence is an excellent starting point for surveying the optical properties of glass. Used, as shown in the figure, in conjunction with the index at 0.5893 μm, it indicates the change in index with wavelength in this wavelength region.

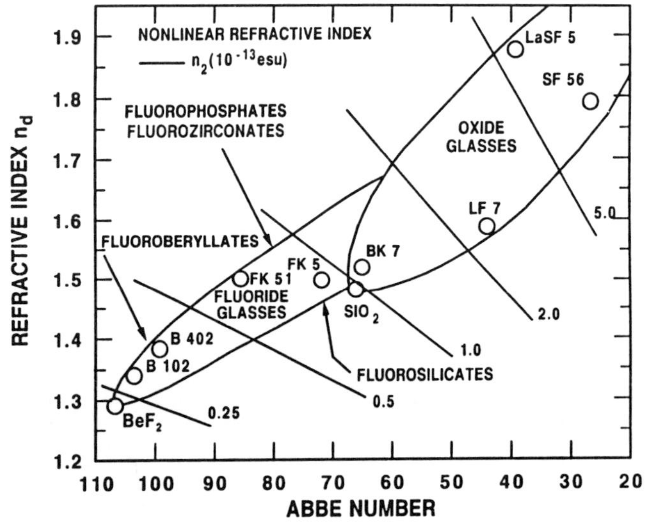


Figure 11. Abbe' diagram showing regions of various glass compositions.

It is evident from the figure that fluoride glasses have higher ν values than the oxide glasses. A correct conclusion that can be drawn from this is that the metal - fluorine bonds predominant in fluoride glasses have higher energies associated with them than the metal - oxide bonds of the oxide glasses. Using the constringence and $n_D$, such physical interpretations are generally limited to the high energy resonances. Also included in Figure 11 are lines of constant non-linear refractive index, $n_2$. This quantity is useful for determination of certain non-linear optical properties and accessing the potential of laser damage to the glass. It will be discussed in a subsequent chapter.

Other relations defining dispersion such as the example given below, are termed partial dispersions, $P_{(x,y)}$ and are useful in the design of lenses. They further define the dispersion at other optical wavelengths and can be used in the choice of glass combinations for lenses which minimize chromatic aberrations in optical assemblies.

$$P_{i,h} = \frac{n_i - n_h}{n_F' - n_c'}, \quad (15)$$

where $n_i$ is the index at 0.3650,
$n_h$ is the index at 4046,
$n_F'$ is the index at 0.4799,
$n_c'$ is the index at 0.6438.

Often one needs a dispersion model that can be used to predict the refractive index and calculate the dispersion at any given wavelength. Power series models such as the following form are often used:

$$n = C_0 + C_1\lambda^2 + C_2\lambda^{-2} + C_3\lambda^{-4} + C_4\lambda^{-6}, \quad (16)$$

where $C_0, C_1, C_2, C_3, C_4$ are constants.

These are functional, but a more physically correct model is one which actually takes into account the fundamental material resonances discussed above. Such a model is the Sellmeier relationship: [8,18]

$$n^2 - 1 = \sum_{i=1}^{n} \frac{A_i\lambda^2}{\lambda^2 - l_i^2}, \quad (17)$$

where $A_i$ are oscillator strengths and
$l_i$ are oscillator wavelengths.

The oscillator strengths and wavelengths actually have physical significance. A fit of the Sellmeier relation to refractive index dispersion data can often lead to insights into the structure of the glass and the effect a certain processing modification has on the glass structure. Three Sellmeier terms are sufficient for a very accurate representation of most refractive index data. In fact, the Sellmeier relationship has been found to provide precision superior to polynomial models containing the same number of dependent variables.[22]

In fitting this model to a set of index data, most glass is best addressed by placing two of the three oscillator terms in the ultraviolet range, uv, and one in the infrared, ir. The author has found that the infrared term tracks in oscillator wavelength the centroid of the fundamental material absorptions in the infrared.[8]

Often it is useful to design a model that permits estimation of the refractive index of a range of glass compositions from a few measurements at endpoint compositions. Surprisingly, a linear first order model is quite accurate for specific wavelengths. However, the Sellmeier model can be applied more effectively here. This can be done by obtaining compositionally dependent linear relationships for the parameters of the Sellmeier model and incorporating these as follows:[8,9]

$$n^2-1 = \sum_{i=1}^{n} \frac{[A_i + x(B_i-A_i)]\lambda^2}{\lambda^2 - [l_i + x(m_i-l_i)]^2}, \tag{18}$$

where $A_i$, and $B_i$ are oscillator strengths of components A and B respectively, and $l_i$ and $m_i$ are oscillator wavelengths of A and B, and x is the mole fraction of B in A.

A key advantage to this approach is that one now has a model for the dispersion of the composition range rather than one that only addresses the refractive index at one wavelength.

A difficulty with the use of the Sellmeier is that a non-linear regression must be used to fit the data to the model. To simplify this and also to obtain relationships that contain parameters that can be more easily used for calculations and predictions of physical properties of materials, a number of investigators have used modified Sellmeier relationships such as the following equation used by Wemple.[23]

$$n^2-1 = [E_d E_o/(E_o^2-w^2)] - E_l^2/w^2, \tag{19}$$

where w = photon energy (eV),
$E_o$ = avg. electronic energy gap (eV),
$E_d$ = electronic oscillator strength (eV),
and $E_l$ = lattice oscillator strength (eV).

**Thermal Dependence of Optical Characterization**

As the temperature of a glass is changed, there is a corresponding variation in the atomic polarizability and the molecular vibrations. These features are principally responsible for the intrinsic optical properties of glasses in the wavelength region between them. An increase in polarizability generally accompanies a temperature increase. This might result in the uv absorptions moving to longer wavelengths and therefore moving closer to the visible region. The result would be increased refractive index and increased optical absorption in the visible. Temperature increases also cause a shift to longer wavelength of the molecular vibrations responsible for fundamental optical interactions in the ir.[1,16] Therefore, temperature increases will cause ir absorptions to be less of an influence on the visible and the refractive index is decreased. The observed resultant change in index is a summation of the uv and ir effects. In most oxide glasses the observed optical effects are an increase in refractive index with temperature and an increase in absorption and scattering. The typical change is $1 \times 10^{-4}$/°C.[1,5]

A unique property of glass is that its structure is a function of the thermal history experienced during preparation.[(24)] Structural variation of a given composition as a result of thermal modification generally changes the optical characteristics of the glass. The molecular vibrations are fixed to the structure to which the glass has equilibrated as a result of its thermal history. This state can be characterized by a temperature at which the glass would have that structure called the fictive temperature. Polarizability or local electronic modifications are not so fixed. The result is that fictive temperature increases cause a decrease in refractive index at a given wavelength in the visible region. As the temperature at which the structure of a glass effectively freezes is increased, the refractive index is generally decreased. In some compositions the change in refractive index as a function of thermal history can be as much as several parts in the third. Thus, it is important when characterizing a glass to define the thermal history.

## REFLECTANCE MEASUREMENTS

Light incident on an interface between two materials of different refractive index will be reflected to some extent from the interface. The amount of incident light reflected is dependent on the quality of the surface and the angle of incidence. Reflection can be specular such as from a polished or mirror-like surface or diffuse such as from a rough surface. The following material concerns measurement of specular reflectance. Fresnel found that the reflectance from a glass surface depended also on the polarization of the light and that the reflectances of the respective polarizations could be determined from the following relations: for light polarized parallel to the surface,

$$R_{\parallel} = \frac{\sin^2(i-r)}{\sin^2(i+r)} \tag{20}$$

for light polarized perpendicular to the surface,

$$R_{\perp} = \frac{\tan^2(i-r)}{\tan^2(i+r)} \tag{21}$$

One takes the average of the above calculations for unpolarized light. It can be determined that for normal incidence the reflectance is given by the relation:

$$R = \left(\frac{n-1}{n+1}\right)^2 \tag{22}$$

Therefore, if one knows the refractive index of a given glass, the reflectance can be determined for that wavelength for all angles of incidence using Snell's law and Fresnel's laws. If the refractive index is not known, then an apparatus such as shown schematically in figure 3 above can be employed. Procedurally, one would first measure the intensity of the incident light when it falls directly on the detector and then allign the apparatus such that the beam is reflected off the sample surface onto the detector. Sample surface preparation requirements are the same here as for the Brewster angle measurement.

## ATTENUATION MEASUREMENTS

As discussed above, light interacts with a material during propagation which determines its velocity, reflection, and attenuation. Attenuation is the opposite of transmission. It is the result of optical absorption and scattering. Absorption converts light into thermal energy in the structure. Through scattering the propagated light is caused to deviate significantly from its original path of propagation in the glass. Absorption can occur through intrinsic absorption of the fundamental electronic and ionic resonance elements of the material and from extrinsic defects such as impurity ions or structural inhomogeneities. The principal extrinsic absorption mechanisms are transition metal ions, rare earth metal ions, water, and structural defects such as broken bonds. Scattering of light can result from intrinsic inhomogeneities in the electronic polarizability or extrinsic defects such as bubbles, inclusions, or surface irregularities. Lines[10] has found that fundamental attenuation, $\alpha$, can be expressed by the following relation.

$$\alpha = Ae^{-a/\lambda} + B/\lambda^{-4}, \qquad (23)$$

where A, a, and B are material parameters and $\lambda$ is the light wavelength.

The nature of the apparatus employed to measure attenuation depends on several factors. One factor is the intensity of the characteristic absorption one desires to measure. For example, searching for the transmission window defined by the characteristic material absorptions does not require extremely high sensitivity compared to searching for the wavelength of lowest loss in a glass for optical fiber transmission. Additionally, one might be interested in separately determining the absorption component of the attenuation or the scattering component. The wavelength of interest will also determine the choice of measurement apparatus.

### Absorption Measurement Methods

The most commonly employed apparatus for measuring attenuation of bulk samples is a spectrophotometer. This instrument compares the transmission intensity of a standard and an unknown with the response being a measure of transmission or optical density. The apparatus consists of a light source, chopper, light directing optics, a monochrometer, and a detector. Typically, spectrophotometers are manufactured to measure certain transmission windows. Commercially available equipment is generally manufactured to measure responses in the range from 0.1 to 2.0 μm on one apparatus and 2.0 through 50 μm on another. This wavelength specificity is a result of detector and monochrometer limitations.

For light that is normally incident on a glass surface, which is the typical orientation employed for attenuation measurement, the first component of attenuation is surface reflection. The fraction of light absorbed by the material is determined by comparing the amount actually injected into the specimen, $I_o$, to the intensity I detected at the end of a given length of travel through the specimen, x. The relations used for quantification of the absorption take the form:

$$\alpha = (\frac{1}{x})\log\frac{I}{I_{o,}} \tag{24}$$

where $\alpha$ is the absorption coefficient generally reported in units of $cm^{-1}$.

Sample preparation consists of optically polishing the sample on two parallel faces through which the beam will pass. This is done to remove surface irregularities which might scatter the incident light and remove impediments to transmission which are of no interest to the investigator. Thickness is an important parameter in the calculation of attenuation and should be measured accurately. It can be increased or decreased to optimize the measurement signal. When one is observing very strong absorptions, it is possible to dilute the sample by mixing the specimen in powder form in a medium that transmits well in the wavelength region of interest. An example of an appropriate immersion material is a low melting salt such as KBr.

Spectrophotometers

Two types of spectrophotometer are commercially available. The single-beam instrument uses a single optical path for measurement of the sample. Quantification and standardization are accomplished through a standard or baseline measurement. To overcome the problem of sample surface reflection, two samples of different lengths with similarly polished faces are required. The sensitivity of this method is a function of the magnitude of difference in the lengths of the samples. These measurements should be made consecutively.

Double-beam spectrophotometers are the other type. These measure the attenuation of the two glass samples simultaneously and provide a comparison of the two as the instrument response. This can either be an ac or dc response, depending upon the nature of the source and arrangement of optics. The double-beam methods are easier to employ and generally more accurate.

For cases in which one desires to measure the attenuation of very low attenuation bulk glass specimens, several techniques have been developed. To the author's knowledge these are not commercially available. One such technique is the two-beam optical bridge. Such an apparatus is shown schematically in figure 12.[25,26] Light from a laser beam is split by a beam splitter and the two beams are then chopped such that synchronized pulses are created where one is passed when the other is blocked.

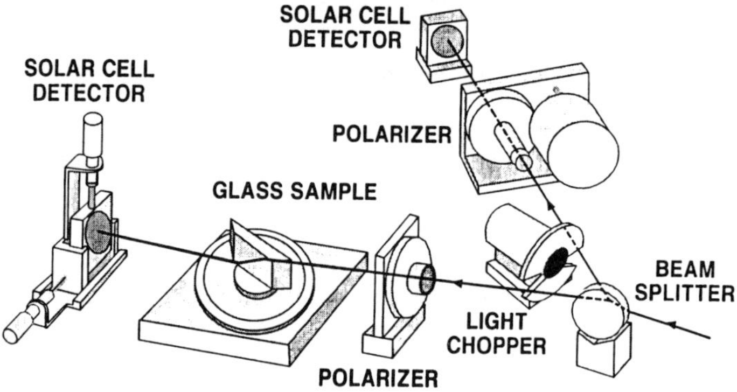


Figure 12. Schematic diagram of a double-beam optical bridge.

The beams next pass through polarizers and onto detectors. The sample is placed in one optical path between a polarizer and detector. The sample must be optically polished on the faces through which the beam passes. These faces should also be parallel. To eliminate the complication of surface reflection, the sample faces are oriented at Brewster's angle with respect to the incident beam as shown in figure 12. Using the polarizer in this light path, the surface reflectivity can be eliminated as an attenuation factor. The other polarizer is used to adjust the attenuation on the reference beam so that the two beams are detected as being of equal intensity when no sample is in the light path. Therefore, with the detectors connected in parallel, the electrical signal is a constant dc value. When the sample is placed in the light path, a small ac signal proportional to the loss of the sample is placed on the dc one. This can be monitored selectively with very high sensitivity, making very accurate measurements of attenuation possible. The optical bridge measures the total sample attenuation, which includes bulk absorption, bulk scattering, and surface scattering. By measuring different thicknesses of the same glass sample, one can isolate the bulk loss effects.

The most critical elements in this measurement are the detectors. Most detectors do not have a uniform response to a given light signal across their surfaces. It is therefore important to orient the detector of the sample beam so that the beam falls on the same spot with and without the sample in place. For this purpose precision movement of the sample beam detector is incorporated into the apparatus.

Calorimetric Methods

Optical calorimetry can be used to determine the bulk absorption loss directly without the confounding effects of surface and bulk scattering and reflectivity.(27) An optical calorimeter is designed as illustrated in figure 13. A collimated beam of monochromatic light, preferably from a high intensity source, impinges on a sample of the glass to be measured, and the temperature rise resulting from absorption of light energy is detected by a thermocouple in contact with the sample surface.

To reduce errors resulting from heating of the sample enclosure from scattered light or ambient changes in temperature and to eliminate the need to determine the heat leakage through sample points of contact or convection, the radiated specimen temperature is compared with that of a specimen of the same glass composition and geometry located within the sample chamber adjacent to the specimen under test.

Measurement sensitivity is principally dependent on the intensity of source radiation. To measure $10^{-3}$ dB/cm one mW of source power is required. The optical absorption can be calculated from the following relation.(27)

$$\alpha = c(\pi r^2)\Delta T/\tau P, \tag{25}$$

where c is the specimen heat capacity,
$\Delta T$ is the equilibrium temperature rise,
r is the radius of the cylinder,
and P is the optical power passing through the sample.

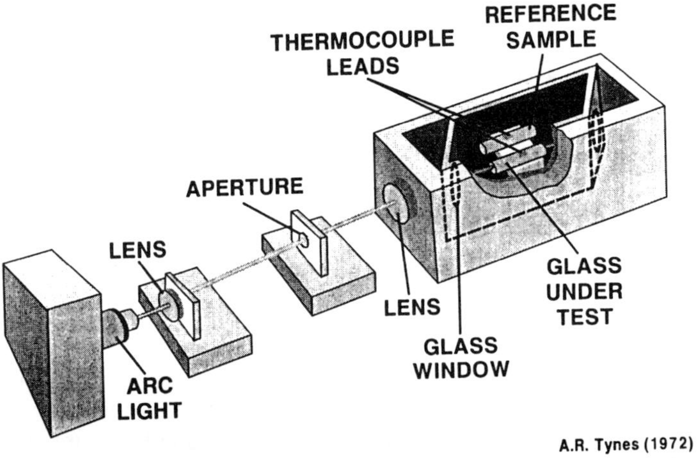


Figure 13. Optical calorimeter for specifically measuring absorption.

The accuracy of this measurement is dependent on apparatus standardization. This can be accomplished with precision Joule heating of an appropriate specimen, for example, a glass tube with an internal precision joule heater. If performed with a sufficiently powerful light source, this method is unsurpassed in attenuation measurement sensitivity. Although the high intensity is necessary, it is important not to overload the specimen with optical energy. Resultant non-linear effects could cause the conversion of light into other wavelengths that exhibit higher attenuation.

**Scattering Measurement Methods**

Scattering loss in glass can be caused by bulk impurities such as dirt particles, bubbles, crystallites, and phase separation. An additional source of scattering is structural inhomogeneity. The scattered light intensity is a function of wavelength and the size of the scattering sites. For scattering sites much smaller than the wavelength of the incident light, the intensity is proportional to $1/\lambda^4$ in wavelength regions far from the fundamental material absorptions. This establishes the baseline for scattering loss and is referred to as Rayleigh scattering.

Scattering loss measurements on bulk samples have become an important part of optical material development in areas examining novel glass compositions such as fluorides for optical transmission. The geometry and size of the scattering sites cause an angular dependence of the scattered light. Analyzing the variation of intensity with scattering angle is often valuable in determining the cause of scattering. However in this section the apparatus described will be limited to examination of the intensity perpendicular to the incident beam.

Scattering can be measured effectively using an apparatus such as the one shown schematically in figure 14.[12] A microscope designed for a camera attachment is fitted with a light detector such as a photomultiplier tube. The polished glass specimen to be measured is placed on the microscope stage.

A microscope objective lens is used to focus a beam of light through a polished face of the specimen under test. Generally, the source is a laser. The microscope is brought to focus on this beam within the sample and the scattered light entering the microscope is measured with the photomultiplier tube. The instrument can be calibrated using specimens of known scattering intensity. Absolute scatter loss is then obtained by comparing the scattered intensity to the incident light intensity. An incident power level of 1 mW is generally sufficient. The entire sampling procedure usually takes no more than a few minutes.

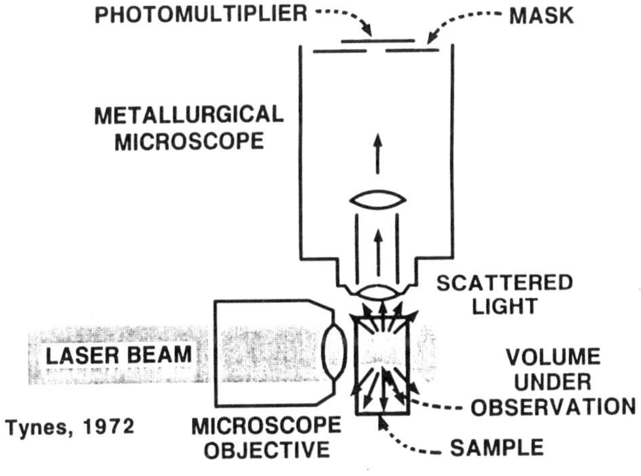


Figure 14. Optical microscope setup for measuring scattering intensity.

### Fiber Loss Measurements

The accuracy of any attenuation measurement is proportional to the transit length of the optical beam within the specimen. For this reason, attenuation measurements on fiber specimens are generally more precise for low attenuation glasses than those performed on bulk specimens. Sample preparation may be more difficult if one does not have access to fiber drawing equipment. In a single material fiber, the guided modes will be reflected off the sample surface. If this surface is not kept pristine, the measurement accuracy will be severely affected. The most common fiber loss measurement is a cut-back technique. This is shown schematically in figure 15.[28,29] In this method, light is launched into the fiber using a microscope objective. The transmitted optical power, $P_A$, detected at the other end is recorded, then the fiber is cut close to the source, and the short fiber remeasured without changing the launch conditions to obtain the launch power, $P_B$.

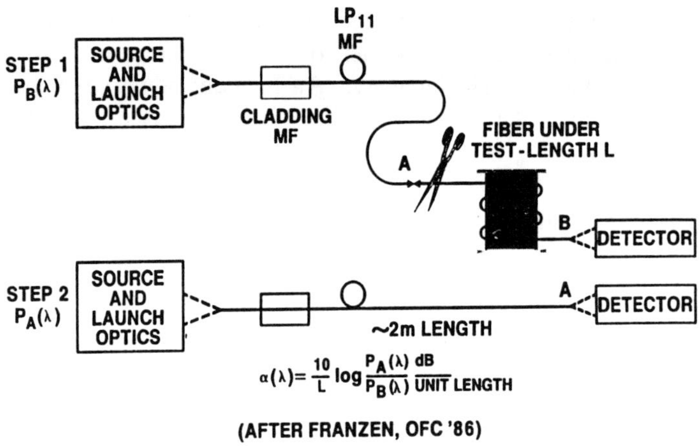


Figure 15. Schematic of a standard optical fiber loss measurement system.

The actual power loss in the fiber is given by the ratio

$$\frac{P_A}{P_B}, \tag{26}$$

which in dB is given by $\alpha(\lambda) = (10/L)\log\frac{P_A}{P_B}$.

While this is a very sensitive measurement method, it is ambiguous. Consider that the loss in a specimen is a function of the length of specimen traversed by the beam. Now, in the case of a fiber, many modes of light exist guided by the critical reflection at the core-clad interface. For reasonable numerical aperature, NA, specimens, the optical path length may be significantly longer than the fiber length. This can be diminished by using a low NA launch. The single material fiber measurement then will generally give higher loss indications than the bulk measurements. While the above measurement problem is bothersome, the real problem is keeping the sample surface clean.

## CONCLUSION

The information in this chapter was compiled to help the reader become familiar with and employ the most effective methods of optical characterization. In this volume, complimenting the material in this chapter, are several chapters covering interactions between glasses and light in other regions of the electromagnetic spectrum. Although the wavelengths are different, the fundamental principles involved are the same. These chapers will help support the readers knowledge of optical characterization.

## REFERENCES

[1] G. W. Morey, The Optical Properties of Glass, 365-441, Properties of Glass, Reinhold, Scranton, Pennsylvania, 1938.

[2] B. G. Bagley, C. R. Kurkjian, J. W. Mitchell, G. E. Peterson, and A. R. Tynes, Optical Characterization of Glass, 212-31, Optical Fiber Telecommunications, S. E. Miller and A. G. Chyenowyth.

[3] D.G. Holloway, Optical Properties of Glass, The Physical Properties of Glass, 88-124, Wykeham, London, 1973.

[4] F. J. T. Maloney, Optical Glass, Glass in the Modern World, 109-28, Doubleday, New York, 1968.

[5] G. W. Mclellan and E. B. Shand, Properties of Glass, Glass Engineering Handbook, 2-1, 2-36, McGraw-Hill, New York.

[6] B. G. Bagley, etal., The Optical Properties of a Soda-Lime-Silica Glass in the Region from 0.006 to 22 eV, J. Non-Crystalline Solids, 22, 423-456, 1976.

[7] G. H. Sigel Jr., Optical Absorption of Glasses, 5-82, Treatise on Materials Science and Technology Vol. 12, Minoru Tomozawa and R. H. Doremus, ed.,

[8] J. W. Fleming, Dispersion in $GeO_2$-$SiO_2$ Glasses, Appl. Opt., 23 (24), 4486-4493, 1984.

[9] J. W. Fleming, Materials for UV and Far IR Transmission, Mat. Res. Soc. Sym. Proc. Vol 88, 149-62, 1987.

[10] M. E. Lines, Scattering Losses in Optic Fiber Materials II. Numerical Results, J. Appl. Phys., 55, (11), 4058-63, 1984.

[11] D. Marcuse, Fundamentals, 55-68, Principles of Optical Fiber Measurement, Academic Press, New York, New York, 1981.

[12] D. Marcuse, Loss Measurements, 197-253, Principles of Optical Fiber Measurement, Academic Press, New York, New York, 1981.

[13] F. A Jenkins and H. E. White, Physical Optics, 191-588, Fundamentals of Optics, 3rd edition, McGraw Hill, New York, New York, 1957.

[14] J. Hecht, Fiber-Optic Measurements, Optics News, April, 7-12, 1987.

[15] P. F. Kerr, Refraction, 59-81, Optical Mineralogy, 4th edition, McGraw Hill, New York, New York, 1977. Academic Press, New York, New York, 1977.

[16] I. Fanderlik, Interaction Between Radiation and Glass, 50-224, Optical Properties of Glass, Elsevier, New York, New York, 1983.

[17] W. L. Bond, Measurement of the Refractive Indices of Several Crystals, J. Appl. Phys., 36, (5), 1674-77, 1965.

[18] D. L. Wood and J. W. Fleming, Computerized Refractive Index Measurements for Bulk Materials at UV, Visible, and IR Wavelengths, Rev. Sci. Instrum., 53, 1, 1982.

[19] M. Debenham, G. D. Dew and D. E. Putland, An Improved Recording Refractometer for Optical Glasses in the Wavelength Range 300 to 2600 nm, Optica Acta, 26, (12), 1487-1503, 1979.

[20] E. Moreels, C. de Greef, and R. Finsy, Laser Light Refractometer, Appl. Opt., 23, (17), 3010-13, 1984.

[21] R. M. Waxler and G. W. Cleek, Refractive Indices of Fused Silica at Low Temperatures, J. of Res. of NBS -A. Phys. and Chem., 75A, (4), 1971.

[22] V. G. Reznik, Dispersion Dependences for Calculating Refractive Index, Sov. J. Opt. Tech., 47, (8), 459-61, 1980.

[23] S. H. Wemple, Material Dispersion in Optical Fibers, Appl. Opt. 18 (1), 31-5, 1979.

[24] J. W. Fleming and J. W. Shiever, Thermal History Dependence of Refractive Index Dispersion of Fused Silica, J. Amer. Ceram. Soc., 62, (9-10), 526, 1979.

[25] D. D. Sell, A Sensitive Spectrophotometer for Optical Reflectance and Transmittance Measurements, Appl. Opt., 9, (8), 1926-30, 1970.

[26] A. R. Tynes, Measuring Loss in Optical Fibers, Bell Laboratories Record, 303-310, November, 1972.

[27] T. C. Rich and D. A. Pinnow, Total Optical Attenuation in Bulk Fused Silica, Appl. Phys. Lett., 20, (7), 264-5, 1972.

[28] A. Jacobsen, N. Neuroth, and F. Reitmayer, Absorption and Scattering Losses in Glasses and Fibers for Light Guidance, J. Am. Cer. Soc., 1, 186-87, 1971.

[29] P. Kaiser, A. R. Tynes, H. W. Astle, A. D. Pearson, W. G. French, R. E. Jaeger and A. H. Cherin, J. Opt. Soc. Amer., 63, (9), 1141-48, 1973.

# CHAPTER 2

## INFRARED AND RAMAN SPECTROSCOPY OF GLASSES

**Bernard Bendow**
TRW Space and Technology Group
One Space Park
Redondo Beach, CA 90278

## INTRODUCTION AND BACKGROUND

This chapter describes two complementary techniques for the characterization of glasses, namely, infrared (IR) spectroscopy and Raman spectroscopy. These non-destructive techniques provide extensive information about the optical, vibrational, chemical and structural properties of gases, liquids and solids and, in particular, glasses and amorphous materials. Raman spectroscopy is sensitive to local structure, and is therefore especially useful for detecting, identifying, monitoring and analyzing spatial variations as well as temporal changes in chemical composition and structure in glasses. An extensive literature is available on the detailed theory and practice of IR and Raman spectroscopies[1-10]. This chapter is intended as an introduction to the subject for the non-specialist, with special emphasis on glasses. The primary goal is to inform readers about the capabilities and limitations of IR and Raman spectroscopies, so they can determine how these techniques may be best applied to their own particular areas of interest.

A wide variety of analytical and spectroscopic techniques are used to probe the excitation spectra of materials and, thereby, their chemical and structural properties. Typically, information is extracted from the material's response to external probes such as electric and magnetic fields, lightwaves, electron beams or molecular beams. Such spectra, many of which are described in detail elsewhere in this volume, depend both on the nature of the material excitations, as well as the coupling coefficients (i.e., interaction strengths) between the probe and the excitations.

Raman and IR spectroscopies use light (photons) to probe material excitations such as phonons (atomic vibrations), plasmons (collective electronic excitations), or magnons (magnetic spin waves). The focus in this chapter is exclusively on vibrational spectra, which are readily probed by both IR and Raman techniques, and which provide the most extensive information about chemical composition and structure. Moreover, in this chapter only spontaneous Raman processes, where the scattering intensity depends linearly on sample thickness, are addressed. Stimulated scattering processes, which are characterized by gain (nonlinear dependence on pathlength), are not addressed here.

While the focus of the present chapter is vibrational spectroscopy, optical spectroscopies which probe electronic transitions in glasses, via absorption and emission of visible and UV light, say, are also useful for characterizing glasses. In addition, the spectra of ions in glass, which span the IR to UV depending on the host-ion combination, provide useful information about local structure and bonding in glasses. Optical spectroscopy of glasses, and spectra of ions in glass, are covered in other chapters in this volume.

**Vibrational Spectra of Materials**

Materials such as glasses possess vibrational modes characteristic of their composition and structure, i.e., of the type of atoms in the material, and their arrangement and bonding. The fundamental modes which vibrate at a single, well-defined frequency are referred to as "normal modes". Mathematically, they are the solutions of the equations of motion for the vibrating atoms, and depend on the number of degrees of freedom, symmetry, and the nature of the interatomic forces.

As indicated above, Raman and IR spectroscopy use photons to probe vibrational modes of matter. In IR spectroscopy, polychromatic light at IR wavelengths is used to couple directly to vibrational modes, either in reflection or transmission; the coupling is via the dipole moment associated with the vibrations. As indicated schematically in Fig. 1, vibrational modes appear as dips in transmission spectra and peaks in absorption and reflectivity spectra. A wide variety of IR spectrometers, operating in both transmission and reflection modes, are available commercially. Generally, they consist of a continuous IR light source, a grating to disperse the transmitted or reflected light, and detectors to sense the light. Variations include multiple beam spectrometers and Fourier transform spectrometers.

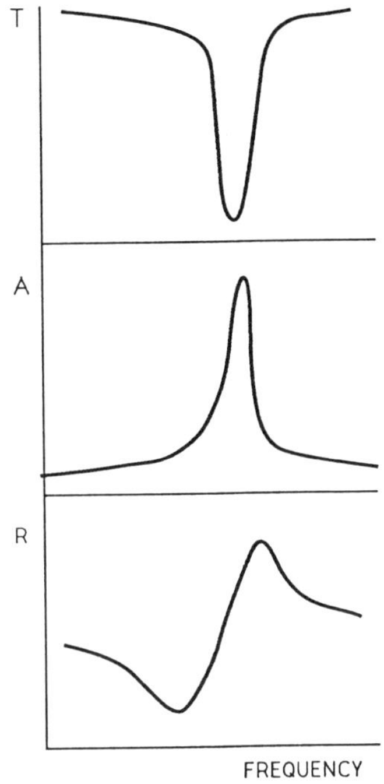


**Figure 1.** Schematic of transmission (T), absorption (A) and reflectivity (R) spectra vs photon frequency, associated with a single infrared-active vibrational mode.

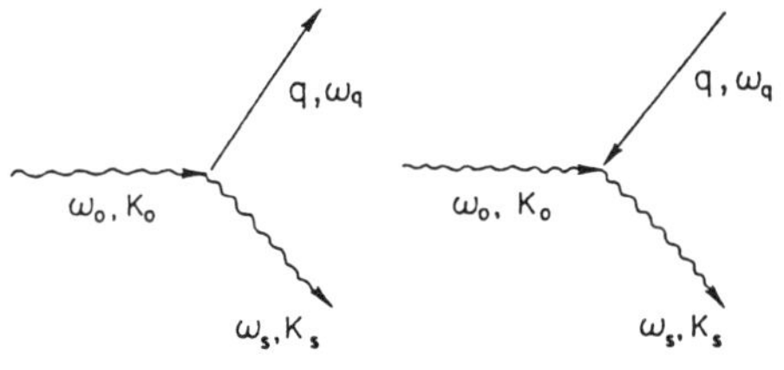


$\omega_s = \omega_0 - \omega_q$
$\boldsymbol{k}_s = \boldsymbol{k}_0 - \boldsymbol{q}$
Stokes

$\omega_s = \omega_0 + \omega_q$
$\boldsymbol{k}_s = \boldsymbol{k}_0 + \boldsymbol{q}$
Antistokes

**Figure 2.** Schematic of first-order Raman processes; a phonon is created or annihilated, and enrgy and momentum are conserved in the process.

In Raman spectroscopy (nearly) monochromatic light, generally in the visible, is used to couple to vibrational modes through the nonlinear polarizability associated with those modes. Visible laser light is generally used because a wide variety of stable, narrow-band laser sources are available; visible light is convenient to work with; and scattering intensities (proportional to the frequency of the light to the fourth power) are enhanced at short wavelengths. The excitation of vibrational modes is manifested by a frequency-shifted spectrum (relative to the incident light frequency) of scattered light from the material.

Spontaneous Raman scattering is a process in which vibrational excitations are either created (Stokes processes), thereby downshifting the scattered light frequency; or annihilated (anti-Stokes), thereby upshifting the frequency. The processes contributing to scattering and the resulting spectra are indicated schematically in Figs. 2 and 3, respectively. As indicated in Fig. 3, the term Raman scattering is usually reserved for scattering by high-frequency, i.e., IR vibrations; elastic scattering is termed Rayleigh, and low frequency ("acoustic wave") scattering is termed Brillouin. Note that these somewhat arbitrary distinctions are based on the magnitudes of the frequency shift, not the type of vibration. For example, both high frequency acoustic as well as optical phonons may contribute to Raman spectra.

A variety of Raman spectrometers are available commercially, combining discrete or continuously tunable laser sources with dispersive elements and detection devices for the scattered radiation. Representative instrumentation comprising a Raman measurement system is indicated schematically in Fig. 4, and illustrates the functional elements common to such systems. The first such element is a source of monochromatic light, usually a visible laser, as indicated in the figure. Visible gas lasers (including Ar, HeNe and Kr) are typical sources; dye lasers are also used as continuously tunable sources for Raman spectroscopy. The laser beam is usually focused into a narrow volume of the sample, positioned in a chamber designed to facilitate collection of scattered light at desired angles (often 90°) from the scattering volume; typical geometries are indicated in Ref. 1. For high resolution spectroscopy, single mode lasers with very narrow spectral bandwidths are advantageous. In addition to the optics which focus the light into the sample and collect and relay the scattered light to the dispersion element, additional filters and polarizers are often inserted in the optical train to suppress unwanted signals, and/or select special spectra (e.g., polarized spectra).

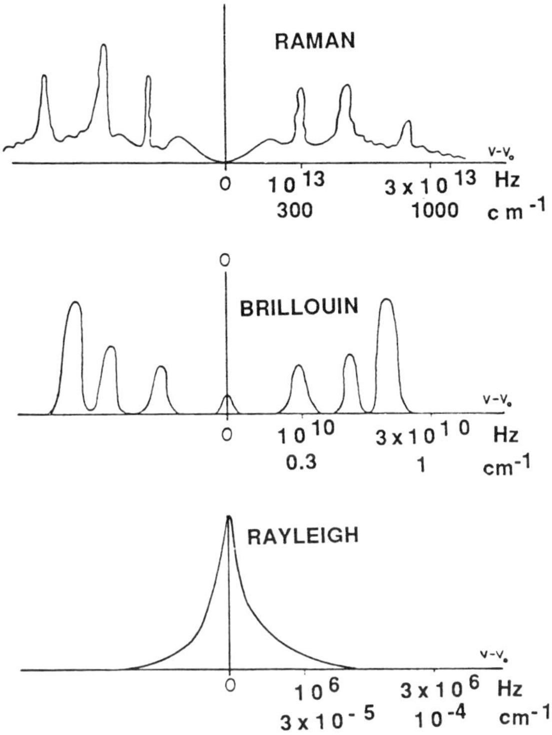


**Figure 3.** Schematic of Raman, Brillouin and Rayleigh scattering spectra in solids, indicating the characteristic frequency ranges for each type of scattering (from H. Cummins, in Ref. 9, pp 3-8).

The scattered light is then input to a dispersive system, usually based on a diffraction grating, such as the double-grating monochromator indicated schematically in Fig. 4. High resolution spectroscopy may require enhancements such as multiple monochromators[1]; also, Fourier transform (interferometric) spectrometers for Raman systems have recently been introduced[2].

As indicated in Fig. 4, the output from the monochromator is provided to a detector, usually a photomultiplier which converts the optical signal to a train of electrical pulses.

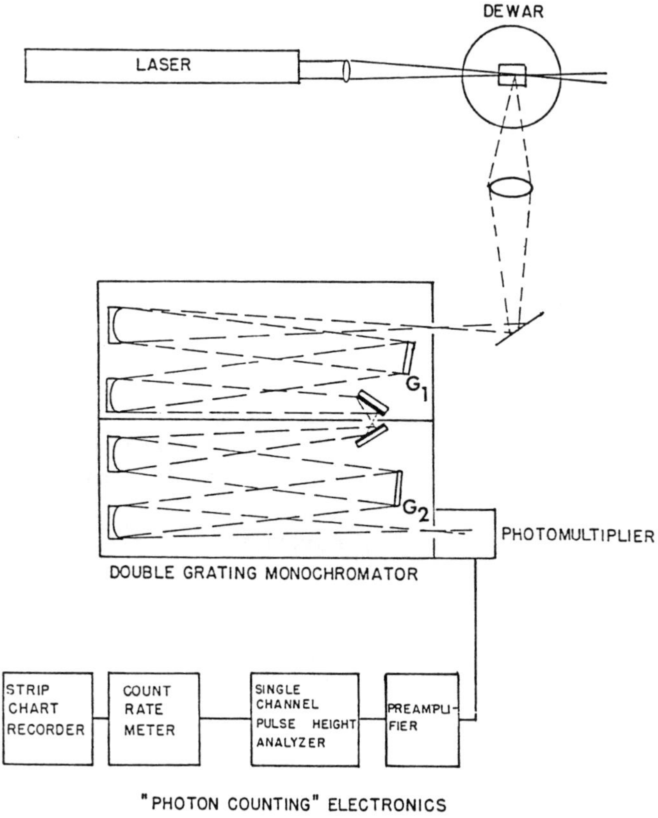


**Figure 4.** Schematic diagram of a representative Raman scattering experimental configuration (from H. Cummins, in Ref. 9, pp3-8).

The highly efficient electron multiplication process in photomultipliers is very suitable for the weak signals accompanying Raman scattering: Usually $10^6$ - $10^8$ incident photons are required (under non-resonant conditions) to generate a single scattered photon. The photomultiplier's output is subsequently processed with "photon counting" electronics consisting of a preamplifier, pulse-height analyzer and count rate meter, as indicated in Fig 4. Systems are also available which use multi-channel detection with image-intensified photodiode arrays or charge-coupled-device detector arrays. Tradeoffs between the various detection systems are discussed in Ref. 2.

The Raman microprobe is a variation of a Raman spectrometer designed to scan samples spatially and to probe small spatial volumes. In essence, the microprobe couples an optical telescope to a conventional Raman spectrometer. Since variations in composition and structure over small spatial scales (microns, say) are often of interest, the microprobe has become a useful adjunct to conventional Raman systems for the study of ceramics and glasses.

**Complementarity of Raman and IR Spectroscopies**

Raman and IR spectroscopy are useful analytical techniques because materials displays unique spectral characteristics (i.e., "signatures") determined by their composition and structure[3]. While at first glance the two techniques might appear to be duplicative, this is generally not the case. Raman and IR spectra usually differ significantly because selection rules governing the coupling of light to particular vibrations differ in the two cases. Fortunately, dominant features in Raman and IR spectra are often complementary rather than redundant, so both are useful for determining vibrational characteristics of materials. In addition, polarized Raman spectra provide additional information about structure not available from IR spectroscopy, as illustrated later in this chapter.

Finally, practical experimental considerations also differentiate between IR and Raman. While IR is simpler in principle, absorption spectroscopy is difficult to carry out in the vicinity of the highly absorbing fundamental vibrations, necessitating reflection spectroscopy along with additional calculations to extract vibrational characteristics from measured data. In Raman, on the other hand, the probe wavelength can usually be selected in a regime where the material is highly transparent. However, Raman measurements generally involve more sophisticated instrumentation and implementation than IR. Also, in certain instances Raman signals may be obscured by broad-band luminescence from trace impurities, which can be difficult to avoid even with tunable sources.

**Scope and Outline of this Chapter**

A detailed theoretical treatment of vibrational characteristics, Raman scattering and IR absorption in condensed matter is beyond the scope of the present chapter. Moreover, the derivation and formulation of selection rules for Raman and IR processes requires extensive use of group theoretical methods[4]. Readers interested in the detailed theory are directed to several excellent books and articles on the subject[1,4,5-9].

The present chapter describes vibrational modes and spectra in physically intuitive terms, with an eye to applications of Raman and IR measurements, and the interpretation of the resulting spectra. Fortunately, key aspects of vibrational spectra can be adequately understood using phenomenological descriptions, for example, representing vibrational modes in terms of (damped) harmonic oscillators. The simplified equations and rules of thumb presented here facilitate the application and interpretation of Raman and IR spectra by the novice, while serving as a point of departure for readers wishing to delve deeper into the present subject.

Glass spectra are best understood in the context of vibrational properties of molecules, crystals and disordered solids. The second section discusses the vibrational modes and associated IR and Raman spectra for these cases, and examines similarities and differences in the spectra. The third section presents a simplified mathematical description of IR response functions and Raman scattering intensities.

The remaining sections focus on the vibrational spectroscopy of glasses. Spectra of common glasses are examined in the fourth section and their implications discussed. The fifth section presents a case study illustrating the application of vibrational spectroscopy to the characterization and interpretation of the properties of a relatively new class of glasses, namely, the heavy metal fluorides. The final section contains a brief discussion of applications of vibrational spectroscopy and some closing remarks.

## VIBRATIONAL SPECTRA OF MOLECULES, CRYSTALS AND DISORDERED SOLIDS

The interpretation of Raman and IR spectra of materials stems from an understanding of vibrational modes and their dependence on composition and chemical structure. Spectral characteristics depend on the types (symmetry) of vibrations and their coupling coefficients to light. In general, light couples directly to a limited number of vibrational modes, as determined by the selection rules discussed below governing Raman and IR processes[4].

In addition to the strong features associated with "allowed" or first order processes, light also couples to a broader range of vibrations via higher order processes, but their contributions to spectra are generally much weaker.

## Vibrational Properties of Molecules and Crystals

Before addressing the vibrational characteristics of disordered solids and glasses in particular, it is instructive to first consider some simpler cases, namely, isolated molecules, ionic crystals, and molecular crystals. The vibrational properties of liquids, which are not treated separately here, are also similar in many ways to those of disordered solids. As discussed below, disordered solids often display a mix of properties, some representative of short-range order, e.g., isolated structural units or molecules in molecular crystals, and others of long-range order, similar to crystalline materials. The following discussion highlights just a few key aspects of vibrations and spectra of molecules and crystals; more detailed treatments of these subjects are available in the literature (e.g., Refs. 10 and 11 for molecules, and Refs. 5 - 7 for crystals).

### Molecular Vibrations and the Anharmonic Oscillator Model

In simple terms, the atoms of a molecule may be considered as masses connected by springs of appropriate stiffness, representing various types of bonds (e.g., ionic, covalent or mixed) between atoms. If the restoring forces exerted by the springs are nearly linear as a function of separation distance, then the vibrations of the molecule will be harmonic, and may be represented as a set of j harmonic oscillators of fundamental frequencies $\omega_j$. Any vibration of the molecule can be expressed as a linear combination of these fundamental oscillations, or normal modes of vibration, which are solutions to the differential equations of motion for the atoms.

In practice, the springs will be nonlinear to some extent, resulting in damping of the motion between atoms and in coupling between the vibrational modes of the idealized system. Thus, the oscillators are no longer purely harmonic, but become "damped" or "anharmonic". Fortunately, it is possible to express the vibrational response of a set of anharmonic oscillators in fairly simple form, as will be indicated later.

A molecule possesses a limited number of distinct vibrational modes, and therefore would be represented as a collection of discrete oscillators of appropriate frequency. Due to anharmonicity, however, the oscillators do not vibrate at just a single frequency but rather some range of frequencies centered about the harmonic value; also, when excited, vibrations persist only for some finite amount of time rather than indefinitely, i.e., are characterized by a damping coefficient.

A simplified example is provided by the case of a linear diatomic molecule of force constant f and atomic masses $m_1$ and $m_2$. In the harmonic approximation, such a molecule possesses a single type of vibration, an antisymmetric stretch mode of reduced mass $\mu$ and frequency $\omega_o$, in which the atoms move in opposite directions, with

$$\frac{1}{\mu} = \frac{1}{m_1} + \frac{1}{m_2}, \quad \omega_o^2 = f/\mu \tag{1}$$

Extending the analysis to molecules containing N atoms indicates that they possess, in general, 3N-6 independent vibrational modes; an important exception, corresponding to the example in the previous paragraph, is the linear molecule, for which there is 3N-5 modes (because one degree of rotational freedom becomes constrained). Derivations and descriptions of the vibrational modes of various simple molecules (e.g., three and four atoms) are widely available in the literature.

In addition to vibrational modes associated with translation of atoms, there are also rotational modes, typically at much lower frequencies, involving rotations of atoms in the molecule. There are also mixed modes consisting of combinations of vibrations and rotations; detailed treatments are presented in Refs. 10 and 11.

The vibrational spectrum of a molecule thus consists of a set of distinct lines corresponding to the vibrational (and rotational) frequencies of the modes. In principle, overtones at frequencies which are integral multiples of the fundamental are also possible, but their contributions to the spectrum are much weaker than the fundamental. Only certain modes of the vibration spectrum are active in Raman, IR or both, as determined by mode symmetries (selection rules). The physical criterion for IR-active modes is that the vibrations result in a net dipole moment. For example, the stretching mode of a diatomic molecule consisting of two oppositely charged ions would be IR-active; in a homopolar molecule with identical atoms (e.g., $O_2$ or $N_2$) the mode would not be IR active. Another example: a linear triatomic molecule has a mode in which one atom stays fixed and the other two vibrate in the same direction; such a mode would not be IR active. Similar considerations apply to rotational modes, which can be IR-active only in molecules with permanent dipole moments.

Since Raman scattering is a nonlinear process involving excitations of virtual electronic states, and because the scattering efficiency depends the polarizations of the incident and scattered light, the selection rules governing Raman processes are more complicated and less intuitive. Nevertheless, some general rules of thumb can be used to predict which modes will be active in Raman. As indicated in the following section, lowest order Raman processes are associated with the first derivative of the polarizability; if the polarizability differs for positive and negative values of the vibrational coordinate, then the mode is Raman-active, and if it is equal for both, then the mode is inactive in Raman. Modes which are symmetric, i.e., in which all the atoms vibrate along equivalent directions, are generally strong in Raman and weak in IR.

An example of this complementarity is provided by centrosymmetric molecules such as $CO_2$, for which Raman-active transitions are forbidden in IR, and vice versa; there are also modes that are forbidden in both Raman and IR. Although the selection rules for molecules are more complicated in general, complementarity between IR and Raman active modes often applies.

**Crystalline Vibrations**

Since a crystal is composed of a very large number of atoms, it could, in principle, possess a continuum of vibrational modes of variable frequencies and wavelengths. However, because the atoms are arranged in a perfectly periodic fashion, only wavelengths corresponding to motions within individual cells of the crystal ("first Brillouin zone" (BZ)) are allowed[5]. Actually, the vibrational spectrum consists of several types of modes or branches, 3N-3 to be exact, where N is the number of atoms in a unit cell. Each branch spans a range of frequencies as a function of the continuous wavenumber **k**.

A linear diatomic chain composed of two different types of atoms (the extension of the linear diatomic molecule to the crystalline case) is an instructive, if not entirely realistic example. This crystal has three types of modes: two acoustic branches, where the frequencies vanish for $\mathbf{k}=0$, and one optical branch with finite frequency at $\mathbf{k}=0$. If the net restoring force acting on each atom is the same as in the diatomic molecule, then the $\mathbf{k}=0$ optical mode corresponds to an oscillator with the same reduced mass and vibration frequency as in the molecule. This result is expected, since the $\mathbf{k}=0$ mode corresponds to the identical motion of atoms in each diatomic unit cell of the crystal.

A more general analysis of the $\mathbf{k}=0$ modes in ionic crystals indicates the existence of both transverse optical (TO) and longitudinal optical (LO) modes, with the longitudinal mode at higher frequency. For homopolar crystals the two optical modes are degenerate, i.e., have equal frequency.

Since the wavenumber of light is very small, light only couples directly to optical phonons near $\mathbf{k}=0$. This restriction is sometimes expressed as conservation of energy and momentum between photons and phonons. Under usual conditions in IR spectroscopy, light couples only to transverse optical modes. Thus, although the vibrational spectrum of a crystal is continuous, the IR spectrum is actually comprised of a limited number of discrete lines, very much like that of an isolated molecule! Of course, the physical origins of the line spectra are different for the two cases: A molecule possesses just a limited number of vibrational modes, some of which are IR-active, so the IR spectrum consists of a few discrete lines. In the crystalline case, the vibrational frequencies span a continuum of values; however, since light only couples to optical modes near $\mathbf{k}=0$, the IR spectrum again consists of discrete lines, similar to the molecular case. In the case of homopolar crystals such as Si, the transverse optical modes are IR forbidden, in analogy to the homopolar molecule case. Analogous comparisons apply to the Raman modes and spectra of crystals relative to those of molecules.

Raman and IR spectra also tend to be complementary for crystals[6] as in the case of molecules. Totally symmetric vibrational modes of the crystal are always Raman-active, and generally strong, while such modes are often absent or weak in IR spectra. For example, in crystals with inversion symmetry, normal modes which are symmetric with respect to the center of symmetry are Raman active but IR-forbidden, whereas antisymmetric modes are IR-active and Raman forbidden. Other examples of complementarity: Homopolar crystals such as Si, where the optical mode is IR-forbidden, but Raman-active; alkali-halide crystals, where the optical modes are IR-active, but Raman-forbidden; and fluorite crystals, where the transverse optical mode is IR-active but Raman-forbidden, while other modes are Raman-active but IR-forbidden. The frequency of the Raman mode in fluorites falls in-between those of the TO and LO modes.

**Molecular Crystals; Internal and External Modes**

So far we've discussed the vibrations of molecules and crystals separately. What happens to the vibrational spectrum when molecules form a periodic arrangement, i.e., become a molecular crystal? Usually, atoms within a molecule bond strongly to each other, but molecules in a crystal bond together only weakly via Van der Waals forces. In this instance, two categories of vibrations can be distinguished[7], internal modes of the molecule, similar to those of the isolated molecule, and external modes, or "lattice vibrations", in which the entire molecule vibrates as a unit. If the intermolecular forces are weak, then the spectrum of internal modes will be only slightly modified (i.e., broadened and shifted) from those of the isolated molecule. Also, the selection rules governing IR and Raman spectra will persist in the crystal to some extent. External modes, on the other hand, will be governed by crystal periodicity and crystalline selection rules, although their relative contribution to the spectrum may be small. Thus, Raman and IR spectra of molecular crystals contain features from both internal as well as external modes.

The persistence in the crystal of modes similar to those of an isolated molecule is an example of the effects of "short range order", i.e., the bonding between atoms within a molecule being much stronger than to atoms outside the molecule. The vibrations of the molecules forming the crystal lattice are governed by crystalline selection rules, and represent the effects of "long range order". Thus, features attributable to both short range order (individual molecule) and long range order (periodicity) are manifested in the vibrational spectra of molecular crystals. Short range order generally dominates in glasses, although longer range order also affects the spectrum, as will be discussed later in this chapter.

**Vibrational Properties of Amorphous Materials**

Amorphous materials, as opposed to crystals, are structurally and sometimes compositionally disordered (i.e., heterogeneous). Although properties such as composition, density, electric moment, and similar quantities may be fixed on average, they nevertheless can vary (sometimes substantially) from point to point in an amorphous material. Generally, despite the variations, correlations exist between nearby atoms, indicative of "short range order"; long range order, i.e., correlations between distant atoms, is weak or absent.

Consider, for example, amorphous as opposed to crystalline silicon. Crystalline silicon is composed of periodically arrange tetrahedral units, with identical bond angles and bond lengths within each unit. In amorphous silicon, however, the bond angles and bond lengths are no longer fixed, but statistically distributed. The distribution is usually narrow, with mean values quite close to those in the crystal. Thus, in the vicinity of any particular silicon atom, the amorphous material appears to be nearly ordered, with a local environment similar to that in crystalline silicon. However, when one proceeds to sites further from the starting point, the deviations from crystalline structure become greater; at some point, the relationship to the starting atom is no longer apparent.

Before discussing vibrational spectra in real disordered solids, it's instructive to consider an extreme case of a hypothetical, "totally disordered" amorphous material where the species of atoms in the material and their locations are totally uncorrelated. Such a material would possess a continuum of vibrations, corresponding to the virtually unlimited number of arrangements of its atoms. Moreover, since the atoms are randomly arranged, there are no universal selection rules governing the coupling of light to the materials. Both the Raman and IR spectrum become continuous and proportional to the density of modes at each frequency ("vibrational density of states") for this hypothetical solid.

In reality, amorphous materials are not "totally" disordered, but possess some form of short range order. In particular, vibrational properties are often attributable, at least in part, to the existence of individual, localized structural units, containing varying numbers, types and arrangements of atoms. Within the units, structural parameters such as bond lengths and strengths may vary randomly, although the average macroscopic properties are fixed. Therefore, the IR and Raman spectra do not simply mimic the vibrational density of states as they would in a perfectly disordered system, as discussed later in this chapter. Rather, the spectrum will be dominated by features associated with each type of unit in the material, and the selection rules apropos to that unit. The statistically distributed structural parameters, plus the interactions between units, causes broadening of the symmetry-allowed spectral lines, and introduces additional spectral features due to previously "forbidden" vibrations. Thus, the spectrum will consist of distinct spectral features characteristic of the individual units, superimposed on a (much weaker) continuous background characteristic of a disordered system of randomly located masses connected by springs of correspondingly variable stiffness.

A related interpretation of the spectral characteristics of amorphous solids considers how crystalline spectra change as the structure deviates from the ideal crystal. In this picture, the vibrational modes are very nearly those of the crystal; the main difference in the amorphous solid is the breakdown in **k**-selection rules, which allows photons to couple with vibrations other than $\mathbf{k}=0$, i.e., throughout the Brillioun zone. The resulting spectrum is therefore a continuum rather than narrow lines. However, the actual shape of the spectrum is determined by the coupling strength as a function of **k**, which depends on the details of the structural disorder and therefore cannot be predicted in general. An example is the disordered version of the diatomic linear chain discussed previously[12]; as opposed to the molecular and crystal cases, a continuum of vibrational modes becomes active in both Raman and IR. Numerical calculations for disordered diatomic chains indicate that the specific from of the vibrational spectrum depends sensitively on the parameters describing the disorder. The significant observation for our purposes is the multiplicity of vibrational modes and the broadening of the spectrum which result from disorder.

In practice, the two simple extremes of total disorder and purely short range order are never realized. Real amorphous solids and glasses possess correlations between atoms and structural units extending over varying scales, a sort of "mid-range order" which may significantly affect selection rules and vibrational spectra. Nevertheless, as will be shown by example later in this chapter, many features of glass spectra can be understood and information deduced using the simplified concepts outlined here.

## IR RESPONSE FUNCTIONS AND RAMAN CROSS SECTIONS FOR SOLIDS

Up to this point, a non-mathematical description of vibrational modes and spectra has been presented. In this section, some simple mathematical formulae for infrared response and Raman scattering intensities are presented and discussed briefly.

### Damped Harmonic Oscillator Model for IR Response in Solids

The propagation of electromagnetic waves in refracting and absorbing media is governed by the frequency ($\omega$)-dependent complex dielectric function $\varepsilon(\omega)$ and its square root, the complex refractive index,

$$\varepsilon^{1/2} = n + i\kappa \;,\; \varepsilon(\omega) = \varepsilon_1 + i\varepsilon_2 = (n^2 - \kappa^2) + 2in\kappa \qquad (2)$$

The dielectric function $\varepsilon$ appears in the expression for the electric field **E** in the medium as:

$$E = E_o \exp[i\omega(e \cdot r \frac{\sqrt{\varepsilon}}{c} - t)]$$
$$= E_o \exp(-\omega\kappa e \cdot \frac{r}{c}) \exp[i\omega(e \cdot r \frac{n}{c} - t)] \quad (3)$$

where **e** is a unit vector in the direction of propagation. The IR response characteristics of a material are usually measured in terms of the reflectivity R (given below for normal incidence) and the absorption coefficient $\alpha$, which are related to n and $\kappa$ by:

$$R = \frac{(n - 1)^2 + \kappa^2}{(n + 1)^2 + \kappa^2} \quad (4)$$

$$\alpha = \frac{4\pi\kappa}{\lambda}, \; |E^2| = |E_o^2| \exp(-\alpha e \cdot r) \quad (5)$$

Explicit expressions can be obtained for $\varepsilon$ vs frequency in simple cases, for example, for a medium characterized by a single anharmonic oscillator of frequency $\omega_o$ and damping $\gamma$. In this case, the dielectric function takes the form[7]:

$$\varepsilon(\omega) = \varepsilon_\infty + \frac{\varepsilon_o - \varepsilon_\infty}{1 - (\omega/\omega_o)^2 - i\,(\gamma/\omega_o)(\omega/\omega_o)} \quad (6)$$

For a crystal with a single IR-active mode, the imaginary part of $\varepsilon$, which represents absorption of photons and excitation of phonons, displays a peak corresponding to the TO mode near $\omega_o$; the width of the peak is proportional to $\gamma$. Similarly, the peak in the imaginary part of $1/\varepsilon$ corresponds to the LO mode of the crystal. The absorption spectrum, which is proportional to $\varepsilon_2$, also has a peak near $\omega = \omega_o$. These peaks may, in general, have complicated shapes because $\gamma$ is a function of frequency, although for simplicity $\gamma$ is often approximated as a constant. The reflectivity R also displays a single peak in the vicinity of $\omega_o$; the explicit expression for R is unnecessarily complicated and need not be given here. Note that in the limit of a perfectly harmonic oscillator, $\gamma$ goes to zero and the absorption spectrum becomes a delta function, i.e., a line which is infinitesimally narrow but infinitely strong. However, the relevant physical quantity for this case, the total line strength obtained by integrating over frequency, is finite and equal to that in the anharmonic case.

In the above example, just a single oscillator was included in the dielectric function. As discussed previously, materials may, in general, possess any number of discrete and/or continuously distributed vibrational modes, and some of these may couple to light via dipole moment interactions. In this instance, the complex dielectric function in Eq. (6) would be generalized[17] to include multiple oscillators of frequencies $\omega_j$, damping coefficients $\gamma_j$, and coupling constants $\rho_j$:

$$\varepsilon = \varepsilon_\infty + \sum_{j=1}^{n} \frac{4\pi\rho_j}{1 - (\omega/\omega_j)^2 - i(\gamma_j/\omega_j)(\omega/\omega_j)} \tag{7}$$

From the above expression, one can directly obtain explicit (albeit somewhat complicated) expressions for the absorption coefficient and reflectivity. The spectrum now contains multiple peaks of varying heights and widths, depending on the values of $\gamma_j$ and $\rho_j$ for each "allowed" oscillator which couples to light. A specific example, in which the dielectric function is parametrized in terms of two damped modes, is presented below in connection with the IR spectrum of fluorozirconate glasses.

When the modes form a continuum, then the discrete sum in Eq. (7) is replaced by an integral over frequency. This would apply, for example, to the hypothetical "totally" disordered solid described in the previous section. However, a continuum of vibrational modes need not imply a continuous IR or Raman spectrum. For example, although the vibrational modes of a crystal span a continuum of values, light couples only to optical phonons at $\mathbf{k}=0$, i.e., the BZ center. Thus, in lowest order, the spectrum is comprised of a discrete set rather than a continuum of oscillators. As indicated previously, the IR response functions of amorphous materials contain peaks analogous to those for crystals and molecules, as well as weaker and broader features in which light couples to a continuum of vibrational modes.

**Raman Scattering Cross Section**

The general form of the Raman spectrum was indicated briefly in the first section. In practice, the Raman spectrum tends to be dominated by Stokes (frequency-downshifted) processes, which originate from the ground state. Anti-Stokes processes, which require phonons to be in excited states, are usually negligible because the initial population of phonons, which is governed by the Bose-Einstein distribution, is small at room temperature. As a function of frequency shift, the Raman spectrum displays peaks due to discrete vibrational modes, or features due to continuous modes, depending on selection rules and coupling strengths for Raman processes in a particular geometry.

The physical mechanism responsible for Raman scattering is the dependence of the electronic polarizability on atomic positions[5,6], whether for molecules, crystals or glasses, and the dipole moments accompanying changes in polarizability. Inelastically scattered photons are produced by the interaction of the electromagnetic field of the incident photons with the electronic polarizability. Quantum mechanically, Raman scattering involves a three-step process, namely: Changes in polarizability which result in transitions from ground to "virtual" excited electronic states; creation of "real" phonons via electron-phonon interactions, accompanied by further changes in electronic states; and, finally, return to the electronic ground state. Within a macroscopic description of scattering processes, the polarizability tensor $\alpha_{lm}$ and associated electric moment **M** is expanded in terms of the normal mode displacements $u_j$ as:

$$\alpha_{lm} = \alpha_{lm}^{(0)} + \alpha_{lm}^{(1)} u + \alpha_{lm}^{(2)} u^2 + \cdots \qquad \textbf{(8a)}$$

$$u_j = u_{jo} e^{i\omega t}, \quad \alpha_{kl}^{(1)} = \alpha_{kl,j} = \left(\frac{\partial \alpha_{kl}}{\partial u_j}\right)_{u=o} \qquad \textbf{(8b)}$$

$$M = \alpha E = \alpha^{(o)} E_o e^{i\omega t} + \alpha^{(1)} u_{jo} E_o e^{i(\omega \pm \omega_j)t} + \alpha^{(2)} u_{jo}^2 E_o e^{i(\omega \pm 2\omega_j)t} + \ldots \qquad \text{(8c)}$$

The first term leads to elastic, or "Rayleigh", scattering. The second term, involving the first derivative of the polarizability, leads to the creation and annihilation of phonons. When the phonons are low-frequency acoustic modes, typically in the range of 0.1 - 1 $cm^{-1}$, the process is referred to as Brillouin scattering. When phonons at IR frequencies ($10^2$-$10^3$ $cm^{-1}$) are involved, whether acoustic or optical, then the process is called Raman scattering. The higher order terms in the polarizability give rise to multiphonon scattering processes of various types. Note that the polarizability is a tensor, so the scattering cross section depends on the scattering geometry and scattering angle, and the polarizations of the incident and scattered photons. In fact, the depolarization manifested in the scattered spectrum provides information about the extent of order or local structure in the material, as discussed further below.

Consider the highly simplified case of a cubic, homopolar crystal with one Raman-active phonon, again represented as a single anharmonic oscillator of frequency $\omega_o$ and damping constant $\gamma$. The Stokes scattering intensity per unit length of crystal and per unit solid angle, at right angles to an incident light beam of frequency $\omega_i$, is given by:

$$I = \frac{3\hbar\omega_s^4}{\rho_D c^4 \omega_o} |\alpha^{(1)}|^2 (n_o + 1) g(\omega)$$

$$g(\omega) \sim \frac{\gamma\omega}{[1-(\omega/\omega_o)^2]^2 + (\gamma/\omega_o)^2(\omega/\omega_o)} \qquad \text{(9)}$$

where $\omega_s$ is the frequency of the scattered light, $\omega = \omega_s - \omega_i$ is the frequency shift, $\rho_D$ is the density, and $g(\omega)$ is the same anharmonicity-induced lineshape as in IR processes (proportional to $\varepsilon_2$). The explicit temperature dependence is often removed by dividing measured Raman spectra by the temperature-dependent Bose-Einstein factor $n_o+1$ to obtain so-called "reduced" spectra. For more general cases involving multiple Raman-active oscillators, the scattered spectrum would consist of lines for discrete, and broad features for continuous modes. The Raman scattering intensity is explicitly temperature dependent, in contract to the IR dielectric function. Moreover, the relative strengths of the modes and the overall dependence of the scattering on frequency is more complicated than for the IR case discussed above. In fact, the Raman spectrum displays frequency-dependent features from sources other than phonon anharmonicity, for example, due to photon-phonon interactions in the case of polar crystals (see Ref. 13, for example).

A theory of Raman scattering in glasses was formulated by Shuker and Gammon (Ref. 9, pp 334-338), based on the short correlation lengths in glasses relative to crystals. Their theory demonstrates that first order processes in glasses yield a continuous Raman spectrum, due to fluctuations in the polarizability modulated by the vibrational modes. The spectrum is related, but in general not proportional to, the vibrational density of states. Assuming a simplified spatial correlation function along with several other simplifying assumptions, yields the following expression for the stokes scattering intensity as a function of frequency shift:

$$I(\omega) \sim \omega^{-1}[1+n(\omega)]\sum_{b} c_b g_b(\omega) \qquad (10)$$

where "b" is a vibrational band index; $g_b(\omega)$ is the density of states for band b; and $c_b$ is a factor which depends on the correlation length associated with the modes in band b, as well as the symmetry of the scattering process. Thus, the scattering spectrum consists of a weighted sum of the densities of states of the various bands. Since the weighting factors may differ significantly from band to band, the spectral shape need not be identical to the unmodulated density of states. In principle, anharmonicity broadening should also be included in the above expression for I, but its effect is expected to be small on the continuous spectra applicable to glasses (as opposed to the line spectra of molecules and crystals).

**Raman Depolarization Ratio**

The depolarization ratio is defined as:

$$\rho = I_{\parallel} / I_{\perp} \quad (11)$$

where $I_{\perp}$ and $I_{\parallel}$ are the intensities of the scattered radiation at frequency shift $\omega$ with their plane of polarization perpendicular and parallel to that of the incident beam. Alternatively, the notations HH, VV, VH and HV, where H is horizontal polarization and V is vertical, are often used in the literature to indicate the polarizations of the incident and scattered beams; this notation appears in subsequent figures in this chapter.

For isolated molecules (such as gases at low pressure, where no preferred orientation exists), it can be shown that $\rho$ assumes values between 0 and 3/4, depending on the symmetry of the mode: For totally symmetric modes, $\rho$ is less than 3/4, corresponding to partial polarization of the scattered light. For all modes of lower symmetry (i.e., not totally symmetric), $\rho$ equals 3/4 and the scattered light is totally depolarized. Thus, the depolarization spectrum provides information about mode symmetries and therefore the symmetry of the molecule. Similar considerations pertain to crystals, although the interpretation of the depolarization spectrum is greatly complicated by the dependence of $\rho$ on the angle of the scattered light as well as the orientation of the crystal[7]. The upside to this complexity is the considerable amount of information about mode symmetries and crystal structure which can be extracted from depolarization spectra.

The depolarization spectra of amorphous solids[14] display certain trends similar to molecules. In particular, for any isotropic disordered solid, $\rho$ again lies between 0 and 3/4, and only symmetric modes can display $\rho$'s less than the maximum value (generally, $\rho < 0.1$ for highly symmetric modes). Thus, a highly disordered solid with no identifiable structural units would be expected to display a flat, featureless depolarization spectrum. On the other hand, if the amorphous solid possesses identifiable structural units and structural heterogeneity, then the depolarization spectrum will display minima corresponding to symmetric or nearly symmetric modes of such units. These features will, of course, be broadened due to anharmonicity, interactions between units and structural heterogeneity, as they are in the Raman spectrum. Thus, the frequency dependence of the depolarization spectrum provides information about disorder, local symmetry and anharmonicity in amorphous solids.

## VIBRATIONAL SPECTRA OF GLASSES

The term "glasses" usually refers to amorphous materials which can be synthesized into bulk objects of various sizes and shapes. Most common glasses are insulators which are transparent over an extended range of frequencies, but there are also metallic and semiconducting glasses. Silicates are perhaps the most prevalent and familiar of all glasses, but there are many other types of oxide as well as nonoxide glasses, including phosphates, borates, chalcogenides and halides, to name a few. Glasses are often described in terms of isolated "identifiable" structural units, which are said to form a "random network", as opposed to the ordered, periodic arrangement of a crystal. Alternatively, more detailed descriptions of glass structure have been formulated in terms of clusters of various types, sizes and symmetries. The interpretation of glass spectra in terms of "simple" identifiable structural units is often a convenient starting point, but as evident from the discussion below, such descriptions are often incomplete and even misleading if taken at face value.

### Vibrational Spectra of Fused Silica

Fused silica, perhaps the most familiar single component glass, provides an instructive example of how deceptively simple glasses can display rather subtle and difficult-to-interpret spectra. The IR reflectivity spectrum of fused silica is illustrated in Fig. 5, and displays two well-separated, strong peaks, along with some subsidiary features (peaks and shoulders). The Raman spectrum, on the other hand, displays one dominant, highly polarized, rather broad peak at low frequency, along with some weak features at higher frequencies, as also indicated in Fig. 5. This behavior contrasts markedly with the more pronounced structure and narrower peaks found for crystalline quartz (see Fig. 6). Some complementarity between IR and Raman spectra of fused silica is evident from Fig. 5, although in this case the secondary IR peak does fall within the same general frequency range as the dominant Raman line.

An interpretation of fused silica spectra in terms of modes of isolated molecular units is very tempting[15]. In particular, consider an isolated $SiO_4$ tetrahedron, which has four modes with frequencies indicated in Fig. 5. These frequencies appear to correspond to positions of principal peaks, as well as some subsidiary peaks in the spectra. For the isolated unit, all four modes are Raman active, while only $\nu_3$ and $\nu_4$ are IR-active. This suggest a possible identification of the two main peaks in the reflectivity spectrum with the $\nu_3$ and $\nu_4$ modes, and one weak feature with $\nu_1$. The main Raman peak appears

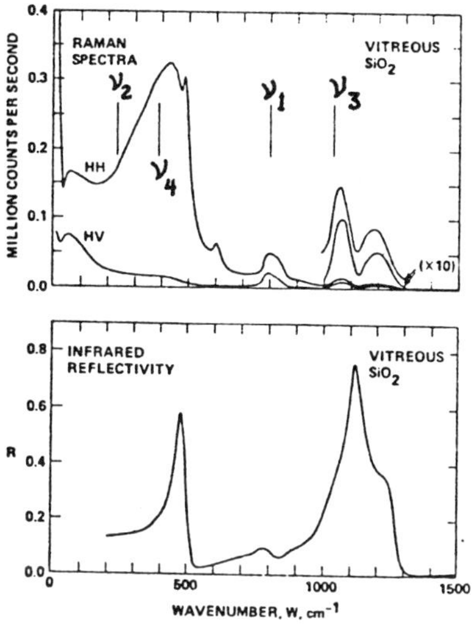


**Figure 5.** IR and Raman spectra of fused silica; the four mode frequencies of an isolated $SiO_4$ tetrahedron are indicated by the vertical bars (from Ref. 15).

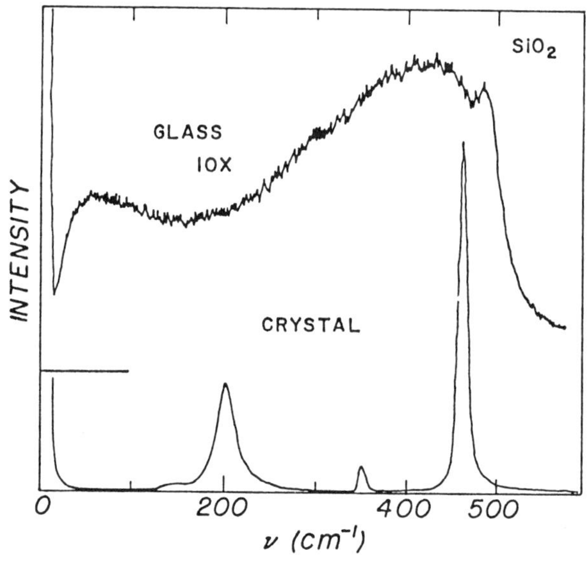


**Figure 6.** Comparison between Raman spectra of crystalline $SiO_2$ (quartz) and vitreous $SiO_2$ (fused silica) (from R. Shuker and R. Gammon, in Ref. 9, pp 334-338).

to be associated with a mixture of both $\nu_2$ and $\nu_4$ modes, and some of the subsidiary Raman peaks appear to correspond to the remaining two frequencies of the isolated unit.

Despite the encouraging correspondences between the observed features of $SiO_2$ spectra and the modes of the isolated tetrahedron, closer scrunity reveals deficiencies with this interpretation. For example, the dominant, highly polarized peak in the Raman spectrum would be expected to be associated with symmetric vibrations in the glass; yet, this peak falls in the vicinity of the $\nu_3$ and $\nu_4$ modes of the tetrahedron, which are not symmetric. On the other hand, the mode $\nu_1$, which is symmetric, should show up as a prominent, highly polarized peak in the glass, but doesn't. And while features appear in the glass spectrum at frequencies corresponding to the four Raman active modes, some are very weak; moreover, additional features not attributable to the four modes are also evident. Further issues surface from the analysis of spectra of other tetrahedrally bonded glasses [16,17] such as $GeO_2$, where the agreement between isolated unit frequencies and spectral features in the glass is not as good as for fused silica. Thus, the apparent correspondence in the case of fused silica between the frequencies of the isolated tetrahedron and features in the glass spectrum may in fact be fortuitous.

A more extensive analysis of spectra for $SiO_2$ and several other tetrahedrally bonded glasses (see Refs. 15 - 17) suggests that the isolated tetrahedron model only accounts for certain limited aspects of observed spectra, such as the appearance of two, well-separated main peaks in the IR spectrum, and the approximate locations of certain peaks in IR and Raman. Nevertheless, the character of the modes is significantly modified in the glass relative to an isolated tetrahedron. For example, the dominant, highly polarized Raman line occuring at low frequencies in these glasses corresponds to a symmetric vibration, rather than the non-symmetric $\nu_2$ and $\nu_4$ modes of the isolated tetrahedron. Similarly, the features in Raman in the vicinity of the $\nu_1$ mode no longer correspond to a symmetric mode. Thus, the isolated tetrahedron does not provide a satisfactory interpretation of the details of the glass spectra, and alternative interpretations are required. For example, more detailed analyses suggest that the most intense Raman mode in these glasses be attributed[18] to a symmetric stretch of the birdging atom (O in the case of fused silica) along the bisector of the angle it makes with its two nearest neighbors.

Further insights into glass spectra are obtained by deriving a complex dielectric function from the reflectivity data, and then plotting $\varepsilon_2$ and $\mathrm{Im}\varepsilon^{-1}$

("energy-loss spectrum") vs frequency, as illustrated in Fig. 7. As discussed previously, peaks in the $\varepsilon_2$ spectrum indicate transverse optical (TO) modes, and peaks in the energy loss spectrum longitudinal optical (LO) modes in polar materials. In this way, one identifies three pairs of LO-TO modes for fused silica from IR data; some of the features in the Raman spectrum may also be attributed to LO modes. Analogous characteristics are manifested in the vibrational spectra of other tetrahedrally bonded single component glasses, including $GeO_2$ and $BeF_2$ (see Refs. 16 and 17), and can be interpreted in a similar manner.

Considerable attention has been given above to the case of fused silica in order to illustrate the factors coming into play in the interpretation of glass spectra, and the various approaches which can be used to assist in the process.

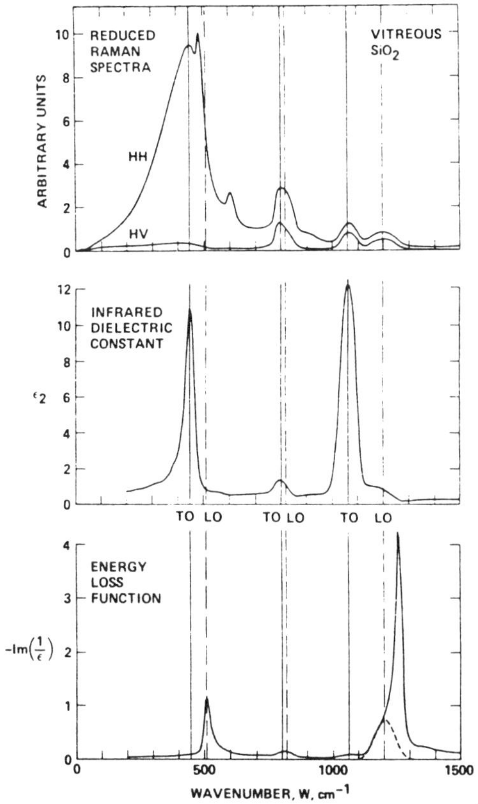


**Figure 7.** Reduced Raman spectra, imaginary part of the IR dielectric constant and energy loss function for fused silica, indicating assignments of three pairs of TO-LO modes (from Ref. 16).

Vibrational spectroscopy does not always provide conclusive information about glass structure, and confirming data from neutron scattering and X-ray spectra, for example, may be required to resolve open issues.

**Vibrational Spectra of Other Single Component Glasses**

Single component chalcogenide glasses such as $As_2S_3$ and $As_2Se_3$ display trends similar to the above[12]. While the spectra for the crystalline versions possess significant structure, the glass spectra are much broader and smoother. Both the IR reflectivity and Raman spectra are dominated by single broad peaks; the Raman peak is highly polarized and at a higher frequency than the IR peak. Some additional smaller peaks are also evident in the Raman spectrum at frequencies lower than the main peak. In this case, the observed spectral features are reasonably well accounted for in terms of two types of identifiable units, namely, $AsX_3$ pyramids and $As_2X$ water-like molecules.

The vibrational characteristics manifested above are typical of single component glasses with one principal type of bond in the glass structure, including certain oxides (e.g., $SiO_2$ and $GeO_2$), halides (e.g., $BeF_2$ and $ZnCl_2$) and chalcogenides (e.g., $As_2S_3$). Such glasses display: (a) Two main peaks in their reflectivity spectra, with the stronger peak generally at the higher frequency; (b) a single dominant, highly polarized peak in the Raman spectrum, generally containing more than 50% of the area under the (reduced) HH spectrum; and (c) a variety of subsidiary features in both spectra. There are generally fewer major peaks in glass spectra relative to their crystalline analogues of the same composition. Moreover, the glass peaks are usually much broader ($>75$ $cm^{-1}$), as expected due to structural heterogeneity and the accompanying breakdown of selection rules in glasses. Glasses such as $B_2O_3$ which contain regular rings are exceptions, and display relatively narrow Raman lines[12,19].

An analysis of the vibrational spectra of single-component glasses[18] indicates that the dominant Raman line in glasses with a single type of bond is attributable to symmetric stretch vibrations of the bridging atoms (O, F, Cl, S) along the bisector of the angle it makes with its two nearest neighbors. The high frequency IR mode represents "rigid-cage" or sublattice-type motion of the bridging atom relative to (Si, Be, Zn, As). As indicated above for $SiO_2$, the IR spectra reveal substantial TO-LO splittings at high frequencies. These split modes also appear as relatively weak and unpolarized lines in the high frequency portion of the Raman spectrum.

Single-component as well as multicomponent glasses may contain more than one type of bond. The Raman spectrum of "simple" glasses containing two different types of bonds, such as alkali oxide glasses, display two highly polarized peaks in Raman, generally of roughly equal strength. These lines are again attributable to symmetric stretch vibrations associated with the two types of bonds involved. Thus a rule of thumb is that the number of intense Raman lines yields the number of well-defined but different bonds in the main framework of the glass structure.

### Vibrational Spectra of Multicomponent Glasses

Multicomponent glasses display a variety of vibrational characteristics depending on the glass structure[12]. In certain multicomponent glasses, identifiable units associated with the individual components persist in the glass structure, and therefore contribute distinct features to the spectrum (depending on the relative percentage of each component in the glass). Additional features due to cross bonding may also occur. Mixed boric-sodium metaphosphate glasses, for example, display prominent features in Raman associated with B-O-B and P-O-P linkages (forming ring structures) for the individual components; intermediate compositions also display additional features attributable to B-O-P bonds[1].

The incorporation of additional components to the glass can significantly modify glass structure and, correspondingly, vibrational spectra. For example, alkali additions (e.g., $Na_2O$) to silicate glasses significantly modify their spectral characteristics and presumably the structure of the glass. In other cases, the glass structure appears to remain dominated, at least over certain composition ranges, by just a single component or small number of components in the glass. Examples will be provided in the following section on heavy metal fluoride glasses.

## CASE STUDY: VIBRATIONAL SPECTRA OF HEAVY METAL FLUORIDE GLASSES

This section examines the spectral characteristics of a relatively new class of glasses, heavy metal fluorides (henceforth "HMFG"), discovered in 1974 by researchers at the University of Rennes in France. These glasses provide an instructive and illuminating example of the application of vibrational spectroscopy to the characterization, and analysis of structure and properties of new glasses. HMFG are also useful examples because they display a rich diversity of spectral characteristics indicative of the variability and complexity of glasses. As vividly demonstrated below, vibrational spectroscopy provides us with an informative glimpse into the properties and structure of complex glasses.

Since their initial discovery, a wide range of HMFG compositions have been synthesized, although the composition ranges for glass forming of a specific set of components tend to be rather narrow. A curious attribute of many of these glasses is that their principal components are not by themselves glass formers; in fact, most practical HMFG compositions contain three or more components. The present discussion is limited to just a few types of HMFG suited for preparation of high optical quality bulk samples; readers interested in more extensive information about HMFG compositions and preparation methods are directed to several recent books and review articles on the subject[20-23].

The spectra of three types of HMFG are considered here:

(a) Fluorozirconte-type glasses, in which Zr or Hf fluoride is the principal constituent, in combination with $BaF_2$ and any number of additional components, typically the fluorides of La, Al, Na or Th.

(b) Rare-earth fluoride glasses, containing fluorides of rare earth elements such as Yb, in combination with the fluorides of thorium, barium and other elements such as Zn and Al.

(c) Transition-metal fluoride glasses, containing fluorides of transition metals such as Zn and Mn, in combination with fluorides of other elements such as Ga, Pb and Al.

**Fluorozirconate Glasses**

The following discussion illustrates how information about glass properties and structure of fluorozirconates can be deduced from their vibrational spectra[24-26]. Consider first, the reflectivity spectra of two fluorozirconate-type glasses based on Zr and Hf fluoride, illustrated in Fig. 8, which are seen to consist of two well-separated peaks. An excellent fit to the reflectivity is obtained using a dielectric function with two damped oscillators, of the form indicated in Eq. (7). The TO and LO modes are then determined from the peaks in $\varepsilon_2$ and $-\mathrm{Im}\varepsilon^{-1}$ vs. frequency, as indicated in Fig. 9.

The absorption coefficient, which can also be calculated directly from the dielectric function, displays a frequency dependence similar to $\varepsilon_2$. Thus, fluorozirconate glasses display two infrared-active modes with discernible TO-LO splitting.

Measurements performed for a variety of other fluorozirconate compositions suggest that tertiary components have a negligible effect on fluorozirconate spectra, although light, high valence metals may affect the spectra of other types of HMFG; an example examined below is rare-earth HMFG containing Al. The dominance of the IR spectrum by two well-separated modes is, in fact, characteristic of many single component glasses.

The polarized Raman spectra of several representative flourozirconate glasses are indicated in Fig. 10, and their depolarization ratios indicated in Fig. 11.

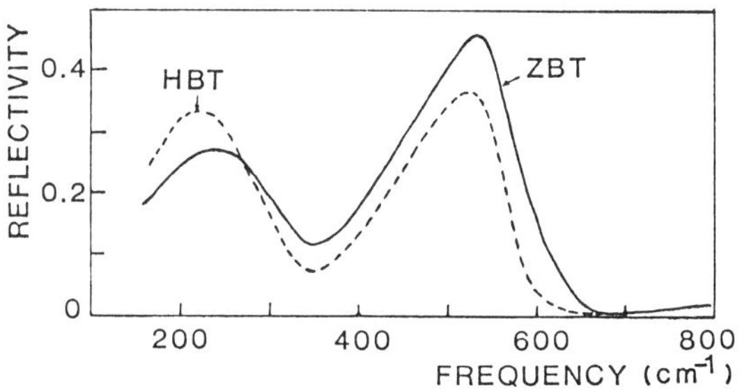


**Figure 8.** IR reflectivity spectra of Zr-Ba-Th fluoride (ZBT) and Hf-Ba-Th fluoride (HBT) glasses (from Ref. 24).

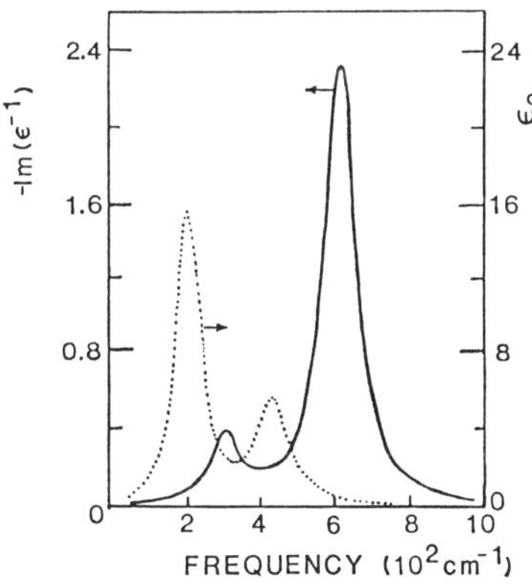


**Figure 9.** Imaginary part of the dielectric constant and energy loss function for HBT glass, calculated from a two-oscillator fit to the measured reflectivity spectrum in Fig.8, indicating assignment of two TO-LO pairs (from Ref. 25).

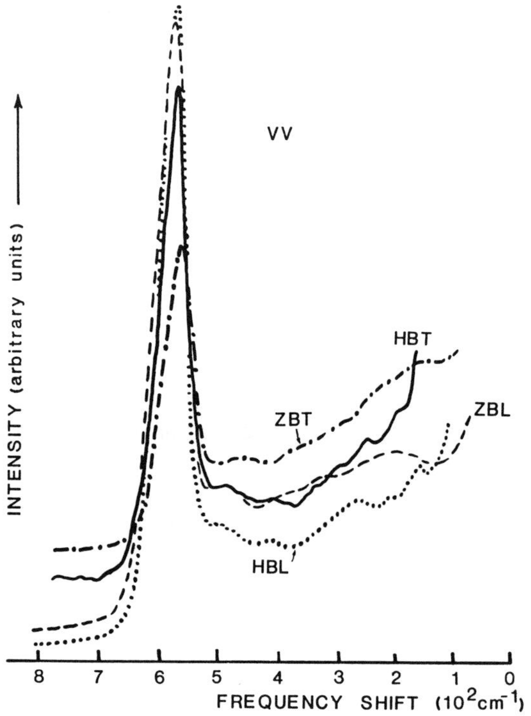


**Figure 10.** Polarized Raman spectra of selected glasses based on Zr and Hf fluorides (B =Ba, L = La, T = Th) (from Ref. 24).

A single, relatively narrow, and highly polarized peak is observed to dominate the spectra, indicative of a single principal bond in the glass. The Raman line falls in-between the high frequency TO and LO modes, which is similar to crystalline fluorides but different from "simple" glasses such as $SiO_2$. This behavior, together with the narrowness of the Raman peak and large TO-LO splitting, suggests that the bonds in fluorozirconates may be ionic. Moreover, the narrow, highly polarized Raman line suggests that the glass contains identifiable units consisting of Zr (or Hf) coordinated with multiple fluorine atoms, and that the Raman mode is associated with a symmetric stretch vibration between Zr (or Hf) and F.

Comparison of glass spectra with those of individual components often

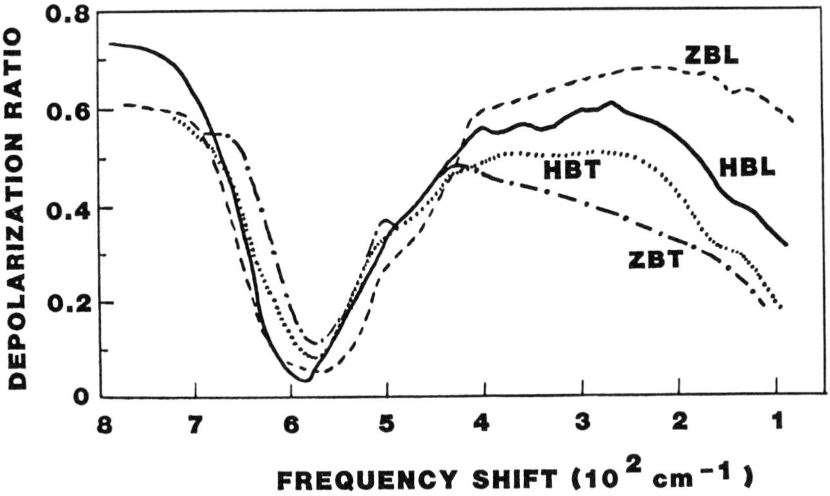


**Figure 11.** Raman depolarization spectrum for the fluoride glasses indicated in Fig. 10.

provides useful insights and additional information. In the case of fluorozirconates, comparisons with the Raman spectra of molten salts of Zr fluorides of various coordination indicates that the coordination number of Zr with F is between 6-8, consistent with results of molecular dynamics simulations and analyses of X-ray data.

Comparisons between spectra of Zr and Hf compositions provide information about relative coupling strengths to light ("effective charges") and values of force constants for Zr vs Hf with F. Also, a comparison of the spectra suggests that the low frequency IR modes are due to Zr/Hf-F bending modes. The dominant effect of $ZrF_4$ on the structure and properties of fluorozirconate glasses is dramatically illustrated by compositional studies of spectra in the Zr-Li-Ba-Th fluoride glass system[27]. This example illustrates the well established technique of using a series of compositionally varying samples, such as indicated in Table 1, to determine the effects of various species and/or components on glass properties and structure.

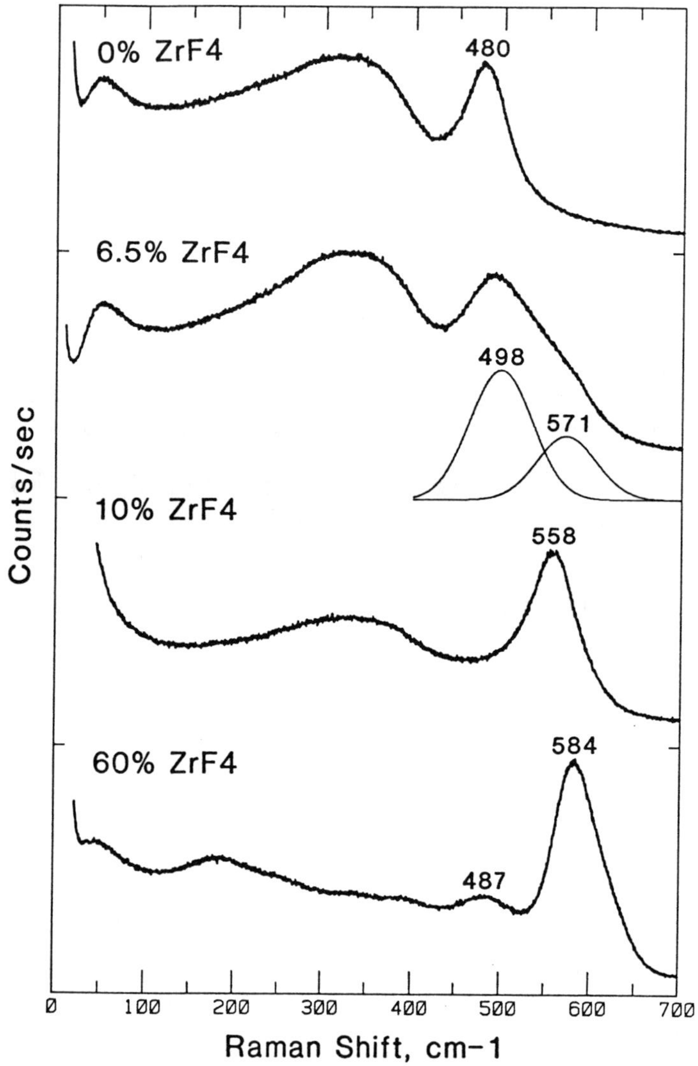


**Figure 12.** Raman spectra of Zr-Li-Ba-Th fluoride glasses containing varying amounts of $ZrF_4$ (from Ref. 27).

**Table 1.** Compositions (Mol%) of Zr-Li-Ba-Th fluoride glasses used for the Raman study in Ref. 27.

| $ZrF_4$ | LiF | $BaF_2$ | $ThF_4$ |
|---|---|---|---|
| 0 | 60 | 10 | 30 |
| 6.5 | 60 | 10 | 23.5 |
| 10 | 60 | 10 | 20 |
| 20 | 60 | 10 | 10 |
| 30 | 45 | 15 | 10 |
| 40 | 25 | 25 | 10 |
| 50 | 20 | 20 | 10 |
| 60 | 20 | 15 | 5 |

The study indicated that for compositions exceeding about 10 mol% $ZrF_4$, both the Raman and IR spectra of the present glasses were dominated by features associated with the $ZrF_4$ component, as illustrated for the case of Raman spectra in Fig. 12. Thus, although $ZrF_4$ is not a glass former in the usual sense, it nevertheless appears to dominate the structure of multicomponent fluorozirconate glasses.

One can proceed much further beyond the analyses presented above, using the detailed features of the glass spectrum to postulate detailed structural models for fluoroziconates (see, for example Ref. 28). Thus, vibrational spectroscopy can be used to extract information at several levels of complexity, some fairly directly and others involving extensive analysis.

**Rare Earth Fluoride Glasses**[24,30]

The IR reflectivity spectra of RE fluoride glases based on Th and Ba fluoride without Al, display two-mode behavior similar to fluorozirconates, as indicated in Fig. 13.

The high frequency peak shifts to lower frequency, resulting in enhanced IR transparency; also, the peak reflectivities are reduced relative to fluorozirconates, indicating lower coupling to light.

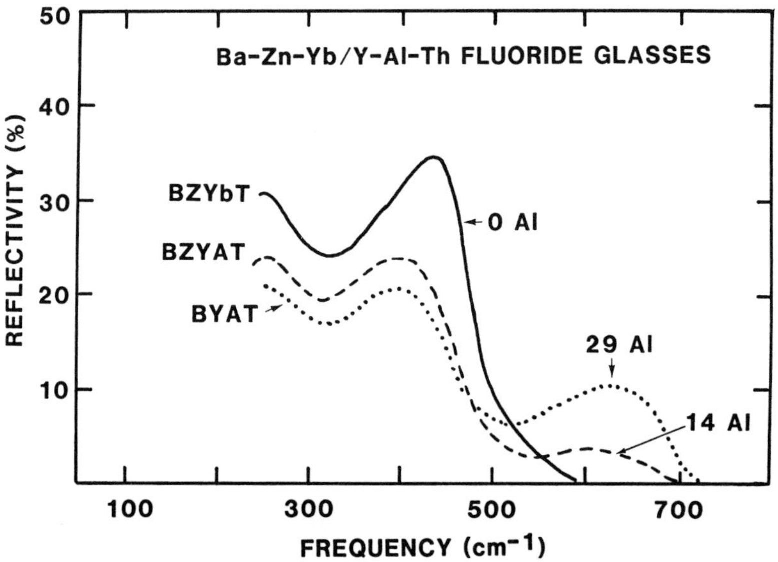


**Figure 13.** IR reflectivity of selected rare-earth fluoride glasses (B = Ba, Z = Zn, A = Al, T = Th) (from Ref. 30).

The addition of certain components to Ba-Th fluoride glasses can significantly influence glass structure, as already suggested by the Zr-Li-Ba-Th system discussed above. In particular, the incorporation of $AlF_3$ in RE fluoride glasses produces a distinct third mode in the reflectivity spectrum, as indicated in Fig. 13. The appearance of this mode suggest possible modification of the glass structure by Al, as will be confirmed by Raman spectra. From a practical standpoint, the additional mode implies degradation of the IR transparency of RE fluoride glasses relative to Al-free compositions, which has, in fact, been verified by IR edge measurements on these glasses.

Further insight into the structure of these glasses may be obtained from their Raman spectra, which display distinct differences from fluorozirconates and "common" single component glasses. As indicated in Fig. 14, the spectra of RE fluoride glasses without Al are dominated by broad, highly polarized peaks in the 200-500 $cm^{-1}$ range, and broad, weakly polarized features at lower

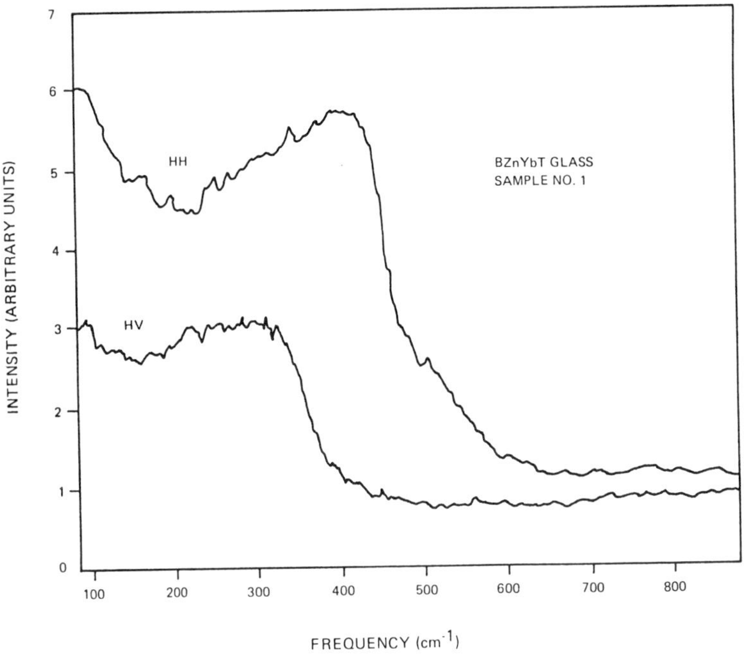


**Figure 14.** Polarized Raman spectrum for Ba-Zn-Yb-Th fluoride glass (from Ref. 29).

frequency. The high frequency peak is attributed to a combination of symmetric modes of the RE ions with F.

When $AlF_3$ is added to the glass, the high frequency peak becomes suppressed and depolarized, as indicated in Fig. 15 for a glass with approximately 15 mol% of $AlF_3$.

With larger amounts of $AlF_3$, the spectrum becomes nearly featureless and depolarized, as illustrated in Fig. 16. As indicated previously, such spectra may be indicative of highly disordered solids without distinct, identifiable structural units. Alternatively, complexes involving RE-F-Al may be formed by the additon of Al to the glass, suppressing the Raman activity of the RE-F mode. The dramatic changes in glass structure associated with Al are also clearly revealed by depolarization spectra, as illustrated in Fig. 17.

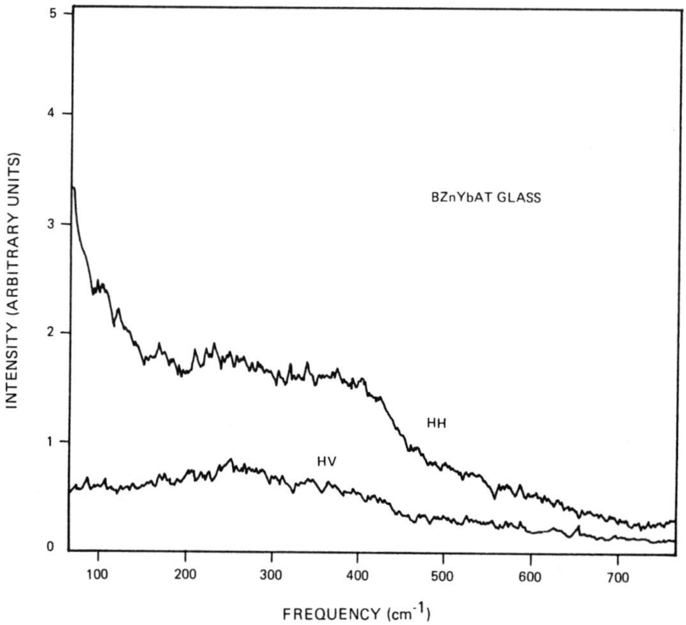


**Figure 15.** Polarized Raman spectrum for Ba-Zn-Yb-Al-Th fluoride glass (from Ref. 29).

Without Al, RE fluoride glasses display deep minima in their depolarization spectrum similar to fluorozirconates. Thus, they presumably contain identifiable units consisting of RE ions with F, although the broader minimum relative to fluorozirconates implies a greater degree of disorder in these glasses. With increasing amounts of Al, the spectrum becomes progressively depolarized, which could be indicative of either increased disorder and/or covalent bonding in the glass.

**Transition Metal Fluoride Glasses**[31]

The TM fluoride glasses display spectral characteristics intermediate to the two families of HMFG discussed above. The IR reflectivity displays two-

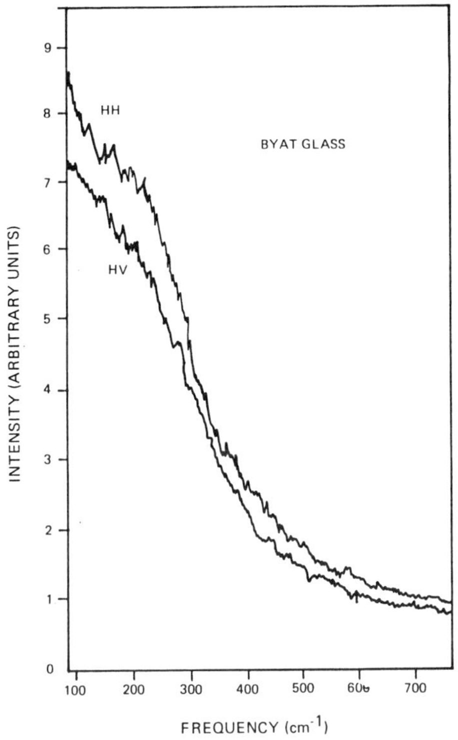


**Figure 16.** Polarized Raman spectrum for Ba-Y-Al-Th fluoride glass (from Ref. 29).

mode behavior similar to fluorozirconates. The Raman spectrum, as illustrated in Fig. 18, is dominated by a single highly polarized peak at high frequency similar to fluorozirconates.

This peak is broader than in fluorozirconates, but much narrower than in RE fluoride glasses. Thus, TM fluoride glasses are believed to contain well-defined structural units, presumably $MF_x$ polyhedra of transition metal ions surrounded by fluorine. The frequency of the dominant Raman peak implies similar coordination numbers to fluorozirconates for TM ions with fluorine. More detailed speculations regarding glass structure[31] derive from comparisons with neutron scattering and EXAFS data.

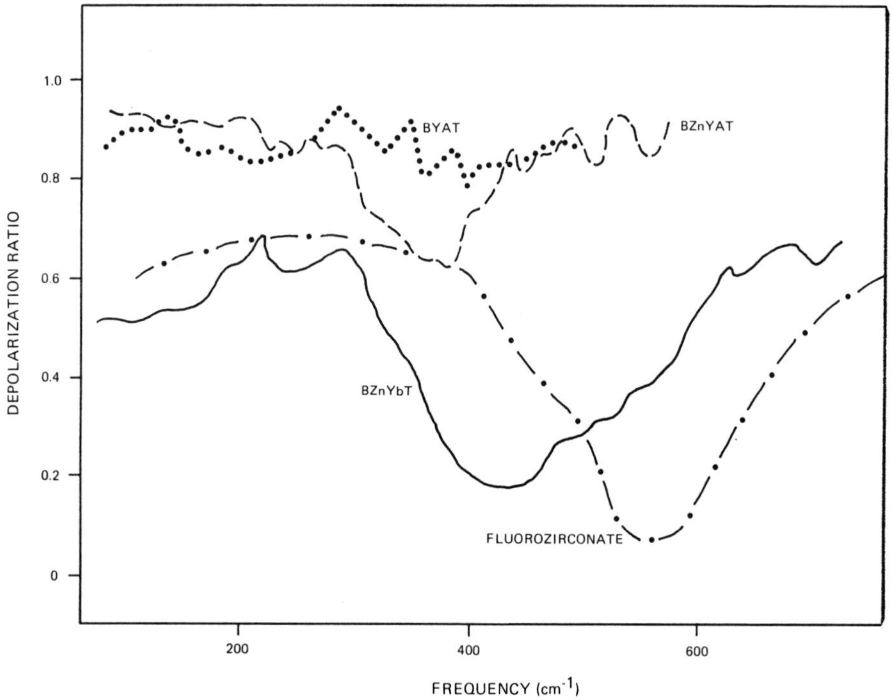


**Figure 17.** Raman depolarization spectra of rare-earth fluoride glasses compared to a representative fluorozirconate glass (from Ref. 29).

The above discussion indicates the usefulness of vibrational spectroscopy in determining the optical and structural properties of new glasses.

## DISCUSSION AND CONCLUDING REMARKS

The above sections have addressed the vibrational modes, and the IR and Raman spectra of glasses, and compared them to molecules and crystals. The spectra of glasses were observed to contain considerable information about

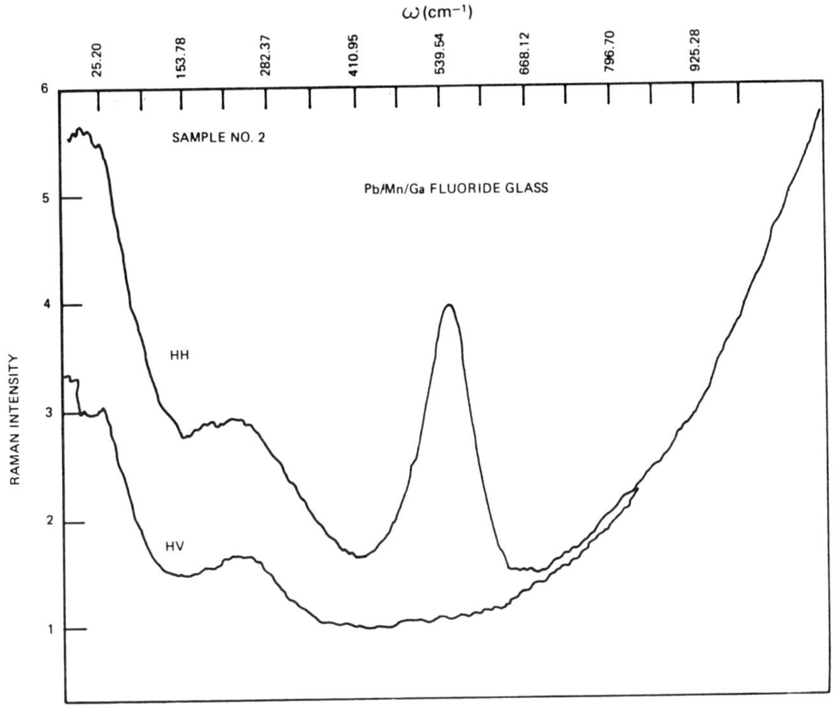


**Figure 18.** Polarized Raman spectrum of a representative transition metal fluoride glass (Pb-Mn-Ga-In fluoride glass) (from Ref. 31).

glass composition, structure and fundamental properties. Thus, vibrational spectroscopy has found many applications[3], ranging from fundamental studies to industrial process control. A few selected examples of applications of IR and Raman spectroscopy are discussed briefly below:

(a) Determination of Vibrational Mode Structure. The frequencies of the fundamental vibrational modes of a material can be obtained from Raman and IR spectra, as illustrated by various examples in this chapter. Moreover, in the case of highly disordered solids, the Raman spectrum may also provide, with certain qualifications, information about the density of vibrational states. On a more detailed level, spectral information can be used to set inputs to models which determine force constants, interatomic potentials, and other fundamental quantities.

(b) Analysis of Material Structure. The vibrational properties deduced from Raman and IR spectra reveal significant amounts of information about symmetry, structure and bonding in materials, including the prevalence of long vs short range order, coordination numbers and other characteristics. For example, narrow spectral features are generally characteristic of crystals, and very broad features of highly disordered solids. The frequencies and number of lines are indicative of certain constituents and types of identifiable structural units in materials.

(c) Identification of Impurities, Imperfections and Multiple Phases. Impurities and defects often display distinctive vibrational signatures, depending on the host material and the specific environment in the host. For example, IR spectroscopy has been used extensively to detect and monitor the presence of water in fluoride glasses, especially in connection with corrosion studies. The presence of multiple phases may also be revealed by features in vibration spectra. Samples can be scanned to determine variations in properties as a function of position; use of tightly focused, short-wavelength laser sources enables scanning with submicron resolution. Thus, vibrational spectroscopy provides a non-destructive means of probing composition and impurity content from sample to sample, as well as within samples.

An example of the use of Raman spectroscopy to analyze inclusions in a rare-earth fluoride glass sample is indicated in Fig. 19. The distinct lines in the inclusion spectrum provided a means of identifying several crystalline species in the inclusion.

(d) Infrared Properties, Including IR Edge and Refractive Index Dispersion. A variety of other IR properties beyond the number, frequencies and widths of the IR active modes, can be conveniently determined from fundamental IR spectra. The high-frequency side of the fundamental IR absorption spectrum can be extrapolated to lower values to determine the approximate position of the IR edge, as confirmed by detailed measurements of IR edge absorption. The effect of additional glass constituents on IR transparency is usually evident from IR reflectivity; for example, the effect of Al fluoride additions on the reflectivity spectrum and transparency of RE fluorides was indicated

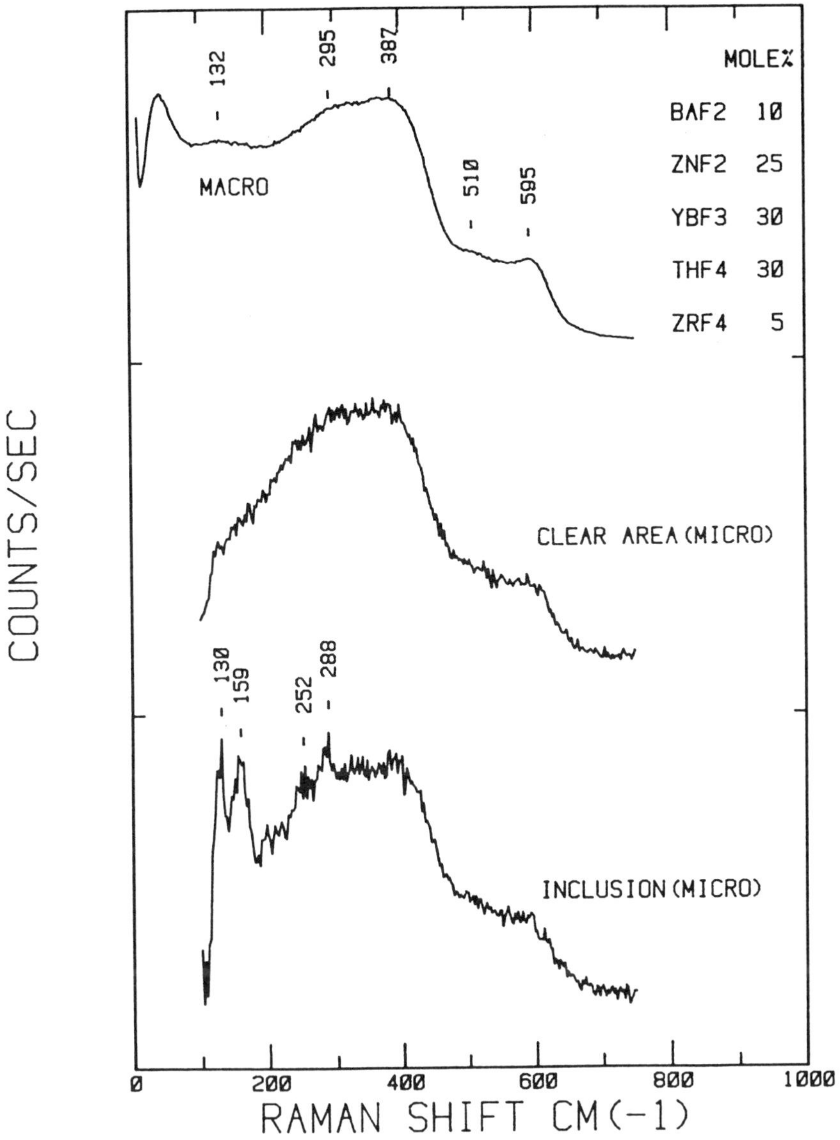


**Figure 19.** Macro-Raman spectrum of a Ba-Zn-Yb-Th-Zr fluoride glass (top), and Raman microprobe spectra of clear area and inclusion in the same sample (from D. Tallant and B. Bendow, "Raman Study of Composition Dependence and Crystallization Phenomena in Heavy Metal Fluoride Glasses" Mat. Sci. For. 6, 533-541 (1985)).

in the previous section.

Another application of IR spectroscopy is the determination of IR dispersion using, for example, the oscillator model in Eq. (7), with parameters deduced from a fit to the reflectivity spectrum.

(e) Process Monitoring. Vibrational spectra provide an efficient means of detecting changes in material composition and structure, through various stages such as fabrication (e.g., formation and annealing), storage, manufacturing into finished parts, and operational use. Examples include nucleation of crystals in glasses, and the adsorption and in-diffusion of impurities in glasses. Phase changes in solids can also be detected and monitored using vibrational spectroscopy, especially Raman. Another example is the use of Raman spectroscopy to study the evolution of gels to densified glasses in the sol-gel processing of glasses.

The above examples are indicative of the broad range of applications in which vibrational spectroscopy is currently in use. The value of vibrational spectroscopy in the study of glasses was aptly expressed by J.C. Phillips[32]:

"Which research tools can we expect to make the greatest contribution to enhancing our understanding of the microscopic morphology of glasses? On the basis of my personal experience, my answer is Raman scattering and electron microscopy. Raman scattering spectra contain an enormous amount of information, but decoding this information is a major task..."

The present chapter has demonstrated how vibrational spectroscopy can be used, at differing levels of detail and complexity, to extract a wide variety of information about glass properties and structure. Hopefully, this chapter has motivated the reader to undertake the major task referred to by Phillips, namely, decoding the enormous amount of information contained in vibrational spectra.

## REFERENCES


1. D.A. Long, Raman Spectroscopy (McGraw-Hill, NY, 1977).
2. J.M. Lerner and F. Adar, "*Raman Microprobing Offers New Potential*", Laser Focus World, 25, 73-82 (1989).
3. A. Fadini and F.M. Schneppel, Vibrational Spectroscopy Methods and Applications (Ellis Horwood Ltd, Chichester, UK, 1989).
4. J.L. Birman, "*Theory of Crystal Space Groups and Infrared and Raman Processes of Insulating Crystals*," Handbuch der Physik XXV/2b (Springer Verlag, Berlin, 1974).
5. H. Bilz, "*Vibrational Infrared and Raman Spectra of Non-Metals*", Handbuch der Physik XXV/2d (Springer Verlag, Berlin, 1984).
6. R. Loudon, "*The Raman Effect in Crystals*," Advances in Physics 13, 423-482 (1964).
7. S.S. Mitra, "*Infrared and Raman Spectra Due to Lattice Vibrations*", in Optical Properties of Solids (Plenum Press, NY, 1969) pp 333-451.
8. G.B. Wright, ed, "*Light Scattering Spectra of Solids*" (Springer Verlag, NY, 1969).
9. M. Balkanski, ed, "*Light Scattering in Solids*" (Flammarion Sciences, Paris, 1971).
10. S. Mizushima, "*Raman Effect*", Handbuch der Physik, Vol. XXVI, S. Flugge, ed (Springer Verlag, Berlin, 1958).
11. M. Weissbluth, Atoms and Molecules (Academic Press, NY, 1978).
12. J. Wong and C.A. Angell, Glass Structure by Spectroscopy (Marcel Dekker, NY, 1976), Chapter 7, pp 409-507.
13. R. Loudon, "*Polaritons, Raman Scattering, E-O Effect and Parametric Amplification*", in Light Scattering Spectra of Solids, op cit, pp 25-42.
14. S.A. Solin, "*Optical Probes of Vibrational Excitations and Structure*", in Structure and Excitations of Amorphous Solids (Amer. Inst. Phys., NY, 1976) pp 205-214.
15. F.L. Galeener and G. Lucovsky, "*Longitudinal Optical Vibrations in Vitreous $SiO_2$*", in Structure and Excitations of Amorphous Solids, G. Lucovsky and F.L. Galeener, eds (Amer. Inst. Phys., NY, 1976), pp 223-228.
16. F.L. Galeener and G. Lucovsky, "*Longitudinal Optical Vibrations in Glasses: $GeO_2$ and $SiO_2$*", Phys. Rev. Lett. 37, 1474-1478 (1976).
17. F.L. Galeener, G. Lucovsky and R.H. Geils, "*The Raman and Infrared Spectra of Vitreous $BeF_2$*", Sol. State Commun. 25, 405-408, 1978.
18. F.L. Galeener, "*Band Limits and the Vibrational Spectra of Tetrahedral glasses*", Phys. Rev. B19, 4297-4297 (1979).
19. F.L. Galeener, G. Lucovsky and J.C. Mikkelsen, "*Vibrational Spectra*

*and the Structure of Pure Vitrious $B_2O_3$*", Phys. Rev. B22, 3983-3990 (1980).
20. M.G. Drexhage, "*Heavy Metal Fluoride Glasses,*" in Treatise on Materials Science and Technology, Vol. 26, Glass IV, M. Tomozowa and R. Doremus, eds (Academic Press, NY, 1985), pp 151-243.
21. T. Katsuyama and M. Matsumara, "*Infrared Optical Fibers*" (Adam Hilger, Bristol, 1989).
22. P.W. France, M.G. Drexhage, J.M. Parker, M.W. Moore, S.F. Carter, and J.V. Wright, "*Fluoride Glass Optical Fibers*" (Blackie/CRC Press, Glasgow, 1990).
23. I.D. Aggarwal and G. Lu, eds, Fluoride Glass Optical Fibers (Academic Press, NY, 1991).
24. B. Bendow, P. Banerjee, M.G. Drexhage, J. Goltman, S.S. Mitra and C.T. Moynihan, "*Comparative Study of Vibrational Characteristics of Fluorozirconate and Fluorohafnate Glasses,*" J. Am. Ceram. Soc. 65, C-8-C-9 (1982).
25. B. Bendow, M.G. Drexhage, P.K. Banerjee, J. Goltman, S.S. Mitra, and C.T. Moynihan, "*Vibrational Spectra of Fluorohafnate Glass,*" Solid State Commun. 37, 485-89 (1981).
26. B. Bendow, P.K. Banerjee, and M.G. Drexhage, "*Applications of Raman Scattering Spectroscopy to Halide Glasses*", in SPIE Proceedings, Vol. 362, 1983.
27. D.R. Tallant, K.J. Ward, B. Bendow, and M. Poulain, "*Vibrational Characteristics and Structure of Heavy Metal Fluoride Glasses Containing Thorium and Zirconium*", Mat. Sci. For. 19-20, 625-632 (1987).
28. R.M. Almeida and J.D. Mackenzie, "*Vibrational Spectra and Structure of Flurozirconate Glasses*", J. Chem. Phys. 74, 5954, 5954-61 (1981).
29. B. Bendow, P.K. Banerjee, J. Lucas, G. Fonteneau, and M.G. Drexhage, "*Polarized Raman Scattering of Rare-Earth Fluoride Glasses*", J. Am. Ceram. Soc. 68, C-92-C-95 (1985).
30. B. Bendow, P.K. Banerjee, M.G. Drexhage, O.H. El-Bayoumi, S.S. Mitra, C.T. Moynihan, D. Gavin, G. Fonteneau, J. Lucas, and M. Poulain, "*Effect of Aluminum Content on IR Characteristics of $BaF_2/ThF_4$-Containing Glasses,*" J. Am. Ceram. Soc. 66, C-64-C-66 (1983).
31. B. Bendow, P.K. Banerjee, S. Mitra, C. Jacoboni, and R. DePape, "*Vibrational Spectroscopy of Pb/Mn/Ga Fluoride Glasses,*" J. Am. Ceram. Soc. 67, C-136-C-138 (1984).
32. J.C. Phillips, "*The Physics of Glass*", Physics Today 35, 27-33 (1982).

# CHAPTER 3

## NMR ANALYSIS OF GLASS STRUCTURE*

J. F. Emerson and P. J. Bray
Brown University
Providence, RI

## INTRODUCTION

Nuclear magnetic resonance (NMR) studies of glasses have been carried out for over thirty years. The studies have extracted stuctural information from the interactions of particular nuclei (*e.g.* $^{1}H$, $^{7}Li$, $^{11}B$, $^{19}F$, $^{27}Al$, $^{29}Si$) with their environment. The interactions are those between nuclear magnetic dipole moments, those between nuclear electrical quadrupole moments and electric field gradients at the nuclear site, and chemical shifts generated by the particular electron distribution about a resonating nucleus. Much of the earlier work has involved the use of continuous wave (CW) spectrometers and relatively low magnetic fields ($<$ 20 kilogauss), but more recent studies have employed high-field superconducting magnets and high-power pulsed spectrometers which yield a transient response that can be Fourier transformed into the standard plot of power absorption versus frequency. Pulsed NMR is particularly useful for studies of the two nuclear relaxation times, $T_1$ and $T_2$, which characterize the dynamic interaction of nuclei with their environment. Studies of $T_1$ as a function of temperature and frequency can distinguish between possible relaxation mechanisms in glasses (*e.g.* ion hopping) and obtain activation energies [1,2]. Growing use of magic-angle sample spinning (MASS) techniques have made possible such accomplishments as the separation of $^{29}Si$ responses from silicons bonded to different numbers of bridging and nonbridging oxygens and/or different numbers of aluminum neighbors [3,4] and the separation of signals from 4-coordinated and 6-coordinated aluminums [5]. Several useful reviews of pulsed NMR studies of glasses, including MASS studies, have appeared recently [6-8].

Since the interactions of a nucleus with its environment have different dependencies on external field strength, it is often useful to complement the information obtained from high-field pulsed NMR spectra with that from low-field studies. The present paper will concentrate on the effects of quadrupolar, chemical shift, and dipolar interactions on NMR spectra and how both high-field pulsed (with and without MASS) and low-field CW, NMR spectroscopy

⋆ Research supported by the National Science Foundation under Grant DMR 8504735 can be used to infer glass structure.

## ZEEMAN INTERACTION

When a nucleus possessing spin angular momentum - and thus a magnetic dipole moment - is subjected to a magnetic field, the otherwise degenerate energy levels of the nucleus are split into 2I+1 energy levels, where $\vec{I}$ is the nuclear spin vector of magnitude $\sqrt{I(I+1)}$; the scalar $I$ is the nuclear spin which is an integer or half-integer. Figure 1a depicts the Zeeman energy levels for a nucleus of spin I=3/2 where $m$ is the magnetic quantum number $(-I \leq m \leq +I)$ and increases from -I in integer steps. The separation between levels is

$$\Delta E = \gamma \hbar H \tag{1}$$

where $H$ is the strength of the magnetic field, $\gamma$ is the gyromagnetic ratio of the nucleus, and $\hbar$ is Planck's constant, $h$ divided by $2\pi$. When the nucleus is subjected to electromagnetic radiation and the frequency $\nu_0$ is adjusted so that

$$\Delta E = h\nu_0 \tag{2}$$

then the quanta of radiation are absorbed and transitions between neighboring energy levels are induced. The resulting NMR absorption spectrum is a single sharp peak at $\nu_0$ (Fig. 1b). The structural information, however, is obtained from the effects on the spectrum of the other interactions of a nucleus in a solid.

## DIPOLAR INTERACTION

Dipole-dipole interactions represent the effect of neighboring nuclear magnetic moments on the probe nucleus. The effect on the NMR absorption lineshape is an overall broadening (Fig. 1c), the extent of which depends on the magnitudes of the magnetic moments, $\vec{\mu} = \gamma\hbar\vec{I}$ and their spatial separation, $r$, and not on the value of the external magnetic field. The dipolar contribution to the full-width at half-maximum, $\delta$, can be calculated using Van Vleck's method of moments.

$$\delta = 2.36\sqrt{M_2^I} \tag{3}$$

where $M_2^I$ is the "second moment" for a nuclear spin in the presence of like spins, I, and unlike spins, S. Van Vleck showed [9] that for polycrystalline and glassy samples the second moment can be written

$$M_2^I = \Delta_{II} + \Delta_{IS} \tag{4}$$

where

$$\Delta_{II} = \frac{3}{5}\gamma^4\hbar^2 I(I+1)\sum\frac{1}{r^6}$$
$$\Delta_{IS} = \frac{4}{15}\gamma_I^2\gamma_S^2\hbar^2 S(S+1)\sum\frac{1}{r^6}. \qquad (5)$$

Equations (3)-(5) can be used to determine interatomic separations in solids.

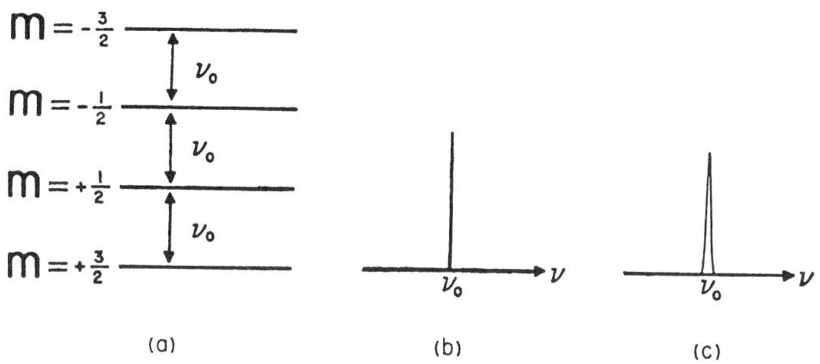


**Fig. 1. (a) Energy levels for a spin 3/2 nucleus in the presence of a magnetic field. (b) The resultant NMR absorption spectrum is a single peak at the frequency $\nu_0$. (c) Dipolar interactions broaden the resonance symmetrically so that it is not the delta-function response depicted in (b).**

## CHEMICAL SHIFT INTERACTION

The motion of the atomic electrons induced by the applied magnetic field will cause a small shift in the magnetic field experienced by the nucleus. The resulting change in the resonant frequency is termed the chemical shift and is measured in parts per million (ppm). The energy of interaction can be written as the coupling of the spin vector $\vec{I}$ with the magnetic field $\vec{H}$ via the chemical shift tensor, $\boldsymbol{\sigma}$:

$$E_{CS} = \gamma\hbar\vec{I}\cdot\boldsymbol{\sigma}\cdot\vec{H}. \qquad (6)$$

The observed spectral frequency is

$$\nu = \nu_0(1 - \sigma_{zz}), \tag{7}$$

where $\nu_0$ is the resonant frequency of the bare nucleus and $\sigma_{zz}$ is the component of the shift tensor along the direction of the external field. Rewriting this in terms of the diagonal components of the chemical shift tensor in its principal axis system, $\sigma_1$,$\sigma_2$, and $\sigma_3$, gives

$$\nu = \nu_0[1 - \sigma_{iso} - \sigma_{ax}(3cos^2\theta - 1) - \sigma_{aniso}(sin^2\theta cos2\phi], \tag{8}$$

where

$$\begin{aligned} \sigma_{iso} &= \frac{1}{3}(\sigma_1 + \sigma_2 + \sigma_3) \\ \sigma_{ax} &= \frac{1}{6}(2\sigma_3 - \sigma_1 - \sigma_2) \\ \sigma_{aniso} &= \frac{1}{2}(\sigma_2 - \sigma_1). \end{aligned} \tag{9}$$

Since the frequency, $\nu$, depends on $\theta$ and $\phi$, the polar angles of the field $\vec{H}$ as referred to the principal axes of the chemical shift tensor, the spectrum of a glassy sample (or powdered crystalline sample) will be the envelope of responses for all values of $\theta$ and $\phi$. The resulting NMR spectrum is a powder pattern, for which representative examples are shown in Fig. 2.

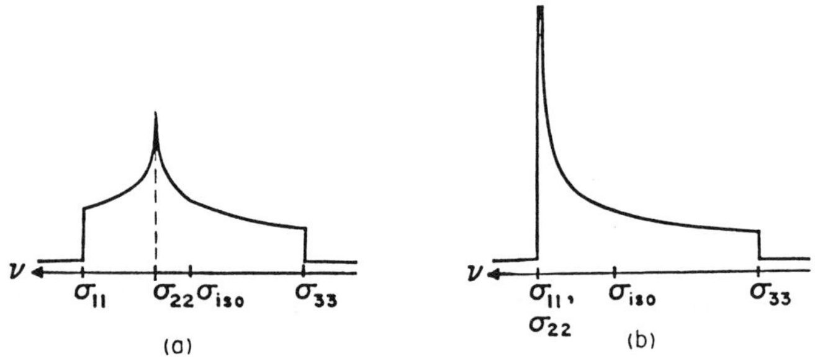


Fig. 2. Chemical shift powder patterns for (a) asymmetric and (b) axially symmetric anisotropies. In practice, the shoulders and divergences are rounded by dipolar interactions and distributions of chemical shifts.

## QUADRUPOLAR INTERACTION

Nuclei with spin 1 or greater possess an electric quadrupole moment, denoted eQ, which provides a measure of the departure from spherical symmetry of the nuclear charge distribution. The quadrupole moment interacts with an electric field gradient (EFG) present at the nuclear site. The energy levels of the nucleus depend on the orientation of the charge distribution with respect to the principal axes of the EFG. Since closed shells of electrons do not generate an EFG, and since the EFG falls off as $1/r^3$, the major contribution to the EFG arises from bonding electrons. Hence, the quadrupolar interaction is a very sensitive probe of chemical bonding and local atomic arrangements. When the quadrupole interaction is treated as a first-order perturbation to the Zeeman energy levels, the transition frequencies become:

$$\nu_{m\rightarrow m-1} = \nu_0 - \nu_Q \frac{(m-1/2)}{2}[3cos^2\theta - 1 - \eta sin^2\theta cos2\phi] \qquad (10)$$

where

$$\nu_Q = \frac{3Q_{cc}}{2I(2I-1)} \qquad (11)$$

and $Q_{cc}$ is the quadrupole coupling constant:

$$Q_{cc} = \frac{eQV_{zz}}{h}. \qquad (12)$$

The asymmetry parameter, $\eta$, is a measure of the departure of the EFG from cylindrical symmetry and is given by

$$\eta = \frac{V_{xx} - V_{yy}}{V_{zz}} \qquad 0 \leq \eta \leq 1 \qquad (13)$$

where $V_{xx}$, $V_{yy}$, and $V_{zz}$ are the diagonal components of the EFG tensor expressed in its principal axis system. The variables $\theta$ and $\phi$ are the polar angles of the magnetic field as referred to the principal axis system of the EFG tensor. In a glass or polycrystalline sample, all orientations of the principal axis system will exist and the resulting lineshape is a powder pattern. Quadrupole effects on the NMR spectrum, to first-order, are independent of the magnitude of the external magnetic field. Figure 3 is an example of a powder pattern due to a first-order quadrupole interaction for a spin 3/2 nucleus. Locating the positions of the shoulders and divergences in an uncomplicated spectrum experimentally determines the values of $Q_{cc}$ and $\eta$, the parameters conventionally measured in NMR spectroscopy.

When the strength of the quadrupole interaction is such that second-order effects must be included, the change in the NMR spectrum for a glass or polycrystalline powder is a broadening that varies inversely with the magnitude of the applied field. Most notably, the central transition ($m = 1/2 \leftrightarrow m = -1/2$), is broadened by an amount that depends on $\eta$ and $Q_{cc}$ (Fig. 4).

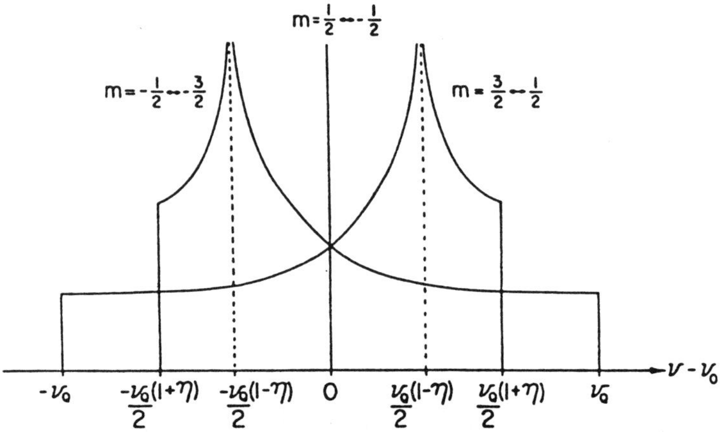


Fig. 3. First-order quadrupolar powder pattern for a system of spin 3/2 nuclei. Note that the central transition ($m = 1/2 \leftrightarrow m = -1/2$) is not affected.

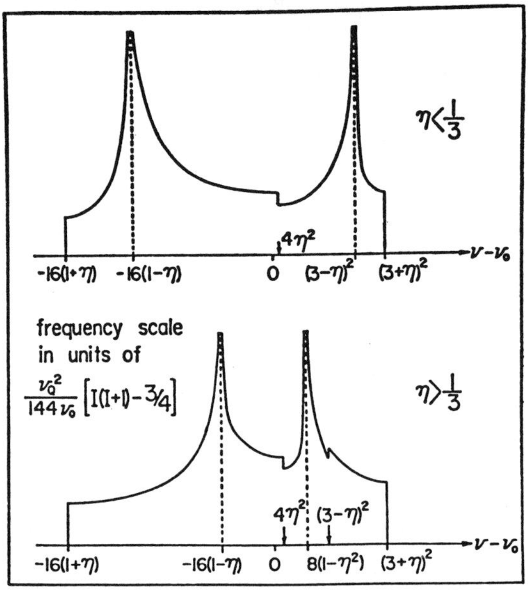


Fig. 4. Second-order quadrupolar powder pattern for the central transition of a half-integral nuclear spin. The two cases shown indicate the lineshapes for $\eta$ less than, or greater than, 1/3. In practice, the shoulders and divergences are rounded by the dipolar interactions and distributions in the components of the EFG.

## EXPERIMENTAL TECHNIQUES

The two basic techniques in NMR, continuous wave and pulsed spectroscopy, involve exciting the nuclear magnetization with radio-frequency (RF) energy and detecting the subsequent response of the spin system. Each method of excitation has advantages and disadvantages depending on the width of the NMR line and the type of information sought. In this section is a brief discussion of the basic principles of these detection methods, especially those which are useful for studying NMR in glasses.

### Continuous Wave NMR

In CW NMR the nuclear magnetization is irradiated at a constant level and either the frequency of irradiation or the magnetic field is swept through the region of the resonance condition given in Eq. (1). In the crossed-coil method [10] the transmitting coil, with its axis perpendicular to the external magnetic field, feeds RF into the sample. The magnetic moments of any nuclei which resonate at that frequency will begin to precess and induce a small RF signal. A reciever coil is placed such that its axis is perpendicular to both the transmitting coil and the external field, thereby recieving only the induced signal. This signal is very small and is usually overwhelmed by random noise. Signal averaging and lock-in detection [11] are the usual means of improving the signal-to-noise (S/N) ratio. One limitation of lock-in detection is that the signal is proportional to the RF power used, but RF levels must be low enough to avoid saturating the spin system and thereby distorting the lineshape. Saturation results when the spin-system cannot sufficiently dissipate energy to the lattice in order to absorb more energy. One way around this problem is to introduce paramagnetic impurities (which can absorb energy from the spin system through electron interactions) such that the glass structure and the absorption lineshape are not affected[12]. Signal averaging when used with lock-in detection further increases the S/N; however, there is an upper limit on how fast one can sweep through resonance without distortion. Non-adiabatic superfast passage (NASP) [13] has been used to solve this problem. In this method the external field is linearly swept through the region of interest in a short time (usually less than 50 msec). Higher RF levels can be used than with lock-in detection since the measured lineshape is not distorted if the resonance is slightly saturated. This, in combination with the rapid accumulation of signals, allows for better S/N ratios in a short time. NASP is limited by the external field sweep range, which is on the order of a few hundred gauss.

Both of these CW techniques are useful for broad lines which commonly occur in glasses due to inhomogeneous broadening mechanisms. NASP is preferable unless the line is broader than a few hundred gauss.

### Pulsed NMR

In pulsed NMR spectroscopy the entire frequency range of interest is irradiated at once by a short intense RF pulse. For a pulse length of $t_p$, with angular frequency $\gamma H_1$ the sample magnetization (the sum of the nuclear magnetic moments) will be rotated from its equilibrium position along the z-axis by the angle

$$\theta_p = \gamma H_1 t_p, \tag{14}$$

as seen in Fig. 5, and will precess about the z-axis. (Recall that a magnetic moment, and hence the magnetization vector for a collection of magnetic moments, will precess about an applied magnetic field with angular frequency $\omega = \gamma H$. The magnetization vector is thus stationary in a frame rotating about the z-axis with that frequency [14].) The components of $\vec{M}$ which lie in the $xy$ plane will induce a signal called the free induction in a reciever coil. Immediately following the pulse the individual spins are aligned but spin-spin interactions (characterized by the time constant $T_2$) cause dephasing of the spins with a reduction in the magnitude of $\vec{M}$. This, along with spin-lattice relaxation mechanisms (characterized by the relaxation time $T_1$) which return the magnetization vector toward equilibrium, diminish the free induction. The decaying signal is termed the free induction decay (FID) and is the one which is accumulated and signal-averaged. The fourier transform (FT) of the real part of the FID gives the NMR absorption spectrum.

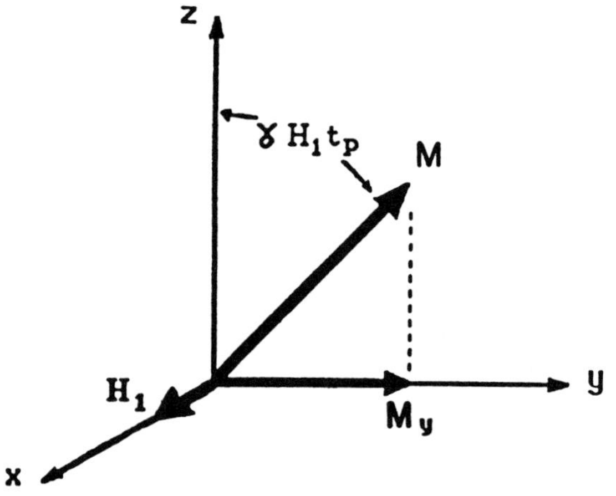


Fig. 5. A pulse of length $t_p$ rotates the magnetization vector from its equilibrium position by an angle $\theta_p = \gamma H_1 t_p$.

The exponential decay of the FID is characterized by a time constant, $T_2^*$, or the apparent $T_2$. It is inversely related to the linewidth of the absorption spectrum as shown in Fig. 6. In general, [14]

$$\frac{1}{T_2^*} \approx \frac{1}{2T_1} + \frac{1}{T_2} + \gamma \delta H_0. \tag{15}$$

In solids $T_1$ is usually much longer than $T_2$; hence, the signal decays before the magnetization has returned to its equilibrium position and the $T_2$ term dominates the decay rate. The last term in Eq. (15) represents broadening due to magnetic field inhomogeneities of extent $\delta H_0$.

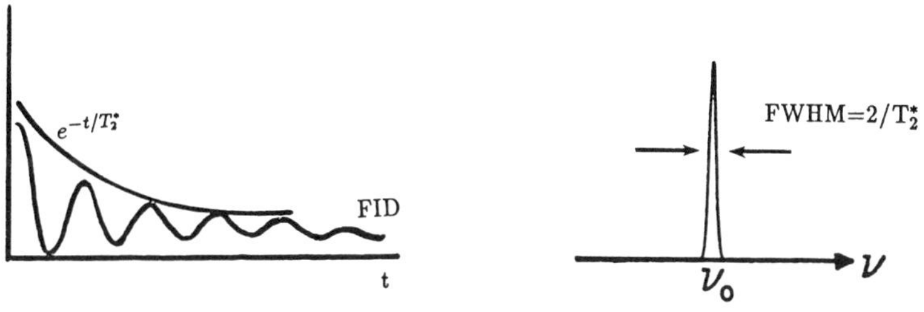


(a) (b)

Fig. 6. (a) The free induction decay (FID) decays with time constant $T_2^*$. (b) Fourier transform of the FID results in an absorption spectrum with a full-width at half-maximum (FWHM) inversely proportional to $T_2^*$.

If the magnetization vector has not returned fully to its equilibrium position along the z-axis, subsequent pulsing rotates the vector to other than its optimal angle. To avoid saturation, the repeated pulse must be delayed by a time on the order of several $T_1$'s. In the study of solids, if the structural information obtained from the lineshape is the objective, rather than the dynamics inferred from relaxation time measuments, $T_1$ times can be effectively shortened by the inclusion of paramagnetic impurities in the sample. Ensuring full relaxation between pulsing is important since nuclei in different structural sites often have different relaxation rates. Even partial saturation will distort the relative spectral intensities due to these sites.

The frequency range of observation is inversely proportional to the pulse length, $t_p$; shorter pulse lengths permit illumination of a broader frequency range. From Eq. (14), a short pulse requires higher RF power to obtain a given tip angle of the magnetization. Pulse lengths are typically on the order of microseconds.

To avoid pulse breakthrough in the FID and to allow for electronic recovery time, data acquisition must be delayed beyond the end of the pulse. For broad lines, when the FID decays rapidly, significant information is lost. Spin-echo techniques can often recreate the early part of the FID which is obscured by reciever dead-time. In this method, (Fig. 7), a 90° pulse is applied at $t = t_0$. (A 90° pulse is defined by Eq.(14) as the pulse length such that $\theta = \pi/2$.) After a certain time interval $\tau$, during which time the spins are dephasing and the FID is decaying, the spin system receives a 180° pulse. This rotates the individual spins 180° around the $x$ axis. The same mechanisms which caused the dephasing of the spins in the $xy$ plane now act to rephase the spins and at a time $2\tau$ later a spin-echo is formed. The echo consists of two FIDs back-to-back and the FT of the second half of the echo gives the absorption spectrum. There have been many pulse sequences designed to observe echos in solids and in liquids and they are described elsewhere [15-17].

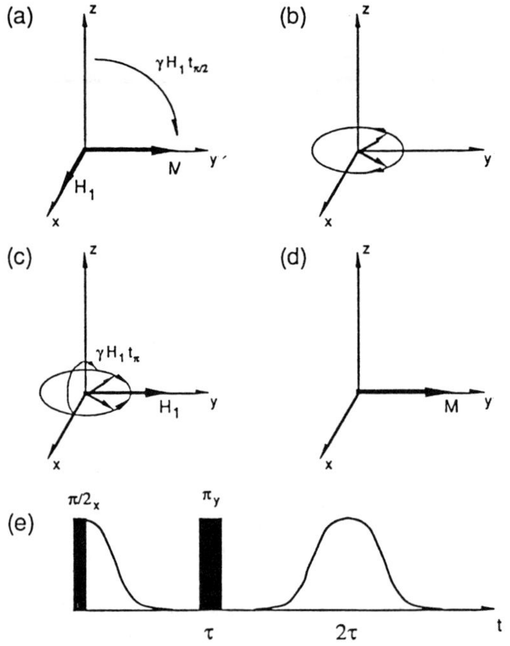


Fig. 7. Formation of a spin-echo. The axes shown represent the rotating frame [14].(a) At $t = 0$ a 90° pulse is applied tipping the magnetization vector into the $xy$ plane. (b) Since the rate of precession varies for the individual magnetic moments, the spins become dephased and the magnetization decays. (c) At time $\tau$ a 180° pulse is applied thereby flipping all the spins 180° about the $y$ axis. (d) The individual magnetic moments then become rephased, restoring the magnetization. (e) The spin-echo pulse sequence and the detected signal.

*Magic-Angle Sample Spinning (MASS)*

Resolving different nuclear sites requires that the broadening for each line be less than the frequency differences between them. Dipolar, chemical shift, and first-order quadrupolar interactions which can contribute to the broadening contain terms which have an angular dependence of $1 - 3cos^2\theta$ where $\theta$ is the angle between the internuclear vector and the external field. For $\theta = 54.74°$ these terms are zero, thus narrowing the overall NMR response. In MASS, the sample is placed in a rotor which spins the sample at this angle relative to the static field. Then any internuclear vector will, on the average, be along the rotation axis and any terms with this angular dependence will average to zero (Fig. 8). To narrow the line effectively, the rate of rotation must exceed the linewidth, otherwise the transverse magnetization decays faster than the time it takes to complete one revolution [18].

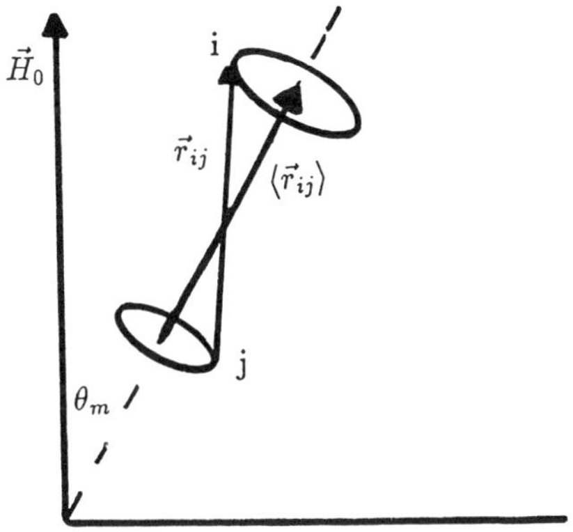


Fig. 8. Sample spinning about an axis at an angle $\theta_m$ with respect to the external field brings the average positions of the nuclei $i$ and $j$ (and the average internuclear vector, $\langle\vec{r}_{ij}\rangle$) to lie along that axis.

## $^{29}Si$ NMR IN ALKALI SILICATE GLASSES

Since $^{29}Si$ is a spin 1/2 nucleus, it has no quadrupole moment and the shape and width of the NMR spectrum are determined by chemical shift and dipolar effects only. In silicates, the basic structural unit is denoted $Q^n$, where

Q refers to silicon in four- fold coordination with oxygen and $n$ is the number of bridging oxygens. In general, as alkali oxide is added to vitreous silica ($SiO_2$), $Q^n$s with fewer bridging oxygens are formed and the magnitude of the chemical shift becomes smaller (the silicon nucleus is less shielded). Figure 9 depicts the $^{29}Si$ NMR powder pattern expected for each of these units.

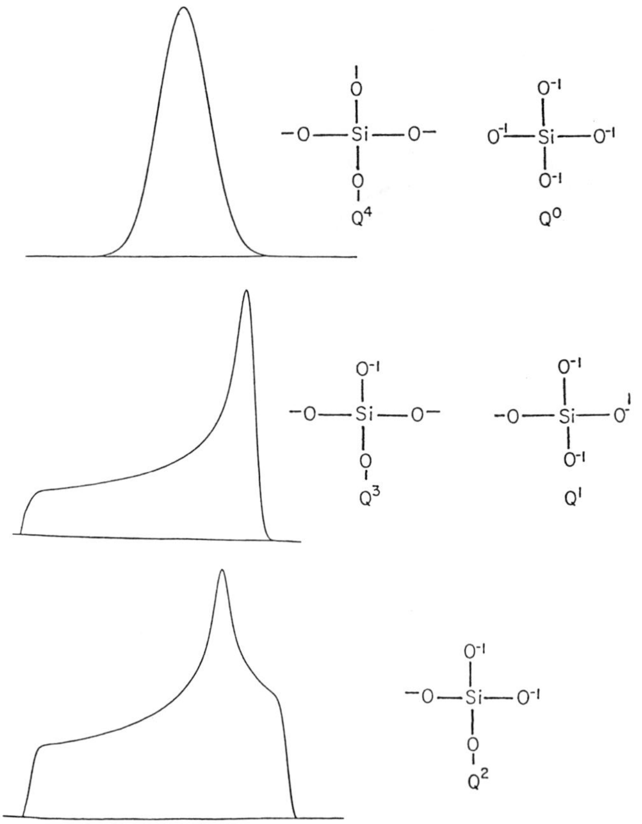


**Fig. 9. NMR powder patterns for the $Q^n$ structural units, including dipolar broadening. An $O^{-1}$ is a non- bridging oxygen.**

MASS effectively reduces dipolar broadening and chemical shift anisotropies thereby enabling measurement of the isotropic chemical shift. However, in some instances the resolution of sites partially depends upon the anisotropies [19,20], in which case static studies at a field strength high enough to reduce the effect of dipolar broadening are more informative than MASS. Figure 10 is a comparison of MASS and static spectra for potassium silicate glasses [21]. It would be concluded from the MASS spectrum for potassium disilicate that only one $Q^n$ site was present, while the static spectrum for the same sample resolves two sites. A third $Q^n$ unit is required by cation and non- bridging oxygen conservation and is found by computer simulation of the data. Computer simulation of the static $^{29}Si$ NMR lineshapes has been used to obtain the chemical shift tensor values for each of the structural units present in alkali di-, meta-, and tetra-silicate glasses [22]. Figure 11 shows the computer-generated lines for potassium and lithium disilicate glasses along with the values for the chemical shift tensors for the structural units present.

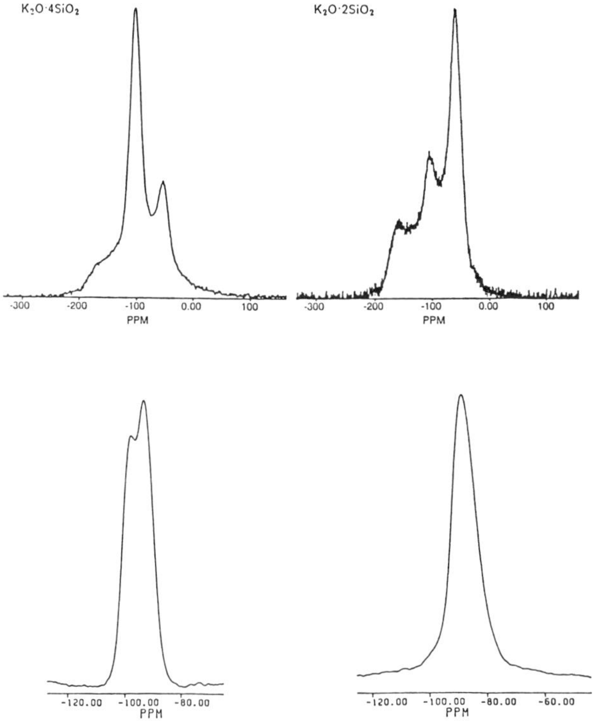


Fig. 10. Static (upper) and MASS (lower) $^{29}Si$ NMR spectra of potassium tetrasilicate (left) and potassium disilicate (right) glasses. The scale is in ppm referenced to the $^{29}Si$ response in tetramethylsilane (TMS).

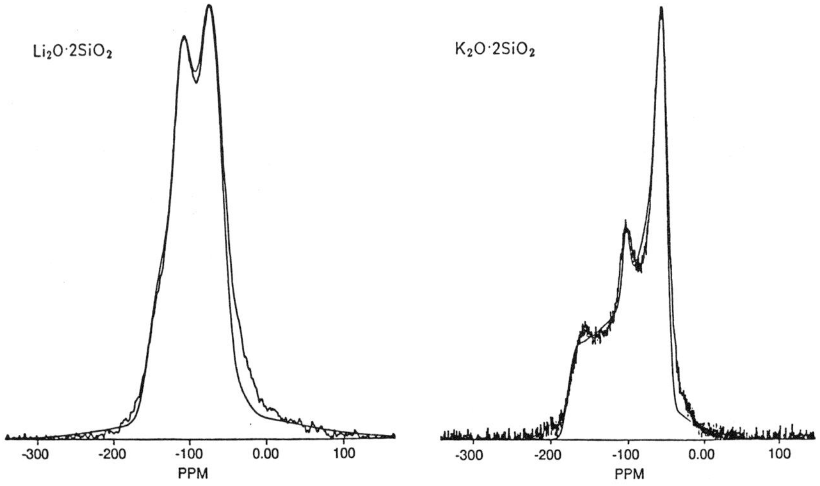


| $Q^n$ | $\sigma_{iso}$ (ppm) | $\sigma_1$ | $\sigma_2$ | $\sigma_3$ |
|---|---|---|---|---|
| $Q^4$ | -107 | -107 | -107 | -107 |
| $Q^3$ | -93 | -63 | -63 | -153 |
| $Q^2$ | -85 | -25 | -75 | -155 |

| $Q^n$ | $\sigma_{iso}$ (ppm) | $\sigma_1$ | $\sigma_2$ | $\sigma_3$ |
|---|---|---|---|---|
| $Q^4$ | -103 | -103 | -103 | -103 |
| $Q^3$ | -95 | -53 | -53 | -179 |
| $Q^2$ | -86 | -27 | -59 | -172 |

**Fig. 11. NMR spectra for lithium and potassium disilicate glasses. The smooth lines are computer-simulations obtained with values (in ppm relative to TMS) for the shift tensors as shown in the tables below each spectrum.**

## $^{27}$Al NMR IN ALUMINO-SILICATE GLASSES

The $^{27}$Al (I=5/2) NMR spectrum for many crystalline compounds and glasses reflects the influence of both quadrupolar and chemical shift interactions. Use of high magnetic fields reduces the size of the quadrupolar effects, since the second-order effect on the central transition depends on $(Q_{cc})^2/\nu_0$, and increases the effect of chemical shifts. The high-field, combined with MASS and a high-powered pulsed spectrometer, yields resolved responses for Al in 4, and 6-coordination with oxygen in some polycrystalline and (nonvitreous) amorphous compounds [23]. With spinning rates in excess of 7 kHz, 4, 5, and 6-coordinated aluminum sites have been found in $SiO_2{\cdot}Al_2O_3$ glasses (Fig. 12)[24].

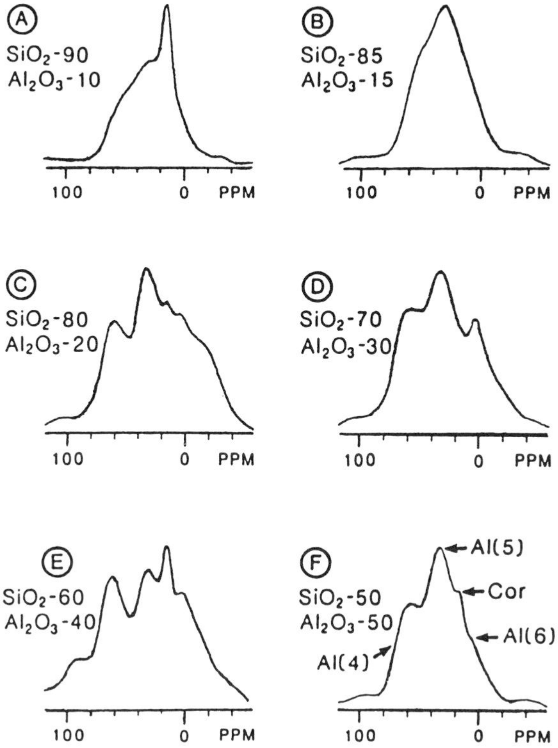


Fig. 12. $^{27}$Al MASS NMR spectra of roller-quenched $SiO_2 \cdot Al_2O_3$ glasses (labeled in weight percent). All spectra were obtained at 11.7 Tesla with MASS spinning rates greater than 7 kHz. The designations Al(4), Al(5), and Al(6) refer to the responses due to 4-fold, 5-fold, and 6-fold aluminum. The designation Cor indicates the aluminum response due to undissolved $Al_2O_3$ in the glass. The ppm scale is relative to the octahedral $^{27}$Al response in $AlCl_3$ solution. (From ref.[24])

## $^{11}$B NMR IN BORATE GLASSES

As shown by previous NMR studies of modified borate glasses [25], addition of modifier oxide to the borate network effects the formation of negatively charged borate units consisting of borons in three- or four-coordination with oxygen. The amounts and types of borate structural units (Fig. 13) vary with the modifier oxide content. Borate glasses are often characterized by the fraction of boron atoms in four-coordination, ($N_4$), the fraction in asymmetric units of three-coordination, ($N_{3A}$), and the fraction in symmetric three-coordination, ($N_{3S}$).

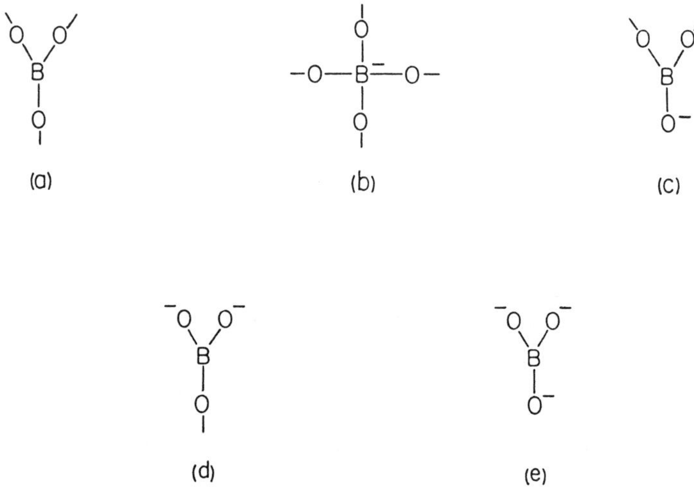


**Fig. 13. Structural units found in modified borate glasses. (a) trigonal $BO_{3/2}$ (b) tetrahedral $BO_{4/2}$ (c) asymmetric trigonal $BO_{2/2}O^-$ with one NBO (d) asymmetric trigonal $BO_{1/2}O_2^{-2}$ with two NBOs (e) symmetric trigonal $BO_3^{-3}$ with three NBOs.**

$^{11}$B is a spin 3/2 nucleus and thus it is subject to quadrupolar interactions. The magnitude of the interaction at field strengths of 1 Tesla or higher is such that it can be treated with second-order perturbation theory. Spectra obtained for a large number of borates show that $Q_{cc}$ has a range from 2.4 to 2.9 MHz for $^{11}$B in all $BO_3$ units regardless of the numbers of bridging and nonbridging oxygens (NBO). However, $\eta \leq 0.2$ for the (relatively) symmetric cases of zero or three NBOs, but $\eta > 0.4$ for one or two NBOs.

It should be noted that responses for the satellite transitions where $Q_{cc}$ is large would be spread over very broad regions to the left and right of the observed central transition and are normally not detected unless special efforts are made to do so [26,27].

Figure 14 shows the expected $^{11}$B derivative spectrum (i.e. $dP/d\nu$ versus $\nu$ where $P$ is the power absorption and $\nu$ is the frequency) from a sample containing boron atoms in distinct environments. By matching computer simulations to the observed response, one can extract $N_4$, $N_{3S}$ and $N_{3A}$, along with values of $Q_{cc}$ and $\eta$ for the latter two cases. (The parameters $Q_{cc}$ and $\eta$ can be obtained for four-coordinated borons in some cases by working at low frequencies [28] for which second-order effects become apparent, or by observing the satellite transitions arising from the $BO_4$ units [29].) Alternatively, $N_4$ can be determined by using the area method for absorption spectra or by integrating derivative spectra as shown in Fig. 15. If rapid MASS (>6 kHz) techniques are employed, the $^{11}B$ NMR responses for $BO_3$ and $BO_4$ units are separated by isotropic chemical shifts, thus determining $N_4$ [30]. Figure 16 displays $N_4$ values obtained for sodium borosilicate glasses [31].

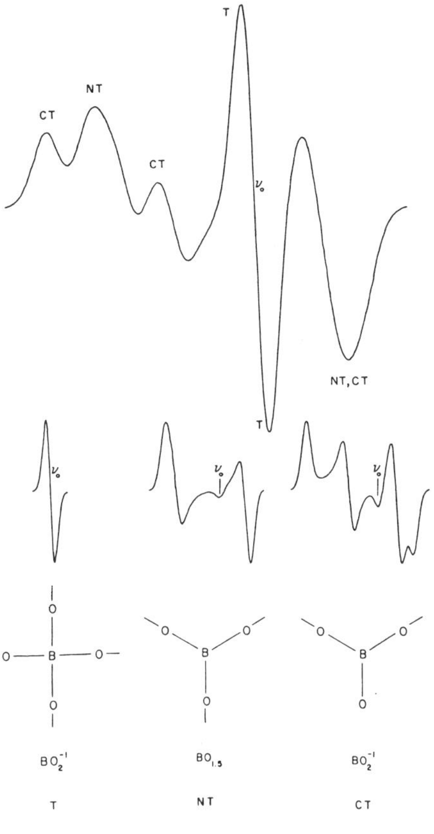


Fig. 14. Derivative spectrum of a dipolar broadened and second-order quadrupolar broadened $^{11}B$ NMR response. The spectrum is comprised of a weighted sum of spectra due to different sites. The features in the spectrum are labeled T (tetrahedral), NT (neutral trigonal), and CT (charged trigonal), corresponding to the boron units as shown below the spectrum.

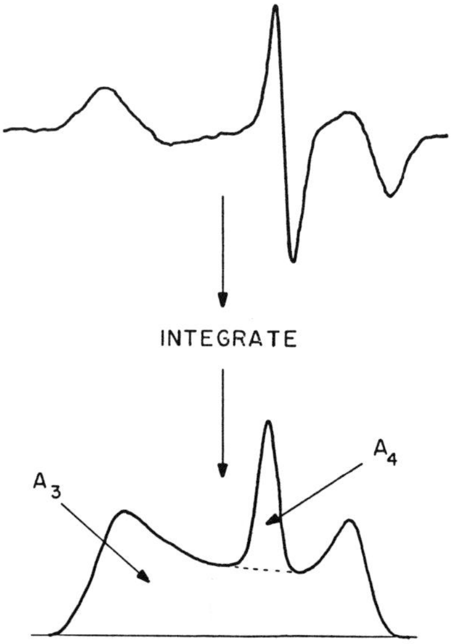


Fig. 15. Determination of $N_4$ by the area method for a typical $^{11}B$ NMR spectrum. $N_4 = A_4/(A_3 + A_4)$.

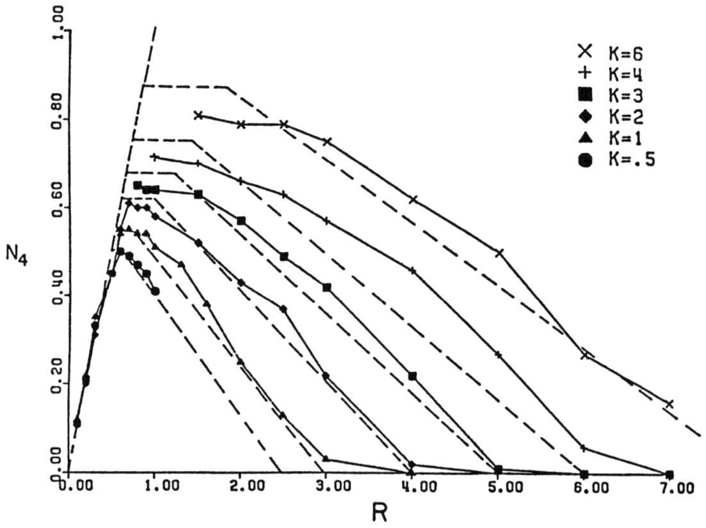


Fig. 16. $N_4$ versus $R$ for sodium borosilicate glasses shown for different $K$ families. $R$=mol%$Na_2O$/mol%$B_2O_3$, $K$=mol%$SiO_2$/mol%$B_2O_3$. The dashed lines are the prediction based on the model of Dell et al. [31].

Data and insights obtained from both $^{10}B$ (spin 3) and $^{11}B$ have been used to construct models for binary alkali borate glasses [32,33] and for glasses in the systems $PbO \cdot B_2O_3 \cdot SiO_2$ [34], $Na_2O \cdot B_2O_3 \cdot SiO_2$ [35,36] and $CdO \cdot B_2O_3 \cdot GeO_2$ [37]. The $^{10}B$ NMR spectrum for $B_2O_3$ glass is shown in Fig. 17 [32]. The observed spectrum agrees with one computer-generated using $Q_{cc}$=0.55 MHz and $\eta$=0.2 MHz. These numbers are in the range of values found for $^{10}B$ for symmetric 3-coordinated borons (see the NT inset in Fig. 14). The work supports the views of Jan Krogh-Moe [38] in which the glasses are composed of the structural groupings found in the crystalline compounds of the glassforming system. [33].

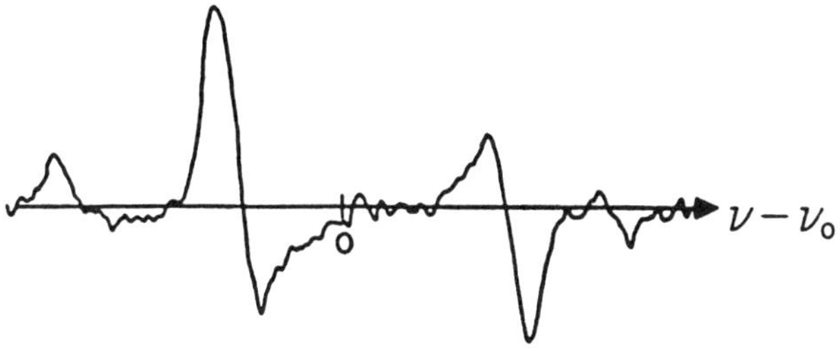


Fig. 17. The $^{10}B$ NMR spectrum observed for $B_2O_3$ glass (derivative of the absorption response). $\nu_0 = 7$ MHz.

## $^6Li$ AND $^7Li$ NMR SPECTRA OF MIXED-ALKALI SILICATE GLASSES

When distinct nuclear sites are not resolved by NMR, information can often be gained from linewidth studies. As discussed in the introduction, the NMR linewidth is due to three main interactions - the dipolar, chemical shift, and quadrupolar. All of these depend on the local nuclear environment and each has a different dependence on the magnitude of the external magnetic field. NMR studies performed at different field strengths, and used in conjunction with studies of different nuclear isotopes, can be significant sources of structural information.

The mixed-alkali effect in glasses is a phenomenon wherein partial substitution of one alkali for another leads to markedly non-linear changes in various properties of the glass. The effect was first reported in 1883 [39] and many theories and models have been proposed to explain this phenomenon; however, a completely successful structural model is still lacking. The work on the mixed-alkali effect had been thouroughly reviewed by Isard [40] and Day [41].

The most successful theories explaining the marked decrease in electrical conduction in these glasses assume an interaction between unlike alkali cations [42-47]. However, in a recent study of lithium-potassium mixed-alkali silicate glasses [22], $^6$Li and $^7$Li NMR studies gave evidence for like-alkali cation pairing. The linewidths (at 7.1 Tesla), as measured by full-width at half-maximum, (FWHM), as a function of alkali composition are shown in Figs. 18 and 19. There is about a fifty percent departure from linearity in each case. For both isotopes the linewidth is maximum when the alkali content is 20-40% potassium. The $^7$Li linewidths at low-field (≈9 kG) demonstrated the same dependence on compostition as at high-field. In addition, the magnitudes of the linewidths were neither significantly broadened or narrowed, being within 10% of the values at high field.

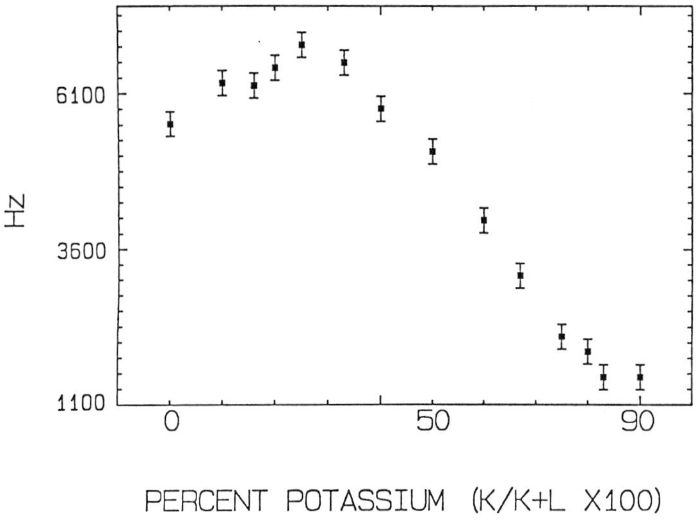


**Fig. 18. Full-width at half-maximum of the $^7$Li lines in the lithium-potassium disilicate glasses.**

In interpreting the results of Fig. 18, recall that the magnitude of the dipolar interaction depends on the square of $\gamma$, the spin, I, of the neighboring nuclei, and is independent of the strength of the applied magnetic field. Since $\gamma$ for $^7$Li (natural abundance 93%) is about an order of magnitude larger than $\gamma$ for $^{39}$K (the most abundant potassium isotope, 93%) it is expected that the dipolar broadening would decrease in the mixed-alkali glasses as potassium is substituted for lithium. The general downward trend in the FWHM with increasing potassium seen both at high-field (Fig. 18) and at low-field (not shown) indicates that the $^7$Li linewidth is predominatly dipolar in nature.

At low field the chemical shift interaction is reduced in magnitude. First-order quadrupolar effects remain unchanged at low field and second-order quadrupolar effects are increased. $^7$Li is a spin 3/2 nucleus and only the second-order effects on the central transition are observed. The broadening of the central transition at low-field, which goes as $(Q_{cc})^2/\nu_0$, is expected to be very small for reported values of $Q_{cc}$ for $^7$Li [48]. The low-field results confirm the dipolar downward trend in the FWHM and further, suggest that the departure from linearity is also predominantly dipolar. Silicon and oxygen are unlikely candidates for dipolar broadening since neither have magnetic isotopes of sufficient natural abundance. From Eq. (5) it is seen that lithium-lithium pairing increases dipolar broadening while lithium-potassium pairing decreases it. Evidently like-cations are paired in these glasses.

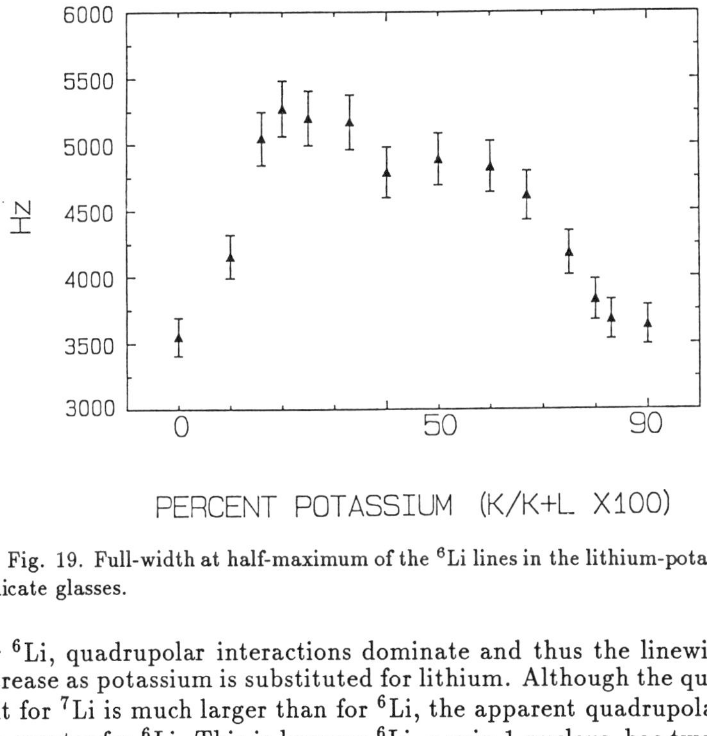


Fig. 19. Full-width at half-maximum of the $^6$Li lines in the lithium-potassium disilicate glasses.

For $^6$Li, quadrupolar interactions dominate and thus the linewidth does not decrease as potassium is substituted for lithium. Although the quadrupole moment for $^7$Li is much larger than for $^6$Li, the apparent quadrupolar broadening is greater for $^6$Li. This is because $^6$Li, a spin 1 nucleus, has two satellite transitions ($m = +1 \leftrightarrow m = 0$ and $m = 0 \leftrightarrow m = -1$) which are observed in first order, and no central transition. Chemical shift effects are expected to be the same for both isotopes. Thus the non-linearity in the FWHM as a function of alkali composition must also be due to increased dipolar broadening, confirming the presence of lithium-lithium pairing.

## REFERENCES


[1] E. Göbel, W. Müller-Warmuth, H. Olyschläger, and H. Dutz, J. Mag.Res. **36**,371 (1979).

[2] M. Rubinstein, Phys.Rev. **B14**,2778 (1976).

[3] E. Lipmaa, M. Magi, A. Samoson, G. Engelhardt, and A.R. Brimmer, J. Amer. Chem. Soc. **102**,4889 (1980).

[4] B.H.W.S. de Jong and Ch.M. Schramm, EOS. **62**,1070 (1981).

[5] D. Mueller, W.Gessner, H.J.Behrens, and G. Scheller, Chem.Phys.Lett. **79**,59 (1981).

[6] E. Oldfield and R.J. Kirkpatrick, Science **227**,1537 (1985).

[7] G.L.Turner, R.J. Kirkpatrick, S.H.Risbud and E. Oldfield, Am. Ceram. Soc. Bull. **66**,656 (1987).

[8] R.J. Kirkpatrick, K.A. Smith, S. Schramm, G.L.Turner, W.H. Yang, Ann.Rev.Earth Planet. Sci. **13**,29 (1985).

[9] J. H. Van Vleck, Phys.Rev. **74**,1168 (1948).

[10] A. Abragam, *The Principles of Nuclear Magnetism*, Oxford University Press (1961).

[11] R.E. Simpson, *Introductory Electronics for Scientists and Engineers*,Allyn and Bacon, Inc., Boston (1974).

[12] F. Bucholtz, and P.J.Bray, J.Non-Cryst.Sol. **54**,43 (1983).

[13] S.L. Segal, R.B. Creel, D.R. Torgenson, J.Mol.Struc. **111**,79 (1983).

[14] E. Fukishima and S.B.W. Roeder, *Experimental Pulse NMR. A Nuts and Bolts Approach*, Addison-Wesley, Reading,MA (1981).

[15] E.L. Hahn, Phys.Rev. **80**,580 (1950).

[16] J.S. Waugh, L.M. Huber, U.Haeberlen, Phys.Rev.Lett. **20**,180 (1968).

[17] M. Mehring, *High Resolution NMR Spectroscopy in Solids*,Springer-Verlag, Heisenberg (1978).

[18] E.R. Andrew, Prog.NMR Spectroscopy **8**,1 (1971).

[19] J.F.Stebbins, Nature **330**,465 (1987).

[20] S.L.Chan, L.F.Gladden and S.R. Elliot, J.Non-Cryst.Sol. **106**,413, (1988).

[21] P.J. Bray, J.F. Emerson, D. Lee, S.A. Feller, D.L.Bain, and D.A.Feil, to appear in J. Non-Cryst. Sol. (1991).

[22] J.F. Emerson, P.E. Stallworth, and P.J. Bray, J.Non-Cryst.Sol. **113**,253 (1989).

[23] D. Müller, G. Berger, I.Grunze, G.Ladwig,E. Hallas, and U. Haubenreisser, Phys. Chem. Glasses **24**,37 (1983).

[24] S.H.Risbud, R.J. Kirkpatrick, A.P.Taglialavore, and B.Montez, J. Am. Ceram. Soc. **70**,C-10 (1987).

[25] S.A. Feller, W.J.Dell, and P.J. Bray, J. Non-Cryst. Sol. **51**,21 (1982).

[26] S.G. Bishop and P.J. Bray, Phys. Chem. Glasses **7**,73 (1966).

[27] P.W.France and M. Wadsworth, J. Mag. Res. **49**,48 (1981).

[28] M. Leventhal and P.J. Bray, Phys.Chem.Classes **6**,113 (1965).

[29] D.E. Hintenlang, Ph.D. Thesis, Brown University (1984).

[30] G.L. Turner, K.A. Smith, R.J. Kirkpatrick, and E. Oldfield, J.Mag.Res. **67**,544 (1986).

[31] W. J. Dell and P.J. Bray, J.Non-Cryst.Sol. **58**,1 (1983).

[32] G.E. Jellison,Jr. and P.J. Bray, Solid State Commun. **29**,517 (1976).

[33] Y.H. Yun and P.J. Bray, J. Non-Cryst. Sol. **44**,227 (1981).

[34] K.S. Kim, P.J. Bray, and S. Merrin, J.Chem.Phys. **64**,4459 (1976).

[35] Y.H. Yun and P.J. Bray, J. Non-Cryst.Sol. **27**,363 (1978).

[36] W.J. Dell and P.J. Bray, J.Non-Cryst.Sol. **58**,1 (1983).

[37] R.V. Mulkern, S.J. Chung, P.J. Bray, G.D. Chryssikos, D.E. Turcotte, and W.M. Risen, J.Non-Cryst.Sol. **85**,69 (1986).

[38] J. Krogh-Moe, Phys.Chem.Glasses **3**,101 (1962).

[39] G. Tomandl and H.A. Schaeffer, J. Non-Cryst.Sol. **73**,179 (1985).

[40] J.O. Isard, J. Non-Cryst.Sol. **1**,235 (1969).

[41] D.E. Day, J.Non-Cryst.Sol. **21**,343 (1976).

[42] O.V. Mazurin, *Structure of Glass*, Vol. 4, Consultants Bureau, New York (1965).

[43] J. Sakurai and K. Ooka, Toshiba Rev. **23**,913 (1968).

[44] W.A. Weyl and E.C. Marboe, *The Constitution of Glasses* Vol. II, Interscience, New York (1962).

[45] J.R. Hendrickson and P.J.Bray, Phys.Chem.Glasses **13**,43,107 (1972).

[46] G.deMarchi and P.Mazzoldi, J.Non-Cryst.Sol. **104**,211 (1988).

[47] R.Terai, H.Wakabayashi, and H.Yamanaka, J.Non-Cryst.Sol. **103**,137 (1988).

[48] J.F. Hon and P.J. Bray, Phys.Rev. **110**,624 (1958).

# CHAPTER 4

## MÖSSBAUER SPECTROSCOPY IN GLASSES

**I. Joseph, L. D. Pye**
Institute of Glass Science and Engineering
New York State College of Ceramics
Alfred University
Alfred, New York 14802

## INTRODUCTION

Mössbauer spectroscopy is an analytical technique that utilizes the recoil free emission and resonant absorption of gamma rays(1) by nuclei bound in solids. For a free nucleus, emission and absorption of $\gamma$–rays are accompanied by large enough recoil energies so that resonance is not observed. Mössbauer(2) discovered that in solids, under certain conditions, the recoil energies will essentially be zero and resonance will be observed.

Since the energies of the gamma ray emitter (source) and absorber (sample) have to be matched, the number of elements that can be studied using this technique is limited. Two of the elements that have been studied extensively are iron and tin. Information has been obtained about the coordination, bonding, and valence state of these elements in glass. In addition, quantitative determination of the elements in different oxidation states has been achieved.

## BACKGROUND THEORY

The Mössbauer effect has been discussed in detail elsewhere(3-6). The theory of Mössbauer spectroscopy and its application to the study of glass is given in the excellent review article by Kurkjian(7).

The isotope most commonly used in Mössbauer spectroscopy experiments is $^{57}Fe$. The following discussion therefore will be based on the emission and absorption of $\gamma$–rays by $^{57}Fe$. The energy level diagram for this isotope is given in figure 1. The transition of interest occurs from the first excited state to the ground state. This transition has a life time ($\tau$) of $\cong 1 \times 10^{-7}$ seconds and energy of 14.4 keV. According to the uncertainty principle we may write:

$$\tau \cdot \Gamma \geq \hbar \quad (1)$$

where, $\hbar = \frac{h}{2\pi}$; h = Planks constant

and, $\Gamma$ = the half-width of the fluorescence peak.

From equation (1), the theoretical minimum half-width of the 14.4 keV $\gamma$-ray peak is approximately $5 \times 10^{-9}$ eV. Since the half-width of the peak is very small, it is possible to distinguish minor changes in the energy level of the nucleus. Although changes external to the nucleus (coordination, valence, bonding) have very little influence on the nuclear energy levels, the very high resolution of this technique makes it possible to distinguish these changes.

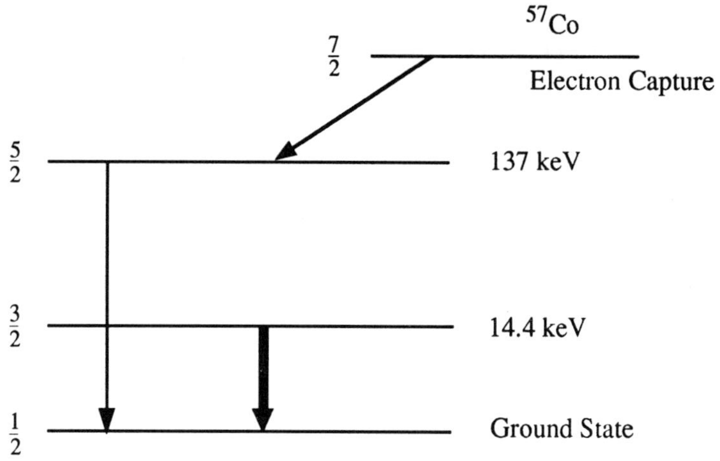


Figure 1. Energy Level Diagram for $^{57}Fe$.

With a stationary source ($\gamma$–ray emitter) and absorber (sample) resonant absorption is observed only when the transitions from the first excited state to the ground state in both the emitter and absorber are at exactly the same energy. The host matrix in which the $^{57}Fe$ isotope is embedded is normally not the same for the source and the sample. Also, a single-line Mössbauer spectrum is observed only when the degeneracy of the ground and excited states of the nucleus is preserved ($^{57}Fe$ isotope in a cubic non-magnetic host). In the presence of electric or magnetic fields, the degeneracy of the ground and excited states is removed either partially or completely. Due to these reasons, it is necessary to scan over a range of energies to obtain the entire Mössbauer spectrum. For $^{57}Fe$ Mössbauer spectroscopy a range of approximately $1 \times 10^{-6}$ eV is necessary. A Doppler shift in the γ–ray energy can be achieved by moving the source or the sample at a velocity, v.

$$\Delta E = \pm E_\gamma \frac{v}{c} \tag{2}$$

| | | |
|---|---|---|
| where, $\Delta E$ | = | change in energy |
| $E_\gamma$ | = | energy of the γ–ray (14.4 keV) |
| c | = | velocity of light ($3 \times 10^8$ m/s) |
| and v | = | velocity of the source or the sample |

To achieve a Doppler shift in energy of $1 \times 10^{-6}$ eV, it is necessary to move the source or the sample at a velocity of approximately 20 mm/s. For experimental convenience, normally the sample is held stationary and the source is moved.

For the Mössbauer effect to be observed it is necessary that the recoil energy accompanying the γ–ray emission and absorption be sufficiently small (less than the line width of the γ–ray emission and absorption process), and that a sufficient number of recoil free processes do take place. The fraction of total processes that are recoilless is related to the energy of the γ–ray, the mass of the emitter and absorber, and the mean square vibrational amplitude of the nucleus. The energy balance for the emission of a γ–ray can be written as follows:

$$E_\gamma = E_e + D - R \tag{3}$$

| | | |
|---|---|---|
| where, $E_\gamma$ | = | energy of the emitted γ–ray |
| $E_e$ | = | nuclear excitation energy (14.4 keV for $^{57}Fe$) |
| D | = | Doppler shift in energy |
| and R | = | recoil energy on the emitting nucleus |

A similar equation can be written for the absorber nucleus. The magnitude of the Doppler shift in energy can be calculated using equation (2). The recoil energy depends on the energy of the γ–ray, and the effective recoil mass.

$$R = \frac{E_\gamma^2}{2mc^2} \tag{4}$$

where, m = recoil mass
and c = velocity of light.

For a free $^{57}Fe$ nucleus with a mass of $9.3 \times 10^{-23}$ g, the recoil energy is approximately $10^{-3}$ eV. Since the recoil energy is orders of magnitude greater than the line width, resonance absorption will not be observed in this case. Theoretically, the recoil can be suppressed by rigidly fastening the nucleus to an infinite mass. For all practical purposes, this criterion is fulfilled for a nucleus in a solid lattice. During the emission or absorption of γ–rays by a nucleus embedded in a solid, the recoil mass (m in equation 4) is the entire mass of the crystal (or glass) rather than the mass of the atom. Hence, the magnitude of R will be small compared to the line width of the 14.4 keV γ–ray peak, provided the size of the particle in which the nucleus is embedded is large enough. For a glass of density 2.5 g/cm$^3$, 20 nm cubic particles will give a recoil energy of approximately $5 \times 10^{-9}$ eV (same as the line width).

As seen above, the Mössbauer effect will be observed only if the recoil energy is small compared to the line width. Since recoil energies depend on the γ–ray energies, only those transitions with an excitation energy of less than 150 keV are normally used in the experiments. When the recoil energy is smaller than that required for lattice excitation and the displacement or removal of atoms from the lattice sites, resonance absorption is observed. A certain fraction (f) of the total process will be recoilless and are called zero phonon processes. The fraction of zero phonon processes depends on the energy of the γ–ray and the mean square vibrational amplitude of the nucleus.

$$f = \exp\left[-\langle x^2 \rangle k^2\right] \tag{5}$$

where, $\langle x^2 \rangle$ = mean square vibrational amplitude of the nucleus
and k = wave vector = $\frac{2\pi}{\lambda}$
λ = wavelength of the γ–ray.

Since the vibrational amplitude of the nucleus increases with temperature, the fraction of zero phonon processes will be larger at low temperatures. Also, a larger value of the Debye temperature for the solid will result in a larger f value.

**Isomer Shift (IS)**

Complete energy resonance between identical types of stationary γ–ray emitting and absorbing nuclei is attained when both nuclei are embedded in identical matrices and the emitter and absorber are at the same temperature. But, in most systems, the emitter and absorber are in different environments and this leads to a small change in the transition energies[8]. This energy difference is termed the isomer shift[9] (chemical shift) and is due to alteration in the energy difference between ground and excited states because of coulombic interaction between the nucleus and the electrons. This change in energy can be calculated using the equation given below.

$$\Delta E = \frac{2\pi}{5} Ze^2 \left[ r_e^2 - r_g^2 \right] \left[ |\Psi_s(0)|_a^2 - |\Psi_s(0)|_s^2 \right] \tag{6}$$

where, $r_e^2$, $r_g^2$ = mean square charge radii of the excited and ground states.

$|\Psi_s(0)|^2$ = S-electron density at the nucleus in the source and absorber (*s* and *a* refer to source and absorber).

$Ze$ = nuclear charge

From equation (6) it can be seen that the isomer shift has contributions from both nuclear and atomic components. If the excited state nuclear radius is larger than the ground state radius, a positive chemical isomer shift indicates an increase in S-electron density in going from source to absorber. The excited state nuclear radius of $^{57}Fe$ is larger than the ground state radius. For $^{151}Eu$, the excited state radius is smaller than the ground state nuclear radius. Therefore, during the interpretation of Mössbauer data (change in the ionic or covalent nature of a chemical bond) it is necessary to have a thorough knowledge of the isotope used in the measurement. Also, the isomer shift is a change rather than an absolute energy value. This necessitates that the isomer shift be reported with respect to a reference material.

**Quadrupole Splitting (QS)**

The quadrupole splitting observed in Mössbauer spectroscopy is a result of the interaction of the nuclear electrical quadrupole moment with the electric field gradient at the nucleus. The energy level diagram for the $^{57}Fe$ nucleus in the presence of an electric field gradient is given in figure 2.

The spin quantum number of the ground state of the nucleus is $\pm\frac{1}{2}$ and the nucleus has no quadrupole moment. The excited state has a spin of $\pm\frac{3}{2}$, and in the presence of an electric field gradient the degeneracy of the excited state is partially removed, thus leading to two transitions separated by an energy $\Delta E_Q$. When the electric field gradient is axially symmetric, $\Delta E_Q$ can be calculated using the equation given below.

$$\Delta E_Q = \frac{1}{2} e^2 q Q \qquad (7)$$

where, eq = electric field gradient
and Q = nuclear quadrupole moment

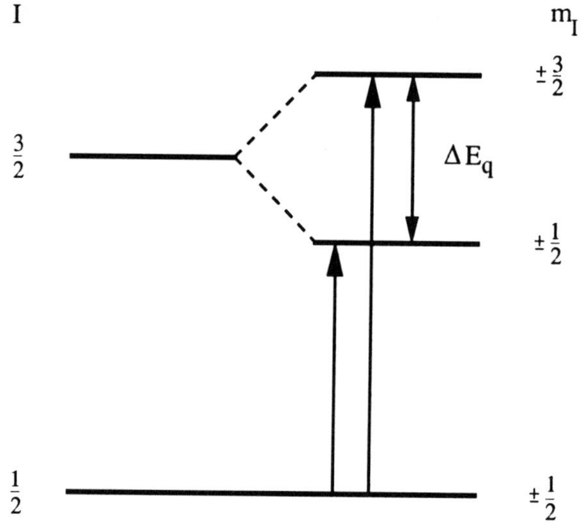


Figure 2. Quadrupole Splitting of $^{57}Fe$.

In $^{57}Fe$ Mössbauer spectra where quadrupole splitting is observed, the isomer shift is measured between the zero point of the reference material and the midpoint of the two peaks in the spectra. The environment and the electronic configuration of the nucleus will contribute to the presence and the magnitude of quadrupole splitting. Since trivalent iron has a half filled 3d shell, the electrons will have no contribution to the electric field gradient at the nucleus. The appearance of quadrupole splitting for trivalent iron therefore is an indication of the distortion of the tetrahedral or octahedral coordination of the absorber atom. Both the electrons and the ligands contribute towards quadrupole splitting in divalent iron.

### Magnetic Hyperfine Splitting

The hyperfine splitting observed in Mössbauer spectra can be due to any one of three reasons. One cause is the nuclear Zeeman effect. In the presence of a magnetic field, the degeneracy of both the ground and excited states is completely removed, producing 2I+1 (I is the nuclear spin quantum number) energy levels each. For the $^{57}Fe$ nucleus, this leads to two and four energy levels for the ground and excited states, respectively. Selection rules allow a total of six transitions(10) that are illustrated in figure 3.

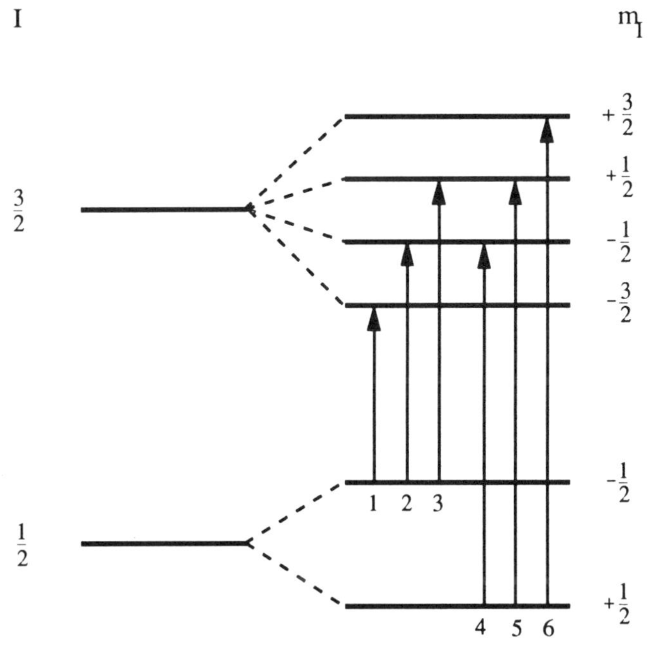


Figure 3. Magnetic Hyperfine Splitting of $^{57}Fe$.

In the studies of iron-containing glasses, the appearance of hyperfine splitting has been used to detect the presence of crystalline phases which are magnetic(11) (e.g., $Fe_3O_4$ spinels). Also, the crystallization temperature of these phases can be determined by collecting Mössbauer spectra over a range of temperatures. Another application of this phenomenon is in the determination of the Nèel and Curie temperatures of entire samples, or secondary phases in the matrix. In addition, if the magnetic domains in the sample are randomly oriented and since the ground state nuclear dipole moment of the absorber nucleus is known(12), the effective magnetic field at the absorber nucleus can be calculated from the magnitude of the hyperfine splitting.

Study of very small particles of $\alpha$-$Fe_2O_3$ showed that the presence of hyperfine structure in the Mössbauer spectrum depended on particle size and temperature, rather than on the bulk magnetic ordering temperature(13) (superparamagnetic behavior). In these cases hyperfine structure is observed when the spin-lattice relaxation time is comparable to or longer than the nuclear Larmor precession time(14). Another reason for the appearance of hyperfine structure is the magnetic field due to unpaired electrons in a paramagnetic ion(15). This will be observed in the Mössbauer spectrum if the electronic relaxation time is long compared to the nuclear lifetime.

## EXPERIMENTAL PROCEDURE

The essential components of a setup to collect Mössbauer spectra are the $\gamma$–ray source, the sample with the $\gamma$–ray absorbing isotope, a device to move the source or sample at a specified velocity, and a data collection system. A typical setup to measure Mössbauer spectra in the transmission mode is given in figure 4. Again, since iron is the most commonly studied element, the following discussion will concentrate on a data collection technique using the $^{57}$Fe isotope.

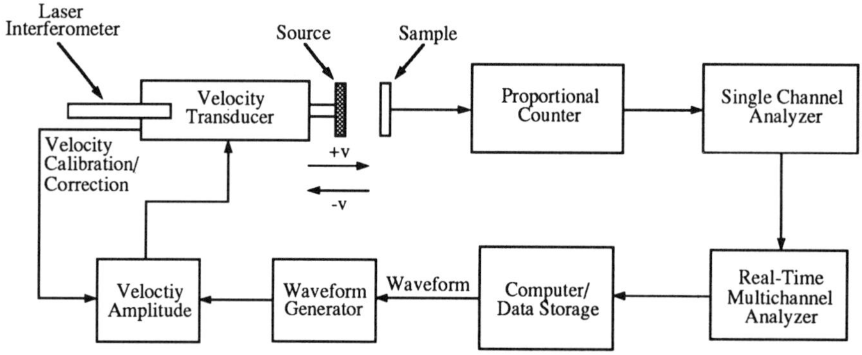


Figure 4. Schematic Diagram of a Mössbauer Spectrometer.

The $\gamma$–ray source consists of $^{57}$Co embedded in a cubic, non-magnetic host. The source (emitter) is prepared by uniform deposition of $^{57}$Co on a matrix foil and subsequent diffusion of the isotope into the matrix at high temperature in a controlled atmosphere[16]. The commonly used matrix materials are Cr, Cu, Pd, Pt, Rh, etc. The choice of the matrix depends on the particular application and the intensity of the γ–ray source. As a matrix material, chromium has undesirable magnetic properties at certain temperatures, whereas copper has a low solubility for iron and cobalt. Due to the high density and resulting electronic absorption, only low intensity sources can be prepared using platinum as the matrix. Rh and Pd have desirable properties if a high resolution detector is used to eliminate signals due to X-rays. Other commercially available sources include $^{119}Sn$ and $^{151}Eu$. In the selection of the ideal isotope-matrix combination, the main parameters of interest are the intensity of the source, line width, and the operating temperature range.

Preparation of the samples for Mössbauer spectroscopy studies is fairly simple. In samples containing large amounts of iron, the glass can be prepared from commercially available iron-containing compounds, because natural iron contains approximately 2.19 wt.% $^{57}Fe$. For very small concentrations of iron it is necessary to use a raw material enriched in $^{57}Fe$. Both solid and powdered samples of the glass have been used with success. The ideal thickness of the sample depends on the concentration of $^{57}Fe$ in the sample, the intensity of the source, and the type of data collection system which is used.

The source is moved through the required range of velocities using a velocity transducer. Automation of Mössbauer spectrometers has allowed the movement of the source according to specific waveforms with the required velocity. This feature facilitates the collection of data in specific velocity ranges, thus reducing the time for data collection. Built-in laser interferometers can be used in the calibration of the velocity of the source to an accuracy of 0.01%[17].

The emission spectrum of the $^{57}Co$ γ–ray source contains peaks at positions other than the 14.4 keV transition. Therefore, before collection of the Mössbauer spectrum, the 14.4 keV γ–ray region is selected. Subsequently, data are collected in real time using a multichannel analyzer and fed to a storage device. Proportional counters are commonly used in the detection of γ–rays. If studies are to be conducted at different temperatures, the necessary equipment (furnace, Dewar, temperature controller) can be added to the system.

As mentioned earlier, the isomer shift is a measure of the energy difference in the transitions between the γ–ray emitter and absorber. Since the isomer shift changes with the host matrix, it is necessary to calibrate for the zero velocity channel and the velocity per channel (energy change per channel). Normally an iron foil is used to calibrate the spectrometer, and all subsequent isomer shift values are reported with reference to this standard. Also, since all the relevant nuclear parameters for iron are known, the magnitude of the hyperfine splitting obtained using the iron foil can be used to calculate the effective magnetic field at the nucleus for other samples.

## ANALYSIS OF MÖSSBAUER SPECTRA

A typical $^{57}Fe$ Mössbauer spectrum of metallic iron is given in figures 5. The metallic iron foil gives a well defined six line spectrum which can be fitted using Lorentzian line shapes. From the fit, the peak locations and the intensity of the peaks can be calculated. The six peaks identified in the Mössbauer spectrum of metallic iron (figure 5) correspond to the six transitions illustrated in figure 3. Peak locations are normally reported in terms of velocity units with respect to the zero velocity point.

Analysis of the Mössbauer spectra of glasses is more complicated. Line widths of the peaks in the spectra depend on the environment of the atoms, absorber thickness, unresolved hyperfine splitting, etc. In addition, compared to the crystals, there is a wider distribution in the site parameters for the $\gamma$–ray absorbing ion in glasses, and this leads to broadening of the lines. The Mössbauer spectra of a series of glasses containing different amounts of $Fe^{2+}$ and $Fe^{3+}$ are given in figure 6. The Mössbauer spectrum of an iron-containing glass along with the computed fit to the raw data is given in figure 7. The sample contains both divalent and trivalent iron and the spectrum exhibits doublets due to quadrupole splitting. In this sample, $Fe^{3+}$ and $Fe^{2+}$ are in tetrahedral and octahedral coordinations, respectively.

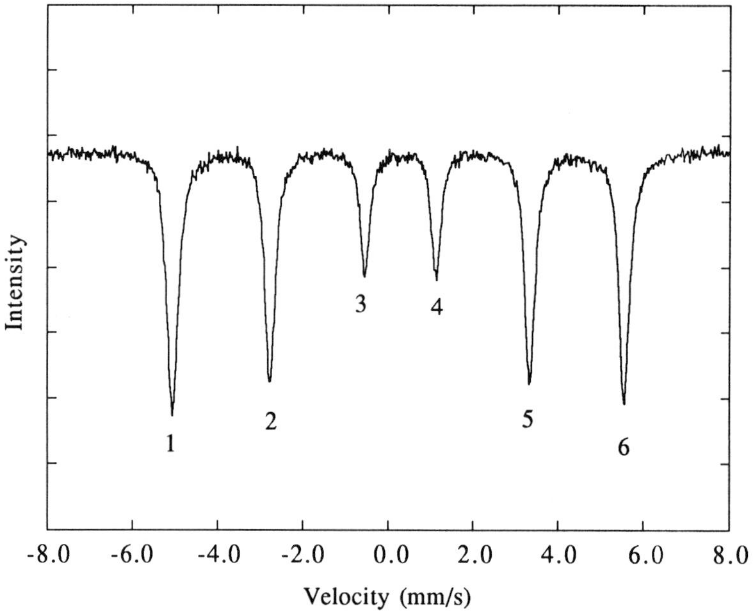


Figure 5. $^{57}Fe$ Mössbauer Spectrum of Metallic Iron Foil (SRM 1541).

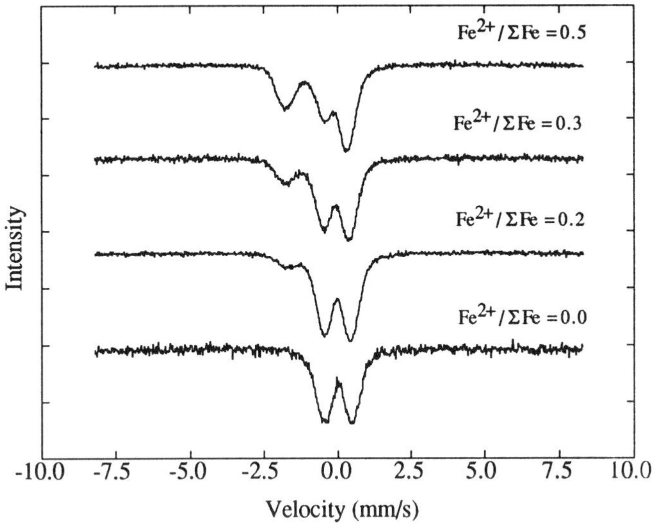


Figure 6. $^{57}$Fe Mössbauer Spectra of a Series of Glasses.

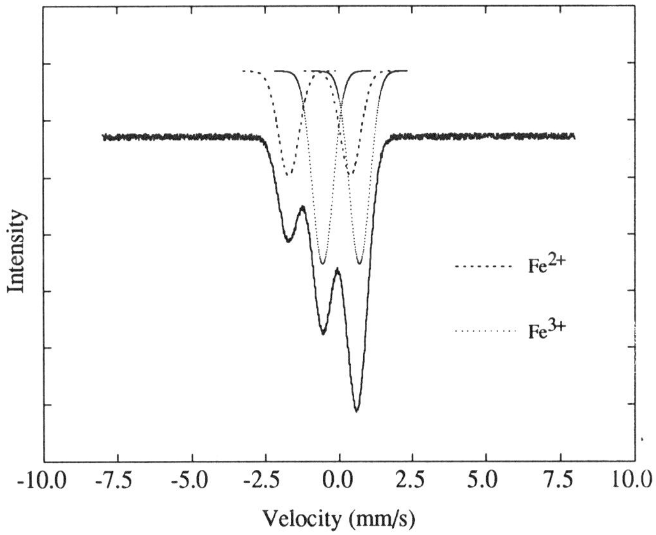


Figure 7. $^{57}$Fe Mössbauer Spectrum of a Partially Reduced Glass.

There are four possible configurations for iron in glass; $Fe^{2+}$ and $Fe^{3+}$ in tetrahedral and octahedral sites. Therefore, a spectrum similar to the one given in figure 7 can be a combination of up to 8 peaks (4 doublets). Though the assumption that the raw Mössbauer spectrum is composed of a large number of peaks may lead to a better fit, the results may not represent the actual distribution of iron in the glass. In those cases where the Mössbauer isotope is suspected to exist in all possible configurations, it is necessary to confirm the results of Mössbauer spectroscopic study using other techniques. In most glass compositions that have been studied, iron has been detected in only two or three configurations.

Normally, line shapes are assumed (Lorentzian, Gaussian-Lorentzian) and peaks are fitted to obtain the best fit. This technique is successful only if the concentrations of the different species are large enough to obtain well defined peaks in the raw Mössbauer spectrum. Studies by Virgo, *et al* (18) concluded that fits to spectra with doublets of Lorentzian shape are inadequate for describing the cumulative envelope, but are accurate in the determination of the peak locations. Models that accounted for the correct distribution of quadrupole splittings provided the best fit for the cumulative envelope. Also, during data analysis certain constraints are imposed on the quadrupole doublets (equal area, equal line width, etc.) which may not necessarily be correct. For example, $Fe^{2+}$ doublets in some glass samples tend be asymmetric(19) due to the Goldanskii-Karyagin effect(20) (directionally dependent f factor).

Levitz, *et al.* (21) analysed $^{57}Fe$ Mössbauer spectra in terms of a two-parameter distribution including variations in isomer shift and quadrupole splitting (electric field gradient). Though this method gave a good description of the cumulative envelope, when the experimental pattern consisted of symmetric Lorentzian doublets, more than one set of parameters gave satisfactory solutions for obtaining the full spectrum. The authors concluded that the constraints used in the analysis of the spectrum should reflect a physically acceptable model for the material being studied.

The method proposed by Massiot(22) involves the shape comparison of physical spectra. Mössbauer spectra of partially reduced glasses are analysed based on the assumption that their shapes are a combination of the spectra of fully oxidized and highly reduced samples. Using this method of analysis they were able to obtain good correlation with chemical analysis on the fraction of iron in each valence state. This method has the advantage of not assuming a structural model in data analysis, but requires spectra of oxidized and highly reduced samples of a glass composition similar to the one being analyzed.

Analysis of the spectra becomes more complicated for other Mössbauer isotopes like $^{151}Eu$ where quadrupole splitting leads to a number of energy levels. In these cases, the observed Mössbauer spectrum will be the result of a large number of overlapping peaks. For most glass forming systems, other techniques are required to confirm the results of Mössbauer spectroscopic studies.

## $^{57}$Fe MÖSSBAUER STUDIES OF GLASSES

In the analysis of glass samples by Mössbauer spectroscopy, the main parameters of interest are the isomer shift (IS), quadrupole splitting (QS) and magnetic hyperfine splitting. From values of the IS and QS, information can be obtained about the coordination and valence of the Mössbauer isotope in the glass. For some of the more commonly used isotopes, values of the IS and QS in various glass compositions have been reported in the literature. The IS and QS for the Mössbauer isotope in different valence states and coordinations can be obtained by measuring the Mössbauer spectra of crystalline materials (with known crystal structures) that contain the specific isotope. If a suitable reference material or data base is not available, it is necessary to combine Mössbauer spectroscopy with other analysis techniques to get information about the environment of the isotope in the glass. The values of IS and QS for a few Mössbauer isotopes in different glass compositions are given in Table 1.

The majority of Mössbauer studies on glasses deal with the behavior of iron ($^{57}$Fe) in silicates, borates and phosphates. Other isotopes which can be used in Mössbauer studies of glasses include $^{40}$K, $^{73}$Ge, $^{107}$Ag, $^{119}$Sn, $^{121}$Sb, $^{125}$Te, $^{133}$Cs, $^{151}$Eu, etc. There are a number of review articles which deal with the research in this area(7, 23-26).

### Borates and Borosilicates

Mössbauer spectroscopic studies on pure borates are limited due to the low solubility of $Fe_2O_3$, and $SnO_2$ in $B_2O_3$. Sekhon and Kamal(27) studied $B_2O_3$ glasses containing 0.5 to 5.0 wt% $Fe_2O_3$ enriched in $^{57}$Fe. They observed the presence of $Fe^{2+}$ in the amorphous phase, whereas $Fe^{3+}$ was present only in crystallites of $Fe_2O_3$.

In pure $B_2O_3$ glass, boron is in triangular coordination and boroxyl rings are the basic units in the network(28). Addition of modifier ions to the glass results in the conversion of $BO_3$ to $BO_4$ units. Since this conversion leads to changes in the properties of the glass, there is considerable interest in the study of the structure of these glasses with varying modifier ion contents. The following general observations can be made about the behavior of iron in alkali borates.

1. Both divalent and trivalent iron are present in these glasses and the relative amount in either oxidation state depend on the composition, starting raw materials and the conditions used in the preparation of the glasses.

2. The isomer shift and quadrupole splitting values indicate that trivalent iron is in tetrahedral coordination and acts as a network former ($FeO_4$ units).

3. In alkali borates containing large amounts of $Fe_2O_3$, crystallites of $Fe_2O_3$, $Fe_3O_4$, and a garnet-type phase form as secondary phases. In addition to the

$Fe_2O_3$ content, the amount of alkali present also has an influence on the precipitation of these crystals.

4. Addition of modifier ions which increase the nonbridging oxygen content of the system leads to a decrease in the isomer shift values for the $Fe^{3+}$ ion.

5. Glasses containing smaller concentrations of $Fe_2O_3$ exhibit superparamagnetic behavior.

Study of potassium iron borate glasses containing up to 40 mol% $K_2O$ by Nishida, *et al.* (29) showed a decrease in the IS and QS for $Fe^{3+}$ with increase in the alkali content above 20 mol%. These changes were attributed to the formation of nonbridging oxygens and an increase in the symmetry of the coordination around the $Fe^{3+}$ ion. In glasses containing less than 10 mol% alkali, Mössbauer spectra and X-ray diffraction results showed the presence of crystalline $\alpha$-$Fe_2O_3$.

Tanaka, *et al.* (30) studied sodium iron borate glasses and observed a decrease in the IS for $Fe^{3+}$ with increase in alkali content. Quadrupole splitting for $Fe^{3+}$ increased with alkali content up to 30 mol% alkali, and further increase in alkali content led to a decrease in the QS. This was taken as an indication of the detection of the boron anomaly using Mössbauer spectroscopy. This is in agreement with the work by Raman, *et al.* (31) on alkali iron borates. However, Mössbauer spectroscopic studies of sodium iron borate glasses by Kamal, *et al.* (32) did not show any sharp changes in the IS or QS values in the boron anomaly region. They reported the presence of $Fe^{3+}$ ions in both tetrahedral and octahedral coordinations.

Another study of sodium borates containing $Fe_2O_3$ (0.5–9 mol%) was conducted by Pye, *et al.* (33) and Belkhiria(34). They observed hyperfine splitting in the room temperature spectra of samples containing less than 1.5 mol% $Fe_2O_3$. Belkhiria attributed this observation to long spin relaxation times and Fermi contact interaction of paramagnetic $Fe^{3+}$ ions in octahedral coordination. Transmission electron microscopy revealed the presence of amorphous iron rich clusters in samples containing more than 1.5 mol% $Fe_2O_3$. The isomer shift values for $Fe^{3+}$ decreased with increase in the $Fe_2O_3$ content of the glass.

Belkhiria(34) conducted detailed studies of a potassium borate glass containing 3 mol% $Fe_2O_3$, prepared under different partial pressures of oxygen. Both the IS and QS for $Fe^{2+}$ decreased with increase in the $Fe^{2+}$ content of the glass, but the values indicated that $Fe^{2+}$ is in octahedral coordination in all the samples. Hyperfine splitting was observed in the spectrum of a sample which contained 80% of iron in the $Fe^{2+}$ state. This was attributed to isolated paramagnetic $Fe^{3+}$ ions dispersed in the glass.

A study of lead iron borates containing up to 50 mol% $Fe_2O_3$ was conducted by Burzo, *et al* (36). Large concentrations of iron are required in these glasses because of the absorption of γ–rays by Pb. Both divalent and trivalent iron were detected in the glasses and their relative concentrations depended on the composition. Spectra were fitted assuming two different sites each for $Fe^{2+}$ and $Fe^{3+}$. Hyperfine splitting was not observed in any glass. This combined with results of IR spectroscopic studies showed that up to 50 mol% of $Fe_2O_3$ can be incorporated in the glass matrix without clustering or crystallization of secondary phases. The isomer shift for $Fe^{3+}$ and $Fe^{2+}$ in both sites decreased with increase in $Fe_2O_3$, whereas the QS increased. Assuming that the recoil free fraction (f factor) is the same for $Fe^{2+}$ and $Fe^{3+}$, the number of Fe ions in each valence state was calculated from the areas of the corresponding Mössbauer peaks and the number of $Fe^{3+}$ ions was found to increase linearly with total $Fe_2O_3$ content. Sekhon, *et al.* (37) also studied lead borates containing iron. They observed $Fe^{3+}$ doublets in samples containing less than 20 wt% $Fe_2O_3$ and evidence of hyperfine splitting with higher amounts of iron. Magnetic susceptibility measurements on lead borate glasses containing 10–20 wt% $Fe_2O_3$ showed the presence of iron in a superparamagnetic state. Study of lead borate glasses containing $GeO_2$ showed that the amount of iron in the divalent state decreased with decrease in temperature and increase in the amount of $GeO_2$(38).

Nishida, *et al.* (35) studied the effect of halide ions on the structure of potassium iron borate glasses containing 10-40 mol% $K_2O$. Addition of halide ions ($F^-$, $Cl^-$, $Br^-$) to glasses with $K_2O \geq 15$ mol% led to a sharp decrease in the IS and QS values for $Fe^{3+}$ which led to the conclusion that halides occur as nonbridging ions. The halides are in interstitial sites in glasses when the alkali content is less than 15 mol%. Another study(39) concentrated on silver borates containing 20 mol% AgBr and 1 mol% $Fe_2O_3$ enriched in $^{57}Fe$. Iron was found to exist as $Fe^{3+}$ in a tetrahedral environment. With $Ag_2O$ content of more than 11-12 mol% the IS and QS decreased with increase in $Ag_2O$ content. This was attributed to the formation of nonbridging bromine and oxygen in boron and iron tetrahedra. Results of the Mössbauer study were consistent with changes in the glass transition temperatures of these glasses. The glass transition temperature varied with $Ag_2O$ content with a maximum at 12 mol%.

Barium borates containing up to 60 mol% $V_2O_5$ were studied by Bansal, *et al.* (40). In this system $Fe^{3+}$ was observed in both tetrahedral and octahedral sites. The relative amount of $Fe^{3+}$ in each site depended on the composition of the glass. In pure barium borates, $Fe^{3+}$ was present in both sites, whereas in barium vanadates, $Fe^{3+}$ was detected only in octahedral sites.

The structural role of iron in borosilicates has been studied by a number of workers. A considerable amount of work has been done on simulated nuclear waste glasses using $^{57}Fe$ Mössbauer spectroscopy. One of the main objectives of this work was the quantitative determination of the amount of divalent and trivalent iron in the glasses and is discussed in a later section. Studies of a simulated nuclear waste glass by Music, *et al.* (41) concluded that the valence and coordination of iron in these glasses depend on the composition, method of preparation, and total iron content. Another study by De Grave, *et al.* (42) on a similar system indicates that $Fe^{3+}$ and $Fe^{2+}$ are most likely in tetrahedral and octahedral coordinations, respectively. However, the shape of the $Fe^{2+}$ peaks indicate at least two distinct local symmetries.

Nishida, *et al.* (43) studied potassium borosilicates containing iron. The isomer shift and QS for $Fe^{3+}$ in tetrahedral coordination decreased with increase in $K_2O$ in the region of 8 -18 mol% alkali. This was attributed to formation of nonbridging oxygen in $BO_4$ and $FeO_4$ tetrahedra. Decreases in the IS and QS at high alkali concentrations were attributed to nonbridging oxygen formation in $SiO_4$ and $FeO_4$ tetrahedra. In glasses containing high amounts of alkali, substitution of $SiO_2$ by $B_2O_3$ resulted in an increase in the concentration of nonbridging oxygen. For glasses with an $SiO_2/B_2O_3$ ratio of 2, all of the $Fe^{3+}$ were in tetrahedral coordination and the IS and QS values were a little smaller than those for alkali borates. The half-width of the peaks decreased with an increase in the amount of $K_2O$.

## Silicates

A considerable amount of work has been reported on Mössbauer spectroscopic studies of iron in silicate glasses. The major results of the studies in these glasses are listed below:

1. Both divalent and trivalent iron are present in silicate glasses and the concentrations in each valence state depend on the glass composition, amount of iron, melting temperature, and oxygen partial pressure.

2. $Fe^{2+}$ ions normally occupy octahedral sites and act as network modifiers.

3. The majority of $Fe^{3+}$ ions occupy tetrahedral sites, but normally a fraction of trivalent ions are in octahedral sites. The fraction of $Fe^{3+}$ in each site depends on the $Fe^{3+}$ to total iron ratio.

4. The $Fe^{2+}$ doublet tends to be asymmetric in a number of silicate glasses.

5. $Fe^{3+}$ exhibits superparamagnetic behavior in silicate glasses.

One of the earliest $^{57}Fe$ Mössbauer studies on glasses was conducted by Kurkjian, *et al.* (14) on a sodium trisilicate composition. They observed $Fe^{2+}$ in octahedral coordination. The large line width of the peaks in the spectrum was attributed to distortion of the Fe sites in glass. Study of sodium silicates by a number of workers showed that $Fe^{3+}$ occurs in both tetrahedral and octahedral sites.

Both $Fe^{2+}$ and $Fe^{3+}$ were observed in calcium iron silicates by Iwamoto, et al (44). For large $Fe_2O_3$ concentrations, all the $Fe^{2+}$ occupied octahedral sites, whereas at low concentrations of $Fe_2O_3$, $Fe^{2+}$ was detected in both octahedral and tetrahedral coordinations. $Fe^{3+}$ was detected in both sites.

Levy, *et al.* (45) studied $SiO_2 \cdot Na_2O \cdot CaO$ glasses containing iron and found that the ratio of divalent to trivalent iron increased with increase in melting temperature, decrease in basicity of the glass, and by the replacement of $Na_2O$ by CaO. $Fe^{3+}$ was observed in both tetrahedral and octahedral sites and the ratio of trivalent iron in either site was not affected by the replacement of $Na_2O$ by CaO. $Fe^{2+}$ was assumed to occupy only octahedral sites.

Hematite dissolution and devitrification in float glass was studied by Brown, *et al* (46). They also measured the diffusion coefficient of $Fe^{3+}$ in the glass. The authors concluded that in this system the f factors for $Fe^{3+}$ and $Fe^{2+}$ are 1.11 and 0.8, respectively. The f factors were calculated based on $Fe^{2+}$ and $Fe^{3+}$ concentrations determined using chemical analysis and Mössbauer spectroscopy. Other studies on similar glass forming systems report good agreement in the quantitative determination of iron in different valence states by the same two techniques(47,48). In the latter studies, the f factors for divalent and trivalent iron were assumed to be equal.

Virgo and Mysen(18), and Mysen(49) conducted detailed studies of the structural role of iron in alumino iron silicates containing sodium, calcium and magnesium. The doublet assigned to $Fe^{2+}$ was asymmetric in nature and the peak widths of divalent and trivalent iron were 2-3 times that in the crystals. The isomer shift values for $Fe^{3+}$ in tetrahedral coordination varied from 0.19–0.3 mm/sec. $Fe^{3+}$ in octahedral coordination showed an IS $\cong$ 0.5 mm/sec and for $Fe^{2+}$ in the same coordination the IS was > 1.0 mm/sec. For $Fe^{3+}/\Sigma Fe > 0.5$, the IS values for $Fe^{3+}$ varied from 0.22 to 0.33 indicating tetrahedral coordination. For $Fe^{3+}/\Sigma Fe < 0.5$ the IS increased with a decrease in the amount of $Fe^{3+}$ indicating that the fraction of $Fe^{3+}$ in octahedral coordination increases.

**Phosphates**

One of the early studies on iron phosphates conducted by Kurkjian, *et al.* (14) demonstrated that the glass shows paramagnetic behavior up to 1.2 K. Both divalent and trivalent iron were observed in the samples. At low temperature and low iron concentration, hyperfine structure was observed in the Mössbauer spectrum due to large electron spin relaxation times.

Lin Yunfei, *et al.* (50) studied iron phosphate and sodium iron phosphates and observed both $Fe^{2+}$ and $Fe^{3+}$ in the samples. $Fe^{3+}$ was observed in both tetrahedral and octahedral coordinations and $Fe^{2+}$ only in octahedral coordination. Increase in the amount of $Fe_2O_3$ increased the fraction of $Fe^{3+}$ in fourfold coordination and also the fraction of iron in the $Fe^{2+}$ state. Study of potassium iron phosphate glasses by Nishida, *et al.* (51) showed that $Fe^{2+}$ and $Fe^{3+}$ both occur only in octahedral coordination. $Fe^{2+}$ decreased linearly with increase in the $K_2O$ content of the glass.

Taragin, *et al.* (52) collected $^{57}Fe$ Mössbauer spectra of alumino phosphate glasses from room temperature to 450°C. Both $Fe^{2+}$ and $Fe^{3+}$ were observed only in octahedral sites. The isomer shifts for $Fe^{2+}$ and $Fe^{3+}$ decreased with increase in temperature and this change was attributed to second-order Doppler shifts. In general, line widths of the peaks in the Mössbauer spectra of phosphates were narrower than those for borosilicates.

A study of calcium iron phosphates by Sanad, *et al.* (53) also showed the presence of both $Fe^{2+}$ and $Fe^{3+}$ in the samples. Iron in both valence states was identified only in octahedral sites, but the authors do not rule out the possibility that a small fraction of the iron may be in tetrahedral sites.

Results of Mössbauer spectroscopy studies of sodium iron fluorophosphate glasses by Menil, *et al.* (54) also showed evidence of iron in both $Fe^{2+}$ and $Fe^{3+}$ states in octahedral coordination. However, electron paramagnetic resonance (EPR) studies suggest that a small fraction of $Fe^{3+}$ could be in tetrahedral coordination. In these glasses $Fe^{3+}$ and $Fe^{2+}$ act as network formers and modifiers, respectively.

### Zirconium Fluoride

Zirconium fluoride glasses containing iron have been studied by Bogomolova, *et al* (55). Iron was added as $FeF_3$ (0.02 - 0.5 wt.%) to $58ZrF_4 \cdot 6LaF_3 \cdot 32BaF_2 \cdot 4AlF_3$ glasses. Most of the iron was present in the $Fe^{2+}$ state in octahedral coordination. EPR studies showed the presence of iron in the $Fe^{3+}$ state also.

Kawamoto, *et al.* (56) studied $ZrF_4$ glasses containing $FeF_3$ (2 and 0.5 mol%). Iron was detected in octahedral coordination in both divalent and trivalent states. The ratio of $Fe^{2+}/Fe^{3+}$ decreased with decrease in the $ZrF_4$ content of the glasses.

### Quantitative Analysis

One of the applications of Mössbauer spectroscopy is in the determination of $Fe^{2+}/\Sigma Fe$ values in glasses. This ratio can be used as an indication of the redox state of the glass. Some natural glasses and certain glass compositions of commercial importance contain a number of transition elements and this can lead to errors in the determination of the percentage of iron in each redox state by chemical methods. Therefore analysis by $^{57}Fe$ Mössbauer spectroscopy is especially advantageous in these cases. If the recoilless fraction (f factor) is the same for $Fe^{2+}$ and $Fe^{3+}$, the integrated areas of the respective peaks will be proportional to the concentration of iron in each valence state.

Labar and Gielen[57] determined the $Fe^{2+}$ content of sodium iron silicate glasses containing different concentrations of $Fe_2O_3$ by Mössbauer spectroscopy and other methods (infrared and visible absorption spectroscopy). For large concentrations of $Fe_2O_3$ (> 4 mol%) both methods gave similar $Fe^{2+}/\Sigma Fe$ values. For small concentrations of $Fe_2O_3$, $Fe^{2+}/\Sigma Fe$ values obtained by Mössbauer spectroscopy were much higher than those obtained by chemical methods.

In the work by Goldman, *et al.* [58] on simulated nuclear waste glasses (multi-component borosilicates), $Fe^{2+}/\Sigma Fe$ were determined from the Mössbauer spectra of the glass samples. They found that the fraction of iron in the $Fe^{2+}$ state, calculated from Mössbauer spectroscopy data, was 1.2 times that obtained by chemical analysis. Studies on similar glasses by Karraker, *et al.* [48] gave good agreement between the results of chemical analysis and Mössbauer spectroscopy for $Fe^{2+}/Fe^{3+}$ values in the range of 0.2 - 2.4. They concluded that the difference obtained by Goldman, *et al.* is probably due to an error in the chemical analysis. Hunter, *et al.* [59] also obtained one-to-one correlation between the results of chemical analysis and Mössbauer spectroscopy.

Mysen, *et al.* [47] determined $Fe^{2+}/\Sigma Fe$ values for a series of glasses by chemical analysis and Mössbauer spectroscopy and compared the results. In these glasses $Fe_2O_3$ varied from 2.2 to 34.7 wt% and $Fe^{3+}/\Sigma Fe$ varied from 0.2–0.96. Within the limits of experimental error, no difference was observed between the results obtained using the two techniques. Integrated areas of the peaks in the Mössbauer spectra were determined using both Lorentzian, and Gaussian-Lorentzian line shapes for the doublets. The line shapes did not affect the areas of the peaks appreciably. If the low velocity components of the ferrous and ferric doublets coincide, for the determination of the integrated areas of the component doublets, a three-line fit can be used instead of the four-line fit. $Fe^{3+}/\Sigma Fe$ calculated in this manner also gave good correlation with the results of chemical analysis. However the error in the quantitative determination of $Fe^{3+}$ by Mössbauer spectroscopy was large for $Fe^{3+}/\Sigma Fe$ values less than 0.1.

## MÖSSBAUER STUDIES USING OTHER ISOTOPES

Mössbauer spectroscopic studies in glasses have been conducted using a few isotopes other than $^{57}Fe$. The isotopes discussed here are $^{119}Sn$, $^{121}Sb$, and $^{151}Eu$.

### $^{119}$Sn Mössbauer Spectroscopy

Paul, *et al.* [60] studied the behavior of tin in borates containing up to 58 mol% SnO. Careful control of the furnace atmosphere was necessary to obtain $SnO{\cdot}B_2O_3$ glasses. Addition of up to 20 mol% SnO to $B_2O_3$ resulted in the formation of tetrahedral boron. SnO concentrations in excess of 20 mol% lead to the formation of NBO in addition to tetrahedral boron. With the addition of SnO, the IS for $Sn^{2+}$

increased up to 20 mol% SnO and then decreased, whereas QS did not show any appreciable change. In high SnO glasses, ionicity of $Sn^{2+}$–O bonds decreased with increase in SnO. Values of the IS and QS for $Sn^{2+}$ were larger in the glass when compared to tetragonal SnO crystal.

Eissa, *et al.* (61) studied alkali (Li, Na, K) borates containing $SnO_2$ and found that all the tin occurs in octahedral coordination in the $Sn^{4+}$ state. With increase in the ionic radii of the alkali ion, IS increased and QS decreased. With increase in the total alkali content, the IS, QS, and the line width of the peaks in the Mössbauer spectra decreased. The asymmetry observed in the absorption lines was attributed to the Goldanskii-Karyagin effect.

Evstropev, *et al.* (62) studied silicate, borate, germanate, and phosphate glasses containing tin, melted in an oxidizing atmosphere, using $^{119}Sn$ Mössbauer spectroscopy. Tin was present in the $Sn^{4+}$ state in silicates, borates, and germanates and in the $Sn^{2+}$ state in phosphates.

Mössbauer spectroscopic studies of the crystallization of chalcogenide glasses of the composition $xSn_{0.5}.Se_{0.5}.(100\ x)As_{0.1}.Ge_{0.3}.Se_{0.6}$ were conducted by Cheng, *et al* (63). These glasses were crystallized using Zr as a nucleating agent. In the presence of Zr, $GeSe_2$ and SnSe crystal phases were observed, whereas in the absence of Zr only SnSe crystals were observed. Mössbauer spectra were collected at 77 K using powdered glass samples. Tin was present in the $Sn^{2+}$ state and the IS decreased with increase in the amount of crystals.

## $^{151}$Eu Mössbauer Spectroscopy

Aluminosilicate glasses and various oxide crystals containing Eu were studied(64) by Mössbauer spectroscopy using $^{151}Sm$ in $Sm_2O_3$ as the source. The isomer shift of $Eu^{3+}$ in crystals varied with Eu–O bond lengths and coordination number of $Eu^{3+}$. In glasses, the IS increased with increase in the $Eu_2O_3$ content, and coordination of $Eu^{3+}$ varied from 8 to 12. The IS of $Eu^{3+}$ was measured in crystals in which $Eu^{3+}$ has coordinations of 6, 8, 9, and 12. Due to the reduced covalency of the Eu–O bond, the isomer shift decreased with increase in coordination number.

Fluorozirconates containing Eu were studied by Coey, *et al* (65). From the differences in coordination and bonding they concluded that $Eu^{3+}$ are network formers, whereas $Eu^{2+}$ are network modifiers. From the IS values of −15.5 to −12.9 mm/s for $Eu^{2+}$, Eu–F bond lengths were calculated to be in the range of 2.4 to 2.8 Å. The broadness of the $Eu^{2+}$ bands in the absorption spectrum was attributed to unresolved quadrupole interactions, and variations in the coordination. With decrease in temperature, the recoilless fraction (f factor) for $Eu^{2+}$ increased more rapidly than that for $Eu^{3+}$.

Winterer, *et al.* (66) conducted studies of borates containing 0.1–33 mol% EuO. Most of the Eu existed as $Eu^{2+}$, but small amounts of $Eu^{3+}$ also were detected. Mössbauer spectra of samples containing small amounts of EuO (< 2 mol%) showed paramagnetic hyperfine structure.

## $^{121}$Sb Mössbauer Spectroscopy

Cremers, *et al.* (67) studied the Mössbauer spectra of the chalcogenide glass forming system Ge·Sb·Se using $Ba^{121m}SnO_3$ (metastable $^{121}$Sn in $BaSnO_3$ matrix) as the source. Only $Sb^{3+}$ was observed in the glasses. In the selenium rich glasses, Sb atoms were coordinated by three Se atoms. By comparing the Mössbauer spectra of crystalline $Sb_2Se_3$ and the glass, it was concluded that in this region, pyramidal $SbSe_{3/2}$ structural units with $GeSe_{4/2}$ groups dominate the glass network. The magnitude of the negative isomer shift decreased (decrease in S-electron density at Sb nucleus) with increase in Se content at constant concentration of Ge. Glasses containing low concentrations of Se may contain $Sb_2Se_{4/2}$ units, glassy Sb and Ge, and GeSb alloys in addition to the structural units mentioned above.

Quantitative determination of trivalent and pentavalent Sb in silicate glasses by $^{121}$Sb Mössbauer spectroscopy was conducted by Stahlberg, *et al* (68). The spectra contained one quadrupole broadened line due to $Sb^{3+}$ and a narrower line due to $Sb^{5+}$. One Lorentzian peak was fitted to $Sb^{5+}$, and based on the allowed quadrupole transitions, $Sb^{3+}$ was fitted using eight Lorentzian lines. $Sb^{3+}/Sb^{5+}$ values were computed as the ratio of the integrated areas of the respective peaks in the spectrum. The accuracy of the measurement technique was checked by determining this ratio in crystalline $Sb_2O_4$. At 4.2 K, the ratio of the integrated areas gave the correct results for $Sb^{3+}/Sb^{5+}$ values. At higher temperatures, this technique yielded higher values for the amount of antimony in the pentavalent state.

## Other Mössbauer Studies

Mössbauer spectroscopy has been used extensively in the study of amorphous metals(69). Though of limited use in the study of their structure, this technique provides considerable information about the crystallization behavior of these materials. Gamma-ray transmission Mössbauer spectroscopy in combination with conversion electron Mössbauer spectroscopy (study of electrons dislodged from the material by the $\gamma$-ray) allows simultaneous study of the surface and the bulk of amorphous metals.

Capillary Mössbauer spectroscopy is another technique useful in the study of liquids trapped in the pores of a suitable carrier like a silicate glass(70). Previously, Mössbauer spectroscopy studies were conducted on liquids by quenching them at a fast rate to obtain an amorphous material. In many organic systems, the change in temperature may lead to large changes in the properties and phase composition of the system. Instead, using Mössbauer spectroscopy, it was possible to study liquids trapped in the pores of a special porous silicate glass at temperatures above the freezing point of the liquid. However, it is necessary to take sufficient care to distinguish between atoms adsorbed on the carrier surface and those in the bulk.

## SUMMARY

Mössbauer spectroscopy has been used successfully in the study of a number of glasses, especially those containing Fe and Sn. Information has been obtained about the structure, crystallization behavior, diffusion coefficients, and magnetic properties of the glasses. The isomer shift and quadrupole splitting values have been used to determine the coordination of the Mössbauer isotope in the glass.

Table 1. Mössbauer Parameters for Some Isotopes in Glasses.

| Composition | Isotope | Valence | CN | IS | QS | Reference |
|---|---|---|---|---|---|---|
| Phosphates | $^{57}Fe$ | $Fe^{3+}$ | 6 | 0.18 - 0.50 | 0.36 - 0.88 | 25, 54 |
| Phosphates | $^{57}Fe$ | $Fe^{2+}$ | 6 | 1.06 - 1.60 | 1.48 - 3.20 | 25 |
| Borates | $^{57}Fe$ | $Fe^{3+}$ | 4 | 0.17 - 0.34 | 0.49 - 1.09 | 25 |
| Borates | $^{57}Fe$ | $Fe^{3+}$ | 6 | 0.41 - 0.41 | 0.47 - 1.38 | 25 |
| Borates | $^{57}Fe$ | $Fe^{2+}$ | 6 | 0.61 - 1.03 | 1.01 - 2.08 | 25 |
| Silicates | $^{57}Fe$ | $Fe^{3+}$ | 4 | 0.20 - 0.32 | 0.73 - 1.40 | 25, 18 |
| Silicates | $^{57}Fe$ | $Fe^{3+}$ | 6 | 0.30 - 0.54 | 0.44 - 1.55 | 25, 18 |
| Silicates | $^{57}Fe$ | $Fe^{2+}$ | 4 | 0.77 - 0.99 | 1.02 - 2.26 | 25 |
| Silicates | $^{57}Fe$ | $Fe^{2+}$ | 6 | 0.98 - 1.24 | 1.84 - 2.67 | 25, 18 |
| Fluorozirconate | $^{57}Fe$ | $Fe^{3+}$ | 6 | 0.44 - 0.49 | 0.48 - 0.64 | 56 |
| Fluorozirconate | $^{57}Fe$ | $Fe^{2+}$ | 6 | 1.25 - 1.37 | 2.24 - 2.42 | 56 |
| Borate | $^{119}Sn$ | $Sn^{2+}$ | 4 | 0.71 - 1.38 | 1.52 - 1.92 | 60 |
| Chalcogenide | $^{119}Sn$ | $Sn^{4+}$ | 6 | 3.29 - 3.54 | 0.58 - 1.04 | 63 |
| Aluminosilicate | $^{151}Eu$ | $Eu^{3+}$ | 8-12 | 0.30 - 0.50 | - | 64 |
| Fluorozirconate | $^{151}Eu$ | $Eu^{3+}$ | 8-12 | 0.06 | - | 65 |
| Fluorozirconate | $^{151}Eu$ | $Eu^{2+}$ | 8-12 | -14.18 | - | 65 |
| Ge·Sb·Se | $^{121}Sb$ | $Sb^{3+}$ | - | -11.4 - -14.2 | 7.50 - 11.5 | 67 |
| Silicate | $^{121}Sb$ | $Sb^{3+}$ | - | -12.0 | - | 68 |
| Silicate | $^{121}Sb$ | $Sb^{5+}$ | - | 1.34 | - | 68 |

The IS for both $Fe^{2+}$ and $Fe^{3+}$ are relative to metallic iron absorber.
The IS for $Sn^{2+}$ is relative to alpha tin absorber.
The IS for $Sn^{4+}$ is relative to $BaSnO_3$ absorber.
The IS for $Eu^{3+}$ (Ref. 64) is relative to $EuF_3$ absorber.
The IS for both $Eu^{3+}$ and $Eu^{2+}$ (Ref. 65) are relative to $SmF_3$ source.
The IS for both $Sb^{3+}$ and $Sb^{5+}$ (Ref. 68) are relative to $Ba^{121m}SnO_3$ source.

Ranges of the IS and QS reported for a few glasses are given in Table 1 (the IS and QS data for phosphates, borates and silicates has been compiled in the review article by Dyar[25]). As can be seen from these data, there is some overlap in the IS and QS values for $Fe^{2+}$ and $Fe^{3+}$ ions in tetrahedral and octahedral coordinations.

Since the changes in the nuclear energy levels of the isotope are caused mainly by the nearest neighbors, Mössbauer spectroscopy is a technique that is ideal for the study of the structure of glasses which lack long range order. However, this lack of order gives rise to a distribution of site parameters for the Mössbauer isotope which leads to the broadening of the peaks in the spectrum. Peaks in the Mössbauer spectra of glasses are 2–3 times broader than those of crystals. In the analysis of the Mössbauer data, the distribution of site parameters has to be accounted for in order to obtain a fit that describes the cumulative envelope of the spectrum. Since detailed information about the structure of glass is not available, the constraints that are imposed during the fit may not necessarily reflect the real physical model. In most cases it is necessary to use other techniques to verify the results obtained by Mössbauer spectroscopy.

A number of Mössbauer isotopes exist in more than one valence state and attempts have been made to determine the fraction in each valence state. If the recoil free fraction (f factor) for the isotope in both valence states is the same, the integrated areas of the respective peaks will be proportional to the concentration of the isotope in each valence state. $Fe^{2+}/\Sigma Fe$ values have been determined at room temperature in a number of glasses and give good agreement with the results of chemical analysis[47,48]. $Sb^{3+}/Sb^{5+}$ values agree with the results of chemical analysis at low temperature (4.2 K) but show large variations at higher temperatures[68]. It has been reported[71] that in certain minerals the recoil free fraction for $Fe^{3+}$ is larger than that for $Fe^{2+}$. Therefore, the assumption that the recoil free fraction for $Fe^{2+}$ and $Fe^{3+}$ is the same must be tested for specific glass compositions at different temperatures.

Iron occurs in both divalent and trivalent states in glasses. The fraction of total iron in each valence state depends on the glass composition, starting raw materials, melting temperature, partial pressure of oxygen, concentration of iron etc. In phosphates, both $Fe^{2+}$ and $Fe^{3+}$ generally exist in octahedral coordination. In borates, $Fe^{2+}$ exists in octahedral coordination whereas $Fe^{3+}$ exists in both tetrahedral and octahedral coordinations. In fluorozirconate glasses both $Fe^{3+}$ and $Fe^{2+}$ were detected in octahedral coordination with fluorine. In silicates, $Fe^{3+}$ and $Fe^{2+}$ have been reported in both octahedral and tetrahedral coordinations. The fraction of Fe in each site depends on the glass composition, $Fe_2O_3$ content, and $Fe^{3+}/Fe^{2+}$ ratio.

**Acknowledgements**

The authors would like to thank Ms. T.V.Palmiter and Mr. D.M.Korwin for their assistance in the preparation of this manuscript. The comments and suggestions of an anonymous reviewer are greatly appreciated.

## REFERENCES


1. Moon, P.B., "Resonant nuclear scattering of gamma-rays: Theory and preliminary experiments," *Proc. Phy. Soc., (London) A* **64** 76-82 (1951).
2. Mössbauer R.L., "Nuclear resonance fluorescence in $Ir^{191}$ for gamma rays," *Z. Physik,* **151** [2] 124-43 (1958).
3. Wong, J., Angell, C.A., **Glass:Structure by Spectroscopy** 97-128, Marcel Dekker, New York, 1977.
4. Frauenfelder, H., **The Mössbauer Effect,** Benjamin, New York, 1962.
5. Mössbauer, R.L., "Recoilless nuclear resonance absorption," *Ann. Rev. Nucl. Sci.,* 123-52 (1962).
6. Gruverman I.J., (Editor) **Mössbauer Effect Methodology** Vol. I-VI, Plenum Press, New York 1965-70.
7. Kurkjian, C.R., "Mössbauer spectroscopy in inorganic glasses," *J. Non Cryst. Solids,* **3** 157-194 (1970).
8. Kistner, O.C., Sunyar, A.W., "Evidence for quadrupole interaction of $Fe^{57m}$ and influence of chemical binding on nuclear gamma-ray energy," *Phys. Rev. Letters,* **4**[8] 412-15 (1960).
9. Walker, L.R., Wertheim, G.K., Jaccarino, V., "Interpretation of $Fe^{57}$ isomer shift," *Phys. Rev. Letters,* **6**[3] 98-101 (1961).
10. Preston, R.S., Hanna, S.S., Heberle, J., "Mössbauer effect in metallic iron," *Phys. Rev.,* **128**[5] 2207-18 (1962).
11. Bandyopadhyay, *et al.*, "Magnetic properties of a basalt glass and glass-ceramics," *J. Non-Cryst. Solids,* **40** 353-368 (1980).
12. Ludwig, G.W., Woodbury, H.H., "Magnetic moment of $Fe^{57}$," *Phys. Rev.,* **117**[5] 1286-87 (1960).
13. Kündig, W. *et al.*, "Some properties of supported small $\alpha$-$Fe_2O_3$ particles determined with the Mössbauer effect," *Phys. Rev.,* **142**[2] 327-33 (1966).
14. Kurkjian, C.R., Buchanan, D.N.E., "Mössbauer absorption of $Fe^{57}$ in inorganic glasses," *Phys. Chem. Glasses,* **5**[3] 63-70 (1964).
15. Wertheim, G.K., Remeika, J.P., "Mössbauer effect hyperfine structure of trivalent $Fe^{57}$ in corundum," *Phys. Letters,* **10**[1] 14-15 (1964).
16. "Mössbauer Sources," **Amersham Catalogue**, Amersham Corporation, Illinois, USA.
17. Spijkerman, J., Ranger Scientific, Inc., Texas, USA, Private communication.
18. Virgo, D. Mysen, B.O., "The structural state of iron in oxidized vs. reduced glasses at 1 atm: A $^{57}Fe$ Mössbauer study," *Phys. Chem. Minerals,* **12** 65-76 (1985).
19. Eibschutz, M., Lines, M.E., "Observation of second-order quadrupole shift in Mössbauer spectrum of amorphous YIG (yttrium iron garnet)," *Phys. Rev. B,* **25** 4256-4259 (1982).
20. Goldanskii, V.I. *et al.*, "Investigation of the Mössbauer effect in tin compounds," *Dokl. Akad. Nauk SSSR,* **147**[1] 127-30 (1962).
21. Levitz, P., *et al.*, "A two-parameter distribution analysis of Mössbauer spectra in noncrystalline solids using general inversion method," *J. Phys. E,* **13**[4] 427-32 (1980).

22. Massiot, D., "Shape comparison of physical spectra: application to Mössbauer spectra of silicate glasses," *J. Non-Cryst. Solids*, **69** 371-80 (1985).
23. Coey, J.M.D., "Amorphous solids: a review of the applications of the Mössbauer effect," *J. de Physique*, **35** C6-89 - C6-105 (1974).
24. Müller-Warmuth, W., Eckert, H., "Nuclear magnetic resonance and Mössbauer spectroscopy of glasses," *Physics Reports (Review Section of Phys. Letters)*, **88**[2] 91-149 (1982).
25. Dyar, M.D., "A review of Mössbauer data on inorganic glasses: the effect of composition on iron valency and coordination," *Am. Mineral.*, **70** 304-316 (1985).
26. Sekhon, S. S., Kamal, R., "The applications of Mössbauer spectroscopy to the study of inorganic oxide glasses," *Phys. Chem. Glasses*, **29**[4] 157-67 (1988).
27. Sekhon, S. S., Kamal, R., "Mössbauer study of $Fe_2O_3$ crystallite formation in $B_2O_3$ glass," *J. Appl. Phys.*, **49**[6] 3444-45 (1978).
28. Krogh-Moe, J., "The structure of vitreous and liquid boron oxide," *J. Non-Cryst. Solids*, **1** 269-84 (1969).
29. Nishida, T., Takashima, Y., "Mössbauer spectroscopic study of the formation of non-bridging oxygen in the potassium borate glasses," *J. Non-Cryst. Solids*, **37** 37-43 (1980).
30. Tanaka, K., Hirao, K., Soga, N., "Mössbauer study of sodium iron borate glasses," *J. Non-Cryst. Solids*, **85** 228-232 (1986).
31. Raman, T. *et al.*, "Mössbauer studies in the glass system $Na_2O{\cdot}B_2O_3{\cdot}Fe_2O_3$," *J. Appl. Phys.*, **46**[2] 972-76 (1975).
32. Kamal, R. *et al.*, "Mössbauer spectra in the $xNa_2O{\cdot}(1-x-y)B_2O_3.yFe_2O_3$ glass system," *J. Non-Cryst. Solids*, **53** 227-33 (1982).
33. Pye, L.D., Belkhiria, N., Montenero, A., "Mössbauer studies of iron soda borate glasses," *J. Non-Cryst. Solids*, **95&96** 359-364 (1987).
34. Belkhiria, N., **Redox, Properties and Structure of Iron-containing Soda Borate Glasses**, Ph.D. Thesis, Alfred University, Alfred, NY (1987).
35. Nishida, T. *et al.*, "Mössbauer and ESR studies on the behavior of halide ions in potassium borate glass," *Phys. Chem. Glasses*, **24**[4] 88-92 (1983).
36. Burzo, E., Ardelean, I., "Mössbauer effect study of iron cations in lead borate glasses," *Phys. Chem. Glasses*, **20**[1] 15-20 (1979).
37. Sekhon, S.S., Kamal, R., "Mössbauer study in the glass system $PbO{\cdot}2B_2O_3{\cdot}Fe_2O_3$," *J. Non-Cryst. Solids*, **28** 189-192 (1978).
38. Burzo, E., Ardelean, I., Ursu, I., "On the physical properties of $20Fe_2O_3 .80[3B_2O_3{\cdot}(1-x)PbO{\cdot}xGeO_2]$ glasses," *J. Mater. Sci.*, **15**[3] 581-93 (1980).
39. Nishida, T., Ogata, M., Takashima, Y., "Application of Mössbauer spectroscopy to a structural study of superionic conducting silver borate glasses," *Phys. Chem. Glasses*, **29**[1] 22-25 (1988).
40. Bansal, T.K., Mendiratta, R.G., "Study of barium borate glasses containing $V_2O_5$ and $Fe_2O_3$ using Mössbauer spectroscopy," *J. Non-Cryst. solids*, **86** 13-21 (1986).
41. Music, S., "$^{57}$Fe Mössbauer effect in borosilicate glasses," *J. Mater. Sci. Letters*, **8**[5] 537-9 (1989).

42. De Grave, E., Stalios, A.D., Van Albloom, A., "Influence of the heat treatment on the Mössbauer spectrum of a simulated nuclear-waste glass," *J. Nuclear Materials*, **171** 189-197 (1990).
43. Nishida, T., Hirai, T., Takashima, Y., "Mössbauer spectroscopic study of nonbridging oxygen atoms in potassium borosilicate glasses," *Phys. Chem. Glasses,* **22** [4] 94-98 (1981).
44. Iwamoto, N., Tsunawaki, Y., Nakagawa, H., "Investigation of calcium-iron-silicate glasses by the Mössbauer method," *J. Non-Cryst. Solids*, **29** 347-356 (1978).
45. Levy, R.A., Lupis, C.H.P., Flinn, P.A., "Mössbauer analysis of the valence and coordination of iron cations in $SiO_2$-$Na_2O$-CaO glasses," *Phys. Chem. Glasses*, **17**[4] 94-103 (1976).
46. Brown, I.W.M., "A Mössbauer study of hematite dissolution and devitrification in float glass," *J. Non-Cryst. Solids*, **50** 233-252 (1982).
47. Mysen, B.O., Carmichael, I.S.E., Virgo, D., "A comparison of iron redox ratios in silicate glasses determined by wet-chemical and $^{57}$Fe Mössbauer resonant absorption methods," *Contrib. Mineral. Petrol.*, **90** 101-106 (1985).
48. Karraker, D.G., "Ferrous/ferric ratio in DWPF glass: Chemical and $^{57}$Fe Mössbauer determinations," *Adv. Ceram. Mater.*, **3**[4] 337-40 (1988).
49. Mysen, B.O., "Redox equilibria and coordination of $Fe^{2+}$ and $Fe^{3+}$ in silicate glasses from $^{57}$Fe Mössbauer spectroscopy," *J. Non-Cryst. Solids*, **95&96** 247-254 (1987).
50. Lin, Y. *et al.*, "Structural study of iron in phosphate glass," J. *Non-Cryst. solids*, **112** 136-141 (1989).
51. Nishida, T., Shiotsuki, T., Takashima, Y., "Mössbauer spectroscopic study of gamma-ray irradiated potassium phosphate glasses," *J. Non-Cryst. Solids*, **43** 123-128 (1981).
52. Taragin, M.F., Eisenstein, J.C., Haller, W., "Mössbauer study of $Fe^{57}$ in aluminophosphate glass," *Phys. Chem. Glasses*, **13**[5] 149-152 (1972).
53. Sanad, A.M. *et al.*, "Mössbauer, structural, and magnetic studies of glasses of the CaO-$P_2O_5$-$Fe_2O_3$ system," *Phys. Chem. Glasses*, **30**[1] 27-29 (1989).
54. Menil, F. *et al.*, "Sodium-iron fluorophosphate glasses: Part 2: EPR and Mössbauer resonance study," *J. Non-Cryst. Solids*, **34** 209-221 (1979).
55. Bogomolova, L.D. *et al.*, "Some spectral investigations of $ZrF_4$ based glasses containing iron impurity," *J. Non-Cryst. Solids*, **91** 203-208 (1987).
56. Kawamoto, Y., Nakashio, K., "Characterization of trace amounts of iron in $ZrF_4$ based optical fiber glasses," *Mat. Res. Bull.*, **21** 779-785 (1986).
57. Labar, Ch., Gielen, P., "A spectroscopic determination of ferrous iron content in glasses," *J. Non-Cryst. Solids*, **13** 107-119 (1973/74).
58. Goldman, D.S., Bewley, D.E., "Ferrous/ferric Mössbauer analysis of simulated nuclear waste glass with and without computer fitting," *J. Am. Ceram. Soc.*, **68**[12] 691-95 (1985).
59. Hunter, R.T. *et al.*, "Determination of the $Fe^{2+}/Fe^{3+}$ ratio in nuclear waste glasses," *J. Am. Ceram. Soc.*, **72**[6] 943-47 (1989).
60. Paul, A. *et al.*, "Infrared and $^{119}$Sn Mössbauer spectra of tin borate glasses," *Phys. Chem. Glasses*, **18**[6] 125-27 (1977).

61. Eissa, N.A., Shaisha, E.E., Hussiun, A.L., "Study of alkali borate glass containing tin by the Mössbauer effect," *J. Non-Cryst. solids,* **16** 206-218 (1974).
62. Evstropev, K.S. *et al.*, "Mössbauer study of tin-containing glasses," *Izv. Akad. Nauk SSSR, Neorg. Mater.*, **6**[4] 781-84 (1970).
63. Cheng, J.J. *et al.*, "Mössbauer investigation of tin-containing chalcogenide glasses and glass-ceramics," *J. Non-Cryst. Solids*, **45** 47-56 (1981).
64. Tanabe, S., Hirao, K., Soga, N., "Mössbauer spectroscopy of $^{151}Eu$ in oxide crystals and glass," *J. Non-Cryst. Solids*, **113** 178-184 (1989).
65. Coey, J.M.D., McEvoy, A., Shafer, M.W., "Mössbauer study of europium in fluorozirconate glasses," *J. Non-Cryst. Solids*, **43** 387-392 (1981).
66. Winterer, M. *et al.*, "Paramagnetic hyperfine structure in $^{151}Eu$ Mössbauer spectra of $Eu^{2+}$ ions in borate glasses," *J. Phys. C: Solid State Phys.*, **20** [32] 5389-99 (1987).
67. Cremers, H. *et al.*, "$^{121}Sb$ Mössbauer studies of glasses in the system Ge-Sb-Se," *Phys. Chem. Glasses*, **30**[3] 79-82 (1989).
68. Stahlberg, B. *et al.*, "Combined electrochemical and Mössbauer studies of the $Sb^{3+}/Sb^{5+}$ equilibrium in a silicate glass-forming melt," *Glastech. Ber.*, **61** [12] 335-40 (1988).
69. Gonser, U., Limbach, C.T., Aubertin, F., "Mössbauer spectroscopy in amorphous metals: Failures and successes," *J. Non-Cryst. solids*, **106** 395-98 (1988).
70. Burger, K., " Capillary Mössbauer spectroscopy (CMS)–a new method for the study of liquids fixed in a suitable carrier," *Spectrochim. Acta, Part A* **43A** [9] 1105-10 (1987).
71. Whipple, E.R., **Quantitative Mössbauer Spectra and Chemistry of Iron**, Ph.D. Thesis, MIT, Cambridge, MA (1973).

# CHAPTER 5

## X-RAY PHOTOELECTRON SPECTROSCOPY OF GLASS

**Carlo G. Pantano**
Department of Materials Science and Engineering
The Pennsylvania State University
University Park, PA 16802 USA

## ABSTRACT

The fundamental basis of x-ray photoelectron spectroscopy (XPS) is presented with an emphasis on those aspects of the method that are important in the analysis of glasses. The capabilities of the technique for characterizing specific features of the chemical structure and surface chemistry of glass are reviewed. The issues associated with the compositional analysis of glass surfaces are discussed in some detail. Finally, examples of the use of XPS in glass science are presented; these include: the study of (i) non-bridging oxygen in glass, (ii) hydrogen surface treatments of lead-silicate glass, (iii) nitridation of silica sol/gel thin-films, and (iv) organic adsorbates on glass fibers.

## INTRODUCTION

X-ray photoelectron spectroscopy (XPS) is an analytical technique that is widely used for chemical characterization of solids. Figure 1 illustrates a typical arrangement for XPS. The specimen is irradiated with monochromatic x-rays to cause the emission of photoelectrons and Auger electrons. The energy distribution of the ejected electrons is measured to obtain the photoelectron spectrum. An example is presented in Figure 2. Each of the peaks represent the photoelectrons or Auger electrons ejected from elements in the specimen surface. The energies of the various peaks are used to identify the elements present in the sample surface, while their relative intensity provides information about the composition. Often, the *energy* values of the electrons are slightly different in a compound or multicomponent material relative to their elemental state. These small *chemical shifts* are indicative of local chemical environment, bonding, or oxidation state of the specific elements detected.

In general, XPS provides information about the elemental composition and their chemical state. It is best known for its ability to provide surface analyses because the ejected electrons originate within the outermost few atomic layers of the specimen surface. Nevertheless, it is also possible to obtain information about the bulk.

ESCA, *electron spectroscopy for chemical analysis*, is a more general term which is often used synonymously with XPS. It emphasizes the chemical aspects of the technique, and does not exclude analysis of the Auger electrons that are always present in the photoelectron spectrum. In fact, the combined analysis of the Auger electrons, and the photoelectrons, can be most advantageous in the evaluation of insulators.

It is, of course, possible to generate photoelectron spectra with any photon excitation of the specimen. The x-ray irradiation used in XPS enhances sensitivity to the core-level electronic states, and thereby, to the atomic composition of the material. The related method of ultraviolet photoelectron spectroscopy (UPS) is based upon ultraviolet irradiation of the specimen. This enhances sensitivity to the valence-band electronic structure of the materials. Most recently, the use of synchrotron light sources has further expanded the capabilities of photoelectron spectroscopy for probing electronic structure and bonding configuration. To date, these methods have been applied to only a limited extent in glass research [1].

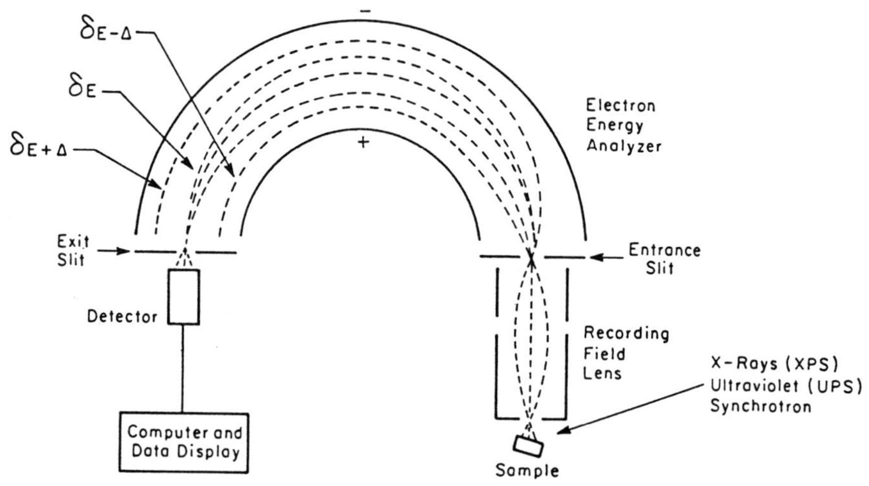


Figure 1. Schematic representation of the typical experimental arrangement for XPS; all components are contained in vacuum.

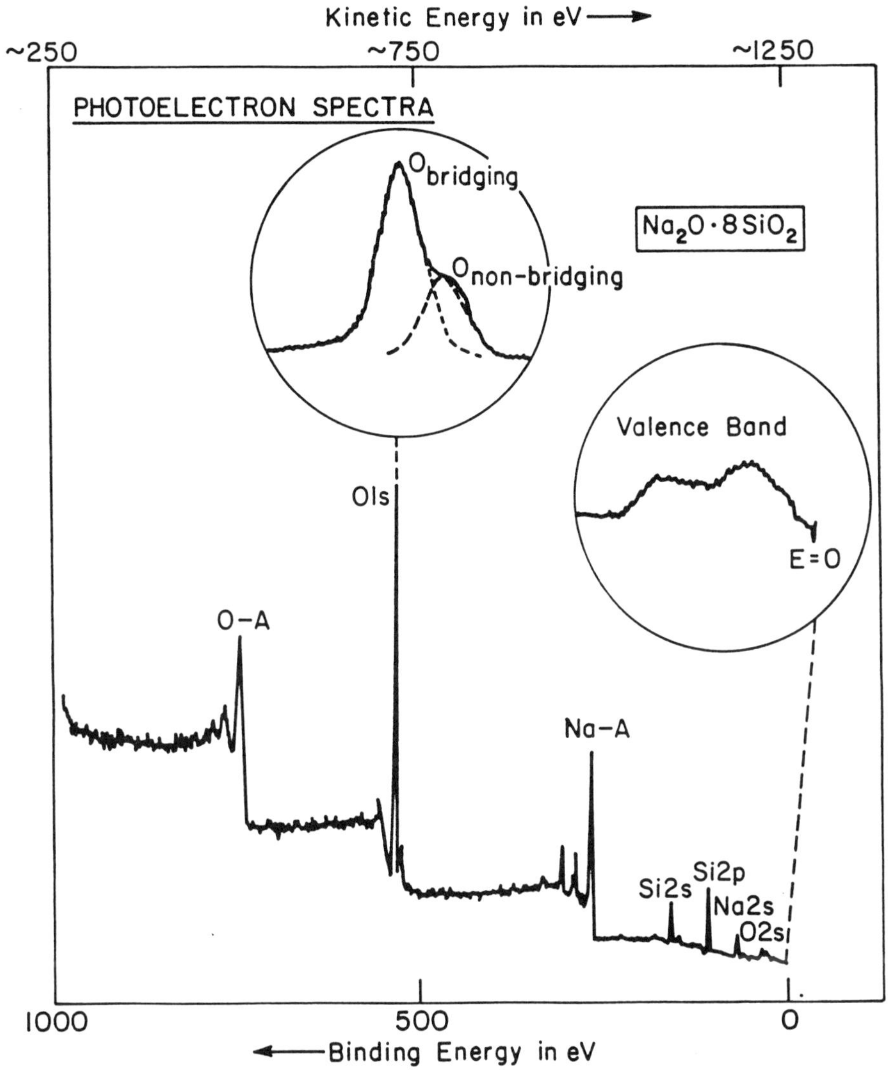


Figure 2. A typical XPS spectrum for the vacuum fracture surface of a glass; the insets are high resolution spectra of the Ols and valance band regions.

XPS can be applied in many different ways to study or analyze glass [2]. The most important and routine applications of XPS in glass research include analysis of the surface composition, surface chemical state, or bulk chemical structure. More specifically,

- elemental surface composition,
- organics on surfaces,
- non-bridging oxygen concentration in the bulk,
- oxidation state of cations on the surface or in the bulk,
- valence band density of states.

The purpose of this chapter is to describe only the basic aspects of XPS. The important issues in spectral interpretation will be explained using specific examples in glass research. The interested reader can find more detailed discussions of photoemission theory, instrumentation and technique in the BIBLIOGRAPHY.

There are a few general characteristics of XPS that are probably best presented here at the beginning. First, XPS *cannot detect hydrogen*; in some cases, indirect information about hydrogen can be inferred through chemical shifts, but not routinely. This is an intrinsic limitation of photoelectron spectroscopy that greatly influences the study of glass where hydrogen, hydroxyls and water are often present. Second, the method is intrinsically sensitive to the outermost .5-5 nm of the glass surface; hence, *XPS is a surface analytical technique*. But if a glass specimen is fractured in ultra-high vacuum, the composition of the fracture surface can be equal to the bulk composition, and in many instances, the non-bridging oxygen content and oxidation state of this surface may be equivalent to the bulk. It is in this sense that XPS is used to analyze the bulk characteristics of glasses. Third, any spectroscopic technique that relies upon the detection of charged particles is susceptible to *charging problems* when applied to insulators. Although XPS is less prone to charging problems than other electron or ion spectroscopies, problems do exist. It is important to understand the origin and nature of the charging in XPS because it can influence the results.

## PHOTOELECTRON EMISSION

The orbital model of the atom in Figure 3 illustrates the fundamental basis of XPS. The absorption of an incident photon (x-ray or ultraviolet) causes the ejection of a photoelectron. The kinetic energy of the photoelectron, KE, is simply

$$KE = h\upsilon - BE \quad (1)$$

where $h\upsilon$ is the energy of the incident photon and BE is the binding energy of the emitted electron in the atom. The binding energy of an electron is characteristic of the specific atom, and in this way, photoelectron spectroscopy provides elemental identification. And since any occupied electron state in the atom can be ejected, the entire spectrum of photoelectrons is a representation of the electronic density of states.

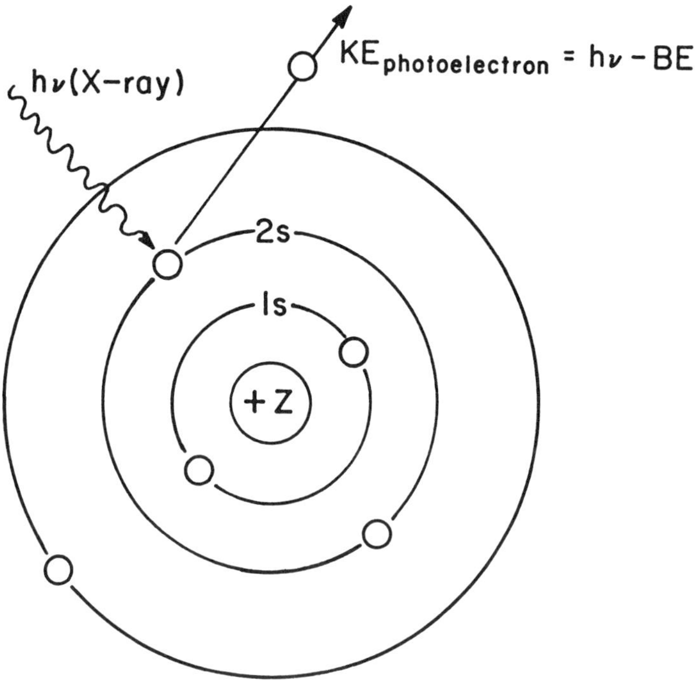


Figure 3. Orbital model of the x-ray photoemission process.

The photoelectrons distinguish the occupied electronic states only on the basis of energy. There are states corresponding to each of the principal quantum numbers 1s, 2s, 3s, .... . In the case of the p, d and f levels, Russell-Saunders coupling creates two distinguishable energy states in each principal shell $2p^{1/2}$, $2p^{3/2}$, $3p^{1/2}$, $3p^{3/2}$, ... , $3d^{3/2}$, $3d^{5/2}$, $4d^{3/2}$, $4d^{5/2}$, ... , and $4f^{5.2}$, $4f^{7/2}$, $5f^{5/2}$, $5f^{7./2}$, .... . There are selection rules and cross-sections that define the relative probability of photoemission in each state. And of course, photoemission cannot occur unless hυ is greater than the binding energy of the electron (typically, AlKα @ 1487 eV or MgKα @ 1254 eV are used in XPS). In general, therefore, the x-ray photoelectron spectrum is dominated by the electrons that occupy the outermost levels of the atom, e.g., the 1s electron in the case of oxygen, the 2p electrons in the case of Si, the 4f electrons in the case of Pb, etc.

The photoemission process is, in reality, more complex than equation (1) would suggest. The binding energy calculated according to equation 1 assumes the frozen orbital model (Koopman's theorem). In reality, there is some relaxation of the electrons during the photoemission process; i.e.,

$$KE = h\upsilon - BE - E_{Relaxation} \tag{2}$$

This is not an issue in routine compositional analyses because it can be accounted for through referencing and calibration. On the other hand, it can be of concern in the characterization of chemical and electronic structure.

In the case of solids, and especially insulators, the referencing of photoelectron binding energy is more complicated than the free atom models represented in eqns. 1 and 2. Figure 4 shows the energy level diagram for an atom in the solid state. The photoemission process, itself, is not affected by the solid state. But it is more sensible and practical to reference the binding energy to the Fermi level (rather than to the vacuum level). Using this reference scheme, the kinetic energy of the ejected photoelectron is reduced by the work function of the specimen, $\phi_s$. Thus,

$$KE = h\upsilon - BE_f - \phi_s \tag{3}$$

The kinetic energy of the photoelectron is perturbed again upon crossing the surface barrier, $\phi_a$, of the analyzer/detector. If the specimen and the analyzer/detector are in electrical contact, this perturbation is the contact potential between the specimen and the analyzer/detector ($\phi_s$-$\phi_a$); thus,

$$KE' = h\upsilon - BE_f - \phi_a \tag{4}$$

In the routine application of XPS, the complications suggested in equations (2), (3) and (4) are eliminated through calibration. That is, standards have been used to establish tables of self-consistent binding energies and/or kinetic energies [see B6]; essentially,

$$E_{kinetic} = h\upsilon - E_{binding}. \tag{5}$$

The assumptions are simply that the Fermi levels of the specimen surface and the analyzer/detector are equilibrated, the relaxation energy ($E_R$) is a constant for an element (independent of of the material composition or chemical structure), and that the analyzer/detector work function, $\phi_a$, is constant. In practice, the situation is further simplified because there is always one peak in the spectrum (usually carbon) that can be used to calibrate the binding energy scale. In fact, calibration of the measured spectrum using the C1s is an almost universal practice. Here, it is assumed that so-called adventitious carbon (essentially carbonaceous surface contamination found on virtually any specimen that has been exposed to the ambient atmosphere) is either hydrocarbon or amorphous carbon, and therefore, is assigned a binding energy of 285 eV [3-4]. Of course, these assumptions must be

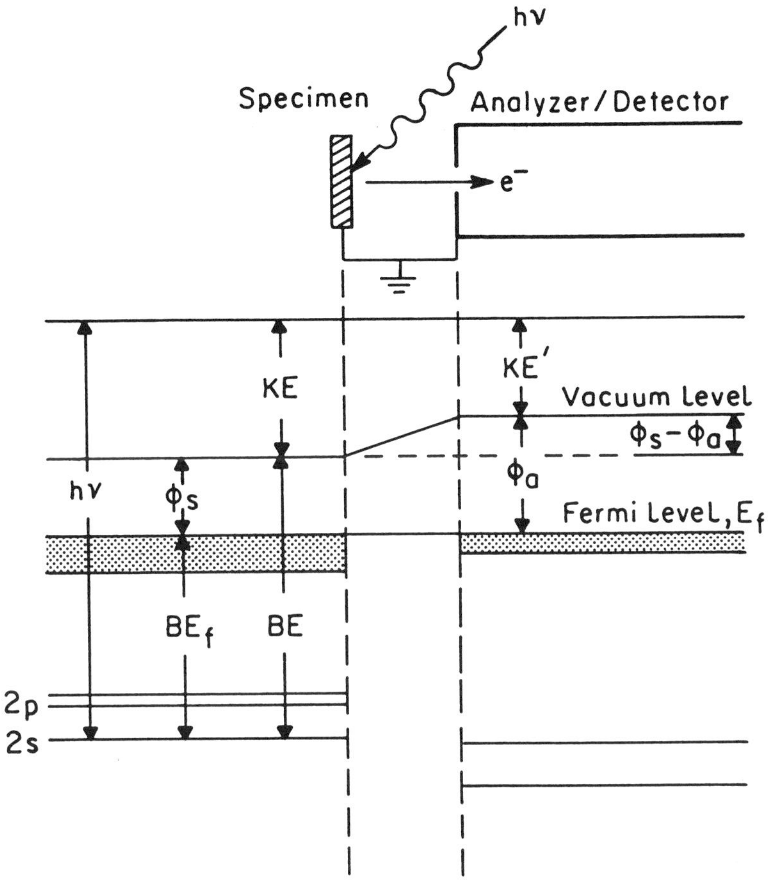


Figure 4. Energy-level references for XPS when the Fermi levels of the specimen and the analyzer/detector are equilibrated.

reconsidered in those cases where quantitative interpretation of a binding energy, or a small binding energy shift, is of interest. This is especially true in the analysis of insulators.

In the case of insulators, the specimen can *charge* due to the ejection of photoelectrons, Auger electrons, and low energy secondary electrons. This positive charging of the surface must be stabilized to obtain a useful XPS spectrum. This is usually achieved through the use of a low energy electron flood gun, although other methods have been proposed (see Appendix 2 in B5). But even if the surface potential is stable, the Fermi levels of the analyzer/detector and specimen may not be in equilibrium. Thus,

$$E_k = h\upsilon - E_b \pm E_{charge}. \quad (6)$$

This surface potential influences all of the electrons uniformly, and so in routine analyses, the measured spectrum can be recalibrated using the carbon 1s line.

In those studies where chemical state identification is of specific interest, referencing to carbon may not be an acceptable approach. Here, it is necessary to define a more reliable reference. One method is to choose a line associated with a specie in the glass structure; essentially, an internal reference. The assumption is that the binding energy of this reference specie is independent of the composition or chemical states of the other constituents of the glass (e.g., Si 2p in many silicate glasses). Unfortunately, this is not always possible because of the collective effect of bonding and network structure upon the electron binding energies of all the atoms. Alternatively, one can implant an internal reference; for example, $N_2$ molecules can be implanted at very low energy (<500 eV). In this way, the N1s binding energy in $N_2$ can be used for calibration. The shortcoming of this approach is the possible introduction of ion beam damage, and the need for an in-situ source of low energy ions in the XPS system. It does not provide an absolute measure of the binding energy, but it does provide a reliable determination of binding energy shifts. Still another approach is to deposit a small metallic dot (usually Au) on the glass surface, and use it to pin the Fermi level of the spectrometer to the glass surface. The Au dot is grounded to the spectrometer, and the Au line that appears in the spectrum of the glass is used for calibration. This method is widely employed in research studies where precise determination of binding energy values are required; it is not routine, and is subject to careful interpretation [5-9]. In general, one must recognize real limitations in precise determinations of binding energy or chemical shifts in the case of insulating glasses.

## AUGER EMISSION

After an electron vacancy is created due to the photoemission process (see Figure 3), a de-excitation process occurs within the atom to fill the vacancy. The de-excitation can occur *either* through an x-ray emission *or* through an Auger electron emission. The relative probability of these two processes depends upon the incident x-ray energy and the atomic number of the atom. During the typical XPS analysis (where soft x-rays are used), Auger emissions are prevalent – especially in low z elements – and these electrons are necessarily included in the photoelectron spectrum.

Auger electron emission – shown schematically in Figure 5 – corresponds to the decay of an outer electron (e.g., an L-shell electron) to fill the electron vacancy (e.g., a K-shell vacancy) and then transfer of the energy difference ($E_K$-$E_L$) to another electron (e.g., another L-shell electron); hence, the nomenclature KLL Auger electron. If, alternatively, an L-shell electron was vacated due to the photoemission, an LMM Auger electron could be created. The kinetic energy of the Auger electron is characteristic of the atom; i.e.,

$$E_{KLL} = E_K - E_{L1} - E_{L2} - \phi_a \qquad (7)$$

There are tables of Auger electron energies for all the elements in the periodic table, and these can be used in routine chemical analyses [see B6].

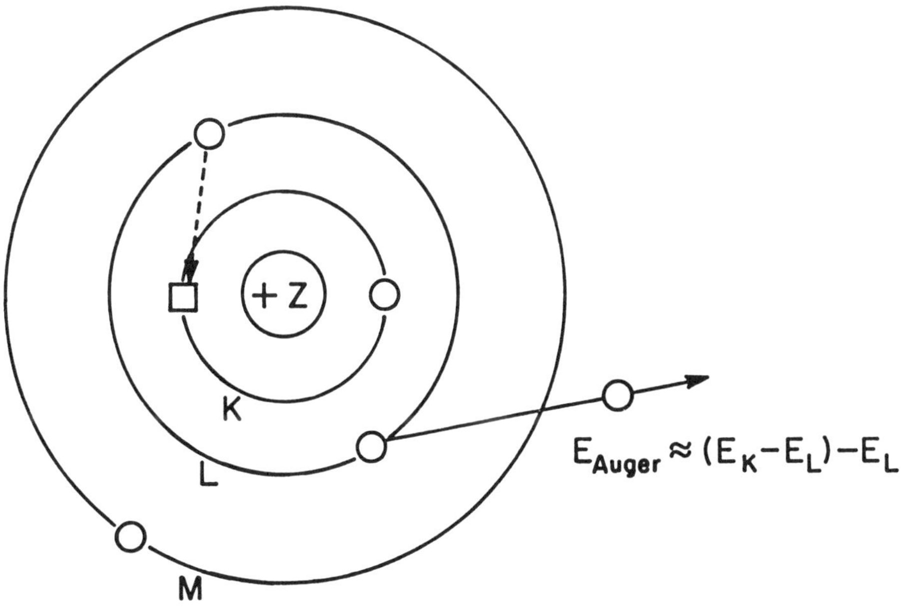


Figure 5. Orbital model of the electron transitions associated with de-excitation of a core-level vacancy (KLL) by the Auger emission process.

It is important to note that is is the *measured kinetic energy* of Auger electrons that is characteristic of the atom where it originates, while in the case of the photoelectrons, it is the *calculated binding energy* that is characteristic; i.e., the measured kinetic energy of any Auger electron is independent of the x-ray used in the analysis ($h\upsilon$), while in the case of photoelectrons, the measured kinetic energy depends upon $h\upsilon$. Thus, one way to distinguish between the Auger electron and photoelectron lines in the spectrum is to change the x-ray source. The photoelectron lines will change in kinetic energy while the Auger electron lines will always appear at their characteristic kinetic energy. In practice, though, experience and the width of the line are more often used in the proper assignment of Auger lines; Auger lines are often broader in energy than photoelectron lines.

One of the most interesting uses of the Auger lines that appear in the XPS spectrum is in the definition of *Auger parameter*. The Auger parameter, $\alpha$, is the difference in kinetic energy between any Auger line and any photoelectron line for a specific element; i.e.,

$$\alpha = KE_{Auger} - KE_{Photoelectron} \tag{8}$$

This definition of Auger parameter is shown schematically in Figure 6. Ideally, the lines chosen would be directly related through a specific transition (e.g., Si 1s and Si KLL), but this is not required. In practice, it is calculated using the most intense lines in the spectrum . The Auger parameters are found to be unique to a specific compound, and in fact, have been tabulated for a wide range of substances including silicates and aluminosilicates. They are especially useful in the case of insulators because the value of $\alpha$ is insensitive to charging and referencing. Table 1 shows, for example, a compilation of data for silicon compounds.

## CHEMICAL AND STRUCTURAL EFFECTS

The binding energy of an electron depends not only on the atomic number, but also upon the electronic structure of the atom and its surroundings. Thus, it is sensitive to the chemical bonding or structure, the oxidation state and the local chemical environment of the atom. In practice, this effect is termed a *chemical shift*; that is, a shift in the binding energy relative to the elemental state. Table 1 shows the Si2p binding energy in various Si compounds. The chemical shift is of the order of a fraction to a few eV. Thus, it is necessary to measure the photoelectron spectra at high resolution in order to extract chemical information. In general, the ability to distinguish chemical states is not apparent in a survey scan of the type shown in Figure 1. The inset in Figure 1 shows the high resolution scan for O1s. Note the presence of two peaks at this binding energy. This implies the presence of two chemical forms or states of oxygen. In this case, the bridging and non-bridging oxygens in this glass are represented.

Table 1. Peak Positions and Auger Parameters ($\alpha_{Si}$) for a Variety of Si Compounds [from 10].

| Sample | Formula | Si 2p BE (eV) | Si KLL KE (eV) | $\alpha_{Si}$ (eV) |
|---|---|---|---|---|
| *silicon* | Si | 99.6 | 1616.4 | 462.4 |
| *silicon carbide* | SiC | 103.9 | 1610.2 | 460.5 |
| *silicon nitride* | $Si_3N_4$ | 102.0 | 1611.7 | 460.1 |
| *silicon dioxide* | $SiO_2$ | 102.6 | 1609.6 | 458.6 |
| *zinc silicate* | $ZnSiO_3$ | 108.1 | 1603.7 | 458.2 |
| *silica gel* | $SiO_2 \cdot xH_2O$ | 107.0 | 1604.3 | 457.7 |

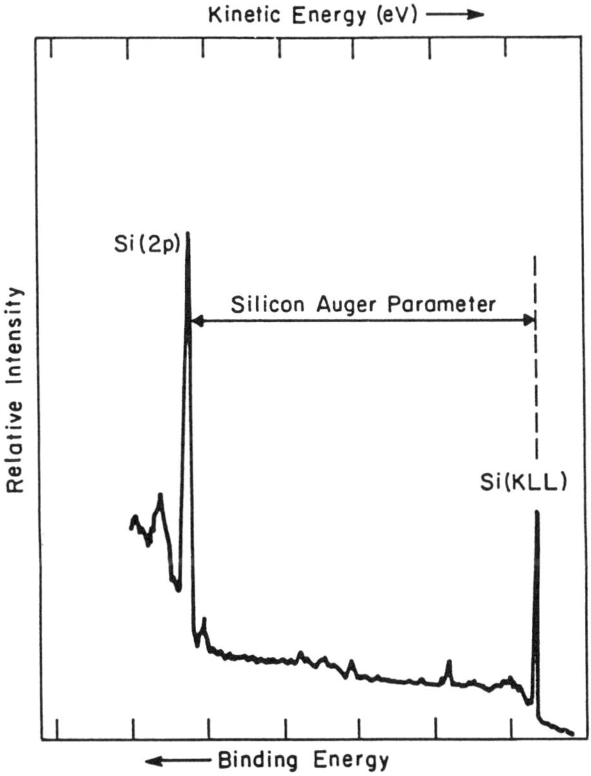


Figure 6. Spectral description of the Auger parameter ($\alpha$) for Si; in this case, the Si(KLL) transition was excited with bremsstrahlung radiation.

Many theories and approaches have been developed to understand and interpret binding energy shifts. The simplest and most basic idea can be realized using the charged shell model. It assumes that increases or decreases in the electron charge density in the outer bonding orbitals causes complementary changes in the binding energy of the inner core levels. Thus, the Si 2p binding energy is higher in SiC, $Si_3N_4$ or $SiO_2$, than in Si due to the electronegativity of the bonding environment; i.e., the electron charge density near the Si atom is decreased, and thereby the Si 2p core level electron is more tightly bound. Similarly, the binding energy of the O 1s in a non-bridging site (NBO) is less than in a bridging site because the higher charge density at the NBO shields the inner-level 1s electron and thereby lowers its effective binding energy. This simple, qualitative concept is useful for understanding the origin and sign of the shift. But the magnitude of the shift depends upon many factors including relative electronegativity of the bonded species, coordination number, relaxation effects, etc. In the case of the silicon compounds, SiC, $Si_3N_4$, and $SiO_2$, wherein the coordination number is always four, the Si 2p shift scales directly with the electronegativity of the anion.

More detailed discussions of the origin and qualitative interpretation of chemical shifts can be found elsewhere (see BIBLIOGRAPHY). In practice, the interpretation of photoelectron binding energies is based upon tabulated values of chemical shifts (and Auger parameters) that have been measured in specific compounds [see B6]; Table 1 is an example. Alternatively, specific standards – or a range of compounds – can be prepared and/or analyzed for reference.

One must recognize that although *in theory* the binding energy of an electron must reflect any change in its local chemical environment, the associated chemical shift may or may not be observed. In some cases, the shift is too small to be detected because of limited resolution in the measurement. Sometimes, an increase in the instrument resolution can be of benefit, but in many instances, the shift in core level states is not significant. In other cases, effects of relaxation can offset the binding energy shift; e.g., changes in the binding energy due to bond electronegativity may cause corresponding changes in $E_R$ (see equation 2). In the specific case of glass, some important chemical shifts that cannot be detected in routine XPS analyses include $Ag^+$ vs Ag, $Sn^{+2}$ vs $Sn^{+4}$ and $O^{-2}$ vs OH. Another unfortunate limitation in glass research concerns the measurement of transition metal and rare-earth ion oxidation equilibria. Although the associated shifts in many variable-valence oxides are significant [2.23], their low concentration in most glasses limits detection.

Nevertheless, XPS has proven to be exceedingly beneficial for studying the chemical structure of glass; specifically, non-bridging oxygens in silicate, aluminosilicate and phosphate glasses, bonding in fluoride and chalcogenide glasses, nitridation of oxide glasses, and redox equilibria in glass. And in many of these studies, it has been the bulk chemical structure that has been studied. In these cases, it has been assumed that a clean, vacuum fractured glass creates a surface that exposes the bulk structure of the glass. The uncertainty in this approach concerns the fact that clean surfaces often reconstruct in vacuum [11], and so, the surface is not an ideal projection of the bulk structure. On the other hand, it has been observed that there is reasonable agreement between the measured NBO concentration and the bulk composition of simple glasses. This agreement could be due to the limited sensitivity of XPS to the outermost monolayer where the reconstruction is confined. In spite of these uncertainties, XPS *is* sensitive to the network chemical structure of a glass – especially BO and NBO – and is necessarily an important tool for the study of these characteristics.

## SURFACE SENSITIVITY

The sampling depth in XPS is determined by the *inelastic mean free path* (imfp) of the photoelectrons. This number, often called the *escape depth*, represents the most probable distance the electron can travel in the solid without an energy loss. Any photoelectron or Auger electron that suffers an energy loss before it escapes the solid surface no longer possesses its characteristic energy, and thereby, contributes only to the background signal at an energy, E, which is less than its original energy, $E_{ki}$. This is reason for the steps in the background at $E_{ki}$ in Figure 2. More precisely, the imfp ($\lambda$) characterizes the degree of attenuation of electrons due to inelastic scattering in a homogeneous layer or surface film of thickness x; i.e.,

$$\frac{I_2}{I_1} = e^{-x/\lambda} \qquad (9)$$

Thus, photoelectrons that originate more than a distance $\lambda$ from the surface have less than a 63% chance of escape without experiencing an inelastic scattering event. Another way to think of $\lambda$ is to recognize that it represents the depth over which 63% of each photoelectron signal originates. A reasonable representation of the *sampling depth of analysis* would be $3\lambda$, where 95% of the detected signal originates.

The imfp of an electron is a function of its kinetic energy, and of the matrix in which it travels. The spectrum in Figure 2 shows the typical range of electron kinetic energy; i.e., 100-1000 eV. In this range, the imfp is of the order .5-5 nm in oxides. The range of imfp is greater in organics (up to 10 nm), and less in metals (.5-3 nm).

It is important to recognize that each peak in the spectrum is due to electrons of different kinetic energy. Thus, the imfp of O1s in Figure 2 is ~2.0 nm, while for Si2p it is ~4.0 nm. This represents a difference in the sampling volume for O and Si It is also important to note that equation 9 is valid only for a homogeneous surface. If the surface is covered with adsorbates or thin layers ($x<\lambda$), the expression must be modified to account for the non-uniformity; that is, the dependence of $\lambda$ upon the substrate material and the surface film/adsorbate material. These issues are of critical nature in any quantitative analysis, and will be discussed further in the next section.

## COMPOSITIONAL ANALYSIS

XPS is probably the most suitable method to measure the surface composition of glass. XPS is less susceptible to charging problems than most of the other surface analysis techniques, it does not introduce any beam damage, and quantitative analysis of the photoelectron spectra is reasonably straightforward.

Equations 10 and 11 define the intensity of the detected photoelectron signal for a specific photoemission line, n, in element i,

$$dI_{i,n} = I_o G \int_{x=0}^{x=\infty} \sigma_{i,n} N_i e^{-x/\lambda} dx \tag{10}$$

$$I_{i,n} = I_o \sigma_{i,n} \lambda_{i,n} N_i G \tag{11}$$

where $I_o$ is the incident x-ray flux, $\sigma_{i,n}$ is the cross-section for photoionization of level n in element i by an x-ray of energy $h\upsilon$, $\lambda_{i,n}$ is the imfp for the emitted photoelectron (which is a function of its kinetic energy and the matrix), $N_i$ is the fraction of atoms i in the sampling volume, and G is an instrumental parameter that accounts for the area of irradiation, the solid angle of the detector, and the transmission of the spectrometer. Equation 11 assumes that the composition of the specimen surface is uniform and homogeneous over the effective sampling volume.

The most common procedure for routine quantitative analysis is through the use of elemental sensitivity factors, $S_i$. These factors are obtained by measuring $I_i$ for a pure element of i. Since $N_i$ is unity in this case, the measurement gives

$$S_{i,n} = I_o \sigma_{i,n} \lambda_{i,n} G \tag{12}$$

These factors can then be used in the analysis of unknown species; e.g., to determine the ratio of two elements i and j in the surface of a specimen

$$\frac{N_i}{N_j} = \frac{I_{i,n}/S_{i,n}}{I_{j,n}/S_{j,n}}, \tag{13}$$

or the atomic concentration

$$N_i = \frac{I_{i,n} S_{i,n}}{\sum\limits_{\substack{\text{all elements} \\ i,j,k\ldots}} I_{i,n}/S_{i,n}} \tag{14}$$

The use of the published elemental sensitivity factors [B6] is reasonably accurate in routine XPS analysis of glass surfaces; but the preferred method is to obtain these factors through the use of oxide or compound standards. Figure 7 shows a calibration curve for the N1s signal in various nitride standards. The standards were a series of oxynitride glasses (whose nitrogen contents were determined independently through spectrochemical analysis), and both $Si_2N_2O$ and $Si_3N_4$ crystals. The ratioing of the N1s intensity provides an internal reference. The slope of this line provides a sensitivity factor that can be used for quantitative surface analysis of N in nitrided glass surfaces or thin films [2.20]. In principal, a calibration curve of the type shown in Figure 7 can be obtained for any species provided that the standards are available, and further, that clean surfaces can be prepared (e.g., by fracture of the glasses in vacuum). The advantage of this procedure over the use of elemental sensitivity factors is that the matrix dependence of the impf is included. In general, the imfp of photoelectrons in oxide glass matrices are greater than those in the pure elements.

The compositions calculated according to equation 11 – regardless of the method used to obtain the sensitivity factors – assume a homogeneous mixture of the detected elements in the sampling volume of the surface. The thickness of this 'homogeneous layer' is defined by the inelastic mean free path of the photoelectrons used to calculate the composition. In most glasses, this range corresponds to about 4 or 5 nm. This procedure for the quantitative analysis of XPS data is routinely employed, even though it is unlikely that a homogeneous surface layer exists. Most real surfaces – especially glasses – exhibit a strong photoelectron signal due to the presence of (hydro)carbon surface contamination. It is important that this be considered in the interpretation of these analyses.

It is common in many analyses to simply 'ignore' the detected carbon signal and calculate the glass composition using the signals for the 'glass species' only. This approach will give a reasonable measure of the glass composition if: (i) the carbonaceous overlayer itself is uniform and homogeneous, (ii) it does not contain any of the species present in the glass, and (iii) the attenuation of the various substrate signals are equivalent in this layer. In most real glass surfaces, these conditions are very unlikely. The carbonaceous layer most certainly contains oxygen, and the inelastic mean free paths for the various photoelectrons (in the overlayer) may differ by up to 30%.

To demonstrate these points, consider the following experiment. A glass rod of known composition is fractured in air where a layer of carbonaceous contamination forms. It is then subjected to a quantitative XPS analysis. The data in Table 2 summarizes the results. In this case, the true glass surface composition is the bulk composition because the surface was created by fracture. The composition calculated including carbon has implicitly assumed that the carbon is mixed homogeneously within the sampled volume of the glass surface. Obviously, this is not correct in the case of these fracture surfaces. The calculation performed by excluding the carbon signal does not eliminate the error because the carbonaceous overlayer has non-uniformly attenuated the various photoelectrons ejected from the underlying glass substrate. Another source of the error exhibited in Table 2 is due

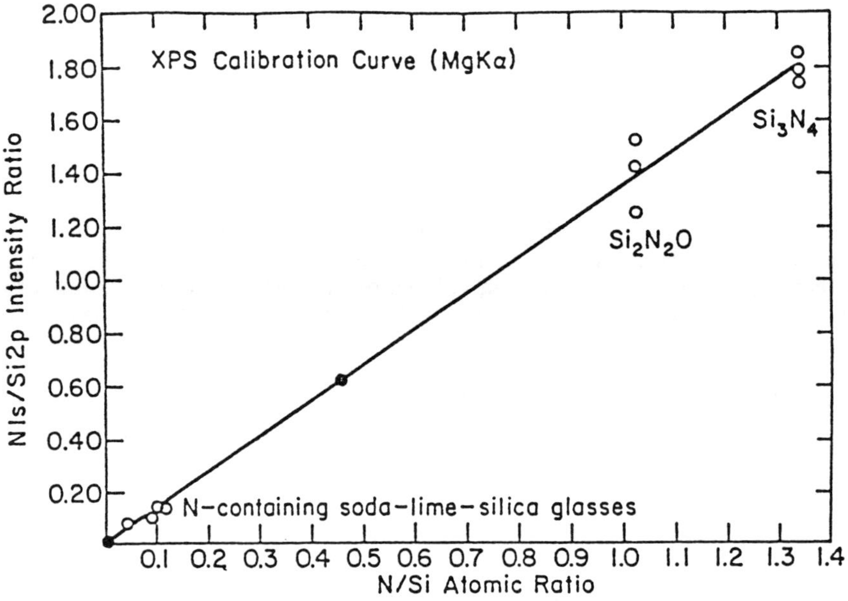


Figure 7. The calibration curve for quantitative analysis of the Nls/Si2p intensity ratio; the N/Si concentration ratios in the glass standards were measured spectrochemically..

Table 2. Glass Surface Compositions in Atomic Percent.

| | C | O | Si | Ca | Al | B |
|---|---|---|---|---|---|---|
| measured composition including carbon | 8.9 | 65.0 | 13.1 | 6.1 | 4.1 | 2.8 |
| measured composition excluding carbon | - | 71.3 | 14.4 | 6.7 | 4.5 | 3.1 |
| true glass surface composition | - | 62.2 | 19.7 | 8.7 | 5.8 | 3.6 |

to the fact that the carbonaceous layer contains some oxygen. Thus, the oxygen content is overestimated and the metal-ion contents are underestimated in all cases. This problem concerning the interpretation of glass surface composition has no easy solution. In many studies, sputter-etching of the surface is used to remove the layer of carbon. This is not considered an acceptable approach, though, because sputtering can drastically alter the composition of glass surfaces due to differential sputtering and beam-induced migration [12]. It is strongly recommended that compositional analyses be performed in the absence of any ion-beam sputtering. This necessarily means that the problems asociated with the presence of carbon be recognized and considered in the data treatment and interpretation.

There are situations where it is possible to make specific assumptions about the nature of the carbonaceous surface, and quantitatively treat the data accordingly. If a uniform overlayer of carbonaceous contamination is assumed, equation 11 can be modified, so that

$$I'_i = I^o_i e^{-t/\lambda} \qquad (15)$$

where t is the thickness of the overlayer, $\lambda$ is the imfp of the photoelectrons through the carbonaceous overlayer, and $I^o_i$ is the intensity of the photoelectrons in the absence of the overlayer (t=0). One can also write

$$I'_c = I^\infty_c (1 - e^{-t/\lambda}), \qquad (16)$$

where $I^\infty_c$ is the intensity of the C 1s photoelectron in an infinitely thick layer ($t \gg \lambda$) of carbonaceous contamination. $I^o_i$ can be obtained using a clean vacuum fracture surface or by extrapolation of a set of data for glasses with overlayers of varying thickness. $I^\infty_i$ can also be obtained by extrapolation, but for a direct measurement, a pure carbon, hydrocarbon or other 'overlayer' reference standard must be chosen to represent the contamination. The use of this overlayer model is exceedingly useful in those cases where it is known, a priori, that an overlayer exists – e.g., in the study of organic adsorbates on glass surfaces – and where a useful reference standard for the overlayer can be prepared.

But in many cases, a discrete overlayer may not be present on real glass surfaces. The melt-formed surfaces of glass (e.g., fibers, bottles, tubing, etc.) may react with $CO_2$, $H_2O$, or hydrocarbons at elevated temperatures where their diffusion into the glass structure is possible. In the polishing and chemical treatment of glass surfaces, carbonaceous overlayers containing constituents of the glass are possible. And in a broader sense, glasses always weather in the ambient atmosphere to form hydrated surface layers that contain carbonates. The point is that it is necessary to establish a model of the glass surface to properly interpret the data. The model usually entails some assumptions about the carbonaceous (or other) overlayer or about the glass substrate composition. Moreover, it often requires calibration of the signals using independent measurements of 'overlayer' standards and/or clean glass fracture surfaces. If the XPS measurements can be made as a function of the angle between the specimen and the detector, the overlayer can be better characterized. This variable angle measurement can be quite useful, but it requires flat, smooth samples to be most effective [see B5].

## INSTRUMENTATION

A detailed discussion of the instrumentation used in XPS will not be provided here because many other sources are available [see BIBLIOGRAPHY]. Only a commentary will be provided about the important features and specialized facilities for glass research. The basic apparatus is represented in Figure 1. An ultra high vacuum system, an x-ray source, an electron energy analyzer and a computerized data acquisition system. Some auxilliary items that may be of benefit include an ion sputtering gun, a fracture stage, a variable-angle specimen holder, an electron flood gun, and a metal evaporation source.

The vacuum system, and especially the availability of a fracture stage, are important in glass research because it is common to use XPS – an inherently surface sensitive technique – for bulk characterization. This necessitates an atomically clean surface that is best obtained by fracturing glass rods in vacuum. Clean fracture surfaces can also be used as calibration standards for quantitative surface analysis. A vacuum level better than $10^{-8}$ Pa is required to maintain the fracture surface in a clean state.

In most commercial XPS systems, MgKα and AlKα x-ray sources are standard. Although the routine analysis of most materials, including glass, can be achieved with these soft x-rays, there are a couple of specific applications in glass where higher energy x-rays, and monochromatic x-ray sources, can be advantageous. A higher energy x-ray source – e.g., SiKα, TiKα or ZrLα – creates core holes that lead to additional Auger lines in the XPS spectrum. These Auger lines can be used to obtain Auger parameters that are useful for structural characterization of silicates and for calibration of charging effects. The use of a monochromatic x-ray source improves the resolution of the analysis and this can be of benefit in the measurement of small chemical shifts. This is because the ability to measure small differences in binding energy, $\Delta E_b$, depends upon the linewidth of the incident x-ray, $\Delta(h\upsilon)$ (see equation 1).

## APPLICATIONS OF XPS

### Non-Bridging Oxygens

This example illustrates a bulk analysis of the network chemical structure, specifically non-bridging oxygens, in silicate glasses. Since the bulk structure was of interest, clean surfaces created by (vacuum) fracture were required.

Figure 8 shows high resolution O1s spectra for sodium aluminosilicate glasses with various Al/Na ratios. The relative concentration of bridging oxygen (BO) and non-bridging oxygen (NBO) is readily obtained through deconvolution; the presence of a distinct line due to the NBO is obvious. The ratio of the integrated intensities is equal to the concentration ratio because the sensitivity factors for the oxygen states are virtually identical. This analysis shows that the addition of $Al_2O_3$ eliminates NBO in sodium-silicate glasses. In this study, though, two different bridging oxygen sites were proposed – one due to Si-O-Si (BO1) and the other due to Si-O-Al (BO2) – even though these two BO lines are not resolved in the raw data [2.17]. A higher resolution measurement could provide more direct evidence of the authors' interpretation of the data. In the absence of experimental evidence of two distinct BO lines, they have based their interpretation of the data upon a specific model of the glass structure, and a systematic broadening and shifting of the proposed (BO1 + BO2) components. In general, curve fitting and deconvolution of XPS peaks should be executed with caution.

Figure 9 shows the high-resolution O1s spectra for binary alkali-silicate glasses containing Li, Na, K or Cs. The fact that the ratio of the BO to NBO lines is nearly equal in each glass is consistent with the fact that the glass specimens were of equal molar composition. However, it can be seen that the binding energy of the (NB)O1s electron depends upon the specific alkali specie in the glass.This reveals the dependence of the (NB)O1s binding energy upon the field strength of the associated alkali. This clearly demonstrates the sensitivity of the NBO photoelectron binding energy to its specific environment.

The interested reader can find many other studies of BO and NBO in binary and ternary silicate and phosphate glasses [see 2]. In more complex multicomponent glasses, where a wider distribution of oxygen sites exists, it is more difficult to resolve the BO and NBO lines. Only a broadening and/or a gradual shift of the O1s line is observed [2.33]. The ability to resolve distinct binding energy states for BO and NBO seems to be unique to simple binary or ternary oxides. Similarly, there have been attempts to measure non-bridging fluorine in fluoride glasses, but only a broadening of the F1s line has been reported [2.18].

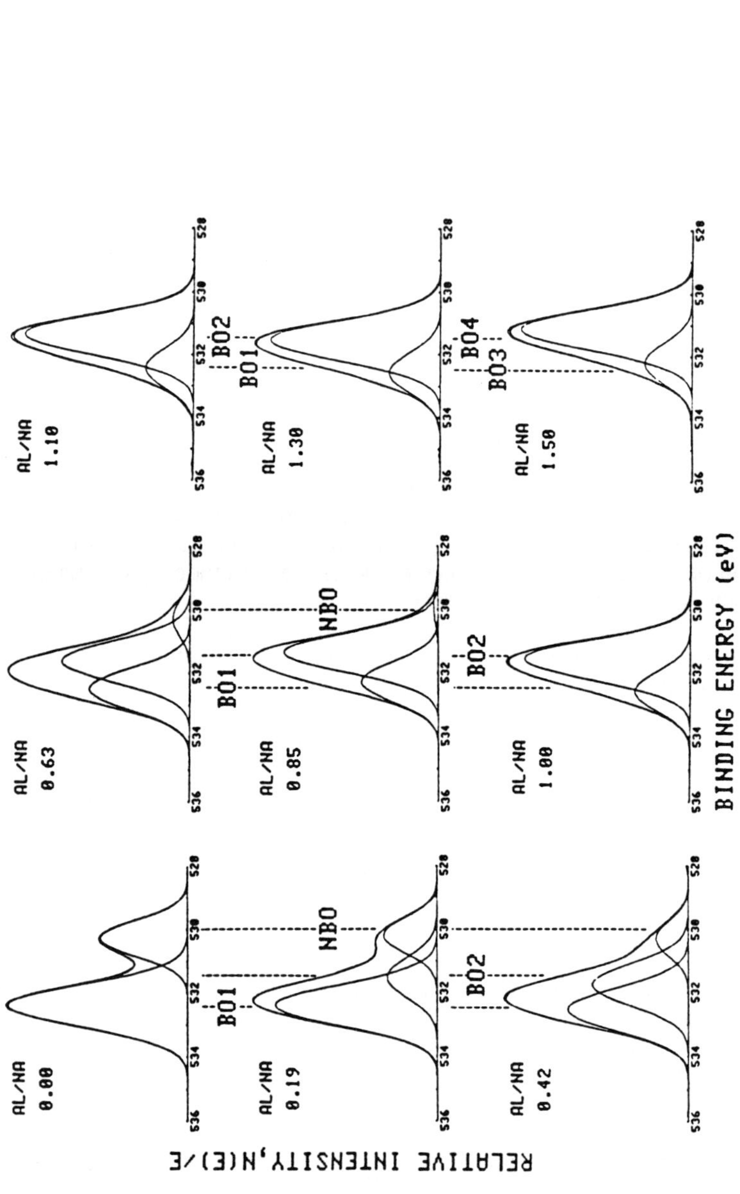


Figure 8. The high-resolution Ols spectra for sodium-aluminosilicate glasses with various Al/Na concentration ratios; each spectra was deconvoluted assuming three Gaussian components of constant width but varying binding energy [from 2.17].

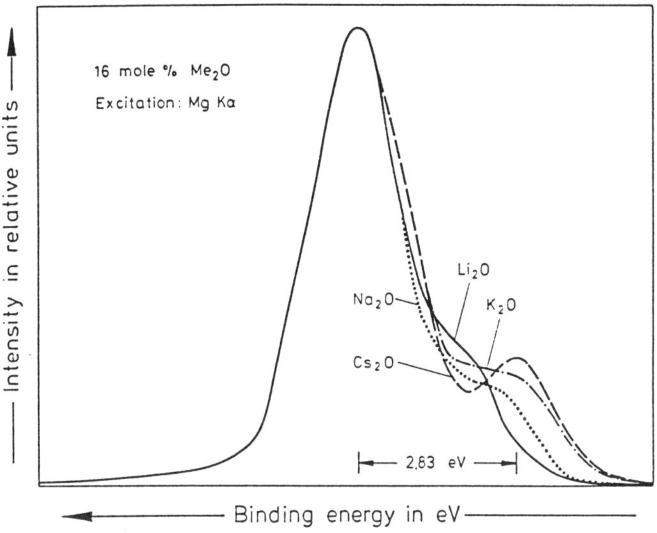


Figure 9. High-resolution Ols spectra of alkali silicate glasses with 16 mol-% $Me_2O$ where Me = Li, Na, K or Cs [ from 2.3].

## Thermochemical Treatment of Glass Surfaces

This example demonstrates the ability to characterize changes in the surface structure and redox state of a glass surface [2.28]. Figure 10 shows the high resolution O1s spectra of two alkali-lead-silicate glass specimens. One is for a vacuum fracture surface, and represents the bulk structure. The presence of two different oxygen species is clearly evident, and presumably, these correspond to BO and NBO. After a surface treatment in hydrogen, elimination of the lower binding energy O1s is observed. Figure 11 shows the high-resolution Pb4f spectra of these specimens. It is most evident that the Pb concentration has been significantly lowered in the treated surface. This probably accounts for the elimination of NBO. After low-energy ion sputtering into the sub-surface of the treated glass, the Pb concentration increases, but it can be seen that the oxidation state of the Pb has been altered. The Pb4f lines have shifted to lower binding energy. Using reference spectra for the identification, it was concluded that these were metallic Pb species. Thus, the hydrogen treatment (at ~500°C) converted the PbO to Pb, and evaporation of Pb at the surface created a silica-rich zone. The

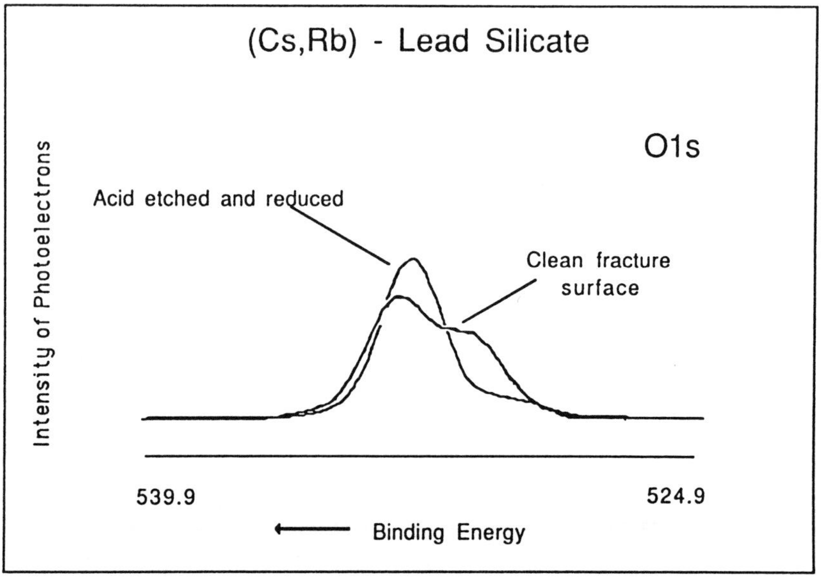


Figure 10. High-resolution Ols spectra of an alkali-lead-silicate glass before and after hydrogen reduction at 485°C.

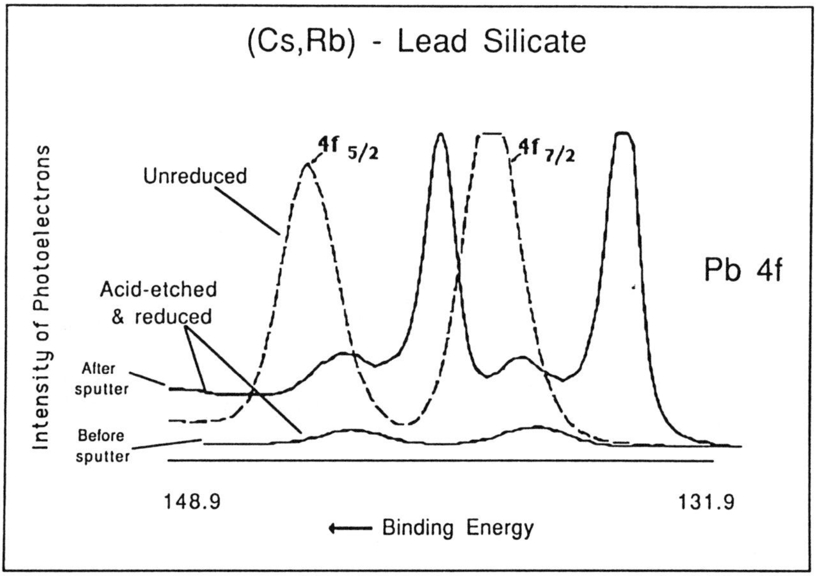


Figure 11. High-resolution Pb4f spectra of an alkali-lead-silicate glass before and after hydrogen reduction at 485°C.

reader should note that, in general, ion sputtering must be used with caution because it can change the composition and chemistry of the surface being analyzed [12]. In this analysis, ion sputtering of the clean fracture surface verified the absence of any perturbing effects of the ion bombardment at low energy.

## Chemical and Compositional Analysis of Sol/Gel Thin Films

This example demonstrates the ability of XPS to provide *quantitative* chemical and compositional analysis of glasses – in this case glass thin films [13]. It also shows that with careful calibration, the binding energy value itself can provide a measure of composition. In this way, a quantitative composition determination can be obtained which is independent of the usual integrated intensity analysis.

Figure 12 shows the high resolution C1s spectra of silica sol/gel films after various heat treatments; the spectra of thermal $SiO_2$ is also shown for reference. The thermal $SiO_2$ exhibits only 'adventitious' carbon due to contamination in the atmosphere, whereas the $SiO_2$ sol/gel films show other chemical states of carbon due to their processing history. These lines were identified using published reference spectra [B6].

Figure 7 showed the calibration curve established for quantitative analysis of nitrogen in the ammonia treated silica films. It was used to determine the nitrogen concentration in films treated at various temperatures. These nitrogen concentrations were then plotted against the Si2p binding energies; this plot is shown in Figure 13. The data reveals that the progressive incorporation of nitrogen influences the Si2p binding energy, and thereby, provides evidence for chemical association of the Si and N in the treated films. The fact that the Si2p binding energy in $Si_2N_2O$ is ~ 102 eV provides strong evidence for the creation of Si-oxynitride species in the nitrided silica gel-film. (A theoretical discussion of this binding energy effect can be found in 2.20.) Figure 13 also provides an additional calibration curve for nitrogen concentration based solely on the Si2p binding energy.

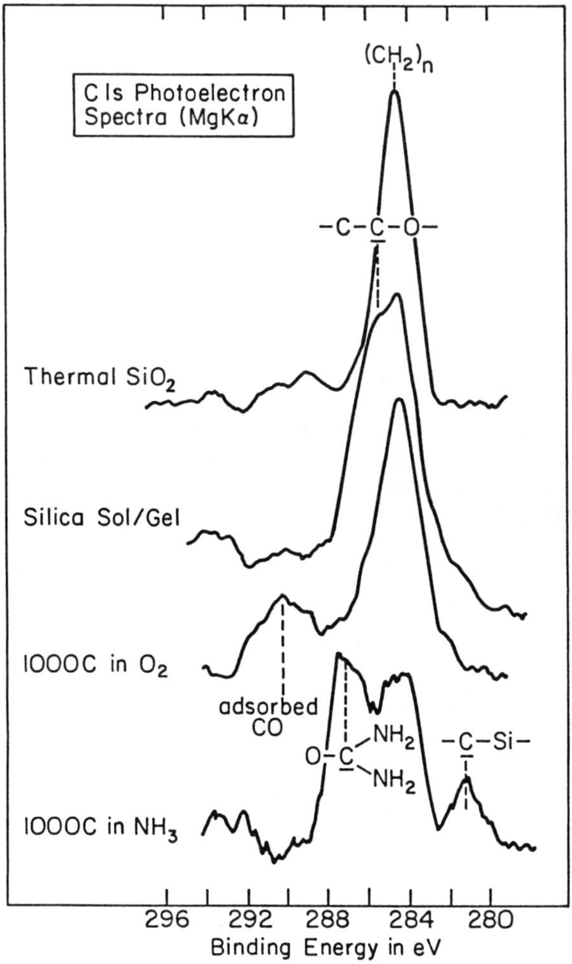


Figure 12. High resolution Cls spectra of an $SiO_2$ reference standard and $SiO_2$ sol/gel thin film before and after thermal processing.

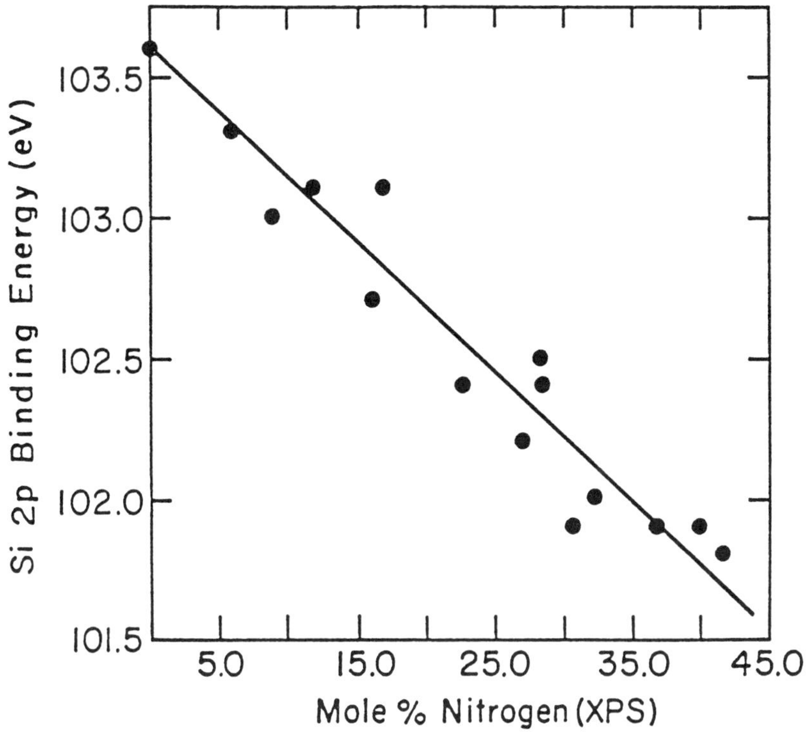


Figure 13. The observed relationship between the Si2p binding energy and nitrogen content of silica sol/gel films treated in ammonia at various temperatures.

## Organic Adsorbates on Glass

XPS is very useful for the detection and chemical analysis of organics on glass surfaces. In comparison to electron-beam and ion-beam methods, the x-ray beam is relatively non-destructive. Moreover, the photoelectron spectra can distinguish the chemical state of the organic species. In this example, glass fibers were analyzed after exposure to five different organic silanes. Untreated fibers were examined for reference. The spectra were quantitatively evaluated using elemental sensitivity factors. A calculation of the fiber surface composition according to equation 14 revealed that all of the treated fibers had considerably more carbon on the surfaces (30-50 a/o) than the untreated control fibers (10-15 a/o). Naturally, this is due to

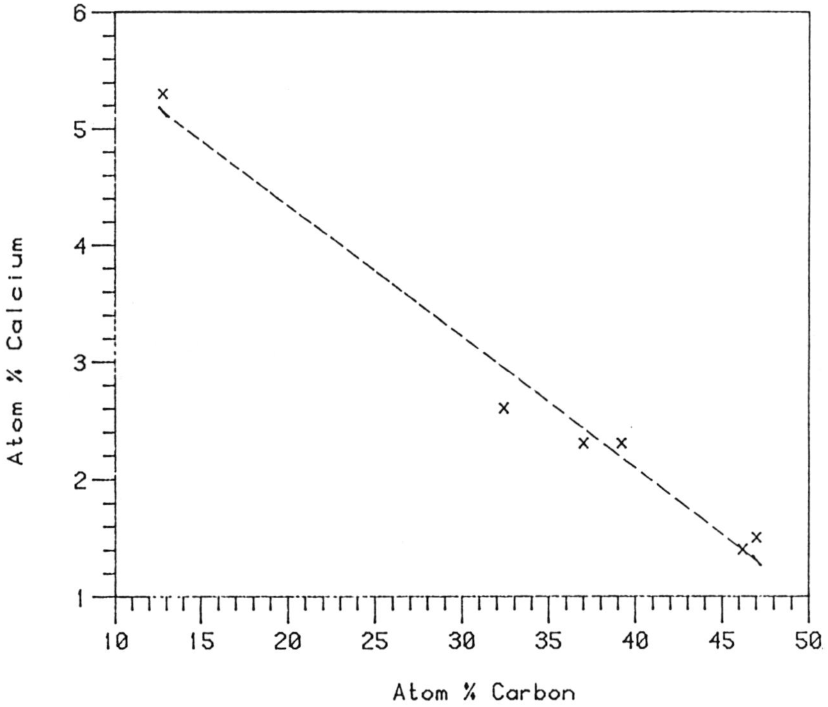


Figure 14. The observed relationship between the calculated Ca and C concentrations (or equivalently, the relative Ca2p and Cls signals) on E-glass fibers treated with organic silanes..

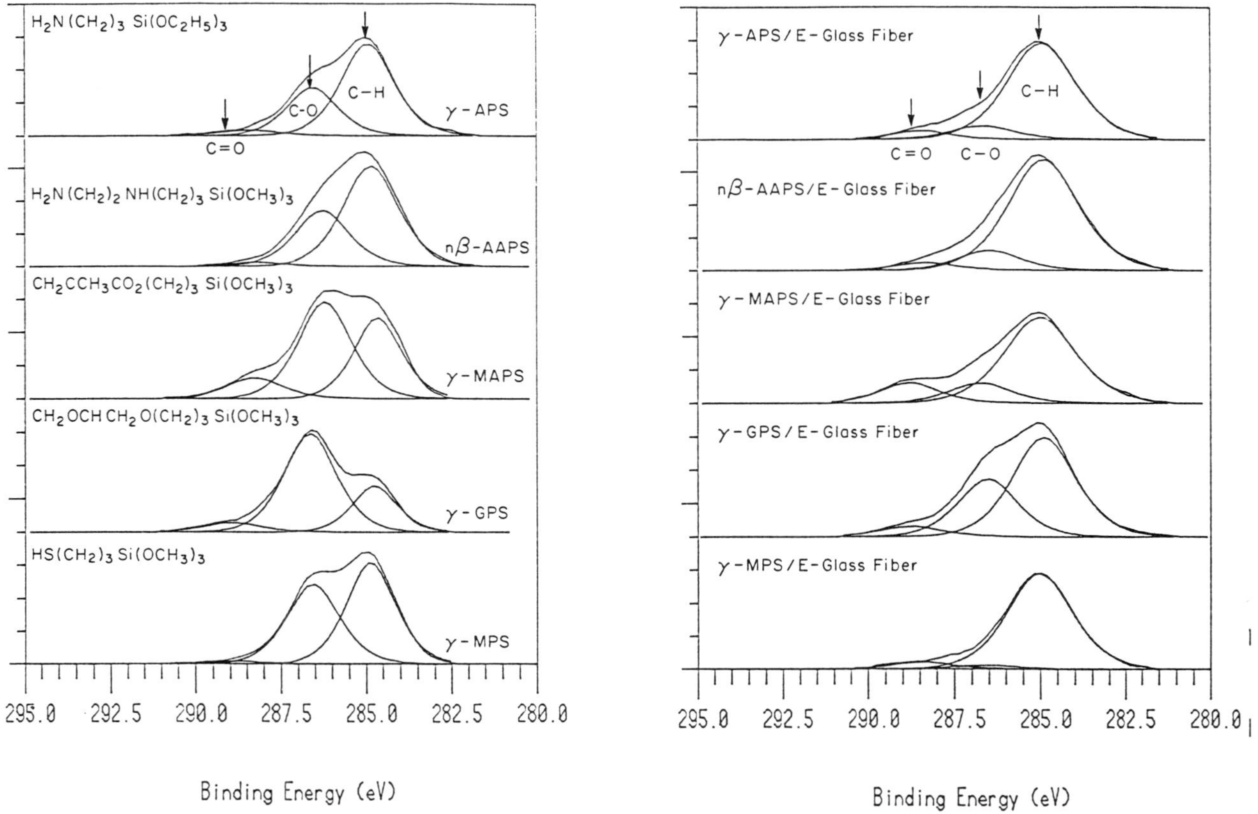


Figure 15. High resolution Cls spectra of organosilanes neat on gold foil (left) and adsorbed to E-glass fibers (right).

adsorption of the organic silanes, and for qualitative characterization, these analyses could be sufficient to show the relative effects of the silane-treatment.

Figure 14 plots the carbon concentrations against the calcium concentrations. The fact that the Ca concentration (associated with the glass fiber) decreases with increasing C concentration verifies that the carbon is associated with an overlayer. But clearly, the *Ca concentrations* are quantitatively meaningless unless they are interpreted using a more representative model of the surface; i.e., the analyzed surface is not a homogeneous mixture of the organic silane and glass. The data in Figure 13 is, in fact, plotted according to Equations 15 and 16. An extrapolation of the line to 0% carbon – that is, no overlayer @ t=0 – gives a value of 6.5% Ca at the glass fiber surface. This is in reasonable agreement with the glass composition (~7.5% Ca). And, the slope of this line allows the thickness of the overlayers to be determined. Assuming a value of $\lambda \approx 2.7$ nm, the overlayers are found to be 1.5-2.5 nm. These thicknesses correlated with the length of the alkyl group in the various silanes [2.30].

In this analysis, it was also possible to examine the chemistry of the adsorbed silane, and compare it to the bulk silane, using the high-resolution C1s spectra. Figures 15a and 15b compares these spectra for all five the silanes. The spectra were deconvoluted to separate C=O, C-O and C-H sites in the silane. The relative decrease in the C-O species in the *adsorbed* silane verifies that hydrolysis of the alkoxy ligands in the silane has occurred.

## SUMMARY

The most common applications of XPS in glass research involve surface analysis; that is, the determination of surface composition and/or the identification of surface impurities or adsorbates. The technique is especially suited to the analysis of organic and metallic species on the surface, although a reasonable quantitative analysis of the glass composition itself can also be achieved. The analysis is limited to species located in the outermost 0.5-5.0 nm of surface whose concentration exceeds .1-1 atomic percent. It should be noted, though, that XPS cannot detect hydrogen. XPS can also be used to characterize features of the chemical structure; that is, non-bridging oxygens, bonding or oxidation state of anions and cations, and valence band density of states. And is those cases where very clean fracture surfaces can be prepared, the spectra can be used to characterize the bulk chemical structure of the glass.

## ACKNOWLEDGEMENTS

The author thanks his colleagues and students for their contributions toward this review: Dick Brow, Lee Carman, Fumio Ohuchi, Bill Tasker, Alan Then and Tom Wittberg.

## BIBLIOGRAPHY

B1. K. Siegbahn, C. N. Nordling, A. Fahlman, R. Nordberg, K. Hamrin, J. Hedman, G. Johansson, T. Bergmark, S. E. Karlsson, I. Lindgren, and B. Lindberg, *ESCA, Atomic, Molecular, and Solid State Structure Studied by Means of Electron Spectroscopy* (Almqvist and Wiksells, Uppsala, Sweden, 1967).

B2. T. A. Carlson, *Photoelectron and Auger Spectroscopy* (Plenum Press, New York, 1975).

B3. P. K. Ghosh, *Introduction to Photoelectron Spectroscopy* (Wiley-Interscience Publishers, New York, 1983).

B4. C. R. Brundle and A. D. Baker, Eds., *Electron Spectroscopy: Theory, Techniques and Applications*, Vol. 1-5 (Academic Press, London, 1978).

B5. D. Briggs and M. P. Seah, *Practical Surface Analysis by Auger and X-ray Photoelectron Spectroscopy* (John Wiley and Sons, New York, 1983).

B6. D. Briggs, Ed., *Handbook of X-ray and Ultraviolet Photoelectron Spectroscopy* (Heydon and Son, London, 1977, 1978).

## REFERENCES

1. See for example:
1.1. D. A. Stephenson and N. J. Binkowski, "X-ray Photoelectron Spectroscopy of Silica in Theory and Experiment," J. Non-Cryst. Solids, 22[2], 399-421 (1976).
1.2. S. R. Nagel, J. Tauc, and B. G. Bagley, "X-ray Photoemission Study of a Soda-Lime-Silica Glass," Solid State Commun., 20[3], 245-9 (1976).
1.3. B. Fischer, R. A. Pollak, T. H. DiStefano, and W. D. Grobman, "Electronic Structure of $SiO_2$, $Si_xGe_{1-x}O_2$, and $GeO_2$ from Photoemission Spectroscopy," Phys. Rev., B15[6], 3193-9 (1977).
1.4. W. Y. Ching, R. A. Murray, D. J. Lam, and B. W. Veal, "Comparative Studies of Electronic Structures of Sodium Metasilicate and $\alpha$ and $\beta$ Phases of Sodium Disilicate," Phys. Rev., B28[8], 4724-34 (1983).
1.5. W. Y. Ching, Y. P. Li, B. W. Veal, and D. J. Lam, "Electronic Structures of Lithium Metasilicate and Lithium Disilicate," Phys. Rev., B32[2], 1203-7 (1985).
2. See for example:
2.1. J. H. Escard and D. J. Brion, "Study of Composition of Leached Glass Surfaces by Photoelectron Spectroscopy," J. Am. Ceram. Soc., 58[7-8], 296-299 (1975).
2.2. N. J. Binkowski, R. F. Heitzenrater, and D. A. Stephenson, "Surface Analysis of a Treated Soda-Lime Glass," J. Am. Ceram. Soc., 59[3-4], 153-7 (1976).

2.3. R. Bruckner, H.-U. Chun, and H. Goretzki, "Photoelectron Spectroscopy (ESCA) on Alkali Silicate and Soda Aluminosilicate Glasses," Glastechn. Ber., 51[1], 1-7 (1978).
2.4. Y. Kaneko, "Application of Photoelectron Spectroscopy to Glasses," Yogyo-Kyokai-Shi, 86[7], 330-3 (1978).
2.5. R. Gresh, W. Müller-Warnuth, and H. Dutz, "X-ray Photoelectron Spectroscopy of Sodium Phosphate Glasses," J. Non-Cryst. Solids, 34, 127-136 (1979).
2.6. J. S. Jen and M. R. Kalinowski, "An ESCA Study of the Bridging to Non-Bridging Oxygen Ratio in Sodium Silicate Glass and the Correlations to Glass Density and Refractive Index,", J. Non-Cryst. Solids, 38-39, 21-6 (1980).
2.7. D. J. Lam, A. P. Paulikas, and B. W. Veal, "X-ray Photoemission Spectroscopy Studies of Soda Aluminosilicate Glasses," J. Non-Cryst. Solids, 42[1-3], 41-8 (1980).
2.8. R. Bruckner, H.-U. Chun, H. Goretski, and M. Sammet, "XPS Measurements and Structural Aspects of Silicate and Phosphate Glasses," J. Non-Cryst. Solids, 42[1-3], 49-60 (1980).
2.9. B.M.J. Smets and T.P.A. Lommen, "The Structure of Germanosilicate Glasses, Studied by X-ray Photoelectron Spectroscopy," J. Non-Cryst. Solids, 46[1], 21-32 (1981).
2.10. B.M.J. Smets and T.P.A. Lommen, "The Incorporation of Aluminum Oxide and Boron Oxide in Sodium Silicate Glasses, Studied by X-ray Photoelectron Spectroscopy," Phys. Chem. Glasses, 22[6], 158-62 (1981).
2.11. T. Sasaki, M. Kawaguchi, M. Yamane, and Y. Suginohara, "On the Quantitative Analysis of $O^{\circ}$, $O^{-}$, and $O^{2-}$ Ions in Binary Silicate by X-ray Photoelectron Spectroscopy," J. Japan Inst. Metals, 45[8], 790-6 (1981).
2.12. K. Matsumoto, "The Analysis of Carbon on the Surface of Glass by ESCA," Yogyo-Kyokai-Shi," 89[1], 19-23 (1981).
2.13. B. W. Veal, D. J. Lam, A. P. Paulikas, and W. Y. Ching, "XPS Study of CaO in Sodium Silicate Glasses," J. Non-Cryst. Solids, 49[1-3], 309-20 (1982).
2.14. Y. Kaneko, H. Nakamura, M. Yamane, K. Mizoguchi, and Y. Suginohara, "Photoelectron Spectra of Silicate Glasses Containing Trivalent Cations," Yogyo-Kyokai-Shi, 91[7], 321-4 (1983).
2.15. R. K. Brow and C. G. Pantano, "Nitrogen Coordination in Oxynitride Glasses," J. Am. Ceram. Soc., 67[4], C72-74 (1984).
2.16. M. Sekita, A. Fujimori, A. Makishima, T. Shimohira, and H. Ohashi, "X-ray Photoelectron Spectroscopy of a Caerium-Doped Lanthanum Aluminosilicate Glass," J. Non-Cryst. Solids, 76, 399-407 (1985).
2.17. G. W. Tasker, D. R. Uhlmann, P.I.K. Onorato, M. N. Alexander, and C. W. Struck, "Structure of Sodium Aluminosilicate Glasses: X-ray Photoelectron Spectroscopy," J. DePhysique, C8[46], C8-273-80 (1985).
2.18. R. M. Almeida, J. Lau, and J. D. Mackenzie, "XPS Studies of Fluorozirconates," Mat. Sci. Forum, 6, 465-470 (1985).
2.19. D. S. Goldman, "Evaluation of the Ratios of Bridging to Nonbridging Oxygens in Simple Silicate Glasses by Electron Spectroscopy for Chemical Analysis," Physics and Chemistry of Glasses, 27[3], 128-133 (1986).

2.20. R. K. Brow and C. G. Pantano, "Compositionally Dependent Si 2p Binding Energy Shifts in Silicon Oxynitride Thin films," J. Am. Ceram. Soc., 69[4], 314-16 (1986).
2.21. G. W. Tasker, "X-ray Photoelectron Spectroscopy of Silicate Glasses," Ph.D. Thesis, Massachusetts Inst. of Tech., 1987.
2.22. R. K. Brow, "Oxidation States of Chromium Dissolved in Glass Determined by X-ray Photoelectron Spectroscopy," J. Am. Ceram. Soc., 70[6], C-129-31 (1987).
2.23. B. W. Veal, J. N. Mundy and D. J. Lam, "Actinides in Silicate Glasses," *Handbook on the Physics and Chemistry of the Actinides*, A. J. Freeman and G. H. Lander, Eds. (Elsevier, 271-309, 1987).
2.24. C. G. Pantano and R. K. Brow, "Hydrolysis Reactions at the Surface of Fluorozirconate Glass," J. Am. Ceram. Soc., 71[7], 577-81 (1988).
2.25. R. Caracciolo and S. H. Garofalini, "X-ray Photoelectron Spectroscopy Analysis of Potassium Adsorption onto $\upsilon$-$SiO_2$: Observation of Reactive Surface Defect Sites," J. Am. Ceram. Soc., 71[7], C346-C349 (1988).
2.26. R. K. Brow, M. R. Reidmeyer and D. E. Day, "Oxygen Bonding in Nitrided Sodium-and Lithium-Metaphosphate Glasses," J. Non-Cryst. Solids, 99, 178-189 (1988).
2.27. R. K. Brow, Y. Zhu, D. E. Day, and G. W. Arnold, "Surface Nitridation of a Phosphate Glass," J. Non-Cryst Solids, 120, 172-7 (1990).
2.28. A. M. Then and C. G. Pantano, "Formation and Behavior of Surface Layers on Electron Emission Glasses," J. Non-Cryst. Solids, 120 178-187 (1990).
2.29. Z. A. Osborne, R. K. Brow, D. R. Tallant, "Short-Range Structure of Fluorine-Modified Phosphate Glass," *Properties and Characteristics of Optical Glass, II*, SPIE Proceedings, 1990 Int. Symp. on Optical and Optoelectronic Apl. Sci. and Eng., San Diego, CA, July 8-13, 1990.
2.30. C. G. Pantano and T. N. Wittberg, "XPS Analysis of Silane Coupling Agents and Silane-treated E-glass Fibers," Surf. and Int. Anal., 15, 498-501 (1990).
2.31. R. K. Brow, R. J. Kirkpatrick, and G. L. Turner, "Local Structure of $xAl_2O_3{\cdot}(1\text{-}x)NaOP_3$ Glasses: An NMR and XPS Study," J. Am. Ceram. Soc., 73[8], 2293-2300 (1990).
2.32. D. Sprenger, H. Bach, W. Meisel and P. Gütlich, "XPS Study of Leached Glass Surfaces," J. Non-Cryst. Solids, 126, 111-129 (1990).
2.33. I. Berthelot and P. Lehuede, "XPS Analysis of Glass: O1s Line Study," Rivista della Staz. Sper. Vetro, n.6 (1990).
3. P. Swift, "Adventitious Carbon - The Panacea for Energy Referencing?" Surf. Interface Anal., 4[2], 47-51 (1982).
4. S. Hohiki and K. Oki, "Problems of Adventitious Carbon as an Energy Reference," J. Electron Spectrosc. Relat. Phenom., 33[4], 375-80 (1984).
5. C. R. Ginnard and W. M. Riggs, "Gold Evaporation for Charge Correction in X-ray Photoelectron Spectroscopy," Anal. Chem., 46[9], 1306-8 (1974).
6. Y. Uwamino, T. Ishizuka, and H. Yamatera, "Charge Correction by Gold Deposition onto Non-Conducting Samples in X-ray Photoelectron

Spectroscopy," J. Electron Spectrosc. Relat. Phenom., 23[1], 55-62 (1981).
7. S. Kohiki, T. Ohmura, and K. Kusao, "A New Charge-Correction Method in X-ray Photoelectron Spectroscopy," J. Electron Spectrosc. Relat. Phenom., 28[4], 229-37 (1983).
8. S. Kohiki and K. Oki, "An Appraisal of Evaporated Gold as an Energy Reference in X-ray Photoelectron Spectroscopy," J. Electron Spectrosc. Relat. Phenom., 36[1], 105-10 (1985).
9. M. J. Edgell, D. R. Baer, and J. E. Castle, "A Test of Biased Referencing for the XPS Analysis of Nonconducting Materials," J. Vac. Sci. Technol., A4[3], 1562-3 (1986).
10. J. E. Castle and R. H. West, "Bremsstrahlung-Induced Auger Peaks," J. Elect. Spec., 18, 355-358 (1980).
11. C. G. Pantano, "Glass Surfaces," Rev. of Solid State Sci., 3[3&4], 379-408 (1989)
12. C. G. Pantano, "Chemical and Compositional Analysis of Glass Surfaces," Rivista della Staz. Sper. Vetro, n.6, 123 (1991).
13. R. K. Brow and C. G. Pantano, "Thermochemical Nitridation of Microporous Silica Films in Ammonia," J. Am. Ceram. Soc., 70[1], 9-14, 1987.

# CHAPTER 6

## ELECTRON SPIN RESONANCE SPECTROSCOPY OF RADIATION-INDUCED DEFECT CENTERS IN INSULATING GLASSES

David L. Griscom
Naval Research Laboratory
Washington, DC 20375

### INTRODUCTION

Electron spin resonance (ESR) spectroscopy generally involves the measurement of the resonant absorption of fixed-frequency microwaves by a system of unpaired electrons, tuned by varying the strength of an externally applied magnetic field. ESR is thus analogous to nuclear magnetic resonance (NMR), which records the resonant absorption of radio waves by nuclei with non-zero nuclear magnetic moments (magnetic nuclei). But whereas many as-quenched insulating glasses contain at least some magnetic nuclei suitable for investigation by NMR, few contain sufficient unpaired electrons for convenient ESR study (the most common exception being cases where transition-group ions are present). However, the situation can be dramatically different when a normally-diamagnetic glass is subjected to ionizing radiations, such as neutrons, protons, energetic electrons, $\gamma$ rays, x rays, or sometimes even UV light. The inevitable result of such irradiations is the creation of paramagnetic defect centers. These defect centers may comprise, e.g., vacancies and interstitials in an otherwise perfect random network. Such displaced atomic species may be created by the radiation or they may be preexisting in the glass, but in either case they are rendered paramagnetic by the trapping of radiation-generated electrons or holes. The concentrations of "unpaired spins" produced in this way seldom exceed 100 per million network atoms. Nevertheless, under favorable conditions, ESR can be called upon to routinely detect numbers of resonators as low as ~1 ppb. ESR is intrinsically more sensitive than NMR by a factor of ~2000 owing to the larger magnetic moment of the electron relative to that of the proton, and further sensitivity gains derive from the high Q of the microwave resonant cavity in which the sample is positioned.

Whereas NMR gives information about the electrostatic field gradients and atomic arrangements at *typical* sites in the glass structure, ESR—when employed to study radiation damage—gives information concerning the quantum mechanical wavefunction of the unpaired spin and the atomic arrangements at *defect* sites. An understanding of these defect centers per se is particularly important because they are generally responsible for the observed optical absorption bands and electrical charge buildups (and are probably the initiators of mechanical changes) which degrade the performances of many common optical and electronic components. Nevertheless, experience has shown that the structures of the radiation-induced defect centers are themselves strongly related to the structure of the undamaged glass, so that an ESR study of an irradiated glass may profitably complement other spectroscopic and diffraction approaches to determining glass structure.

The present chapter will illustrate by a small number of examples how an ESR experiment is performed in irradiated glasses and how certain types of structural information can be derived therefrom. A number of review articles and text books provide discussions of radiation effects in a broader range of glass types and compositions,[1-4] accounts of ESR studies of transition group ions[3,5] and ferromagnetic precipitates[6,7] in glasses, and in-depth treatments of ESR theory and technique[8-10].

## BASIC THEORY

An electron has a quantum mechanical property called *spin*, which is often visualized classically as a rotation about "its own axis" (although strictly speaking this visualization is inaccurate since elementary particles do not have finite dimensions). The spin of a system of one or more electrons is denoted by a dimensionless vector **S** of magnitude $\sqrt{S(S+1)}$, where $S$ is a scalar number which may be an integer, half integer, or zero. Always, for a *single* electron, $S = \frac{1}{2}$. The spin angular momentum is $\hbar\mathbf{S}$, where $\hbar$ is Planck's constant $h$ divided by $2\pi$. Electrons possess a spin magnetic moment given by

$$\mu_S = -g\beta\mathbf{S}, \tag{1}$$

where $\beta \equiv |e|\hbar/2mc$ is the Bohr magneton ($e$ and $m$ are the electron charge and mass, respectively, and $c$ is the speed of light) and $g$ is a dimensionless number. For the free electron, $g = g_0 \equiv 2.0023$.

An unpaired electron on an atom in free space also has an orbital angular momentum $\mathbf{L}\hbar$. The magnitude of **L** is $\sqrt{L(L+1)}$, where $L$ = 1, 2, or 3 for a single electron in a $p$, $d$, or $f$ orbital, respectively. The associated orbital magnetic moment is given by

$$\mu_L = -\beta\mathbf{L} \tag{2}$$

The total electron magnetic moment is given by the sum of Eqs. (1) and (2). However, when placed in a solid, electrons are generally denied the freedom to circulate freely about the central nucleus, with the result that the effective orbital angular momentum is said to be "quenched" ($\mu_L \approx 0$). In such a situation (which pertains to all the cases of interest here), the net magnetic moment can be expressed in the form of Eq. (1), where the small residual orbital contribution is subsumed into the $g$ value—which in turn can be calculated in quantum mechanical perturbation theory.[8, 9] Thus, the "$g$ shift", $g - g_0$, carries information relating to the interaction of the orbital of the unpaired spin with the crystal fields of the host solid.[3,8,9]

The energy of the Zeeman interaction of an electron with an applied magnetic field **H** is given by

$$E_{Zeeman} = - \mu_S \cdot \mathbf{H} = \beta \mathbf{S} \cdot \mathbf{g} \cdot \mathbf{H}, \tag{3}$$

where, in anticipation of anisotropic crystal field influences, the "$g$ value" of eq. (1) has been made a matrix quantity. Quantum mechanics dictate that the measurable projection of **S** on any axis (e.g., the direction of **g·H**) can only take on the values $S$, $S-1, \ldots, -S$. Since $S = \frac{1}{2}$ for all cases of interest in this chapter, Eq. (3) can have only two evaluations for a given orientation of **H** with respect to the principal axes of **g**:

$$E_{\uparrow} = +(1/2)\beta \mathbf{g} \cdot \mathbf{H}, \tag{4a}$$

$$E_{\downarrow} = -(1/2)\beta \mathbf{g} \cdot \mathbf{H}. \tag{4b}$$

These Zeeman energies are schematically graphed as a function of the magnitude of **H** in Fig. 1a. Also illustrated in Fig. 1a is a transition between the lower-energy "spin down" state and the higher-energy "spin up" state induced by bathing the sample in

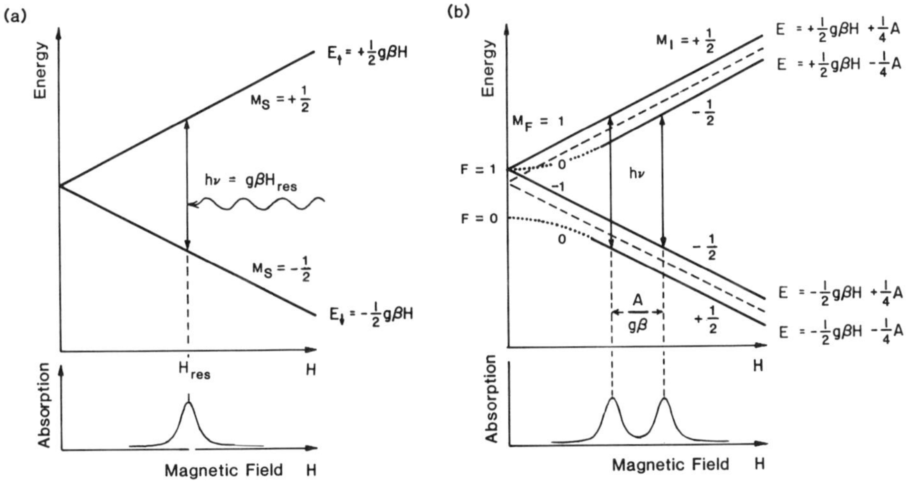


**Figure 1** Energy levels and ESR resonance absorption schemes for (a) the Zeeman interaction of a single electron ($S = \frac{1}{2}$) and (b) the Zeeman and hyperfine interactions for $S = \frac{1}{2}$ and $I = \frac{1}{2}$. (From ref. 3.)

microwaves of frequency $\nu$. The measured absorption spectrum, peaking at $|\mathbf{H}| = h\nu/g\beta$, is schematically drawn in the lower graph of Fig. 1a. It should be noted, however, that the probability of inducing a downwards transition is equal to the probability of inducing one upwards, so that a net absorption is observed only because the population of the upper state is less than that of the lower state in consequence of Maxwell-Boltzmann statistics. Indeed, were it not for the existence of spin-lattice relaxation processes, the performance of a resonance experiment would eventually equalize the upper and lower populations and, consequently, absorption of energy from the microwave field would cease. At sufficiently low microwave powers $P$, spin-lattice relaxation maintains the population difference. Under these conditions the intensity of the resonance is proportional to the strength of the microwave magnetic field $H_1$ at the position of the sample (note that $H_1 \propto \sqrt{P}$). When the power level is raised to the point where this proportionality no longer holds, the spin system is said to be entering "saturation". Since different centers generally have different relaxation times, microwave saturation studies are often useful for separating overlapping spectra.

Without a doubt, from the standpoint of identifying an unknown paramagnetic species and its immediate atomic surroundings, the most useful property of an ESR spectrum can be the presence of hyperfine structure (hfs) arising from the interaction of the magnetic moment of the unpaired electron spin with the magnetic moment of a nearby nucleus. The hyperfine interaction energy can be expressed as

$$E_{hfs} = \mathbf{S}\cdot\mathbf{A}\cdot\mathbf{I}, \tag{5}$$

where $\mathbf{I}$ is the dimensionless nuclear spin vector of magnitude $\sqrt{I(I+1)}$ and $\mathbf{A}$ is the hyperfine coupling matrix. Clearly, hyperfine interactions only occur with isotopes for which $I \neq 0$.

The magnetostatic contributions to the energy of a single electron of spin $S = \frac{1}{2}$ interacting with an applied magnetic field $\mathbf{H}$ and undergoing a hyperfine interaction with a nucleus of spin $I$ can be obtained by summing Eqs. (3) and (5). (Such a sum of energy terms involving the quantum mechanical spin vector $\mathbf{S}$ is called the *spin Hamiltonian.*) These energies are graphed in Fig. 1b under the assumptions that there is a single magnetic nucleus of spin $I = \frac{1}{2}$ and the matrices $\mathbf{g}$ and $\mathbf{A}$ are isotropic ($\mathbf{g} = g\mathbf{1}$ and $\mathbf{A} = A\mathbf{1}$, where $\mathbf{1}$ is the unit matrix). Here, the diverging long-dashed lines recapitulate the Zeeman interaction as taken from Fig. 1a. The unbroken lines show the result of adding in the hyperfine interaction in the limit that $g\beta|\mathbf{H}| >> A$ — a condition which prevails in many common experimental situations. The dotted curves schematically represent the outcome of an exact matrix diagonalization,[9] as must be performed under conditions when the Zeeman interaction is comparable to or weaker than the hyperfine interaction. It can be seen in Fig. 1b that, in the high-field limit, the single resonance line of Fig. 1a splits into a pair of lines separated by $A/g\beta$ (which is the hyperfine coupling constant in units of magnetic field). In general, for a nuclear spin $I$ there will be $2I+1$ hyperfine lines, uniformly spaced in the high-field limit. The positions of these lines (to first order in perturbation theory) are given by the *resonance condition*

$$H_{res}(M_I) = (h\nu/g\beta) + (A/g\beta)M_I, \tag{6}$$

where $M_I$ is the nuclear magnetic quantum number, which may assume values $I, I-1, I-2, \ldots, -I$.

Equation (6) can be adapted to cases where the $g$ values and hyperfine matrix are anisotropic. In the case that **g** and **A** are axially symmetric (and share a common symmetry axis), it is necessary only to substitute the following expressions for $g$ and $A$ into the resonance condition:

$$g = [(g_\parallel^2 - g_\perp^2)\cos^2\theta + g_\perp^2]^{\frac{1}{2}}, \tag{7a}$$

$$A = g^{-1}[(A_\parallel^2 g_\parallel^2 - A_\perp^2 g_\perp^2)\cos^2\theta + A_\perp^2 g_\perp^2]^{\frac{1}{2}}, \tag{7b}$$

where $\theta$ is the angle between the applied field **H** and the symmetry axis of the defect. Equations (7), although specialized to axial symmetry, turn out to pertain to a wide range of radiation-induced defect centers in insulating glasses. The parameters $A_\parallel/g_\parallel\beta$ and $A_\perp/g_\perp\beta$ (measured for $\theta = 0$ and $\theta = \pi/2$, respectively) provide a great deal of structural information about these defects. In many cases it turns out that the ground state wavefunction of the unpaired spin can be expressed

$$|0\rangle = c_s|ns\rangle + c_p|np\rangle + \sum_j c_j|k_j\rangle, \tag{8}$$

where $|ns\rangle$ and $|np\rangle$ are n$s$-state and n$p$-state orbitals centered on the magnetic nucleus (n is the principal quantum number) and $|k_j\rangle$ are ligand orbitals. In such cases, the coefficients $c_s$ and $c_p$ can be determined from the experimentally determined parameters according to the relationships

$$c_s^2 = (A_\parallel + 2A_\perp)/3A_0, \tag{9a}$$

$$c_p^2 = (A_\parallel - A_\perp)/3B, \tag{9b}$$

where the $A_0$ and $B$ are, respectively, the isotropic and anisotropic hyperfine coupling constants for the free atom (or ion). Tabulations of these atomic coupling constants are available in the literature.[2,3,9] For different isotopes (i = 1, 2,...) of the same atom, both $A_0$(i) and $B$(i) are rigorously proportional to $\mu_N$(i)/$I$(i), where $\mu_N$(i) is the nuclear magnetic moment of the i$^{th}$ isotope and $I$(i) is its spin. Thus, with the aid of tabulated values of $\mu_I$(i) and making use of isotopically-enriched samples, the atom giving rise to the hyperfine interaction can be identified unambiguously.

## POWDER PATTERNS

It is clear from Eqs. (6) and (7) that an ESR spectrum in a crystal can generally be expected to exhibit an angular dependence. That is, the value of the resonance field is likely to depend on the angle between the axis of crystalline symmetry and the direction of the applied field **H**. Thus, by comparing the angular dependence of the spectrum with the known crystal symmetries, the exact site of the paramagnetic entity (e.g., substitutional, interstitial, etc.) can often be determined. For a sample consisting of a powder or glass, all possible angular orientations are simultaneously present, thus cutting off this valuable crystallographic information. However, there remains a rich information content in the spectrum which is *not* lost in consequence of this angular averaging. Rather, the effect is amenable to theoretical calculation, with the resulting line-shape envelope being termed a "powder pattern".

The calculation of a powder pattern is schematically illustrated in Fig. 2 for the special case of axial symmetry.

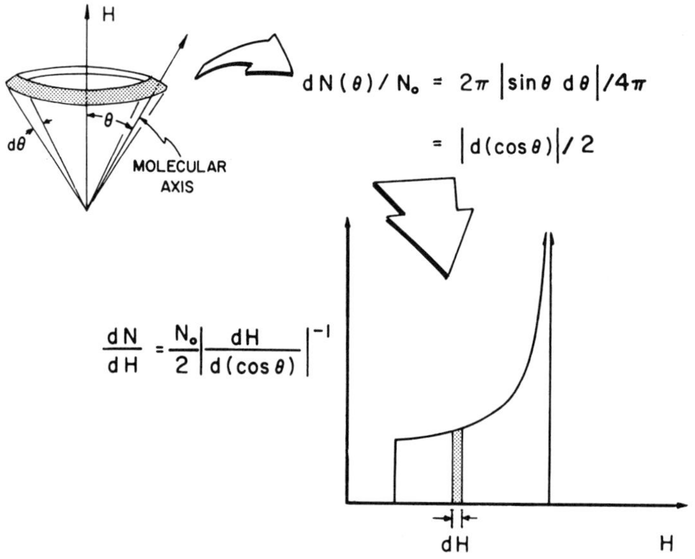


**Figure 2** Transformation of the probability of orientation in a differential element of solid angle dΩ to the probability of resonance absorption in a differential element of magnetic field d$H$ for an axially symmetric paramagnetic center.

If the total number of unpaired spins is $N_0$, the fractional number $dN/N_0$ undergoing resonance in the field interval $H \rightarrow H + dH$ must be equal to the fractional solid angle $d\Omega/4\pi$ for which the resonance condition gives solutions in this interval. (Here $H$ stands for the magnitude of the applied field **H**.) Since axial symmetry is assumed, it follows that $d\Omega = 2\pi|\sin\theta \, d\theta|$. Whence

$$dN/N_0 = d\Omega/4\pi = |d(\cos\theta)|/2. \tag{10}$$

The desired powder pattern $dN/dH$ can now be obtained in "closed form" by dividing both sides of Eq. (10) by the differential element of magnetic field:[11]

$$\frac{dN}{dH} = \frac{N_0}{2} \left| \frac{dH}{d(\cos\theta)} \right|^{-1} \tag{11}$$

Thus, the resonance condition of Eq. (6) can be differentiated with respect to $\cos\theta$ to obtain the desired powder pattern. In the case of no hyperfine interactions (i.e., $I = 0$), $\cos\theta$ can be algebraically eliminated from Eq. (11) after the differentiation is performed, thus making the right-hand side solely a function of $H$. This closed-form solution for $g_\parallel > g_\perp$ is illustrated as the dashed curve in Fig. 3a. Note the "step

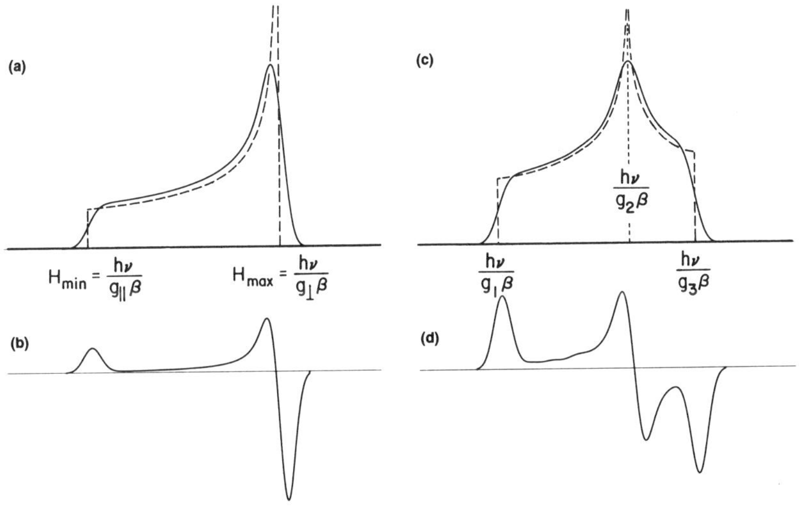


**Figure 3** Computer simulated ESR powder patterns (dashed curves) for the case $S = \frac{1}{2}$ and $I = 0$ with (a) axial and (c) orthorhombic $g$ matrices. Unbroken curves (a) and (c) result from convoluting with a Gaussian broadening function, while curves (b) and (d) are the respective first derivatives.

shoulder" at $H_{min} = h\nu/g_{\parallel}\beta$ and the mathematical divergence at $H_{max} = h\nu/g_{\perp}\beta$. (A similar but reversed pattern would be obtained for $g_{\perp} > g_{\parallel}$). The unbroken curve in Fig. 3a is the result of convoluting the dashed powder pattern with a Gaussian "single-crystal" broadening function. Fig. 3b shows the first derivative of the convoluted powder pattern, which corresponds to the usual experimental mode of observation. It is clear from Fig. 3a and b how the $g$ values, $g_{\parallel}$ and $g_{\perp}$, can be estimated from an experimental spectrum by inspection—and determined with even greater precision by computer line-shape simulation. It should be noted that in the case of non-zero hyperfine coupling, $\cos\theta$ cannot be eliminated from Eq. (11), but powder patterns can still be generated in point-by-point fashion by calculating $dN/dH$ and $H$ both as functions of $\cos\theta$.

Equation (6) can also be cast in a form appropriate for orthorhombic symmetry (all three $g$ values and/or $A$ values unequal), but in this case a simple algebraic expression like Eq. (11) cannot be obtained, and numerical methods[12,13] must be employed from the start. In the absence of a hyperfine interaction, the powder pattern for three unequal $g$ values shows step shoulders at $H_{min}$ and $H_{max}$, corresponding to the maximum and minimum $g$ values, respectively, and a divergence at an intermediate field corresponding to the intermediate $g$ value (see Fig. 3c and d).

## GLASS SPECTRA

The computed spectra of Fig. 3 are strictly applicable only to powdered crystals. From the standpoint of ESR, it is abundantly clear from years of research that glasses differ from powders in a fundamental way, namely, the discrete $g$ values and $A$ values that characterize a crystal must be replaced by statistical distributions in some or all of these parameters to properly reflect the influences of vitreous-state disorder.[14] ESR spectral simulations including such effects are generally performed by numerical methods on a computer.[13]

In Fig. 4a, the dashed curve is a computer simulation[15,16] of an experimental ESR first-derivative spectrum (unbroken curve) commonly observed in many irradiated borate glasses. Part b of this figure schematically illustrates the four overlapping hyperfine component powder patterns derivable from Eq. (6) by setting $M_I(^{11}B) = 3/2$, 1/2, -1/2, and -3/2. In combination, these four components lead to the computed shape. The principal-axis values of the orthorhombic hyperfine matrix are indicated as $A_1$, $A_2$, and $A_3$, and the values employed in the dashed "five-line" simulation are listed in Table 1. As in Fig. 3d, the mathematical divergences tend to correspond to zero-crossings and the step shoulders to positive or negative extrema in the derivative spectrum. However, Fig. 4 serves to illustrate how interferences between overlapping components can sometimes make the fitting process difficult. Figure 4 also brings home the important characteristic of glasses alluded to above: Statistical distributions in spin-Hamiltonian parameters endemic to the amorphous state can profoundly alter the line shapes which would otherwise be predicted for ordinary powdered crystals. In Fig. 4, a broad distribution in $g_3$ values of width $W$ is seen to result in a smearing together and obscuration of the four derivative extrema which would normally have been expected in association with the hyperfine splitting $A_3$. Indeed, this masking effect impeded the correct analysis of the borate-glass spectra for a number of years.

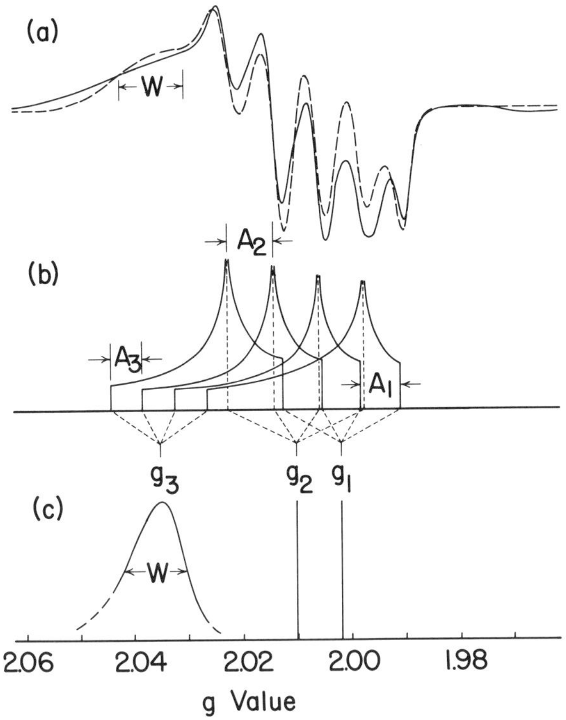


**Figure 4** Analysis of the "five-line-plus-a-shoulder" X-band ESR spectrum of irradiated alkali borate glasses of low alkali contents. (a) Experimental and computed spectra (unbroken and dashed curves, respectively). (b) "Single site" component powder patterns for hyperfine interaction with $^{11}B$ ($I = 3/2$). (c) $g$-value distribution used in simulation of (a). (From ref. 16.)

Table 1. Average $g$ Values and $^{11}B$ Hyperfine Coupling Constants for "Boron-Oxygen Hole Centers" in Alkali Borate Glasses, as Derived from Computer-Line-Shape-Simulation Analysis.[15]

| Spectrum Type | $g_1$ | $g_2$ | $g_3$ | $A_1$(mT) | $A_2$(mT) | $A_3$(mT) |
|---|---|---|---|---|---|---|
| "Five Line"[a] | 2.0020 | 2.0103 | 2.035 | 1.21 | 1.42 | 1.0 |
| "Four Line"[b] | 2.0049 | 2.0092 | 2.025 | 1.12 | 1.29 | 0.8 |

[a]Glass compositions containing ≤25 mol% alkali oxide.
[b]Glass compositions containing ≥35 mol% alkali oxide.

## THE ALKALI BORATES: A CASE STUDY

ESR studies of radiation-induced defect centers in glasses containing boron have been reported as early as 1953,[17] less than a decade after the inception of the ESR technique. The wealth of structural information which has since become available for the alkali borates, coupled with the long history of their study by ESR, renders this material system an excellent medium for framing the principles of the successful application of ESR to glasses and ceramics. The specific family of defect centers to be discussed has been termed the "boron-oxygen hole centers" because competitive trapping experiments[18] have shown the species to be of the trapped-hole type and because the spectra exhibit ubiquitous hyperfine interactions with boron(s) of such small magnitudes as to imply that in all cases the hole must be chiefly localized on (nonmagnetic) oxygens.[15,20]

Figure 4 has already illustrated and analyzed the familiar "five-line-plus-a-shoulder" spectrum observed at X-band frequencies ($\nu \approx 9$ GHz) in irradiated $x\mathrm{Me_2O}$ - $(1-x)\mathrm{B_2O_3}$ glasses for which $x \leq 0.25$ (Me ≡ alkali). Before the advent of the line-shape simulation analysis described above, the "five lines" were believed to have a separate origin from the shoulder in the region marked "*W*". But the successful computer fit of Fig. 4a was only part of the argument for the current interpretation of the spectrum. Indeed, it is often contended that a single spectral simulation such as that of Fig. 4 is not necessarily unique and that equally successful simulations might conceivably be achieved under the constraints of entirely different model hypotheses. In fact, the present interpretation of the borate-glass spectrum is the end product of a long-term process, involving many types of data and parallel lines of reasoning. A critical first step was the original demonstration that the "five lines" observed at X band in the low-alkali glasses were definitely due to a hyperfine interaction with $^{11}$B ($I$ = 3/2, 81 % natural abundance). This demonstration[20] was accomplished by preparing a sample enriched to 96 at% in $^{10}$B ($I$ = 3) which when irradiated exhibited a greater multiplicity of lines, with separations approximately 1/3 of the "five-line" splitting. This change in line spacings exactly followed expectation for hyperfine interactions with boron, since $[\mu_N(^{10}\mathrm{B})/I(^{10}\mathrm{B})]/[\mu_N(^{11}\mathrm{B})/I(^{11}\mathrm{B})] \approx 0.335$.

The next, and perhaps most crucial, step was the performance of the same experiments on polycrystalline compounds falling in the same composition range as the glasses—in this case, $\mathrm{Li_2O{\cdot}4B_2O_3}$ (1:4) and $\mathrm{Li_2O{\cdot}3B_2O_3}$ (1:3). The results for the 1:4 compound $\gamma$ irradiated at 77 K are shown in Figs. 5a and b for the $^{11}$B- and $^{10}$B-enriched samples, respectively.[15] Also shown in this figure are computer simulations employing a common set of principal $g$ values (Table 2) but with boron hyperfine coupling constants constrained to obey the relation $A_i(^{11}\mathrm{B})/A_i(^{10}\mathrm{B}) = [\mu_N(^{11}\mathrm{B})/I(^{11}\mathrm{B})]/[\mu_N(^{10}\mathrm{B})/I(^{10}\mathrm{B})]$. Since neither this ratio nor the known isotopic abundances are adjustable parameters, it becomes much less likely that *both* spectra could be simultaneously fitted with incorrect choices of $g$ and $A$ values than would be the case if only one of the two spectra had been simulated. Moreover, adding considerably to the confidence in the derived spin-Hamiltonian parameters was the similarly successful fit of spectra obtained at a much higher microwave frequency;

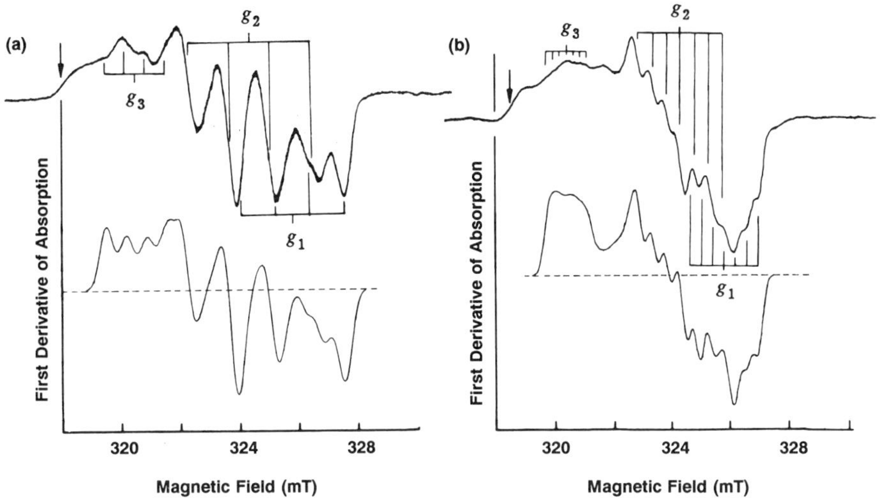


**Figure 5** X-band ESR spectra of $\gamma$-irradiated polycrystalline $Li_2O{\cdot}4B_2O_3$ enriched in (a) $^{11}B$ and (b) $^{10}B$. Upper traces are experimental; lower curves are computer simulations based on the "1:3" (Site 1) parameters of Table 2. (From ref. 15.)

Table 2. $g$ Values and $^{11}B$ Hyperfine Coupling Constants for "Boron-Oxygen Hole Centers" in Alkali and Alkaline-Earth Borate Polycrystalline Compounds, as Derived from Computer-Line-Shape Analysis.

| Spectrum Type | $g_1$ | $g_2$ | $g_3$ | $A_1$(mT) | $A_2$(mT) | $A_3$(mT) |
|---|---|---|---|---|---|---|
| "1:3"[a] (Site 1) | 2.0020 | 2.0103 | 2.035 | 1.22 | 1.44 | 0.68 |
| (Site 2) | 2.0020 | 2.0103 | 2.039 | 1.22 | 1.44 | 1.3 |
| "1:2"[b] | 2.003 | 2.009 | 2.022 | 0.51 | 0.60 | 0.29 |
| "2:1"[c] | 2.0048 | 2.0081 | 2.020 | 1.02 | 1.29 | 0.69 |
| "?"[d] | | 2.0091 | | 1.1 | 1.0 | 0.9 |

[a]Data for $Li_2O{\cdot}4B_2O_3$ (ref. 15).
[b]Data: $Li_2O{\cdot}2B_2O_3$. Hyperfine interactions are with *two equivalent* borons (ref. 22).
[c]Data for $2CaO{\cdot}B_2O_3$ (ref. 25).
[d]Data for both $CaO{\cdot}B_2O_3$ and $2CaO{\cdot}B_2O_3$ following anneal at 200 C (ref. 25).

Figure 6a illustrates the entirely different character of the boron-oxygen hole center spectrum of $Li_2O{\cdot}4^{11}B_2O_3$ at $\nu \approx 35$ GHz. Note how the $^{11}B$ hyperfine "combs" are "pulled apart" in Fig. 6, as compared to Fig. 5. Inspection of Eq. (6) reveals that the spacing between any pair of $g$-value features should be proportional to the microwave

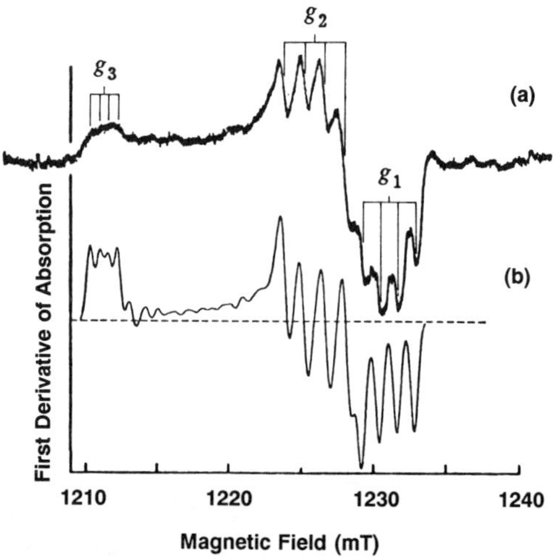


**Figure 6** Ka-band (35 GHz) ESR spectrum of $^{11}B$-enriched polycrystalline $Li_2O{\cdot}4B_2O_3$ $\gamma$-irradiated at 77 K. (a) Experimental. (b) Computer simulation based on "1:3" (Site 1) parameters of Table 2. (After ref. 15.)

frequency $\nu$. And, indeed, the detailed predictions of Eq. (6) are borne out by the computer simulation of Fig. 6b which was accomplished using the same spin-Hamiltonian parameters as were employed in the X-band simulation of Fig. 5a.

The success of the simulations of Fig. 5 in the "five-line" region is striking. Moreover, the structure and overall extent of the "$g_3$ region", though imperfectly simulated here, is seen to depend on boron isotope. (Somewhat improved simulations were achieved[15] by assuming the existence of two chemically distinct sites differing mainly in the values of $g_3$ and $A_3$; see Table 2). Therefore the premise that all parts of the resonance signal arise from a single defect species is well supported by the simulations of the spectra of the polycrystalline compounds.

Since the glass of Fig. 4 and the compound of Fig. 5 have very nearly the same chemical composition while the "five-line" parts of Figs. 4a and 5a are virtually identical, the detailed analyses of the 1:4-compound spectra provided additional support for the interpretation of the glass spectrum schematically laid out in Fig. 4. That is, the region marked "*W*" in Fig. 4 is an integral part of the "five-line" spectrum, this part exhibiting resolved hfs in the polycrystalline materials but becoming smeared out in the glasses due to the influences of vitreous-state disorder.

A spectrum similar to that of Fig. 5a was also obtained for irradiated polycrystalline lithium triborate (1:3), while totally different spectra were recorded for the diborate (1:2) and metaborate (1:1) compounds.[15] Since the 1:4 and 1:3 compounds are known[21] to share a common structural grouping—that is, the triborate ring of Fig. 7—it was

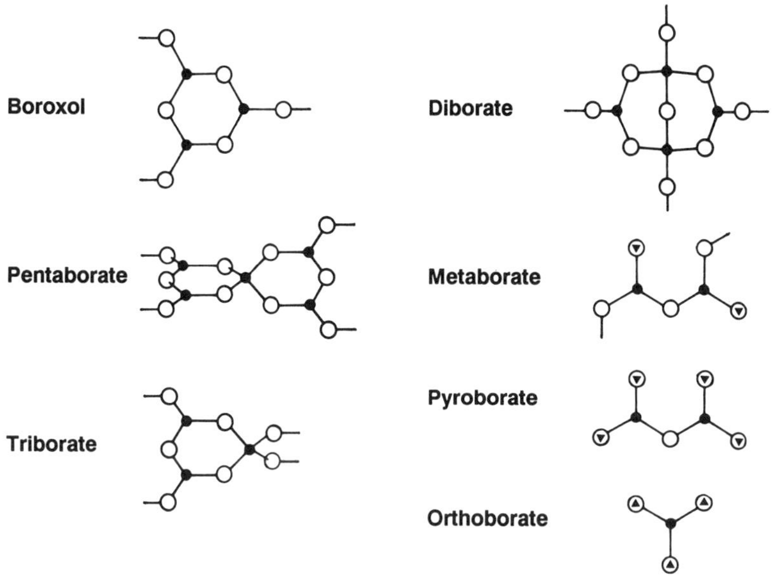


**Figure 7** Structural groupings characteristic of alkali and alkaline-earth borate compounds. Nonbridging oxygens are demarked by triangles. (After ref. 21.)

inferred that the "five-line" spectrum must be associated with this ring. For this reason, the defect giving rise to this spectrum was termed[22] the "1:3" center. The most promising model for this defect is that of a hole trapped in a $2p$ orbital of a single ring oxygen in the triborate structure (Fig. 8a), with the measured hyperfine interaction involving the adjacent four-coordinated boron. (A hyperfine interaction with the nearest-neighbor three-coordinated boron is also expected, but this interaction is presently believed to be too small to be measured experimentally.) An abundance of evidence from unrelated structural studies supports the presence of triborate rings in the low-alkali glasses (for a review, see ref. 23). Thus, the "1:3" center would appear to be an independent indicator of triborate groupings in borate glasses containing $\leq 25$ mol % alkali oxide.

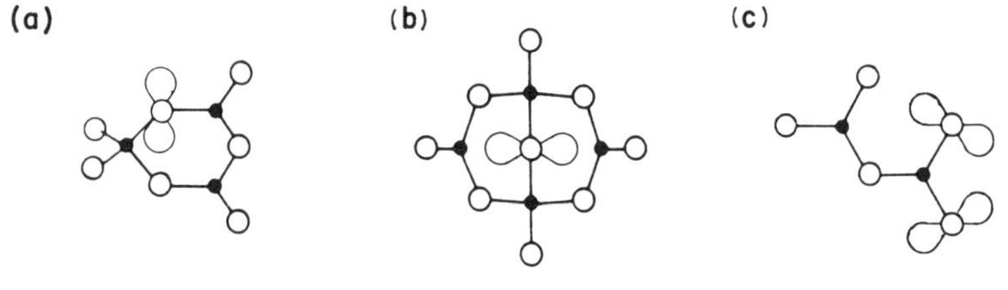


**Figure 8** Postulated models for (a) the "1:3" center, (b) the "1:2" center, and (c) the "2:1" center. The unpaired electron occupies the oxygen $2p$ orbitals indicated.

As illustrated in Fig. 9,[24] the nature of the room-temperature stable, $\gamma$-ray induced spectra in alkali borate glasses changes dramatically as the alkali content is raised above 25 mol%. An almost symmetrical "four-line" spectrum (Fig. 9e) was obtained for a 40%$K_2O$-60%$B_2O_3$ glass after a thermal treatment to remove two peaks indicated by arrows in Fig. 9d. The origin of the "four-line" spectrum was further explored[15] by $\gamma$ irradiating a 50%$Li_2O$-50%$B_2O_3$ glass at 77 K and computer simulating the spectrum observed before warming (Fig. 10). To the uninitiated, the spin-Hamiltonian parameters which effected the computer fit of Fig. 10 might seem to differ little from those derived in simulating the "five-line" spectrum of Fig. 4 (see Table 1). But, in fact, the broader distribution of $g_3$ values and the greater $g_1$ shift, $g_1$-$g_0$, apparent in Fig. 10b vis-à-vis Fig. 4c foreshadow a model for the "four-line" center differing profoundly from that presented above for the "five-line" center.

As in the case of the "five-line" spectrum, the final breakthrough in interpreting the "four-line" spectrum derives from parallel studies of polycrystalline compounds. Only a hint of a "four-line" spectrum was detected in the 1:1 lithium borate compound, and it was entirely absent in the 1:2 compound. Thus, it was tentatively concluded[15] that *none* of the crystalline compounds of the *lithium* borate system contain the structural groupings which serve as trapping sites for the "four-line" center in the glasses.

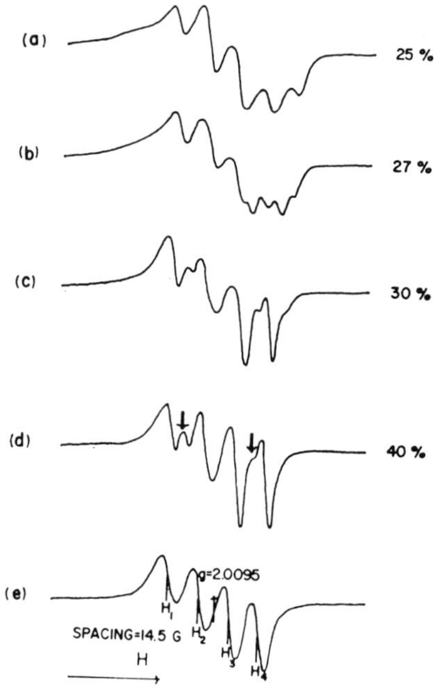


**Figure 9** X-band ESR spectra of a series of potassium borate glasses $\gamma$-irradiated at room temperature. (a) 25 mol% $K_2O$. (b) 27 mol% $K_2O$. (c) 30 mol% $K_2O$. (d) 40 mol% $K_2O$. (e) As (d), but annealed at 200 C for 90 min (From ref. 24.)

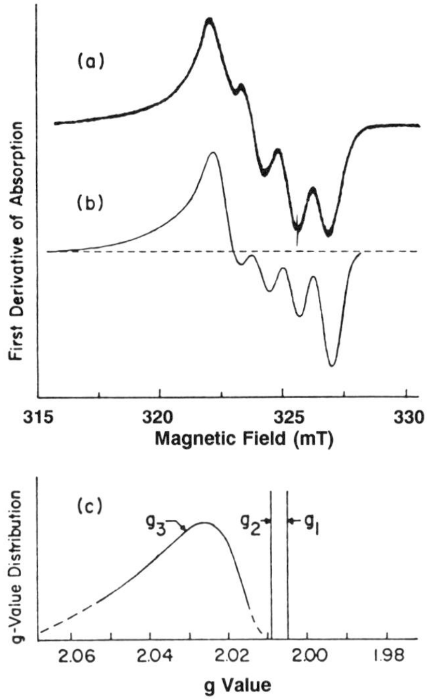


**Figure 10** X-band ESR spectrum of a 50% $Li_2O$-50% $B_2O_3$ glass $\gamma$ irradiated and observed at 77 K. (a) Experimental. (b) Computer simulated based on the $g$-value distribution of (c) and the other "four-line" parameters of Table 1. (From ref. 15.)

Subsequent studies[25] of x-irradiated calcium metaborate (1:1) and calcium pyroborate (2:1) compounds revealed a "four-line" spectrum in the latter (see Fig. 11a) but not the former. The spectra of irradiated calcium pyroborate compounds containing 82% $^{11}B$ and 95% $^{10}B$ were computer simulated (dotted curves in Figs. 11a and 12a, respectively) using spin-Hamiltonian parameters virtually identical with those derived from the simulation of the high-alkali glass spectrum of Fig. 10b (compare the parameters of the "four-line" spectrum in Table 1 with those of the "2:1" spectrum in Table 2). Indeed, the main difference between the "four-line" spectrum in the glasses and the "four-line" spectrum in the "2:1" compound was the requirement of discrete values of $g_3$ to simulate the compound spectra vis-à-vis the broad distribution of $g_3$ values which characterized the spectrum in the glasses (Fig. 10c). This clue that the "four-line" spectrum may be associated with pyroborate groupings was then combined with a theoretical interpretation of the $g$ values (discussed elsewhere[1] in connection with the $HC_2$ defect in alkali silicate glasses) to suggest[25] a model of a hole delocalized over a pair of nonbridging oxygens bonded to the same boron (Fig 8c). It is on this basis that the defect giving rise to the "four-line" spectrum is termed the "2:1" center.

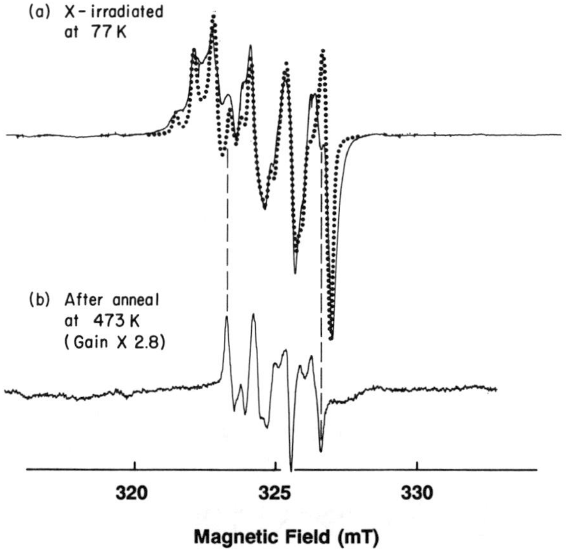


**Figure 11** X-band ESR spectrum of polycrystalline $2CaO{\cdot}B_2O_3$ (82% $^{11}B$) x-irradiated at 77 K. (a) Recorded without warming. (b) After anneal to 200 C. Dotted curve is a computer simulation using "2:1" parameters of Table 2. (After ref. 25.)

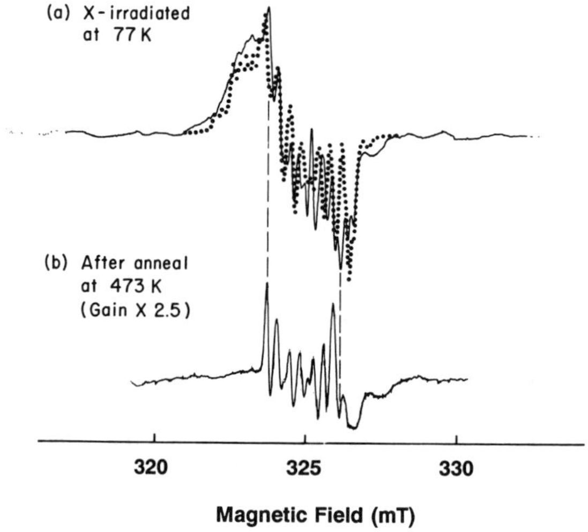


**Figure 12** X-band ESR spectrum of polycrystalline $2CaO{\cdot}B_2O_3$ (95% $^{10}B$) x-irradiated at 77 K. (a) Recorded without warming. (b) After anneal to 200 C. Dotted curve is a computer simulation using "2:1" parameters of Table 2. (After ref. 25.)

For completeness, it is appropriate to comment on the narrower, rather symmetrical four-line spectrum which remains stable in the calcium pyroborate compound after an anneal to 473 K (Fig. 11b). This spectrum is clearly different from the "four-line" spectrum computer simulated in Fig. 11a and tagged as the "2:1" center. As suggested by the vertical dashed lines in Fig. 11, the "narrow four-line" spectrum may be discernable as weak peaks between the first and second and the third and fourth main peaks in the spectrum recorded at 77 K before annealing. [A spectrally similar "narrow four-line" resonance may be present in the alkali borate glasses (see peaks indicated by arrows in the spectrum of Fig. 9d), although in view of their differing thermal stabilities the two manifestations may have totally separate origins.] While no model has yet been postulated for the "narrow four-line" spectrum of Fig. 11b, some data can be cited which constrain the possibilities. First, the four lines are definitely $^{11}B$ hyperfine structure, as verified by $^{10}B$ isotopic substitution (Fig. 12b). Second, the identical spectrum was also observed[25] in calcium *meta*borate after x irradiation and an anneal to 473 K. Thus, the resonance is not necessarily to be associated with a pyroborate unit. Third, a very different resonance has been unambiguously identified in (unirradiated) potassium borate-potassium carbonate ceramics as arising from a hole delocalized over three nonbridging oxygens of an *ortho*borate unit.[26] This eliminates such a model for the present spectrum. Finally, the nearly isotropic spin Hamiltonian fitted to the "narrow four-line" spectrum (labeled "?" in Table 2) seems difficult to reconcile theoretically with a model of a hole trapped on a *single* nonbridging oxygen (see discussion of $g$ values of the $HC_1$ center in ref. 1), which would otherwise be inferred if the defect were to be associated with a metaborate unit. The origin of the "narrow four-line" spectrum therefore remains an enigma.

Returning to the lower-alkali compositions, the spectrum of the "1:2" center observed in the lithium diborate compound was successfully computer simulated[22] by assuming a hyperfine interaction with *two equivalent* borons (see spin-Hamiltonian parameters in Table 2). This outcome led naturally to the model of the "1:2" center as a hole trapped on the oxygen bridging between the two four-coordinated borons of the diborate ring (Fig. 8b). Such a conclusion was strongly supported by the observation[22] of "1:2-type" centers in polycrystalline strontium diborate where all of the borons are known to be four-coordinated. Hyperfine structure attributable to the "1:2" center is not resolvable in the ESR spectra of irradiated alkali borate *glasses*, but computer simulation studies[27] strongly suggest its presence as an underlying component. The quantitative results of these computer decompositions are illustrated in Fig. 13a as a partitioning of the total[28] boron-oxygen hole center intensity into the contributions of "1:3" and "1:2" centers for alkali contents from 5 to 33 mol %. To date, no comparable attempt has been made to separate the contribution of the "2:1" centers, which clearly becomes the dominant species at higher alkali oxide contents (Fig. 9).

The ESR defect-center yields of Fig. 13a can be compared[23] with the numbers of fundamental structural groupings known or postulated[21] to be present in these glasses as illustrated in Fig. 13b. The correlations of the "1:3" and "1:2" centers with the predicted numbers of tetraborate and diborate groups is reasonable, given that the overall hole-trapping efficiency is seen to be a strong function of composition. Presumably the "2:1" center (which is modelled as a hole trapped on two non-bridging oxygens) tracks the nonbridging oxygen (NBO) population up to ~45 mol % alkali oxide. Above this composition, the number of NBOs continues to increase monotonically, while the overall trapped-hole center concentration (presumably comprising mainly "2:1" centers) is seen to fall off—possibly due to the available electron-trapping sites becoming increasingly shallow as the alkali oxide content increases above ~50 mol %. (Note that there must be one trapped electron for each trapped hole.)

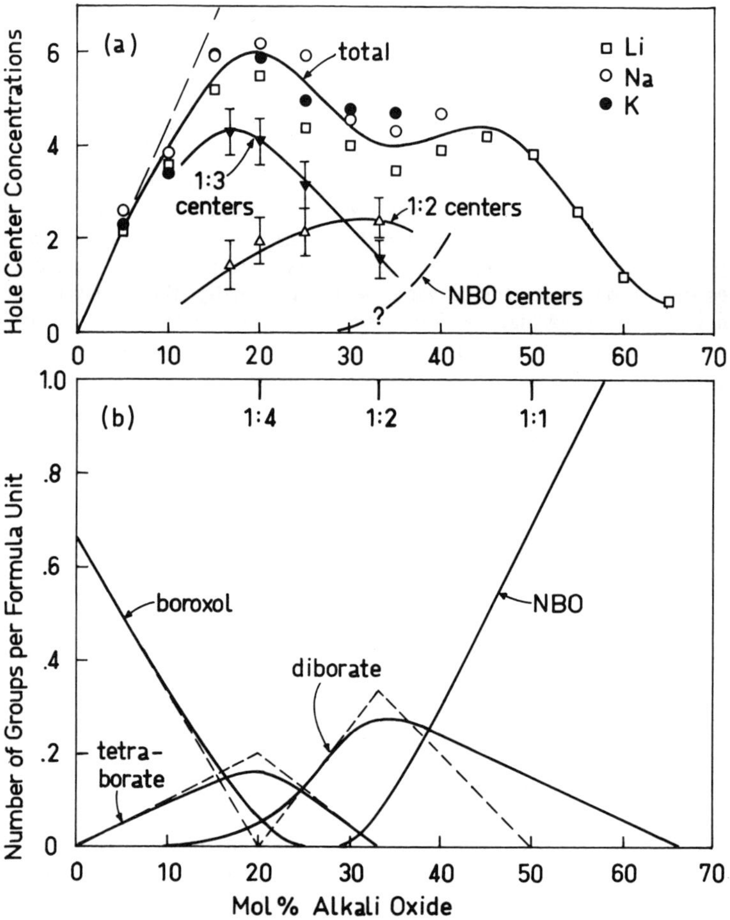


**Figure 13** Comparison of ESR intensity data for radiation-induced trapped-hole centers (a) with predicted numbers of structural groupings (b) in alkali borate glasses. Circles and squares in (a) are *total* boron-oxygen hole center concentrations (ref. 28); triangles indicate partitioning into "1:3" and "1:2" centers by computer simulation (ref. 27). Sources of data for (b) are cited in ref. 23. (From ref. 23).

The present case study illuminates many facets of the performance of an ESR investigation of an irradiated glass system. Perhaps the foremost lesson to be drawn is that some problems do not solve themselves quickly. Often they must by gradually worn down in the course of several interlocking experimental and theoretical assaults over a period of years. Such has been the history of the fundamental trapped-hole centers in the alkali borate glass system. However, given the large number of characteristic vitreous-state ESR spectra which now have been thoroughly analyzed in the literature, the day is fast approaching when ESR will become a more routine analytical tool for the study of defects in glasses. Some recent examples of problem solving based on previously elucidated defect-center spectra are noted in the following section.

## OTHER CASES: A NON-OXIDE GLASS AND A GLASS CERAMIC

One of the two concluding studies can trace its genesis back to the alkali borate work described above. This history began with the study of radiation effects in alkali borate glasses containing alkali halide additions. ESR spectra recorded for these materials were distinct from those observed in halide-free borate glasses and varied dramatically in character when one halogen was substituted for another.[18] In some of the potassium chloride-potassium borate glasses investigated, the new spectrum appeared with a far greater intensity than the usual boron-oxygen hole centers described above. This spectrum, reproduced in Fig. 14b, was analyzed in terms of the theoretical powder patterns for the $Cl_2^-$ molecular ion. These powder patterns (inset to Fig. 14) were originally calculated from Eqs. (6), (7), and (11) in a point-by-point manner for a sequence of values of $\cos\theta$, exactly as described above. It is to be noted that there are two magnetic isotopes of chlorine, $^{35}Cl$ (75 % abundant) and $^{37}Cl$ (25 % abundant), both having nuclear spins $I = 3/2$. The atomic coupling constant for $^{35}Cl$ is larger than that for $^{37}Cl$ by a factor of 1.2. Statistically, $^{35}Cl$-$^{35}Cl$, $^{35}Cl$-$^{37}Cl$, and $^{37}Cl$-$^{37}Cl$ pairs must occur in the ratios 9:6:1. The powder patterns shown in the inset to Fig. 14 were generated[29] only for the most probable case of $^{35}Cl$-$^{35}Cl$ pairs by treating the hyperfine interaction as though it involved a single "merged" nucleus of spin $I = I_1+I_2 = 3$. This is a rigorously valid approach for like nuclei, in so far as the *shapes* of the hyperfine components are concerned. However, the underlying reality that there are two nuclei with $I_1 = I_2 = 3/2$ imposes itself in the relative *intensities* of the components (see, e.g., ref. 8). This is because there is only one way to achieve $M_I = 3$ (i.e., $M_1 = M_2 = 3/2$), while there are two ways to achieve $M_I = 2$ (i.e., $M_1 = 3/2$, $M_2 = 1/2$ or $M_1 = 1/2$, $M_2 = 3/2$), three ways to get $M_I = 1$, and so on, so that the seven hyperfine components corresponding to $M_I = 3, 2, 1, 0, -1, -2, -3$ must be artificially weighted 1:2:3:4:3:2:1 in order to reproduce a true $Cl_2^-$ envelope.

The correspondence of the positions and relative amplitudes of the powder patterns calculated for the $^{35}Cl$-$^{35}Cl$ case with the principal features of the experimental spectrum of Fig. 14b was adequate to postulate the origin of the spectrum as arising from $Cl_2^-$ molecular ions—noting that additional weak structure in the experimental spectrum concentric with the "35-35" lines quantitatively matches the predicted positions of the lines due to "35-37" and "35-35" pairs. However, to complete the analysis, the derivative spectrum of Fig. 14b was subsequently computer simulated[18] by numerical methods (Fig. 14c) using the appropriate resonance condition[30] for the

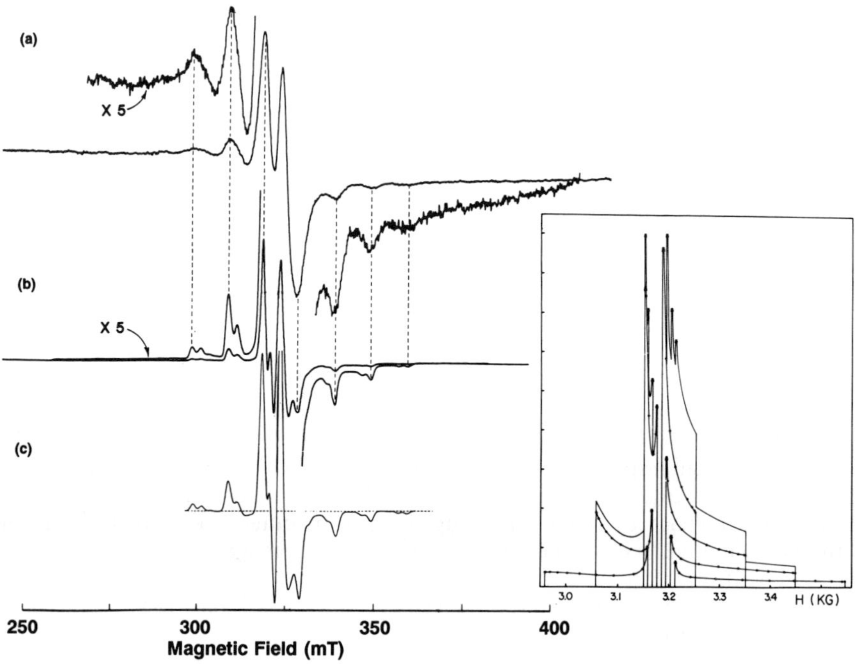


**Figure 14** X-band ESR spectra of $Cl_2^-$ molecular ions (a) in a Cl-doped $ZrF_4$-based glass (ref. 32), (b) in a KCl-potassium borate glass, and (c) as computer simulated (ref. 18). Inset shows calculated powder patterns for $^{35}Cl$-$^{35}Cl$ case.

mixed-isotope pairs and adding all three contributions in the required 9:6:1 ratio. It was noted[18] that the spin-Hamiltonian parameters for $Cl_2^-$ in glass are very close to those for $Cl_2^-$ ($V_K$ centers)[31] in alkali halide crystals (Table 3).

Table 3. $g$ values and $^{35}Cl$ Hyperfine Coupling Constants for $Cl_2^-$ Molecular Ions in Alkali Halide-Alkali Borate Glasses and in Alkali Halide Crystals

| Material | $g_\parallel$ | $g_\perp$ | $A_\parallel$(mT) | $A_\perp$(mT) |
|---|---|---|---|---|
| Glasses[a] | 2.0042 | 2.035 | 10.0 | 0.8 |
| KCl[b] | 2.0010 | 2.044 | 10.1 | 0.9 |
| NaCl[b] | 2.0010 | 2.046 | 9.8 | 0.9 |

[a]LiCl-$B_2O_3$, NaCl-$B_2O_3$, and KCl-$B_2O_3$ (ref. 18).
[b](ref. 31).

The present, rather complete understanding of the $Cl_2^-$ defect in the alkali chloride-alkali borate glasses hinged on a substantial body of work. However, the results of this effort left the door wide open for the immediate identification of an unanticipated radiation-induced ESR spectrum recorded some 16 years later for a number of glasses based on $ZrF_4$ and $HfF_4$ (Fig. 14a).[32] The correspondence of the seven principal peaks of the unknown spectrum with those of the previously analyzed $Cl_2^-$ spectrum (emphasized by the vertical dashed lines), coupled with the fact that the seven-line spectrum was observed in chloride-doped fluoride glasses but not in chloride-free samples, was all that was required to identify the $Cl_2^-$ center in the fluoride glasses. In fact, this ESR spectral signature is now widely used as a diagnostic for chloride impurities in heavy-metal fluoride glasses.[4,32-34]

A useful tip for practitioners of ESR in glasses can be injected at this point: Neither of the experimental spectra of Fig. 14 could have been conclusively identified without observation of *all seven* of the principal hyperfine peaks. It can be seen that this accomplishment required very broad magnetic-field scans obtained at spectrometer gains greatly exceeding those needed to record the strong central part of the spectrum. Indeed, the first report[35] of a spectrum in fluoride glasses similar to that of Fig. 14a did not recognize the weaker satellite lines as being associated with the two strongest peaks, which were consequently described as a "doublet" of unknown origin.

The final case study to be described involves radiation-induced defects in low-thermal-expansion glass ceramics. For example, Zerodur (a product of Schott Glass Technologies, Inc.) is a glass-ceramic material composed of 80% lithium silicate crystals ($\beta$ eucryptite and $\beta$ quartz) and 20% high-silica glass. This material owes its low coefficient of thermal expansion (CTE) to the fact that the CTE of the crystalline and glassy phases are of the same magnitude but are opposite in algebraic sign. Many applications of this and similar low-CTE materials are anticipated (e.g., as mirror substrates for space telescopes) where exposures to ionizing radiations may routinely occur. The questions must then be asked, (i) what are the effects of radiation on the bulk properties and (ii) what are the microscopic origins of the bulk property changes? To answer the second question, several papers[36-38] have addressed the induced ESR spectra and property changes in irradiated Zerodur and other low-CTE glass ceramics.

Figure 15 illustrates the X-band ESR spectra of Zerodur following 100-keV x irradiation. Spectral lines attributable to $Ti^{3+}$ and/or $Zr^{3+}$, oxygen hole centers (OHCs), and $Fe^{3+}$ are recognizable in the traditional first-derivative recording of Fig. 15a (the well-known $Fe^{3+}$ spectrum[39] was present before irradiation). The high-power second-harmonic-mode spectra of Fig. 15b manifest themselves in a form essentially indistinguishable from the undifferentiated absorption curves. Detection at the second-harmonic of the field-modulation frequency while applying high microwave powers is a useful "trick" for obtaining higher signal-to-noise ratios for defect centers which are sensitive to microwave saturation in the normal first-derivative mode.[40] Thus, the spectrum of the $Fe^{3+}$ ion, which is *not* very susceptible to microwave saturation, is clearly observable in Fig. 15a but is almost totally suppressed in Fig. 15b. By contrast, the very broad, but weak spectral "oscillations" seen in Fig. 15a are clearly revealed as the succession of well-defined peaks marked

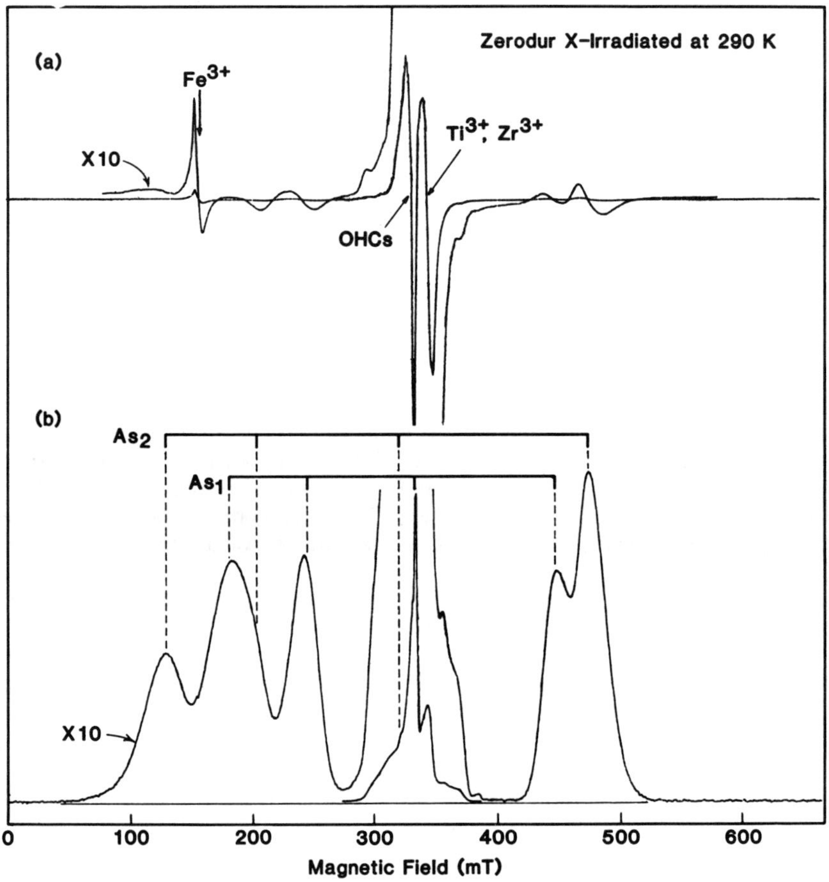


**Figure 15** (a) First-harmonic and (b) high-power second-harmonic X-band ESR spectra of x-irradiated Zerodur. (After ref. 37.)

by the combs in Fig. 15b; these are ascribable to two separate arsenic-related defect centers, $As_1$ and $As_2$ ($I$ = 3/2 for $^{75}As$). It has been shown[37] that the concentration of OHCs correlates well with the fractional increase in bulk density as functions of both x-ray dose and isochronal anneal temperature. A subsequent ESR study[38] of a special series of samples of varying degrees of crystallization ranging from the as-quenched glass of Zerodur composition to the fully ceramed commercial product gave indications of the phases (glassy versus crystalline) in which the various defect centers are primarily located (Fig. 16). In particular, it can be seen that for equal conditions of irradiation the number of OHCs decreases on going from the glass to the glass-ceramic. On this basis it has been suggested[37] that radiation-induced compaction in Zerodur is primarily due to OHC formation (bond breakage) in the glassy phase.

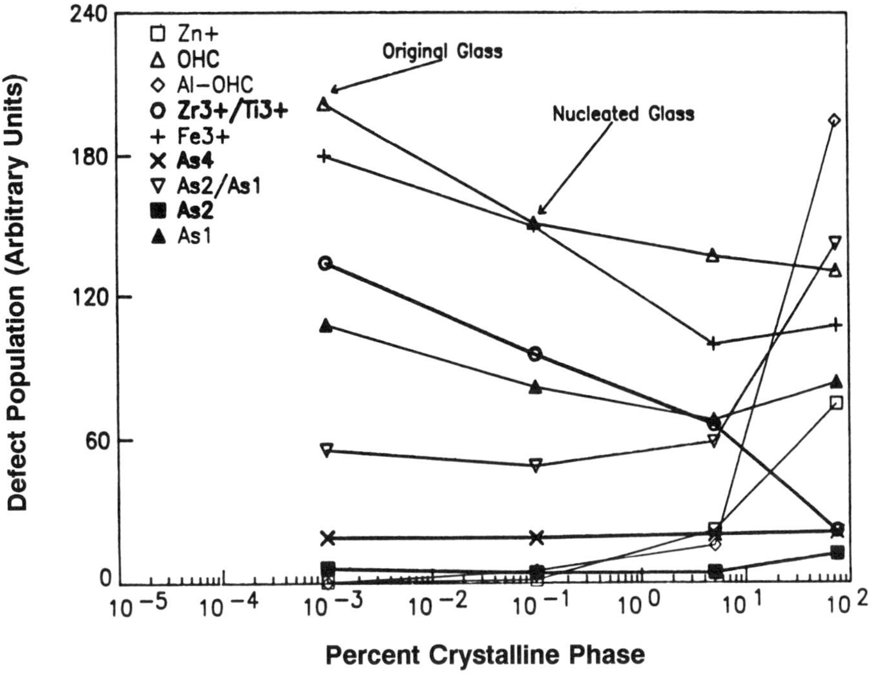


**Figure 16** Relative concentrations of paramagnetic defect centers as a function of the degree of crystallization of a series of identically-irradiated glass ceramics having the composition of Zerodur. The percentage of crystalline phases in the original glass was arbitrarily taken to be $10^{-3}$. (After ref. 38.)

## References


1. D.L. Griscom, J. Non-Cryst. Solids **31**, 241-266 (1978).
2. H. Kawazoe, J. Non-Cryst. Solids **71**, 231-243 (1985).
3. D.L. Griscom, in *Glass: Science and Technology Vol. 4B*, Edited by D.R. Uhlmann and N.J. Kreidl, (Academic Press, Boston, 1990), pp. 151-251.
4. D.L. Griscom and E.J. Friebele, in *Fluoride Glass Optical Fibers*, Edited by I. Aggarwal and G. Lu, (Academic Press, Boston, 1990).
5. H. Kawazoe, H. Hosono, H. Kokumai, J. Nishii, and T. Kanazawa, J. Non-Cryst. Solids **40**, 291-303 (1980).
6. D.L. Griscom, J. Non-Cryst. Solids **42**, 287-296 (1980).
7. D.L. Griscom, J. Non-Cryst. Solids **67**, 81-118 (1984).

8. G.E. Pake and T.L. Estle, *The Physical Principles of Electron Paramagnetic Resonance, 2nd Edition*, (W.A. Benjamin, Inc., Reading, MA, 1973).
9. J.E. Wertz and J.R. Bolton, *Electron Spin Resonance Elementary Theory and Practical Applications*, (McGraw-Hill, New York, 1972).
10. C.P. Poole, Jr., *Electron Spin Resonance: A Comprehensive Treatise on Experimental Technique 2nd Edition*, (John Wiley & Sons, New York, 1983).
11. R.H. Sands, Phys. Rev. **99**, 1222-1226 (1955).
12. J.H. Mackey, M. Kopp, E.C. Tynan, and T.F. Yen, in *Electron Spin Resonance of Metal Complexes*, (Plenum Press, New York, 1969), pp. 33-57.
13. P.C. Taylor and P.J. Bray, J. Mag. Res. **2**, 305-331 (1970).
14. D.L. Griscom and R.E. Griscom, J. Chem. Phys. **47**, 2711-2722 (1967).
15. D.L. Griscom, P.C. Taylor, D.A. Ware, and P.J. Bray, J. Chem. Phys. **48**, 5158-5173 (1968).
16. D.L. Griscom, J. Non-Cryst. Solids **13**, 251-285 (1973).
17. E.L. Yasaitis and B. Smaller, Phys. Rev. **92**, 1068 (1953).
18. D.L. Griscom, P.C. Taylor and P.J. Bray, J. Chem. Phys. **50**, 977-983 (1969).
20. S. Lee and P.J. Bray, J. Chem. Phys. **39**, 2863-2873 (1963).
21. J. Krogh-Moe, Phys. Chem. Glasses **6**, 46-54 (1965).
22. P.C. Taylor and D.L. Griscom, J. Chem. Phys. **55**, 3610-3611 (1971).
23. D.L. Griscom, in *Borate Glasses: Structure, Properties, Applications*, Edited by L.D. Pye, V.D. Frechette, and N.J. Kreidl, (Plenum Publishing Corp., New York, 1978), pp. 11-149.
24. S. Lee and P.J. Bray, J. Chem. Phys. **40**, 2982-2988 (1964).
25. D.L. Griscom and H.M. Kriz, Am. Cer. Soc. Bull. **58**, 381 (1979), (otherwise unpublished).
26. P.C. Taylor, D.L. Griscom, and P.J. Bray, J. Chem. Phys. **54**, 748-760 (1971).
27. P.C. Taylor and P.J. Bray, J. Am. Ceram. Soc. **51**, 234-239 (1972).
28. S. Lee, *Ph.D. Thesis*, Brown University, 1964.
29. D.L. Griscom and P.J. Bray, (unpublished, 1965).
30. D. Schoemaker, Phys. Rev. **149**, 693-704 (1966).
31. T.G. Castner and W. Känzig, J. Phys. Chem. Solids **3**, 178 (1957).
32. D.L. Griscom and D.C. Tran, J. Non-Cryst. Solids **72**, 159-163 (1985).
33. K. Tanimura, W.A. Sibley, M. Drexhage and M. Suscavage, J. Appl. Phys. **58**, 4544-4552 (1985).
34. R.N. Schwartz, M. Robinson, and G.L. Tangonan, Materials Science Forum **19-20**, 275-285 (1987).
35. P.E. Fisanich, L.E. Halliburton, L.N. Feuerhelm, and S.M. Sibley, J. Non-Cryst. Solids **70**, 37-44 (1985).
36. N. Koumvakalis, M.G. Jani and L.E. Halliburton, Appl. Opt. **25**, 4288-4293 (1986).
37. T.-E. Tsai, P.L. Higby, E.J. Friebele, and D.L. Griscom, J. Appl. Phys. **62**, 3488-3490 (1987).
38. T.-E. Tsai, E.J. Friebele, D.L. Griscom, and W. Pannhorst, J. Appl. Phys. **65**, 507-514 (1989).
39. T. Castner, Jr., G.S. Newell, W.C. Holton, and C.P. Slichter, J. Chem. Phys. **32**, 668-673 (1960).
40. D.L. Griscom, Nucl. Inst. & Methods **B1**, 481-488 (1984).

# CHAPTER 7

## INDUCED OPTICAL EFFECTS IN GLASS

**Satyam C. Cherukuri**
David Sarnoff Research Center
Princeton, NJ 08543

**I. Joseph, L. David Pye**
Institute of Glass Science and Engineering
NYSCC, Alfred University
Alfred, NY 14802

## INTRODUCTION

Glass, annealed under ideal conditions and in the absence of external fields, is isotropic to the propagation of light. In isotropic materials the permittivity $\varepsilon$ relates the displacement field D to the electric field E by the equation:

$$D = \varepsilon E \qquad (1)$$

where $\varepsilon = \varepsilon_r \varepsilon_0$; $\varepsilon_r$ is the relative dielectric constant (relative permittivity), and $\varepsilon_0$ is the permittivity of free space. The polarization P is related to E through the dielectric susceptibility $\chi$ via:

$$P = \varepsilon_0 \chi E \qquad (2)$$

with

$$\varepsilon = \varepsilon_0 (1 + \chi) \qquad (3)$$

The index of refraction in isotropic materials at optical frequencies is related to the frequency dependent dielectric constant by the equation:

$$n = \sqrt{\varepsilon_r} \tag{4}$$

Applied electric or magnetic fields, or stress, introduce anisotropy into the glass. To describe the behavior of propagation of light in anisotropic materials, $\varepsilon$ and $\chi$ must be treated as tensors rather than scalars. The most general expression relating each of the three axial components of D to E is[1]:

$$D_i = \varepsilon_0 \textstyle\sum_j \varepsilon_{ij} E_j \qquad (i, j = x_1, x_2, x_3; \text{ principal axes}) \tag{5}$$

or in expanded matrix form:

$$\begin{bmatrix} D1 \\ D2 \\ D3 \end{bmatrix} = \varepsilon_0 \begin{bmatrix} \varepsilon_{11} & \varepsilon_{12} & \varepsilon_{13} \\ \varepsilon_{21} & \varepsilon_{22} & \varepsilon_{23} \\ \varepsilon_{31} & \varepsilon_{32} & \varepsilon_{33} \end{bmatrix} \begin{bmatrix} E1 \\ E2 \\ E3 \end{bmatrix} \tag{6}$$

For noncrystalline materials the dielectric tensor ($\varepsilon_{ij}$) takes the form:

$$\varepsilon_{ij} = \varepsilon_r, \text{ for } i=j \text{ ; and } \varepsilon_{ij} = 0, \text{ for } i \neq j \tag{7}$$

The reciprocal of relative permittivity is referred to as relative dielectric impermeability ($\xi$) and is defined by[2]:

$$\xi_{ij} = \varepsilon_0 \, \partial E_i/\partial D_j \tag{8}$$

By applying stress, or electric, or magnetic fields, the symmetry of $\varepsilon_{ij}$ is relaxed leading to non-equal diagonal elements or non-zero off-diagonal elements in the dielectric tensor. The ultimate symmetry of $\varepsilon_{ij}$ reflects the symmetry of the applied fields. The asymmetry induced by the applied fields leads to the nonlinear optical properties of glass. The following discussion will be limited to the treatment of the behavior of glass in uniaxial applied fields only. The major effects induced in glass by applied fields are given in Table 1. The linear electro-optic (first order) effect is not observed in glasses due to the presence of the center of symmetry. The quadratic magneto-optic effect, also known as the Cotton-Mouton effect, has been observed in glasses but not much data are available in the literature. In this paper, emphasis will be given to the stress-optical, Kerr, and Faraday effects which are the most prominent of all induced optical effects in glass.

Table 1. Induced optical effects in glass.

| FIELD | EFFECT | RESULT |
|---|---|---|
| Stress | Stress-optical Effect<br>Strain-optical Effect<br>(PHOTOELASTIC EFFECTS) | Birefringence |
| Electric | Quadratic Electro-optic Effect<br>(KERR EFFECT) | Birefringence |
| Magnetic | Linear Magneto-optic Effect<br>(FARADAY EFFECT) | Optical Rotation<br>(FARADAY ROTATION) |
| | Quadratic Magneto-optic Effect<br>(COTTON-MOUTON EFFECT) | Birefringence |

## THEORY

### Photoelastic Effect

The relative change of the dielectric impermeability due to applied stress is expressed as[2]:

$$\Delta\xi_{ij} = \pi_{ijkl}\,\sigma_{kl} \tag{9}$$

where $\pi_{ijkl}$ are stress-optical coefficients and $\sigma_{kl}$ is the applied stress. By considerations of symmetry, the principal dielectric impermeabilities can be reduced to:

$$\Delta\xi_{i} = \pi_{ij}\,\sigma_{j} \tag{10}$$

where $\sigma_j$ is the tensile stress. Similarly, the strain-optical coefficients ($p_{ij}$) are expressed as:

$$\Delta\xi_i = p_{ij}\, s_j \tag{11}$$

where $s_j$ is the strain. For small strains Hooke's law is valid between $\sigma_j$ and $s_j$, and $\pi_{ij}$ are related to $p_{ij}$ as[3]:

$$\pi_{ij} = p_{ir}\, c_{rj} \tag{12}$$

where $c_{rj}$ are compliances. For isotropic materials, the following relationship can be derived:

$$\pi_{11} - \pi_{12} = (p_{11} - p_{12}) / 2G \tag{13}$$

where G is the shear modulus.

Consider a beam of monochromatic light travelling along the x2-axis through glass under uniaxial tension along the x1-axis as shown in figure 1. For the two components of light with polarization in x1 and x3 directions, the difference in the refractive indices (birefringence) can be expressed as:

$$\begin{aligned} n_1 - n_3 &= - n^3 (\pi_{11} - \pi_{12}) / 2\sigma \\ &= - n^3 (p_{11} - p_{12})(s_1 - s_2) / 2 \\ &= n^2 (p - q)(s_1 - s_2) \end{aligned} \tag{14}$$

where n is the refractive index free of stress, $s_1$ and $s_2$ are the strains in the $x_1$ and $x_2$ directions, and p and q are the Neuman strain-optical coefficients.

The most commonly used expression for the stress-optical coefficient (C) is:

$$\Delta n = n_{||} - n_{\perp} = n_1 - n_3 = C\sigma \tag{15}$$

where $\Delta n$ is the birefringence.

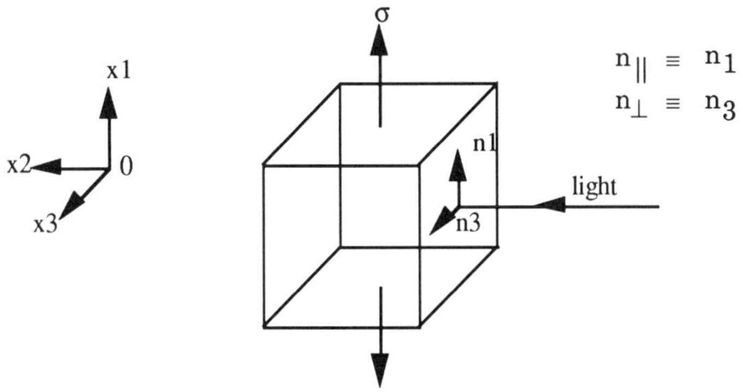


Figure 1. Stress induced birefringence in isotropic materials

From equation (14), C can be expressed in terms of other stress-optical coefficients as[3]:

$$C = -n^3(\pi_{11} - \pi_{12})/2 = -n^3(p_{11} - p_{12})/4G$$
$$= n^2(p - q)/2G \qquad (16)$$

The convention is that the photoelastic effect is positive if $n_1$ (polarization parallel to the tensile stress) is greater than $n_3$ (polarization perpendicular to the stress direction). The anisotropy induced by stress in the electron density distribution can be attributed to two distinct phenomena: 1) relative change of the atomic positions (lattice effect - analogous to crystalline lattice), and 2) deformation of the electron clouds within the atoms (atomic effect). According to Borrelli and Miller,[4-5] the strain optical coefficients are expressed in terms of the atomic and lattice contributions as:

$$p_{12} - p_{11} = 4\pi(n^2 + 2)^2 \sum N_i(\alpha_{11}'^{i} - \alpha_{12}'^{i})/9n^4$$
$$- 2(n^2 - 1)^2/5n^4 \qquad (17)$$

where $N_i$ is the concentration of i ion, and $\alpha_{11}'^{i}$, and $\alpha_{12}'^{i}$ are the elements of the nonlinear polarizability. The first term in equation (17) represents the atomic effect and the second term represents the lattice effect. The lattice contribution gives rise to the negative photoelastic effect and the atomic contribution gives rise to the positive photoelastic effect. In general, the atomic contribution dominates the lattice effect, thus most common glasses are photoelastically positive. High concentrations of heavy-metal atoms such as Pb would lead to dominance of the lattice effect in certain glasses. For those cases where the negative and positive contributions sum to zero, the compositions are referred to as Pockel's glasses in honor of their discoverer.

**Kerr Effect**:

The Kerr effect is often coupled with the photoelastic effect arising from electrostriction induced by the electric field. It is not always easy to separate the individual contributions from the electro-optic and photoelastic effects. The combined birefringence can be expressed as[5]:

$$\Delta n = n_{||} - n_{\perp} = n_1 - n_3 = \lambda K E^2 = K' E^2 \tag{18}$$

where $\lambda$ is the wavelength, E is the applied electric field, and K is the Kerr constant. $n_{||}$ and $n_{\perp}$ are the refractive indices with polarization parallel and perpendicular to the applied electric field respectively. K' , known as absolute Kerr constant, is a product of Kerr constant K and wavelength $\lambda$ and is often represented in esu's. It is evident from equation (18) that the sign of birefringence is not affected by the polarity of the electric field due to the quadratic relationship. A more general expression for birefringence that takes electrostriction into account is given by[5]:

$$n_{||} - n_{\perp} = [(r_{11} - r_{12}) + 2R(\gamma_{11}^{\omega} - \gamma_{12}^{\omega})(\gamma_{11}^{0} - \gamma_{12}^{0}) - (\gamma_{11}^{\omega} - \gamma_{12}^{\omega})\varepsilon^0/8\pi)] E^2 = \lambda K E^2 \tag{19}$$

where $\varepsilon^0$ is the static dielectric constant, $R = 1/2\,(c_{11} - c_{12})$ is the rigidity modulus, c is the elastic tensor, $\gamma^{\omega}$ is the stress-optic tensor, $\gamma^0$ is the electrostrictive tensor, and r is the electro-optic tensor. Assuming an additive relation for the nonlinear polarizability of the participating ions, the electro-optic contribution to the Kerr effect would be given by[5]:

$$(n_{||} - n_{\perp})/E_0^2 = (r_{11} - r_{12}) = (2\pi/n_0)[(n_0^2 + 2)/3]^2 [(\varepsilon^0 + 2)/3]^2 \Sigma N_i (\beta_{11} - \beta_{12})i \tag{20}$$

where $\beta$ is the nonlinear polarizability.

The phase retardation $\delta$ from birefringence for a path length of l is given by:

$$\delta = 2\pi l(n_{||} - n_{\perp})/\lambda = 2\pi lKE^2 \tag{21}$$

Similar to the photoelastic effect, the birefringence can be negative or positive depending on the composition of the glass. Glasses containing Nb or Ta are known to attain negative birefringence[5], whereas typical glasses are positive.

**Faraday Effect**:

The Kerr and photoelastic effects introduce birefringence in glass whereas the linear magneto-optic effect, also known as the Faraday effect, rotates the plane of polarization of a linearly polarized light (Faraday rotation). Distinction should be made between the birefringence and rotation where the former converts a linearly polarized light into an elliptically polarized light whereas the latter simply rotates the plane of polarization of the linearly polarized light. This rotation can also be considered as a special case of elliptical polarization. The angle of rotation $\theta$ is proportional to the applied magnetic field B parallel to the path of light travel, and is given by:

$$\theta = VBl \tag{22}$$

where l is the length of the medium and V is the proportionality constant referred to as the Verdet constant.

The effect can be described along the lines of Fresnel's interpretation of optical activity in which a plane polarized wave is represented as the superposition of two oppositely rotating circularly polarized waves. In the presence of an external magnetic field, the right and left-circularly polarized waves ($E_r$ and $E_l$) travel with different velocities and at the exit recombine to form a rotated linearly polarized wave ($E_p$) as shown in figure 2. This phenomenon can also be considered as the magnetic field induced circular-birefringence.

The effective dielectric tensor in glass in the presence of a magnetic field in the $x_3$-direction (parallel to the propagation of light) can be represented as:

$$\mathrm{eff}(\varepsilon) = \begin{bmatrix} \varepsilon & i\Theta & 0 \\ -i\Theta & \varepsilon & 0 \\ 0 & 0 & \varepsilon \end{bmatrix}$$

$$\Theta = VB\lambda/\pi \tag{23}$$

One of the most important features of the Faraday effect is that the direction of rotation is the same whether the direction of light path is parallel or antiparallel to B[1]. If a wave passes through a Faraday cell, gets reflected and retraces its path, the rotation is 2θ as opposed to the optical activity where the net angle of rotation would be zero.

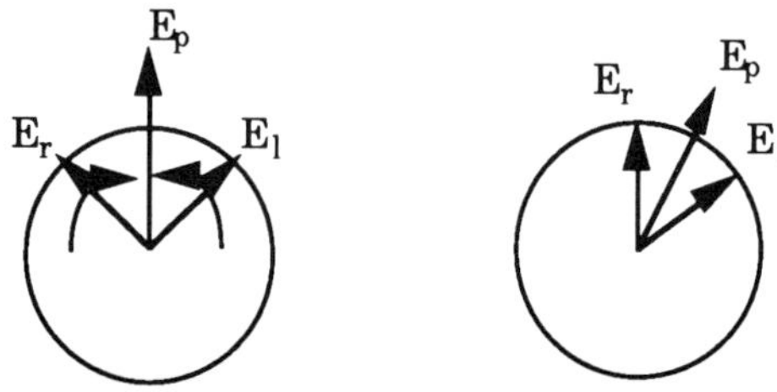


Figure 2. Magnetic field induced circular birefringence

The Faraday rotation of glasses in the absence of transition metal or rare earth ions is due to diamagnetic susceptibility. The diamagnetic rotation, by convention, is considered positive and the paramagnetic rotation is considered negative. Following Van Vleck[6] and Borrelli[7], the appropriate quantum mechanical equation for the Verdet constant is:

$$V = 4\pi N \nu^2 \sum \left[ \frac{A_1\,(n,g)}{\{\nu^2 - \nu(n,g)^2\}^2} + \frac{A_2\,(n,g)}{\{\nu^2 - \nu(n,g)^2\}T} + \frac{A_3\,(n,g)}{\nu^2 - \nu(n,g)^2} \right] \tag{24}$$

where N is atoms/unit volume, ν is frequency, $A_1$, $A_2$, and $A_3$ are related to transition moments, n and g are indices related to ground and excited states, and T is the temperature. The first and third terms represent the diamagnetic contribution to V, and the second term represents the paramagnetic contribution[8]. Becquerel's classical expression for the diamagnetic rotation is given by:

$$V = \gamma \frac{e}{2mc^2} \lambda \frac{dn}{d\lambda} \tag{25}$$

where e and m are the electron charge and mass, respectively, c is the velocity of light, and $\gamma$ is a correction factor, referred to as the magneto-optic anomaly factor, used to bring measured values of V, n, and $\lambda$ into agreement. $\gamma$ should be 1 in an ideal case with no paramagnetic contribution. Attempts have been made to link $\gamma$ to the nature of bonding in the glass-forming and modifier ions[9,10]; however, no strict correlation was found.

## EXPERIMENTAL METHODS

The birefringence induced by the Kerr and photoelastic effects is normally measured using a Babinet compensator. The phase retardation is obtained from the position of the Babinet compensator, and the birefringence is computed using equation (21). A schematic illustration of the birefringence measurement technique is shown in figure 3. In the case of Faraday effect, the rotation angle is directly obtained from the angular position of the analyzer which is adjusted to the extinction position. A schematic diagram of the Faraday rotation apparatus is shown in figure 4. A summary of optical constants and units is given in Table 2.

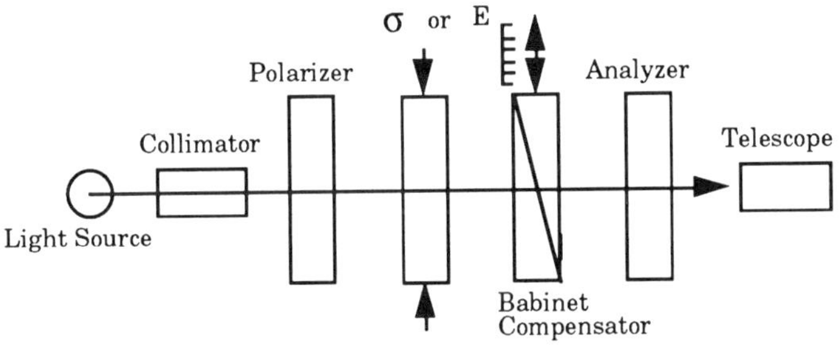


Figure 3. Birefringence measurement apparatus.

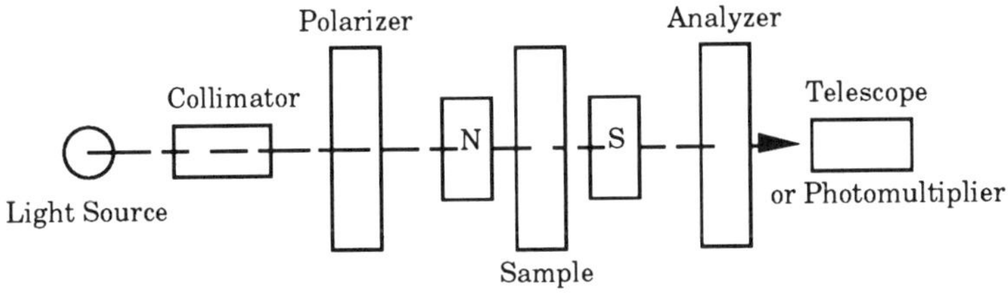


Figure 4. Faraday rotation measurement apparatus.

Table 2. Summary of optical constants and units.

| OPTICAL EFFECT | EQUATION | UNITS |
|---|---|---|
| Photoelastic | $C = \Delta n/\sigma$ | 1 Brewster = $10^{-12} \cdot n^{-1} \cdot m^2$ |
| Kerr | $K = \Delta n/\lambda E^2$<br>$K' = K\lambda$ | $V^{-2} \cdot m$<br>$V^{-2} \cdot m^2$ or esu |
| Faraday | $V = \theta/Bl$ | radians·$Tesla^{-1} \cdot m^{-1}$ |

**DATA**

**Photoelastic Effect**:

The stress-optical coefficient of common glasses typically varies from 2 to 10 Brewsters and decreases and even becomes negative with increase in the density of glass. The stress-optical coefficients of some selected glasses are given in Table 3.

Table 3. Stress-optical coefficients of some selected glasses

| Glass System (mole ratios) | Refractive Index (n) | λ | Stress-optic Coefficient (C in Brewsters) | Ref. |
|---|---|---|---|---|
| $B_2O_3$ | 1.456 | 546.1 (nm) | 11 | 3 |
| $16Li_2O{\cdot}84B_2O_3$ | 1.514 | " | 4.6 | 3 |
| $16Na_2O{\cdot}84B_2O_3$ | 1.498 | " | 4.92 | 3 |
| $16K_2O{\cdot}84B_2O_3$ | 1.488 | " | 5.34 | 3 |
| $36CaO{\cdot}64B_2O_3$ | 1.602 | " | 2.36 | 3 |
| $36BaO{\cdot}64B_2O_3$ | 1.630 | " | 1.82 | 3 |
| $25PbO{\cdot}75B_2O_3$ | 1.672 | " | 2.50 | 3 |
| $50PbO{\cdot}50B_2O_3$ | 1.878 | " | 0.04 | 3 |
| $65PbO{\cdot}35B_2O_3$ | 2.011 | " | -2.15 | 3 |
| $55ZnO{\cdot}45B_2O_3$ | 1.655 | " | 4.35 | 3 |

Table 3. Stress-optical coefficients of some selected glasses (cont'd)

| Glass System (mole ratios) | Refractive Index (n) | λ | Stress-optic Coefficient (C in Brewsters) | Ref. |
|---|---|---|---|---|
| $20Na_2O{\cdot}10Al_2O3{\cdot}70B_2O_3$ | 1.494 | 546.1 (nm) | 4.95 | 3 |
| $49PbO{\cdot}16ZnO{\cdot}35B_2O_3$ | --- | " | 0.10 | 3 |
| $SiO_2$ | 1.458 | " | 3.47 | 11 |
| $40Li_2O{\cdot}60SiO_2$ | 1.548 | " | 2.24 | 11 |
| $25Na_2O{\cdot}75SiO_2$ | 1.491 | " | 2.92 | 11 |
| $30K_2O{\cdot}70SiO_2$ | 1.477 | " | 2.59 | 11 |
| $50ZnO{\cdot}50SiO_2$ | 1.653 | " | 3.51 | 11 |
| $20Li_2O{\cdot}10Al2O3{\cdot}70SiO_2$ | 1.513 | " | 2.80 | 11 |
| $20Na_2O{\cdot}10Al2O3{\cdot}70SiO_2$ | 1.500 | " | 3.13 | 11 |
| $30K_2O{\cdot}10Al_2O_3{\cdot}60SiO_2$ | 1.508 | " | 3.13 | 11 |
| $20Na_2O{\cdot}20B_2O_3{\cdot}60SiO_2$ | 1.524 | " | 2.87 | 11 |
| $50MgO{\cdot}50P_2O_5$ | 1.496 | " | 2.48 | 12 |
| $50CaO{\cdot}50P_2O_5$ | 1.549 | " | 1.73 | 12 |
| $50SrO{\cdot}50P_2O_5$ | 1.555 | " | 1.37 | 12 |
| $50BaO{\cdot}50P_2O_5$ | 1.593 | " | 0.42 | 12 |
| $50ZnO{\cdot}50P_2O_5$ | 1.525 | " | 6.45 | 12 |
| $30Na_2O{\cdot}30TiO_2{\cdot}40SiO_2$ | 3.0 | " | 3.01 | 13 |
| Soda-lime-silica | --- | " | 2.41 | 14 |
| $63BaF_2{\cdot}32Al(PO_3)_3{\cdot}5AlF_3$ | -- | | 0.22 | 16 |
| $62ZrF_4{\cdot}35BaF_2{\cdot}3GdF_3$ | 1.522 | " | -0.03 | 19 |
| $62ZrF_4{\cdot}31BaF_2{\cdot}7LaF_3$ | 1.524 | " | 0.07 | 19 |
| $60ZrF_4{\cdot}27BaF_2{\cdot}3GdF_3{\cdot}10PbF_2$ | 1.545 | " | -0.10 | 19 |
| TI-1173 Chalcogenide | --- | 1-12 μm | 9-14 | 21 |

Dispersion of the stress-optical coefficient with the wavelength of light is shown in figure 5. The coefficient decreases with increasing wavelength of light[17]. The trend with respect to the density of glass is shown in figure 6. The photoelastic effect of the fluoride glasses has been of considerable interest owing to very small birefringence observed under stress[16,19]. The stress-optical coefficients were measured at higher temperatures (in the annealing range) by Fontana[18] and the magnitude was observed to increase with increasing temperature. In the case of high-lead glasses, the stress-optical coefficients are negative and become more negative in the annealing range[18]. These observations were attributed to the ease of both atomic and lattice distortions enabled by the open structures at higher

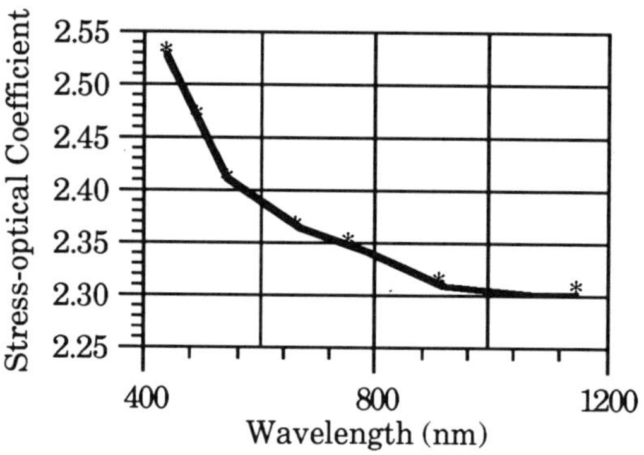


Figure 5. Dispersion of stress-optic coefficient in soda-lime-silica glass. (Data taken from reference 14)

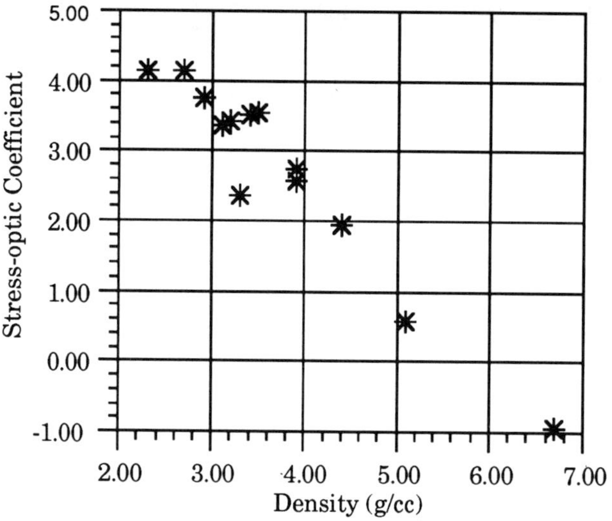


Figure 6. Stress-optic coefficient vs. density in optical glasses. (Data taken from reference 15)

temperatures. Thermo-optical constant, which is dependent on the stress-optical constant, thermal expansion coefficient, elastic modulus, and Poisson's ratio, was computed by Shchavelev and Plutalova[27] for some silicate and phosphate glasses. There is considerable interest in the photoelastic behavior of optical waveguides which may differ considerably from the bulk glasses due to the thermal history and surface stress effects[20-26]. Raftopoulos and Kartalopulos[28] used an interferometry technique to measure the stress-optical coefficient of solid plates whereas Borrelli and Miller[4] used an ultrasonic technique to measure the stress-optical coefficients of optical glasses.

**Kerr Effect**:

The Kerr constants of some selected glasses are shown in Table 4 and are typically on the order of 1-4 x $10^{-12}$ cm/$V^2$ for ordinary glasses. In binary lead-silicates, the Kerr constant decreases with increasing content of PbO[5].

Table 4. Kerr constants of some selected glasses and glass-ceramics

| Glass System (Weight ratios) | Dielectric Constant | λ (nm) | Kerr Constant x $10^{12}$ $(V^2/cm)^{-1}$ (or K' in $10^{-13}$esu)* | Ref. |
|---|---|---|---|---|
| $65.8Nb_2O_5{\cdot}15.4Na_2O{\cdot}18.8SiO_2$ | 25 | 632.8 | -2.8 | 5 |
| $40Tl_2O{\cdot}14K_2O{\cdot}46SiO_2$ | --- | " | 0.3 | 5 |
| $57.5Ta_2O_5{\cdot}12.1K_2O{\cdot}30.4SiO_2$ | 11 | " | -0.57 | 5 |
| $84PbO{\cdot}5SiO_2{\cdot}3Al_2O_3{\cdot}10B_2O_3$ | 15.3 | " | 0.24 | 5 |
| $66PbO{\cdot}30SiO_2{\cdot}2Na_2O{\cdot}2K_2O$ | 10.8 | " | 0.085 | 5 |
| $As_2S_3$ | 7.4 | " | 8.7 | 5 |
| Borosilicate crown 517 | --- | --- | 0.80* | 29,30 |
| Vitreous silica | --- | --- | 0.56* | 29,30 |
| Dense Flint SF 55 | --- | --- | 6.4* | 29,30 |
| Fused silica fiber | --- | --- | 1.9* | 31,32 |
| 25% PbO-borosilicate | --- | --- | 0.0 | 33 |
| $NaNbO_3$ glass-ceramic | 607 | 632.8 | 8000 | 67 |
| $(Na,Sr)NbO_3$ glass-ceramic | - | " | 11500 | 67 |

Glasses containing niobium and tantalum showed negative birefringence. $As_2S_3$ shows a high birefringence in an electric field[5]. Che *et al.*[34] measured DC Kerr effects in glass-organic polymer composites which can be optimized for increasing the Kerr effect. The Kerr effect was studied in chalcogenide glasses[35], semiconductor-doped glasses[36], and glasses containing gold colloids[37]. Hoye and Stell[38] analyzed the birefringence contributions from hyperpolarizability and anisotropic polarizability. Farries and Rogers[39] studied the temperature dependence of the Kerr constant in a silica optical fiber and found that it increases with temperature.

Borrelli and Layton[66-68] measured the electro-optic properties of transparent glass-ceramic systems containing sub-micron ferroelectric crystalline phases. These systems showed a relatively large electro-optic effect owing to the presence of ferroelectric phases. They also exhibit the linear electro-optic effect when poled in DC fields.

**Faraday Effect**:

The Verdet constants of some selected glasses are shown in Table 5. The wavelength dispersion of the Verdet constant of some Hoya glasses is shown in figure 7. The absolute value of the Verdet constant typically decreases with increasing wavelength. The Verdet constants for diamagnetic glasses typically range from 1 to 10 $rad \cdot T^{-1} \cdot m^{-1}$ and are higher in chalcogenide and fluoride glasses. The Verdet constant decreases and reverses sign with increasing additions of paramagnetic ions. The Faraday rotation decreases with increasing temperature[39,54]. Van den Keybus and Grevendonk[41] found that the Faraday rotation reverses sign at certain critical photon energies in chalcogenide glasses. Burkov *et al.*[45,46] found a similar behavior in Tb-doped fluoroborate glasses. Wertheimer[42] measured the rotation of Ce $(PO_3)_3$ glass at liquid-helium temperature and showed that the reciprocal of the Verdet constant extrapolated to zero at absolute zero. Berger and Rubinstein[64] reported similar results for $Ce_2O_3 \cdot 2.67\ P_2O_5$ glass. Daybell *et al.*[48] found similar effects in a Tb-alumino-silicate glass.

Table 5. Verdet constants of some selected glasses

| Glass System | Wavelength (nm) | Verdet Constant ($rad \cdot T^{-1} \cdot m^{-1}$) | Ref. |
|---|---|---|---|
| $20TeO_2 \cdot 80PbO$ | 700.0 | 43 | 7 |
| Pr-Phosphate | 700.0 | -35 | 7 |
| $B_2O_3$ | 632.8 | 2.9 | 8 |
| $SiO_2$ | 632.8 | 4.0 | 8 |
| Alkali-Germanate | 632.8 | 11.5 | 60 |
| 50 mole% Tb-Borosilicate | 632.8 | -45 | 40 |
| IKS-Chalcogenide | " | 29 | 43 |
| Photochromic | " | 5 | 44 |
| Sol-gel derived $SiO_2$ | " | 3.9 | 47 |
| $As_2S_3 \cdot AsI_3$ | 600.0 | 73 | 50 |
| $As_2S_3 \cdot AsBr_3$ | 600.0 | 73 | 50 |
| $37ZrF_4 \cdot 5FeF_3 \cdot 52PbF_2 \cdot 6YF_3$ | 632.8 | 8 | 52 |
| $0.2BaF_2 \cdot 0.325TbF_3 \cdot 0.475NaPO_3$ | 500.0 | 85 | 53 |
| FR-123 Terbium Borate | 633.0 | -71 | 65 |

Stepanov *et al.*[49] studied the magneto-optic properties of $K_2O{\cdot}Al_2O_3{\cdot}SiO_2$ glasses dispersed with magnetic $Fe_2O_3$ particles and measured a strong temperature dependence of rotation at about 253 K attributed to Morin's phase transition in $\alpha$-$Fe_2O_3$. Van Der Ziel *et al.*[51] studied the inverse Faraday effect in which a circularly polarized light induces magnetization in glasses. Stolen[56] and Harms *et al.*[57] studied the Faraday rotation in highly birefringent optical fibers where a significant rotation was achieved by alternating magnetic field with a period equal to the fiber's beat length. Shiraishi *et al.*[59] built a temperature-insensitive Faraday rotator with the Hoya FR-2 glass in which the Verdet constant is stable over a broad temperature range.

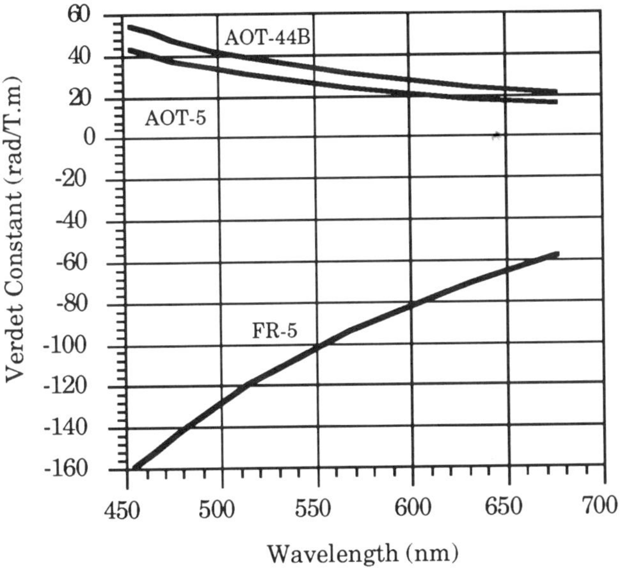


Figure 7. Verdet constant vs. wavelength for some optical glasses. (Data taken from reference 55)

## ZERO-EFFECT GLASSES

As discussed above, all three induced optical effects produce positive or negative birefringence (or rotation) depending on the composition of the glass. In the photoelastic effect, the competing factors are the atomic and lattice polarizations. In the Kerr effect, the contributing factors are atomic and molecular nonlinear polarizabilities, and electrostriction. In the Faraday effect, the intrinsic diamagnetic susceptibility can be compensated or dominated by paramagnetic ions or dispersed ferromagnetic phases. The dominating contribution will dictate the sign of birefringence or rotation. By balancing the competing contributions, it was possible to build zero-effect glasses in each case, within certain temperature and wavelength constraints.

For the photoelastic effect , "Pockels concluded that it would be zero for a glass of the composition $SiO_2$, 22.8; $K_2O$, 1.5; PbO, 75.7, and an attempt was made to make a glass of this composition. The resulting glass, an analysis of which gave: $SiO_2$, 23.64; $Na_2O$, 0.24; $K_2O$, 0.37; CaO, 0.34; PbO, 74.61; $Al_2O_3$+$Fe_2O_3$, 0.8, was not of high quality, and showed a positive birefringence of about one-tenth of the heavier glass. Pockels concluded that a diminution of 0.6 percent in the PbO content would have caused the birefringence to vanish"[33]. Other references to low-birefringence glass systems can be found in the literature. Tauern measured the Kerr effect in a number of Schott glasses and found that "a silicate glass with barium oxide, zinc oxide and potash (O211), and a borosilicate glass with potash and soda (O3832) showed no birefringence"[33]. Borrelli[5] measured the Kerr effect in glasses containing Nb and Ta which showed negative Kerr effect. In the case of the Faraday effect, although there were indications of zero rotation at certain wavelengths,[41,45,46] explicit reference to zero Faraday rotation glasses was made by Cherukuri and Pye[60,61]. They computed the concentrations of paramagnetic rare-earth ions required to produce zero Faraday effect in the diamagnetic alkali germanate glasses for a broad range of wavelengths.

The authors believe that, by carefully balancing various competing influences, it should be possible to create a "triple zero-effect" glass which has zero or negligible photoelastic, Kerr, and Faraday effects simultaneously within certain temperature and wavelength ranges. One likely candidate is a lead-alkali-silicate system containing niobium, tantalum, and rare-earth oxides.

## APPLICATIONS

The optical effects in glass are directly or indirectly used in a wide range of applications. When a glass article is inspected with a polariscope to assess the state of annealing, the photoelastic effect is being used. A thorough understanding of the phenomenon is crucial to a wide variety of optical, electro-optic, and magneto-optic devices and waveguides. For the photoelastic effect, the areas of interest include the processing of optical glasses, residual stress measurements, compressive strengthening of fibers, control of radial refractive index gradient[22], fibers exhibiting low or high polarization birefringence, acousto-optic fiber sensors, fiber Raman lasers, and the fiber gyroscope[20]. The Kerr effect is used in optical fibers to sense electric fields and temperature[39], for self-focusing in optical glasses[29,30], and for modulators[35] and optical shutters for picosecond optical pulses[31,32]. The Faraday rotation has been used or suggested for applications in magneto-optic shutters, light switches, modulators, laser gyroscopes, decoupling devices[9], magnetic field measurement devices[62], and isolators[9]. The cumulative nature of rotation makes glass an attractive candidate for use in Faraday isolators[1,63]. The insight gained into the nature of the glassy state by studying the phenomenon cannot be overemphasized.

**Acknowledgement**

The authors would like to thank Mr. D.M.Korwin for his assistance in the preparation of this manuscript.

**REFERENCES**

1. Nussbaum A., and Phillips R.A., **Contemporary Optics for Scientists and Engineers**, 370-416, Prentice Hall, Englewood Cliffs, New Jersey, 1976.
2. Nye J.F., **Physical Properties of Crystals**, 235-259, Oxford University Press, Ely House, London, 1969.
3. Matusita K., Yokota R., Kimijima T., Komatsu T., Ihara C., "Compositional trends in photoelastic constants of borate glasses," *J. Am. Ceram. Soc.*, **67**[4] 261-5 (1984).
4. Borrelli N.F., Miller R.A., "Determination of the individual strain-optic coefficients of glass by an ultrasonic technique," *Appl. Opt.*, **7**[5] 745-50 (1968).
5. Borrelli N.F., "Electric field induced birefringence in glasses," *Phys. Chem. Glasses*, **12**[4] 93-6 (1971).
6. Van Vleck J.H., **The Theory of Electric and Magnetic Susceptibilities**, Oxford University Press, London, 1932.
7. Borrelli N.F., "Faraday rotation in glasses," *J. Chem. Phys.*, **41**[11] 3289 (1964).
8. Pye L.D., Cherukuri S.C., Mansfield J., Loretz T., "The Faraday effect in some non-crystalline fluorides," *J. Non-Cryst. Solids*, **56**[1-3] 99-104 (1983).
9. Stepanov S.A., "Faraday Rotation in Glasses," *Sov. J. Opt. Technol.*, **41**[3] 179-85 (1974).
10. Cole H., "Magneto-optic effects in glass," *J. Soc. Glass Tech.*, **34** 221 (1950).
11. Matusita M., Ihara C., Komatsu T., "Photoelastic effects in silicate glasses," *J. Am. Ceram. Soc.*, **67**[10] 700-4 (1984).
12. Matusita M., Ihara C., Komatsu T., "Photoelastic effects in phosphate glasses," *J. Am. Ceram. Soc.*, **68**[7] 389-91 (1985).
13. Nissle T.R., Babcock C.L., "Stress-optical coefficient as related to glass composition," *J. Am. Ceram. Soc.*, **56**[11] 596-8 (1973).
14. Sinha N.K., "Normalized dispersion of birefringence of quartz and stress optical coefficient of fused silica and plate glass," *Phys. Chem. Glasses*, **19**[4] 69-77 (1978).
15. Vasudevan T.N., Krishnan R.S., "Dispersion of the stress-optic effect in glasses," *J. Phys. D*, **5**[12] 2283-7 (1972).

16. Galant V.E., "Stress-optical coefficient of fluorophosphate glasses," *Sov. J. Glass Phys. and Chem.*, **5**[5] 543-7 (1980).
17. Sinha N.K., "Dispersion of birefringence in glasses and its measurement with the Babinet compensator using white light," *Phys. Chem. Glasses*, **18**[4] 65-9 (1977).
18. Fontana E.H., "Stress-optical coefficients for glasses in their annealing range," *Am. Ceram. Soc. Bull.*, **64**[11] 1456-8 (1985).
19. Matusita K., Kato H., Komatsu T., Yoshimoto M., Soga N., "Photoelastic effects in some fluoride glasses based on the $ZrF_4$-$BaF_2$ system," *J. Non-Cryst. Solids*, **112**[1-3] 341-6 (1989).
20. Lagakos N., Mohr R.K., El-Bayoumi O.H., "Stress optic coefficient and stress profile in optical fibers," *Appl. Opt.*, **20**[N13] 2309-13 (1981).
21. Szczesniak J.P., Cuddeback D., Corelli J.C., "Stress-induced birefringence of solids transparent to 1-12 μm light," *J. Appl. Phys.*, **47**[12] 5356-9 (1976).
22. Scherer G.W., "Stress-optical effects in optical waveguides," *J. Non-Cryst. Solids*, **38-39**[1] 201-4 (1980).
23. Stockhorst H., Brueckner R., "Stress-optical investigations of glass fibers with linearly and three-dimensionally branched networks," *Phys. Chem. Glasses*, **27**[5] 204-9 (1986).
24. Barlow A.J., Payne D.N., "Stress-optic effect in optical fibers," *IEEE J. Quantum Electron.*, **19**[5] 834-9 (1983).
25. Shibata N., Okamoto K., Tateda M., Seikai S., Sasaki Y., "Modal birefringence and polarization mode dispersion in single-mode fibers with stress-induced anisotropy," *IEEE J. Quantum Electron.*, **QE-19**[6] 1110-15 (1983).
26. Scherer G.W., "Stress-induced index profile distortion in optical waveguides," *Appl. Opt.*, **19**[12] 2000-6 (1980).
27. Shchavelev, O.S., Plutalova N.Y., "Dependence of thermooptical constant Q of silicate and phosphate glasses on their chemical composition," *Sov. J. Glass Phys. and Chem.*, **4**[1] 81-6 (1978).
28. Raftopoulos D.D., Kartalopoulos S.V., "Evaluation of the stress-optical coefficients of transparent solid plates using interferometry," *Appl. Opt.*, **18**[17] 2975-8 (1979).
29. Feldman A., Horowitz D., Waxler R.M., "Mechanisms of self-focusing in optical glasses," *IEEE J. Quantum Electron.,* **9**[11] 1054-61 (1973).
30. Feldman A., Horowitz D., Waxler R.M., "Relative importance of electrostriction and the Kerr effect to self-focusing in optical glasses," *Appl. Phys. Lett.*, **21**[6] 260-2 (1972).
31. Stolen R.H., Ashkin A., "Optical Kerr effect in glass waveguide," *Appl. Phys. Lett .,* **22**[6] 294-6 (1973).
32. Dziedzic J.M., Stolen R.H., Ashkin A., "Optical Kerr effect in long fibers," *Appl. Opt.*, **20**[8] 1403-6 (1981).
33. Morey G.W., **The Properties of Glass**, Reinhold Publishing Corp., New York, 1938.
34. Che, T.M., Carney R.V., Khanarian G., Keosian R.A., Borzo M., "Electro-optical DC Kerr effects and solid state dueteron NMR studies of stable gel-derived glass-organic polymer composites," *J. Non-Cryst. Solids*, **102**[1-3] 280-7 (1988).

35. Klein R.M., "Chalcogenide glasses as passive thin film structures for integrated optics," *J. Electron. Mater.*, **3**[1] 79-99 (1974).
36. Roussignol P., Ricard D., Flytzanis C., "Nonlinear optical properties of commercial semiconductor-doped glasses," *Appl. Phys. A, Solids Surf.*, **A44**[4] 285-92 (1987).
37. Hache F., Ricard D., Flytzanis C., Kreibig U., "The optical Kerr effect in small metal particles and metal-doped glasses: the case of gold," *Proc. SPIE-Int. Soc. Opt. Eng.*, **1127** (Nonlinear Opt. Mater. 2), 115-22 (1989).
38. Hoye J.S., Stell G., "Kerr effect. II. Contributions from hyperpolarizability and anisotropic polarizability," *J. Chem. Phys.*, **88**[9] 5798-804 (1988).
39. Petrovskii G.T., Zarubina T.V., Edel'man I.S., Kim T.A., Smirnova L.A., "Temperature dependence and dispersion of the Faraday effect in glasses based on terbium and cerium oxides," *Sov. J. Opt. Technol.*, **54**[11] 680-2 (1987).
40. Chiba Y., Hori H., Yonenobu K., Yamagishi A., Date M., "High-field Faraday effect of terbium-doped glass above paramagnetic saturation," *Jpn. J. Appl. Phys., Part 2*, **26**[8] L1258-L1259 (1987).
41. Van den Keybus P., Grevendonk W., "Nonresonant Faraday rotation in glassy semiconductors," *Phys. Rev. B*, **33**[12] 8540-5 (1986).
42. Wertheimer M.R., "Faraday rotation in $Ce(PO_3)_3$ glass at liquid-helium temperatures," *J. Appl. Phys.*, **44**[11] 5078-79 (1973).
43. Galanov E.K., Potikhonov G.N., "Magneto-optic properties of IR glasses," *Sov. J. Opt. Technol.*, **43**[12] 729-31 (1976).
44. Peshkov N.M., Kaplun V.A., Romodanov I.S., "Magnetooptical rotation in photochromic glasses," *Sov. J. Glass Phys. and Chem.*, **2**[6] 545-6 (1976).
45. Burkov V.I, Fedorushkov B.G., Vakhrameev V.I., Bratkovskii V.M., Solinov V.F., Kraevskii S.L., Kotov V.A., Khalilev V.D., "Magnetooptical characteristics of fluoroborate glasses activated by Tb," *Sov. J. Glass Phys. and Chem.*, **5**[5] 565-7 (1980).
46. Burkov V.I, Fedorushkov B.G., Bratkovskii V.M., Vakhrameev V.I., Kotov V.A., Landa L.M., Solinov V.F., Khalilev V.D., Semin G.S., "Magnetooptical characteristics of fluorophosphate glasses containing rare-earth ions," *Sov. J. Glass Phys. and Chem.*, **5**[1] 104-7 (1979).
47. Bouaziz J., Bourret D., Sempere R., Regnier J., "Faraday effect of sol-gel derived glasses," *J. Non-Cryst. Solids*, **82**[1-3] 183-9 (1986).
48. Daybell M.D., Overton W.C., Laquer H.L., "Faraday effect at low temperatures in terbium alumina silicate glass," *Appl. Phys. Lett.*, **11**[3] 79-81 (1967).
49. Stepanov S.A., Edelman I.S., Kim T.A., Petrovski G.T., Popov G.V., "Properties of magnetically ordered microparticles in borate glasses," *Phys. Status Solidi A*, **104**[2] 805-13 (1987).
50. Burkov V.I., Dembovskii S.A., Kizel' V.A., Kozlova N.L., Krasilov Y.I., Chernov A.P., "Optical and magnetooptical properties of glass of the systems As-S-I and As-S-Br," *Inorg. Mater. (USA)* , **9**[3] 333-6, (1973).

51. Van der Ziel J.P., Pershan P.S., Malmstrom L.D., "Optically induced magnetization resulting from the inverse Faraday effect," *Phys. Rev. Lett.*, **15**[5] 190-3 (1965).
52. Han J., Cheng J., "New heavy-metal fluoride glasses in $ZrF_4$-$FeF_3$-$PbF_2$-$YF_3$ systems," *J. Non-Cryst. Solids*, **109**[1] 64-68 (1989).
53. Letellier V., Seignac A., Le Floch A., Matecki M., "Magneto-optical properties of heavily rare-earth doped noncrystalline fluorophosphates," *J. Non-Cryst. Solids*, **111**[1] 55-62 (1989).
54. Sato H., Kawase M., Saito M., "Temperature dependence of the Faraday effect in As-S glass fiber," *Appl. Opt.*, **24**[15] 2300-3 (1985).
55. Villaverde A.B., Vasconcellos, E.C.C., "Magnetooptic dispersion of Hoya glasses: AOT-5, AOT-44B, and FR-5," *Appl. Opt.*, **21**[N8] 1347-8 (1982).
56. Stolen R.H., Turner E.H., "Faraday rotation in highly birefringent optical fibers," *Appl. Opt.*, **19**[6] 842-5 (1980).
57. Harms H., Papp A., Kempter K., "Magneto-optical properties of index-gradient optical fibers," *Appl. Opt.*, **15**[3] 799-801 (1976).
58. Brannon P.J., Franklin F.R., Hauser G.C., Lavasek J.W., Jones E.D., "Faraday rotator using ED4 glass," *Appl. Opt.*, **13**[7] 1555-7 (1974).
59. Shiraishi K., Nishino K., Kawakami S., "Temperature-insensitive fiber Faraday rotator," *Appl. Opt.*, **24**[13] 1896-7 (1985).
60. Cherukuri S.C., Pye L.D., "Faraday rotation of rare earth alkali germanate glasses," **Proc. Rare Earths in Modern Science and Technology, Vol. 3**, 465, Plenum, N.Y., 1982.
61. Cherukuri S.C., **Faraday Rotation of Rare Earth Alkali Germanate Glasses**, M.S. Thesis, Alfred University, Alfred, NY (1980).
62. Thompson J.E., Kristiansen M., Hagler M.O., "Optical measurement of high electric and magnetic fields," *IEEE Transactions on Instrumentation and Measurement*, **25**[1] 1-7 (1976).
63. Aplet L.J., Carson J.W., "A Faraday effect optical isolator," *Appl. Opt.*, **3**[4] 544-45 (1964).
64. Berger S.B., Rubinstein C.B., "A comparison of the optical Faraday rotation and magnetic susceptibility of cerous phosphate glass," *J. Appl. Phys.,* **35**[6] 1798-1801 (1964).
65. Weber M.J., **Faraday Rotator Materials**, Lawrence Livermore Laboratory, University of California, Livermore, CA, 1982.
66. Borrelli N.F., "Electro-optic effect in transparent niobate glass-ceramic systems," *J. Appl. Phys.*, **38**[11] 4243-47 (1967).
67. Borrelli N.F., Layton M.M., "Dielectric and optical properties of transparent ferroelectric glass-ceramic systems," *J. Non-Cryst. Solids*, **6** 197-212 (1971).
68. Borrelli N.F., Layton M.M., "Electrooptic properties of transparent ferroelectric glass-ceramic systems," *IEEE Trans.*, **ED-16**[6] 511-14 (1969).

# CHAPTER 8

## NEUTRON AND X-RAY AMORPHOGRAPHY

**Adrian C. WRIGHT**
New York State College of Ceramics at Alfred University
Alfred, NY 14802-1296, U.S.A.

## 1. INTRODUCTION

A knowledge of structure is an important prerequisite for a full understanding of the physical and chemical properties of any material at a microscopic level. However, unlike the situation for crystalline solids where the structures of most simple materials are not in question, the structures of even the simplest single component inorganic glasses such as vitreous silica are still not established and are the subject of considerable debate in the literature. This is a consequence of the fact that glasses are amorphous solids and hence have structures which lack periodicity, symmetry and long range order. Also glasses are normally isotropic on a macroscopic scale, although the same is not necessarily true for amorphous thin films prepared by techniques other than quenching from the melt*. Amorphography is the name given to the branch of science concerned with the determination of amorphous solid structures and their systemmatic classification (c.f. crystallography for crystalline solids).

The existence of periodicity means that, in the absence of defects, a complete, quantitative description of the structure of a crystalline material requires only the detailed specification of a single unit cell. This involves relatively few independent parameters: the lattice parameters ($a_o$, $b_o$, $c_o$, $\alpha$, $\beta$, $\gamma$) and the positional ($x_i$, $y_i$, $z_i$) and thermal displacement parameters for each of the unique atoms as defined by the space group. Diffraction experiments may be performed on single crystal samples yielding three-dimensional data and hence, given good diffraction data over a reasonable region of reciprocal space, it is in practice possible to determine the structure of simple crystalline solids absolutely. As a consequence of the constraints imposed by their periodic nature, most crystalline materials exhibit discrete polymorphism and discrete stoichiometry, any non-

---

* In this chapter the term "glass" will be reserved for materials covered by the A.S.T.M. definition of a glass [1] as "an inorganic product of fusion which has cooled to a rigid condition without crystallising".

Network former ($A_2O_3$) only

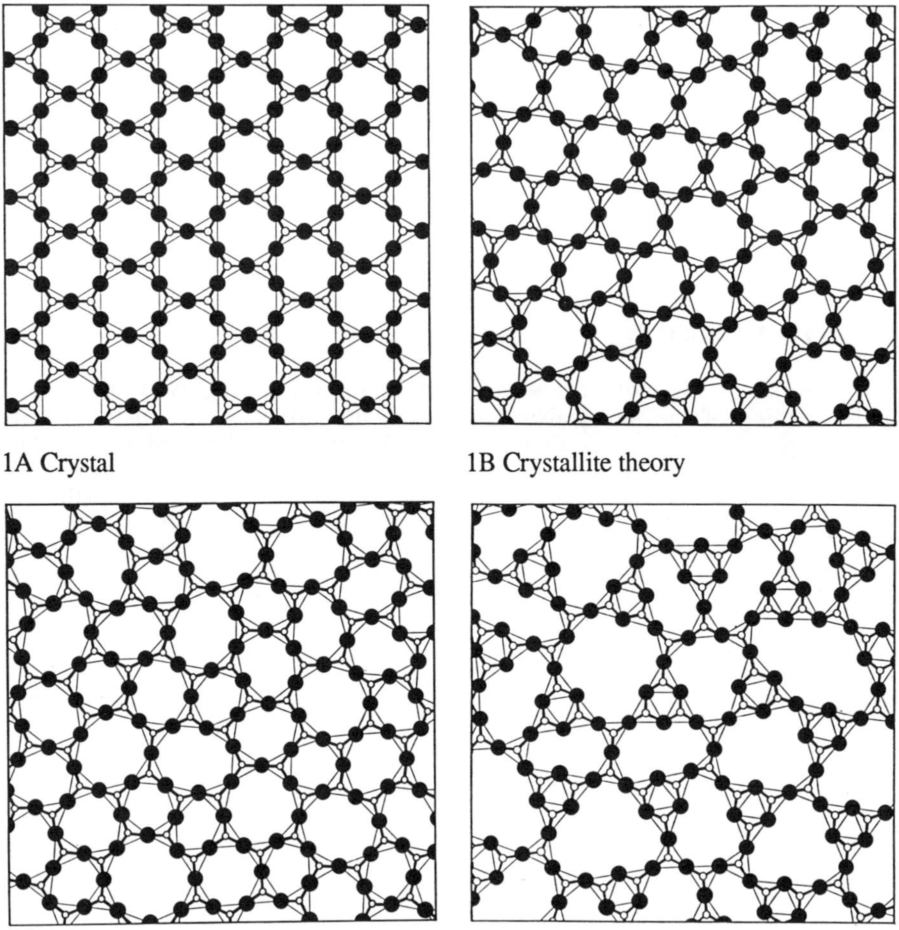

1A Crystal

1B Crystallite theory

1C Random network theory

1D $B_2O_3$ structure containing $B_3O_6$ rings

**Figure 1** Theories of glass structure. o, A atom; ●, O atom and ◍ network modifying cation ($M^{2+}$).

Network former + network modifier (MO)

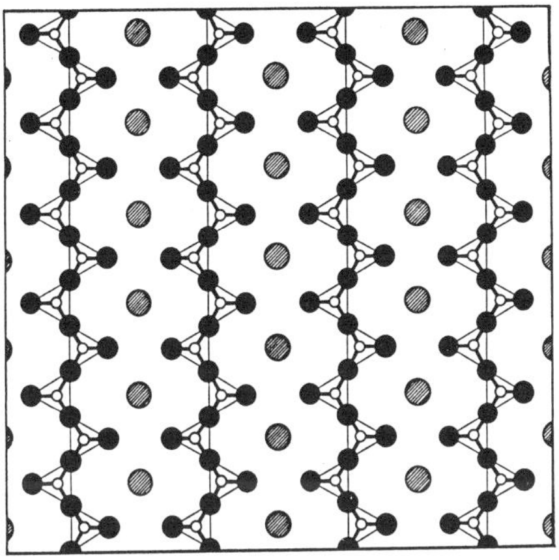

1E Crystal

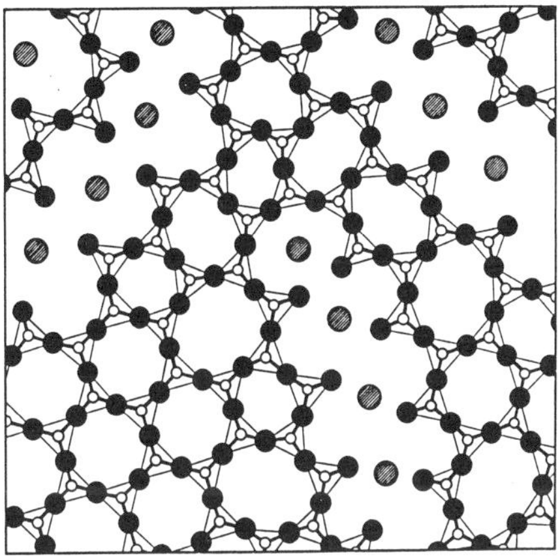

1F Crystallite theory

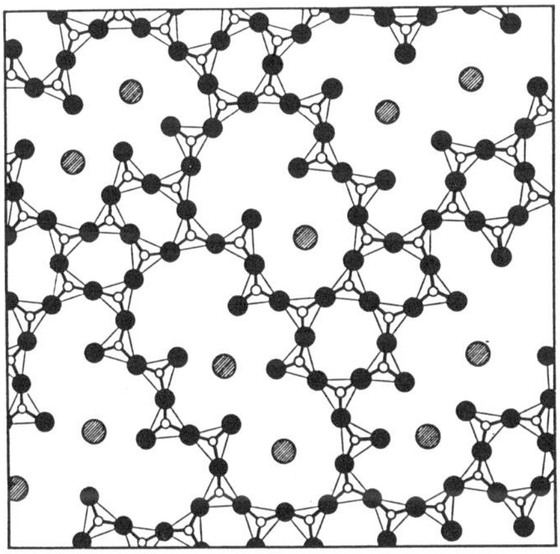

1G Random network theory

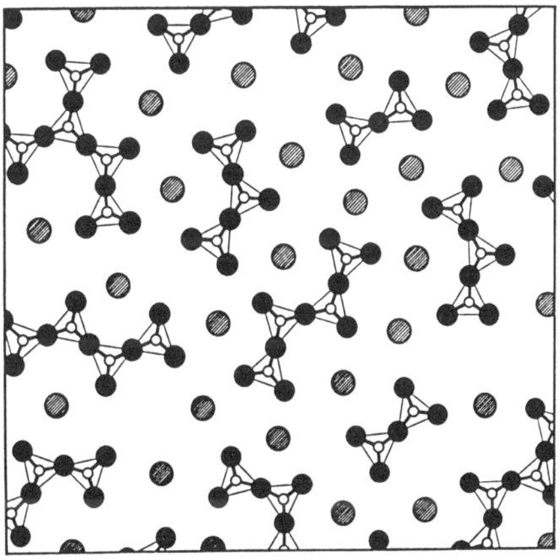

1H Invert glass

stoichiometry which is observed usually being the result of the partial occupancy of well defined atomic sites.

For an amorphous solid, on the other hand, the lack of periodicity and long range order means that to completely define the structure it would be necessary to specify co-ordinates and thermal parameters for every atom present, which is clearly impossible for any real sample. The absence of a simple structural formalism, such as the unit cell used for crystalline materials, means that the maximum that can be extracted from a diffraction experiment on an isotropic amorphous sample is a one-dimensional correlation function (c.f. section 3.6) from which the regeneration of the underlying three-dimensional structure can never be unique. In other words, it is in fact ***impossible*** to uniquely determine the structure of an amorphous solid and it is for this reason that modelling and computer simulation (c.f. Chapter 9) play such an important role in structural studies of amorphous solids and why the choice between possible models involves not only neutron and X-ray diffraction but also a wide range of the other techniques discussed elsewhere in this book. The removal of the constraints imposed by periodicity (c.f. crystalline solids) allows amorphous solids to exhibit structures which are preparation dependent and thus such materials may be said to exhibit continuous polymorphism. For the same reason many amorphous solid systems exist over a continuous range of composition. Hence any structural theory of amorphous solids must be able to accommodate continuous changes in both structure and composition.

For the reasons just outlined, the problems encountered in using diffraction techniques to investigate amorphous solids are somewhat different from those which arise in crystallography. In general, diffraction studies of amorphous solids involve two stages, the actual experiment, culminating in the derivation of a real space correlation function, and the interpretation of this correlation function in terms of a three-dimensional structure. Before discussing these stages in more detail, however, it will first be useful to consider the traditional models for the structure of amorphous solids and then, using these as a basis, to discuss the quantification of amorphous solid structures.

## 2. STRUCTURAL THEORIES AND MODELS

### 2.1 The Random Network Theory

The random network theory was first introduced by Zachariasen [2] to describe the structure of conventional oxide glasses and is illustrated schematically in two dimensions for a pure glass forming oxide $A_2O_3$ in fig. 1C. The corresponding crystalline material is shown in fig. 1A. As pointed out by Zachariasen, the atoms in a glass are linked together by forces essentially the same as those in the corresponding crystals and, if the free energy of the glass is to be comparable to that of the crystal, then it is necessary that the oxygen polyhedra in the glass and crystal are similar. As in the crystals, extended three dimensional networks are formed and the atoms vibrate about definite equilibrium positions, but in the glass the structural units ($AO_3$ triangles in fig. 1C) are linked together randomly to form a non-periodic structure which lacks long range order. It is clear from Zachariasen's original diagram that he expected the structural units in a network glass with a low network forming cation co-ordination number to be as regular as those in the crystal and that, in a three dimensional structure, disorder would be introduced via a distribution of the torsion angles $\alpha_1$ and $\alpha_2$ and the bond angle β as defined in fig. 2. This view is supported by modern diffraction data, but unfortunately subsequent versions of Zachariasen's schematic are much less accurate in that the triangular units have frequently become very distorted.

In multicomponent systems the individual constituents may be classified as network formers or network modifiers, although some intermediate materials can act in either capacity. The addition of a further network former to a single component glass will result in a mixed network of two different structural units. The conventional view of a network modifier, on the other hand, is that it leads to the formation of negatively-charged non-bridging oxygen atoms, as indicated in fig. 3A. Traditionally, the network-modifying cations have been envisaged as occupying nearby holes in the resulting network (c.f. fig. 1G), but modern EXAFS [3] and diffraction evidence [4] suggests that the network modifying cations adapt their local environment to obtain their own desired co-ordination polyhedron as found in related crystalline materials. This modification of the local network-modifying cation environment does not necessarily imply the introduction of regions of crystalline order (crystallites - c.f. section 2.3), but is simply the consequence of radius ratio effects along the lines discussed below for ionic systems.

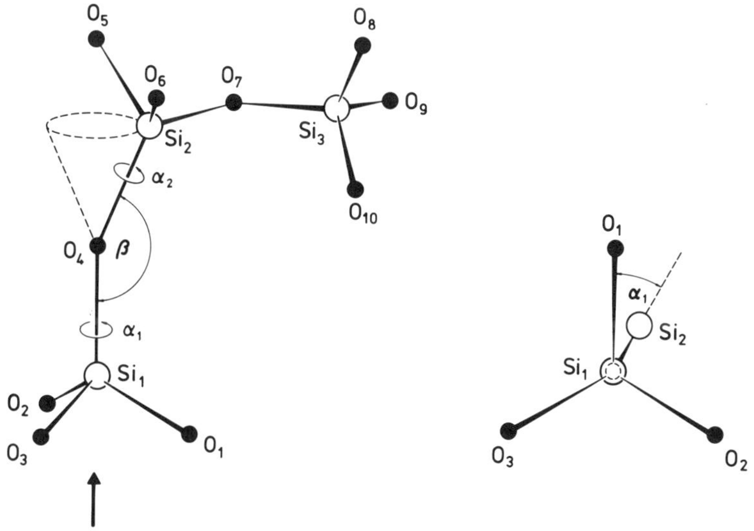


**Figure 2** Definition of the bond angle β and the torsion angles $\alpha_1$ and $\alpha_2$ for vitreous silica.

The above picture is not correct for all multicomponent glass forming systems, particularly when the radius ratio of the network-forming cation is close to that for the transition to a higher co-ordination number. In this case the addition of a network modifier may result in an increase in the network-forming cation co-ordination number, $n_{A(O)}$, such as in the formation of $BO_4$ tetrahedra in borate systems [5] and of $GeO_6$ octahedra in germanate glasses [6] (c.f. fig. 3).

In order to obtain a random network which is relatively distortion free, Zachariasen [2] postulated a set of criteria for glass formation:

(1) An oxygen atom (O) should not be linked to more than two network forming cations (A).

(2) The co-ordination number of oxygen around the network forming cations should be small (i.e. 3 or 4).

(3) Oxygen polyhedra should share corners and not edges or faces.

(4) For a three dimensional network at least three corners of each polyhedron should be shared.

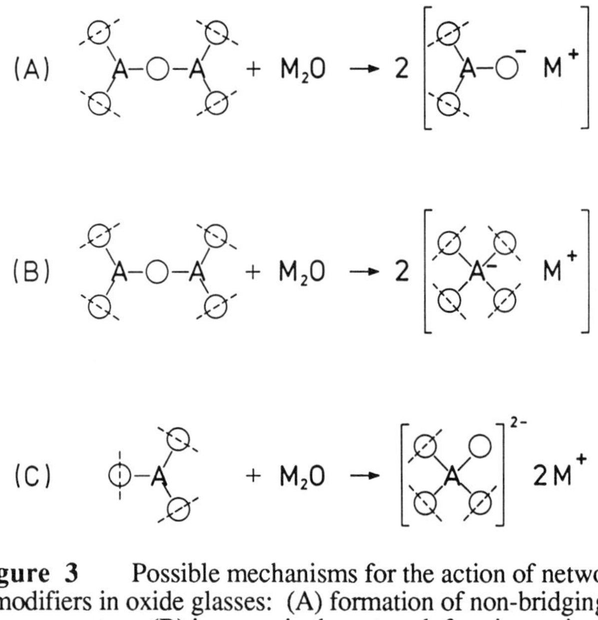


**Figure 3** Possible mechanisms for the action of network modifiers in oxide glasses: (A) formation of non-bridging oxygen atom; (B) increase in the network forming cation co-ordination number and (C) combination of (A) and (B). A circle with a dashed line through it denotes a bridging oxygen atom.

The violation of one of Zachariasen's criteria does not necessarily mean that a glass cannot be formed, merely that it is energetically less favourable than one for which all the criteria are fulfilled. For example, $TiO_6$, $MoO_6$ and $WO_6$ octahedral structural units have all been reported in the glassy state [7,8]. Similarly a three dimensional network is not a necessary prerequisite even for oxide glass formation as demonstrated by the existence of invert glasses [9] (fig. 1H).

Two important points should be noted about the random network theory. First, the perfect (continuous) random network plays the same role in amorphography as does the idealised crystal structure, represented by the unit cell, in crystallography. In real materials there will be "defects" but these will in general be undetectable in an X-ray or neutron diffraction study. Second, the random network theory was formulated to describe conventional oxide glasses quenched from the melt in metastable thermodynamic equilibrium at $T_g$, well below the crystalline melting temperature $T_m$, at relatively slow cooling rates. Thus, although the random network theory has been extended to cover other amorphous network solids such as amorphous semiconductors (e.g. amorphous germanium and hydrogenated amorphous silicon), Zachariasen's arguments, based on free energy considerations, concerning the relationship between the structural units found in the crystal and glass are not necessarily applicable to materials which have been prepared by techniques involving highly non-equilibrium processes such as vapour deposition or glow discharge. For example, the tetrahedral bonding in evaporated amorphous Si and Ge (c.f. later, fig. 6B) is very much more distorted than that in the ambient crystalline modification and both materials contain a much greater fraction of broken bonds than do the corresponding oxide glasses $SiO_2$ and $GeO_2$.

Chalcogenide glasses and the corresponding vapour deposited amorphous solids are in general structurally much more complicated than their oxide counterparts in that even two element systems such as Ge-Se exist over a wide range of composition and contain bonds between like atoms. Models for the numbers of each type of bond present range from the random covalent model [10,11] to complete chemical ordering. The former places no restriction on the bonds which may be formed, the number of each type of bond being determined solely from the 8-N rule and the concentration of each atomic species. A chemically ordered system, on the other hand, contains the maximum number of bonds between unlike atoms. In the vitreous state, the relevant bond strengths are a useful guide to the relative number of each type of bond likely to be formed, but in vapour deposited materials the balance may be significantly affected by the molecules present in the vapour phase. For example vapour deposited $As_{10}S_{13}$ contains many more As-As bonds than the minimum required by stoichiometry (As-As bonds are energetically the least favourable in this system), as a consequence of the presence of $As_4S_4$ realgar molecules in the vapour phase [12] (see section 7.6).

## 2.2 The Locally Ordered Random Packing Model

For metallic and ionic systems with no well-defined directional bonding the concept of an amorphous network is inappropriate and a much more relevant first-order model is a random packing of spheres. In such structures the important concepts are those of radius ratio, packing density and hole filling. These simple ideas are frequently used in a discussion of metallic and ionic crystal structures, but are often neglected when considering those of the equivalent glasses. Brunner [13] has presented an unconventional view of crystalline close packing in which he argues that, given a finite volume, the close packed arrangement results from the requirement that the separation between neighbouring atoms/ions should be a maximum. In this case the spatial distribution of the centres is considered and assumptions of spherical shape and mutual contact are in fact unnecessary. The mutual repulsion idea can clearly be extended to ionic glasses, and to the ion cores in metallic systems, and, in the glass, crystalline close packing will be replaced by random close packing.

In systems where the bonding is at least partially ionic in character, the co-ordination polyhedra present are to a large extent determined by the relative sizes of their constituent ions, usually expressed as the radius ratio, $r_A/r_X$ in which A is the cation and X is the anion. (Anions are in general larger than cations, hence $r_A/r_X<1$). For stability, the cation at the centre of an $AX_n$ polyhedron should fill the

**Table 1. Critical Radius Ratios for Common Co-ordination Polyhedra and Bernal Holes**

| Polyhedron | $(r_A/r_X)_c$ | A-X | | A-(2)X | | Hybrid Orbital |
|---|---|---|---|---|---|---|
| | | $n_{A(X)}$ | $r_{A\text{-}X}/r_X$ | $n_{A(X)}$ | $r_{A\text{-}X}/r_X$ | |
| Line | 0.000 | 2 | — | — | — | sp,dp |
| Triangle | 0.155 | 3 | 1.155 | — | — | $sp^2,ds^2$ |
| Trigonal pyramid | — | 3 | — | — | — | $p^3,d^2p$ |
| Tetrahedron | 0.225 | 4 | 1.225 | — | — | $sp^3,d^3s$ |
| Trigonal bipyramid | — | 5 | — | — | — | $sp^3d,d^3sp$ |
| Dodecadeltahedron | 0.353* | 4 | 1.353 | 4 | 1.860 | — |
| Square | 0.414 | 4 | 1.414 | — | — | $sp^2d,p^2d^2$ |
| Octahedron | 0.414 | 6 | 1.414 | — | — | $d^2sp^3$ |
| Tetragonal pyramid | — | 4 | — | — | — | $d^4$ |
| Trigonal prism | 0.528 | 6 | 1.528 | 3† | 1.992 | $d^4sp$ |
| Archimedean antiprism | 0.645 | 8 | 1.645 | 2† | 2.255 | $d^4sp^3$ |
| Cube | 0.732 | 8 | 1.732 | — | — | — |
| Icosahedron | 0.902 | 12 | 1.902 | — | — | — |

* for atom at centre. A larger atom can be inserted off-centre.
† refers to half octahedron cap.

cavity defined by the anions and there exists a critical radius ratio at which this criterion is just fulfilled, and below which the cation cannot remain in contact with all the anions and the $AX_n$ polyhedron is unstable. Above this value, the $AX_n$ polyhedron will remain stable until the point is reached where extra X atoms can be incorporated. The critical radius ratio $(r_A/r_X)_c$ is given in table 1 for the common co-ordination polyhedra. In general, small radius ratios (n = 3 or 4) tend to result in well defined structural units whereas larger radius ratios (larger values of n) allow a greater latitude for distortion and frequently lead to the presence of more than one type of polyhedron. The effective radius ratio for a given pair of ions is a function of the degree of ionicity of their bond, and small values of n are often associated with considerable covalent bonding character, because of the compatibility of the corresponding polyhedral shapes with common bond hybridization geometries (c.f. table 1). Conversely, at high values of n, there is an increased degree of ionicity in the bonds and the different polyhedra are determined by ionic size (radius ratio) and geometrical packing. There is also an increased probability of both edge and face sharing between adjacent polyhedra since the additional strain energy required is much less.

For a random close packing of equal sized spheres the packing density, $\eta$, is equal to 0.6366 [14], which may be compared to the value of 0.7405 for crystalline close packing. In the latter case the remaining space is distributed between octahedral holes (one per atom) and tetrahedral holes (two per atom) and a number of common binary crystal structures can be generated by inserting appropriately sized cations into the holes in such a close packing of anions (see for example Chapter 2 of ref. [15]). The types of hole in a random close packing of hard spheres have been discussed by Bernal [16] who identifies the five ideal canonical holes shown in fig. 4. Subsequent work by Finney and Wallace [17], however, indicates that for soft spheres the more complex holes disappear and the hole structure can be considered as comprised of distorted octahedra and tetrahedra.

Polk [18,19] has suggested a model for the transition metal - metalloid glasses in which metalloid atoms, M, are inserted into the larger Bernal holes [16] in a random close packed array of transition metal atoms, T. If all such holes are filled the composition is very close to that ($T_4M$) of the deep eutectic which occurs in many of these systems. Optimum glass formation occurs around this eutectic composition and Polk argues that the presence of the metalloid atoms in the holes stabilises the structure. The main objection to Polk's model as originally proposed is that the metalloid atoms are considered to be almost totally passive which is contrary to experience with amorphous solids where the absence of a periodic structure means that an added atom/ion can very much more easily adapt its immediate environment to satisfy its own local bonding/radius ratio requirements, than is the case in crystalline materials. As discussed by Gaskell [20,21] the metalloid co-ordination polyhedron in the corresponding crystalline T-M alloys is the trigonal prism and hence it is likely that metalloid atoms inserted initially into other types of Bernal hole would modify their environment accordingly.

Similar arguments may also be applied to the heavy metal fluoride glasses. Since the $F^-$ species is strongly ionic, it is clear that its electric field cannot be screened until it is surrounded by particles of opposite charge. It is this character which allows the monovalent fluorine ion to form strong cation-anion-cation bridges, leading to the sluggish relaxation behaviour (increase in viscosity) near the freezing point necessary to favour glass formation. The structural differences between fluoroberyllate glasses on the one hand, which have network structures similar to those of silicate glasses, and the heavy metal fluoride glasses on the other, arise simply from the different radius ratios of the glass forming cations involved. As discussed above, the larger radius ratios found in the latter systems

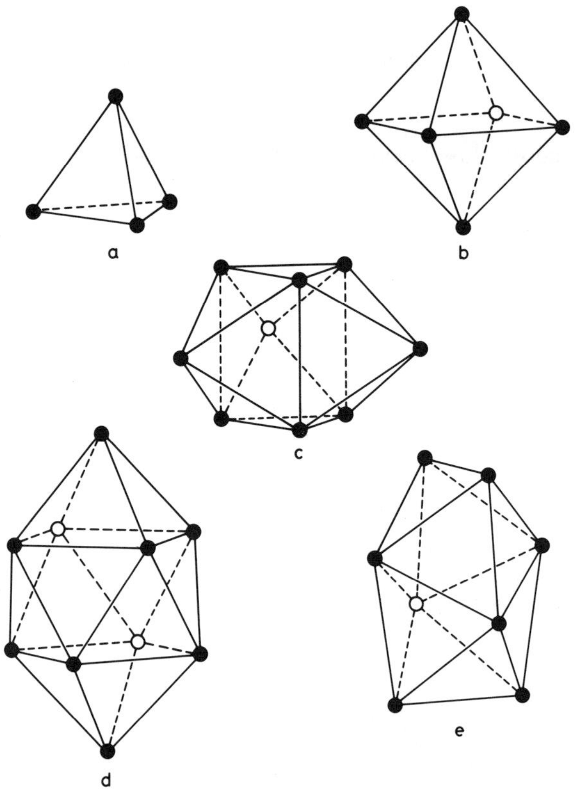


**Figure 4** The five canonical holes of a random close packing [16]. (a) tetrahedron, (b) octahedron, (c) trigonal prism (capped with three half octahedra), (d) Archimedean anti-prism and (e) dodecadeltahedron.

lead to less well defined co-ordination polyhedra and a range of polyhedral types. For example the average co-ordination number $n_{Zr(F)}$ in fluorozirconate glasses is not an integer indicating that not all the $Zr^{4+}$ ions have the same environment, even in pure $ZrF_4$. This result is consistent with the crystalline fluorozirconates which exhibit $ZrF_n$ polyhedra with n equal to 6, 7 and 8 [22]. Glass formation in heavy metal fluoride systems results either from the random packing of different polyhedral types or from the inefficient packing of distorted polyhedra which do not correspond to space-filling geometries. In general, the variation in bond ionicity in heavy-metal fluoride glasses is much less than in multicomponent oxide systems. Hence the classification of constituents as either network formers or network modifiers is not appropriate. It is primarily the absence of a well defined co-ordination number and first co-ordination shell which makes the analysis of random packed structures so much more difficult than their random network counterparts.

## 2.3 Other Structural Theories and Models

There have been a number of other structural theories/models of amorphous solids and the most important of these will now be summarised, since they frequently form a basis for the interpretation of diffraction data. Early workers involved in X-ray diffraction studies of glasses noted that the diffuse halos obtained frequently coincided with prominent powder diffraction peaks for related crystalline materials but were very much broader. This was interpreted as support for the crystallite theory of Lebedev [23] which envisaged glass as an assembly of very small crystals, termed crystallites. It is now generally accepted, however, mainly due to the work of Warren [24], that discrete crystallites do not exist in simple glasses and the modern crystallite theory [25,26] (figs 1B and 1F) argues in terms of fluctuations in the degree of intermediate range order, more highly ordered regions (crystallites), where the atomic arrangement approaches that in related crystalline materials, being interconnected by regions in which the degree of order is somewhat less. For multicomponent systems the crystallite theory leads to clustering and a non-uniform spatial distribution of the individual components. In any random network, constructed from units with well-defined bonding requirements (e.g. the $SiO_4$ tetrahedron), there are bound to be fluctuations in the degree of local order. Hence the essential difference between the crystallite and random network theories lies in the magnitude of these fluctuations, particularly in the range 10-100Å.

Two structural models closely related to the crystallite theory are those involving layers or amorphous clusters in that they both envisage more ordered regions linked together by less ordered connecting material. Amorphous clusters are groups of atoms in a regular but non-crystallographic configuration, frequently based on pentagonal dodecahedra, icosahedra or five-fold symmetry. Cluster size is limited by the fact that the strain energy increases as the clusters become larger. Layer formation has been postulated in various amorphous solids, particularly where the corresponding crystalline polymorphs have layer structures such as in chalcogenide systems (e.g. Ge-X and As-X with X = S or Se) and those containing three-fold co-ordinated atoms [27]. The problem with almost all models of this type, including those based on the crystallite theory, is that they are incomplete because they concentrate on the more ordered regions and fail to specify the interconnecting material. In order to interconnect crystallites, amorphous clusters or regions of roughly parallel layers, without introducing too much strain or giving rise to excess scattering at small scattering vectors (c.f. section 3.4), the interconnecting regions become substantial and hence represent the major fraction of the total volume. In order to demonstrate that such a model is valid, therefore, it is essential to compare a complete model with experiment, which includes the correct volume fraction of interconnecting material.

The structures of amorphous solids containing all or predominantly two-fold co-ordinated atoms may be likened to those of linear organic polymers, although an added complication in the case of S and Se is that these elements can exist as either chains or rings (molecules), the relative numbers of each in the various amorphous forms being both controversial and preparation dependent. Possible first order polymer models, which may be adapted for chain structures, are the random coil [28] and the various amorphous bundled and folded chain models as summarised by Sperling [29]. By analogy with polymers the addition of atoms of higher covalency (e.g. As or Ge) leads to branching and cross-linking as in the chain crossing model [30] which is identical to the "dilute-x" model of Bienenstock and coworkers [10,11].

The final type of amorphous solid structure which should be mentioned is that comprising a random packing of individual molecules. Inorganic molecular amorphous solids are usually, but not always, prepared by vapour deposition as for example in the case of amorphous $CCl_4$ [31]. $As_2O_3$ and $P_2O_5$ are also thought to exist as molecular amorphous solids and at least some vapour deposited chalcogenide thin films contain molecules or molecular fragments [12]. Molecular phosphorus, $P_4$, on the other hand, is known to exist in the liquid state and may be maintained as a supercooled liquid for long periods of time [32]. It is not clear, however, whether the supercooled liquid can be quenched to yield a true molecular glass. A molecular amorphous solid is characterised by having two distinct types of bonding, strong covalent bonds within the molecules and a weaker Van der Waals attraction between molecules. As with molecular liquids the simplest structural model assumes that there is no orientational correlation between adjacent molecules and that the molecules themselves are roughly spherical. For real materials, however, it is necessary to consider such orientational correlations, which are caused by departures of the molecules from spherical symmetry. An extreme case in this regard is amorphous ice where hydrogen bonding leads to a network structure not unlike that of vitreous silica [33].

## 3. THE QUANTIFICATION OF AMORPHOUS SOLID STRUCTURES

Although the concepts of a random network and a random sphere packing are easy to describe qualitatively, there is no way to completely define the resulting structures quantitatively other than by specifying the co-ordinates of every atom present as indicated in section 1. It is therefore necessary in practice to work with a series of statistically averaged distributions of parameters which are the subject of the present section. The parameters themselves are summarised in table 2. Traditionally the structural order within an amorphous solid has been divided into three ranges: short, intermediate (medium) and long. However, particularly for systems with well defined directional bonding, it is more convenient to use the four ranges in table 2 since not only do they relate more directly to the analysis of X-ray and neutron scattering data but also the boundary between short and intermediate range order is not clearly defined in the literature, different authors including a varying number of the range II parameters under the heading of short range order.

**Table 2. Ranges of Order for an Amorphous Solid $A_{1-x}X_x$**

| Range | Order | Characteristic Parameters | Symbol |
|---|---|---|---|
| I | Structural unit (s.u.) or co-ordination polyhedron | Identity | $AX_n$ |
| | | Internal co-ordination number | $n_{A(X)}$ |
| | | Bond length | $r_{A\text{-}X}$ |
| | | Bond angle | $\beta_A$ |
| | | Superstructural unit definition | — |
| II | Interconnection/ relative orientation of adjacent units | Connection mode | C |
| | | Connectivity | c |
| | | Bond angle | $\beta_X$ |
| | | Bond torsion angle | $\alpha$ |
| III | Network topology/ order beyond adjacent unit | Shortest path ring size | m |
| | | Network dimensionality | D |
| | | Topological cluster type | — |
| IV | Longer range fluctuations | Type (composition or density) | — |
| | | Morphology (droplet, spinodal decomposition, etc.) | — |
| | | Radius of gyration | $R_g$ |
| | | Inter-region separation | R |

## 3.1 Range I - The Structural Unit / First Co-ordination Shell

The most important fact to establish for any amorphous solid is the identity of any structural unit(s) or co-ordination polyhedra present. In materials where there is directional bonding, usually characterised by the 8-N rule, this leads to well-defined structural units such as the $SiO_4$ tetrahedron in vitreous $SiO_2$ and the $AsS_3$ pyramid in melt-quenched stoichiometric $As_2S_3$ glass. In more complex systems there may be more than one structural unit and there is also the possibility of discrete combinations of basic structural units in the form of "superstructural units", such as the $B_3O_6$ boroxol group, molecules ($P_4O_{10}$, $P_4Se_3$, $As_4S_4$ etc.) or polyanions. The complete characterisation of a given structural unit or co-ordination polyhedron involves a number of internal parameters including co-ordination numbers and the detailed distribution of bond lengths and angles arising from both static and thermal disorder. Larger units may also require the specification of internal bond torsion angles.

In random packed systems with no clearly identifiable co-ordination polyhedra, the first co-ordination shell surrounding each atom may be analysed in terms of Voronoi polyhedra as illustrated in two dimensions in figs. 5A and 5B. For a monatomic packing, the Voronoi polyhedron (polygon in two dimensions) surrounding any given atom is the smallest polyhedron (polygon) formed by the planes (lines) which perpendicularly bisect the lines joining the atom to its neighbours. Atoms whose Voronoi polyhedra (polygons) share faces (edges) are termed geometric neighbours and the network of dashed lines joining the centres of the geometric neighbours (fig. 5B) is known as the simplicial graph. The geometric structure of such a packing may thus be characterised either by the distribution of Voronoi polyhedron types or by the simplicial graph. An extension of the Voronoi analysis to packings containing more than one size of atom has been proposed by Gellatly and Finney [34] and involves dividing the lines of the simplicial graph using a partitioning method based on radical planes.

## 3.2 Range II - The Interconnection / Relative Orientation of Adjacent Units

For systems in which the structural units or co-ordination polyhedra are bounded by atoms with a co-ordination number of 2, the connection mode is given by the number of atoms, C, which adjacent units have in common. The most frequent arrangement in simple covalent systems is corner sharing (C = 1), although edge sharing (C = 2) does exist in chalcogenide glasses such as $SiSe_2$ [35] and $GeSe_2$ [36]. In molecular systems there are no shared atoms between adjacent molecules (C = 0) while face sharing (C = 3) is only found between polyhedra in materials with little covalent character in their bonding, such that the co-ordination number is high ($n_{A(X)} \gtrsim 8$) and the units themselves are relatively easily distorted [37] (c.f. section 2.2).

The structural unit connectivity (i.e. the number of neighbours connected to an individual structural unit) is single valued for simple structures but in more complex systems there may be a distribution of connectivities. Thus in silicate glasses the network connectivities of the $SiO_4$ tetrahedra present may be determined by $^{29}Si$ Magic Angle Spinning NMR, the various tetrahedral types being denoted $Q_i$ where i is the number of bridging oxygen atoms (see Chapter 3).

(A)

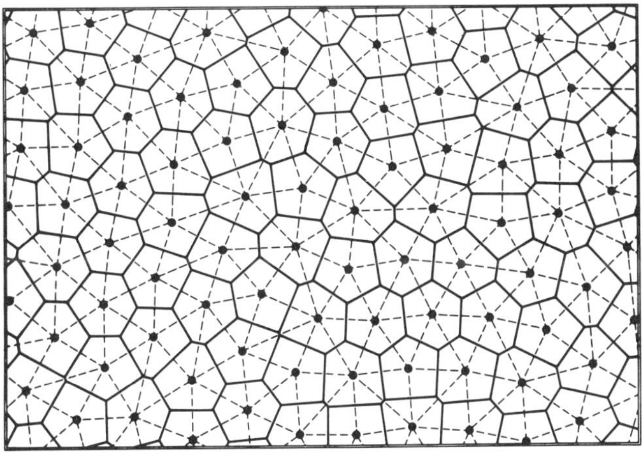

(B)

**Figure 5** Two dimensional analogue of the geometrical analysis of a random close packing. (A) original packing, (B) simplicial graph (-----) and Voronoi polygons (———).

The number of variable parameters required to specify the interconnection and relative orientation of adjacent structural units decreases with increasing C, i.e. the greater the number of shared atoms the lower the number of degrees of freedom. For example, the interconnection of two corner-sharing $SiO_4$ tetrahedra involves one bond angle $\beta$ and two torsion angles $\alpha_1$ and $\alpha_2$ as defined in fig. 2. Note that there may be correlations between $\alpha_1$ , $\alpha_2$ and $\beta$ due to steric effects, particularly at lower values of $\beta$. It is in range II that major disorder first arises for a conventional melt quenched glass in excess of that for the corresponding

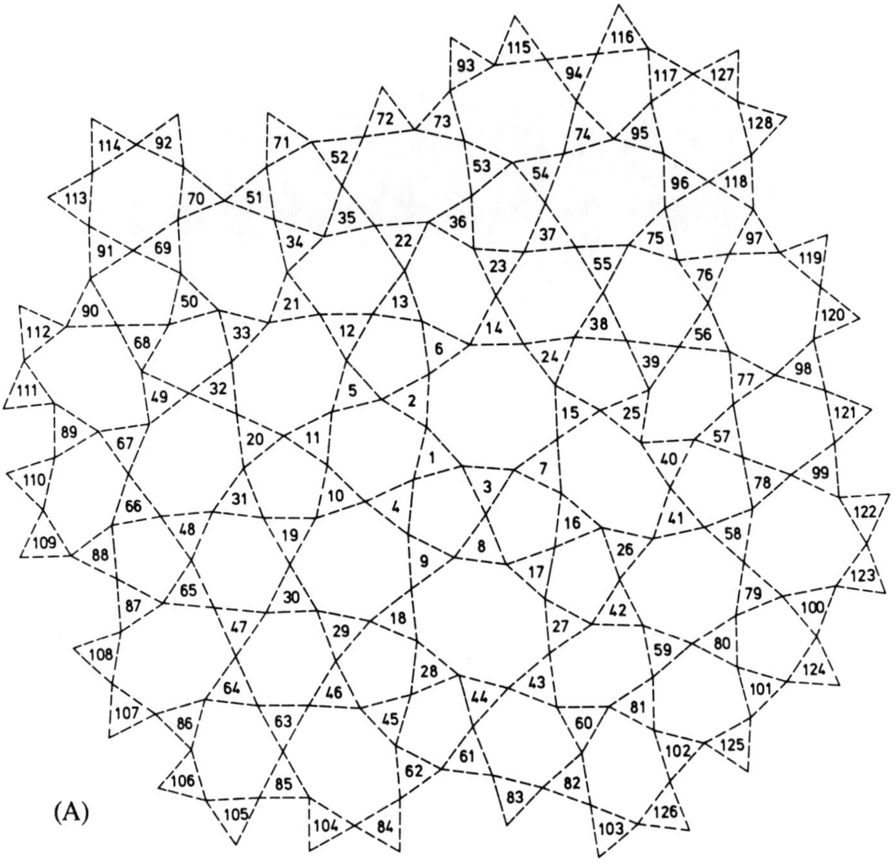


**Figure 6** Decoration of a topological network. The topological units, represented by the dashed triangles in (A), are decorated in (B) to represent an amorphous elemental semiconductor and in (C) to represent a chemically-ordered $A_2X_3$ glass.

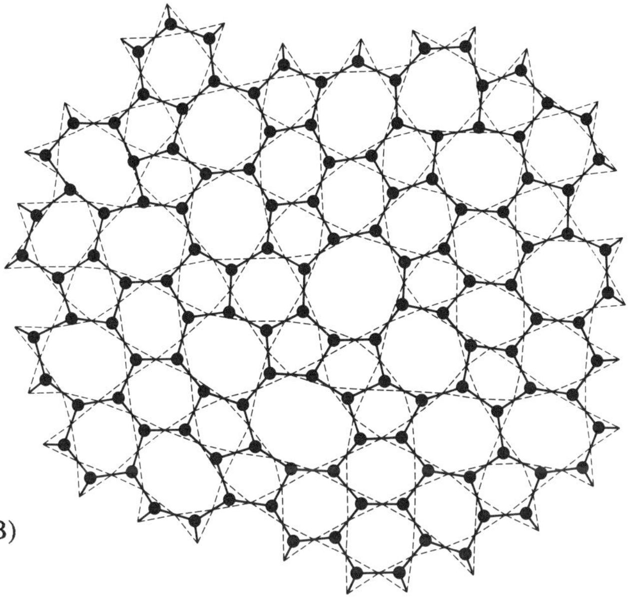

(B)

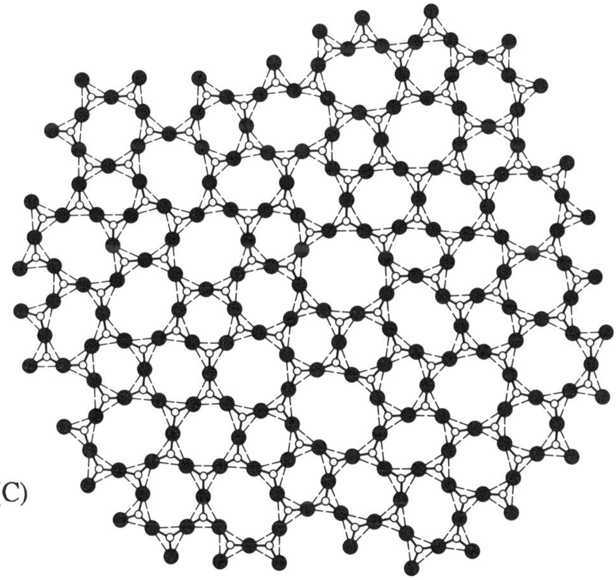

(C)

crystalline material. It is important to realise that the formation of a random network structure on quenching from the melt requires such degrees of freedom, if the network is to be continuous with relatively few broken bonds. Without the necessary degrees of freedom the material will crystallise or else a more non-equilibrium preparation technique is required which can accommodate increased strain and a larger number of broken bonds. In systems containing more than one structural unit it is also necessary to specify which pairs can be interconnected and with what frequency.

### 3.3 Range III - The Network Topology / Order Beyond the Adjacent Unit

A useful concept in discussing the so-called intermediate range order for an amorphous network solid is that of an underlying topological network which can then be decorated using various atomic motifs to represent different amorphous solids [38]. This is illustrated schematically in two dimensions in fig. 6. The topological network, indicated by the corner sharing dashed triangles in fig. 6A, is decorated in fig. 6B to represent an elemental amorphous solid and in fig. 6C a chemically ordered compound of composition $A_2X_3$. With 4-connected units in three dimensions, figs. 6B and 6C would correspond respectively to the group IV amorphous semiconductors Si and Ge and the $AX_2$ tetrahedral network glasses.

The topology of the network in fig. 6A may be completely specified either in terms of the symmetric connectivity matrix

$$c_{pq} = \begin{bmatrix} 0 & 1 & 1 & 1 & 0 & 0 & 0 & 0 & \cdots \\ 1 & 0 & 0 & 0 & 1 & 1 & 0 & 0 & \cdots \\ 1 & 0 & 0 & 0 & 0 & 0 & 1 & 1 & \cdots \\ 1 & 0 & 0 & 0 & 0 & 0 & 0 & 0 & \cdots \\ 0 & 1 & 0 & 0 & 0 & 0 & 0 & 0 & \cdots \\ 0 & 1 & 0 & 0 & 0 & 0 & 0 & 0 & \cdots \\ 0 & 0 & 1 & 0 & 0 & 0 & 0 & 0 & \cdots \\ 0 & 0 & 1 & 0 & 0 & 0 & 0 & 0 & \cdots \\ . & . & . & . & . & . & . & . & \cdots \\ . & . & . & . & . & . & . & . & \cdots \end{bmatrix} \tag{1}$$

or a first neighbour table as shown in table 3. For a model either of these forms may be used although the latter has the advantage of requiring much less computer memory - nN locations instead of $N^2$, N being the number of topological units. The connectivity (co-ordination number) of the pth topological unit is given by

$$n_p = \sum_{q=1}^{N} c_{pq} \quad (2)$$

As with the atomic co-ordinates, it is impossible to specify either $c_{pq}$ or the near neighbour table for any real material and so it is necessary to work with a much simpler statistical average. The quantity usually employed is the ring statistics obtained by shortest path analysis [39], M(m), which have the advantage that the total number of rings per topological unit is well defined [n(n-1)/2 for an n-connected unit] and at large ring sizes tends to zero. The shortest path ring statistics are obtained by taking every pair of vertices on each topological unit in turn and then finding the smallest sized ring, expressed as a number of topological units, in which they are both contained. The shortest path ring statistics for the topological network of fig. 6 are given in fig. 7. If all the topological units in a finite sized model are taken as centre, the percentage ring closure is less than 100%, due to edge effects. The percentage closure for fig. 6 is 78%. Marians and Hobbs [40] have extended these ideas to consider the multiplicity of the shortest path rings through each pair of vertices on a given topological unit and hence formulated the idea of a topological cluster of shortest-path rings which can for example distinguish between the crystalline cristobalite and tridymite structures both of which have only six-membered shortest-path rings.

**Table 3. First Neighbour Table for the Topological Network in Figure 6**

| Topological unit | Neighbours | | |
|---|---|---|---|
| | 1 | 2 | 3 |
| 1 | 2 | 3 | 4 |
| 2 | 1 | 5 | 6 |
| 3 | 1 | 7 | 8 |
| 4 | 1 | 9 | 10 |
| 5 | 2 | 11 | 12 |
| 6 | 2 | 13 | 14 |
| 7 | 3 | 15 | 16 |
| 8 | 3 | 9 | 17 |
| . | . | . | . |
| . | . | . | . |
| . | . | . | . |

No method has yet been devised for measuring M(m). However, more limited information as to the presence of particularly the smaller sized rings can be obtained from the correlation function, vibrational spectroscopy (Chapter 2) and nuclear magnetic resonance experiments (Chapter 3). Gaseous absorption measurements also provide information on ring statistics and as a result of such measurements Shackelford and Brown [41] have suggested that the distribution of ring sizes in a random network glass is log-normal.

Another important parameter which characterises an amorphous network is its dimensionality, D, or more likely its predominant dimensionality since the presence of disorder will lead to occasional fluctuations in D. Thus as an alternative to a truly three dimensional network an amorphous solid may exist as molecules (D = 0), chains (D = 1) or layers (D = 2).

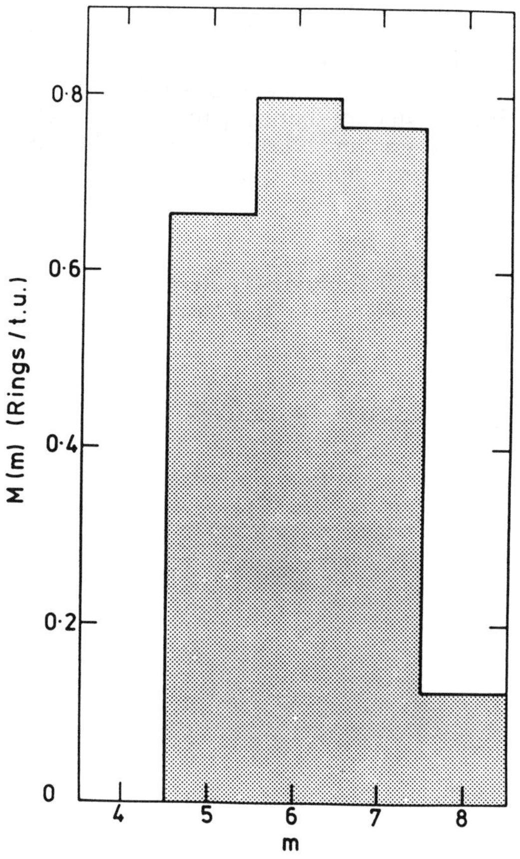


**Figure 7** Shortest path ring statistics for the topological network of fig. 6.

### 3.4 Range IV - Long Range Fluctuations

Although amorphous solids lack long range order there may nevertheless be longer range fluctuations in density and/or composition arising from phase separation, etc. The characteristic distances involved in such fluctuations are much greater than those studied in conventional diffraction experiments and hence require the use of small angle scattering techniques which are beyond the scope of the present chapter. An account of small angle scattering techniques, as applied to amorphous solids, is given in ref. [42].

### 3.5 The Overall Structure

It is important to realise that the order in ranges I to III is correlated. The stereochemistry and regularity of the structural unit will limit the variation of range II parameters that can be accommodated into the structure. Similarly both the range I and range II order determine the extent of topological space which the network can explore. Conversely, for a given set of interatomic potentials the requirement that the static strain energy be minimised means that for a particular topology the range I and II order is fixed. This is especially important in respect of modelling studies of amorphous solids in that if, on relaxation with realistic potentials, a model fails to yield the correct range I and II order then that model must be incorrect. Similarly a model constructed from units which constrain the range I and II order to be incorrect will inevitably lead to the wrong network topology. Examples of both types of incorrect model are given later in sections 6 and 7.

### 3.6 Experimental Parameters

In general the parameters which characterise the order in ranges I to III are not those which are obtained from experiment, diffraction or otherwise. The isotropic nature of most amorphous solids leads naturally to a structural description in terms of a one-dimensional real space correlation function, the various forms of which are illustrated for a monatomic material in fig. 8. The radial density $\rho(r)$ describes on average the number density a distance r from an arbitrarily chosen origin atom and is a gross average over all the atoms in the sample taken as centre. $\rho^\circ$ is the average number density. $\rho(r)$ is essentially a one dimensional form of the crystallographer's Patterson function.

The number of neighbours between r and r + dr is given by

$$g(r)dr = 4\pi r^2 \rho(r) dr \tag{3}$$

in which g(r) is known as the radial distribution function. The correlation function

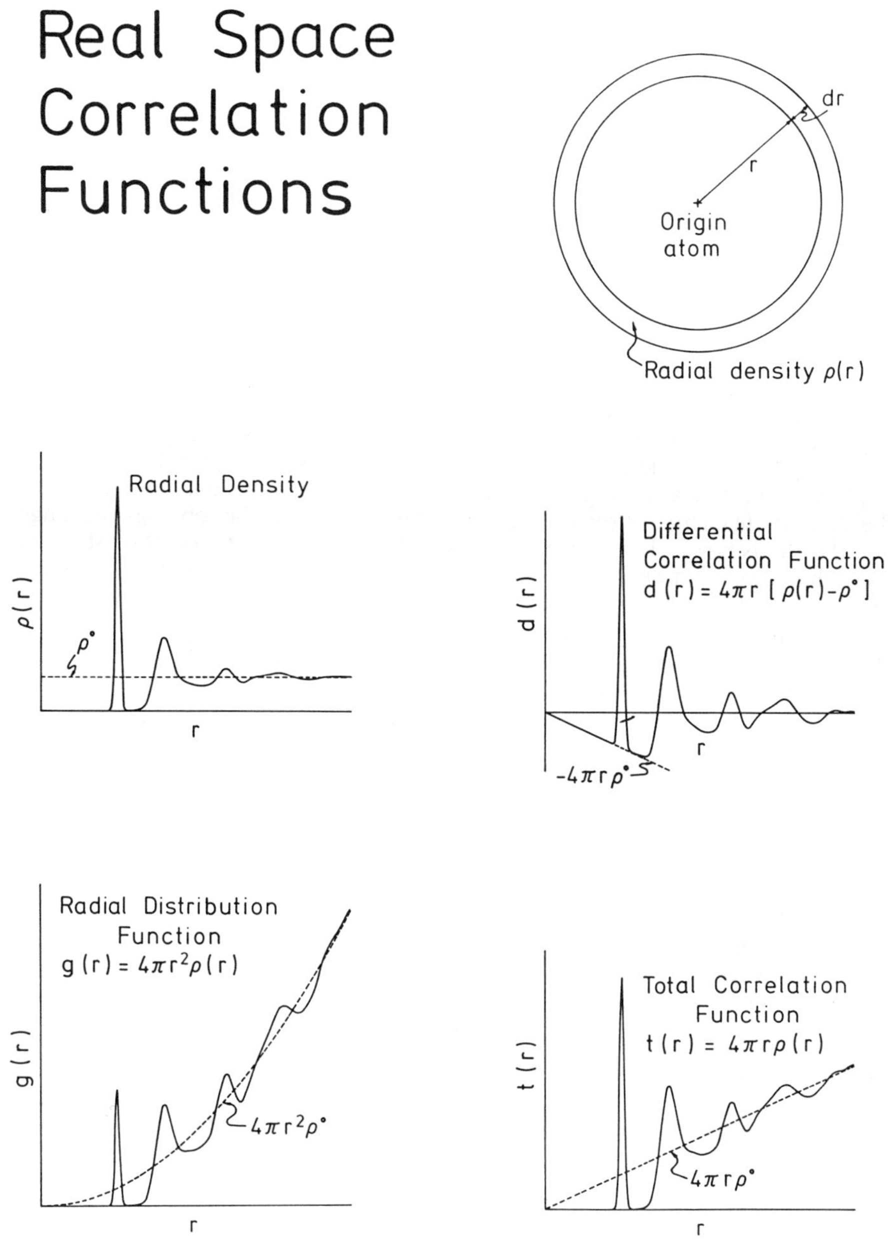


**Figure 8** Real space correlation functions for a monatomic amorphous solid.

$$t(r) = 4\pi r\rho(r) \tag{4}$$

is best used in the analysis of diffraction data since it is in this function that experimental broadening is both symmetric and r-independent. The differential correlation function d(r) is obtained on subtracting the average density contribution

$$t^o(r) = 4\pi r\rho^o \tag{5}$$

from t(r)

$$d(r) = 4\pi r[\rho(r) - \rho^o] \tag{6}$$

For a sample containing n atom types (usually elements, but it may be important to distinguish chemically different species of the same element such as the H atoms in $CH_3OH$), the structure may be described in terms of n(n+1)/2 independent component correlation functions

$$t_{ij}(r) = 4\pi r\rho_{ij}(r) = d_{ij}(r) + t_j^o(r)$$

$$d_{ij}(r) = 4\pi r[\rho_{ij}(r) - \rho_j^o] \tag{7}$$

$$t_j^o(r) = 4\pi r\rho_j^o$$

which specify the distribution of element j atoms about the $i^{th}$ atom in the sample composition unit. Any single X-ray or neutron diffraction experiment measures a linear combination of these components as detailed in the next section.

## 4. THEORY OF NEUTRON AND X-RAY DIFFRACTION BY AMORPHOUS SOLIDS

In a scattering experiment X-rays or neutrons of wavevector $\mathbf{k}_o$ and energy $E_o$ are incident on the sample and are scattered through an angle $2\theta$ with a final wavevector $\mathbf{k}$ and energy E (c.f. fig. 9). Each scattering event may thus be

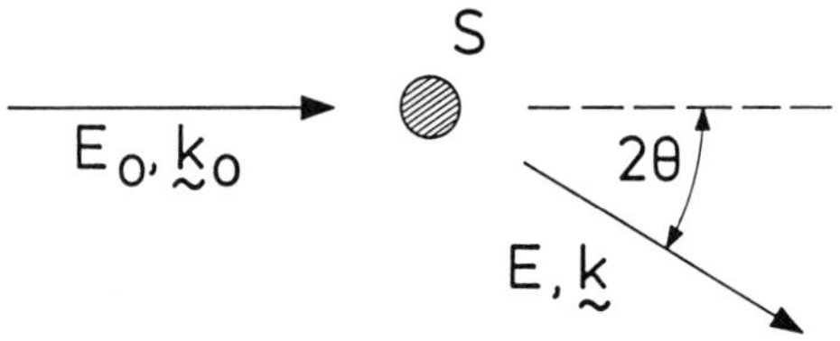


**Figure 9** A scattering event. S is the sample, $2\theta$ the scattering angle, $E_0$ and E the incident and final energies and $\mathbf{k}_0$ and $\mathbf{k}$ the corresponding wavevectors.

characterised by a momentum transfer

$$\hbar\mathbf{Q} = \hbar\mathbf{k}_0 - \hbar\mathbf{k} \tag{8}$$

and an energy transfer

$$\hbar\omega = E_0 - E \tag{9}$$

($\hbar Q$ and $\hbar\omega$ are defined as the momentum/energy transferred to the sample.)

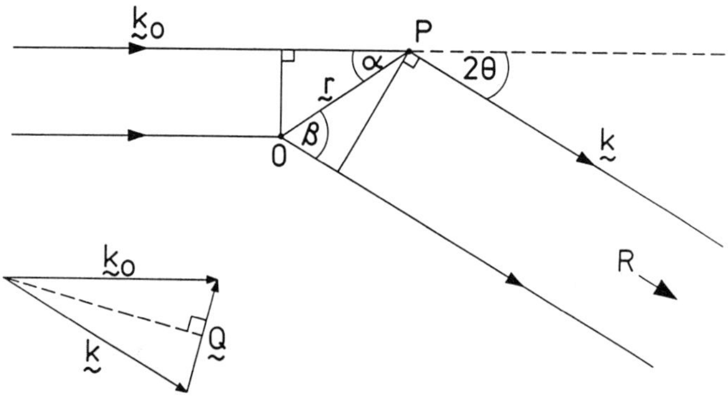


**Figure 10** Scattering geometry. See text for a definition of the various symbols.

The theory of neutron and X-ray diffraction by amorphous solids may be formally derived in terms of the Van Hove [43] space time correlation function $G(\mathbf{r},t)$ and the corresponding scattering law $S(\mathbf{Q},\omega)$ and such a derivation, using the same nomenclature as that below, has been published elsewhere by the present author and co-workers [44-47]. However, in view of the intended readership of this particular book, a much simpler derivation will be given here along the lines of that by Urnes [48] and that for diffraction by crystals found in elementary texts on crystallography.

For a diffraction measurement the relevant value of the scattering vector $\mathbf{Q}$ is that for elastic scattering. In this case the X-ray or neutron wavelength, $\lambda$, is unchanged on scattering and the magnitude of $\mathbf{Q}$ becomes

$$Q = \frac{4\pi}{\lambda}\sin\theta \tag{10}$$

as may be seen from the vector diagram in fig. 10. (Note that $k = k_o = 2\pi/\lambda$). In practice the detector in a conventional amorphous diffraction experiment actually records both the elastic scattering and the inelastic scattering arising from the thermal vibrations (hence the term total diffraction to distinguish this type of measurement from true elastic diffraction [46] in which an analyser is used in the diffracted beam to remove the inelastic component) and thus performs an integration over $\omega$ at a constant scattering angle $2\theta$. As a result the quantity obtained is the instantaneous space time correlation function $G(\mathbf{r},0)$ (c.f. ref. [7]), which may be envisaged as the average of a large number of "snapshots" of the system, one for each neutron or X-ray photon scattered. It is for this reason that this type of measurement can be used to study liquids where the local structure is constantly fluctuating with time.

There are important differences between the scattering of X-rays and neutrons by atoms, as summarised in table 4, which mean that the two techniques should be regarded as complementary rather than as competitive. The much smaller mass of an electron compared to that of the nucleus means that the scattering of X-rays is essentially electronic in character. In a diffraction experiment the incident X-ray wavelength is comparable to the atomic size and hence the atomic scattering amplitude shows the characteristic form factor dependence $f(Q)$ illustrated for Fe in fig. 11. The fall off in $f(Q)$ with increasing Q in practice limits the maximum value of Q to which it is possible to obtain X-ray data from amorphous materials.

For an electron the non-magnetic neutron scattering length $b_e$ is much less than that for a nucleus and hence neutron scattering by non-magnetic systems is predominantly nuclear in origin. Also, since the nuclear size is very much smaller than the incident neutron wavelength the neutron scattering amplitude for an isolated bound atom is independent of Q. Measurements can therefore be made to much higher Q than is currently the case for X-ray scattering. An important consequence

**Table 4. Comparison of X-rays and Neutrons**

| No. Property | X-rays | Neutrons |
|---|---|---|
| *General properties* | | |
| 1 Wavelength | 0.5-2.5 Å | 0.1-20 Å |
| 2 Energy ($\lambda$=1Å) | 12.4 keV | 80 meV (c.f. thermal and magnetic excitations |
| 3 Velocity ($\lambda$=1Å) | $3 \times 10^8$ ms$^{-1}$ | $4 \times 10^3$ ms$^{-1}$ (hence time-of-flight techniques) |
| 4 Production | (a) X-ray tube (characteristic line & continuous spectra)<br>(b) Synchrotron radiation | (a) Nuclear reactor<br>(b) Electron linear accelerator pulsed source<br>(c) Proton spallation pulsed source |
| 5 Detection | (a) Photographic film<br>(b) Proportional counter<br>(c) Scintillation counter | (a) $^{10}BF_3$ or $^3He$ proportional counter<br>(b) $^6Li$ scintillation counter |
| *Scattering by atoms* | | |
| 6 Nature (non-magnetic atom) | Electron | Nuclear |
| 7 Q dependence | Q-dependent scattering factor, f(Q) | Isotropic scattering length, $\bar{b}$ |
| 8 Variation with Z | f(0)=Z | $\bar{b}$ varies irregularly with Z |
| 9 Phase change on scattering | $\pi$ | Usually $\pi$, but some exceptions ($\bar{b}$ negative) |
| 10 Isotopic variation | None | $\bar{b}$ varies for different isotopes |
| 11 Anomalous dispersion | Near an absorption edge | Near an absorption resonance |
| 12 Magnetic scattering | No extra scattering | Additional Q-dependent scattering |
| 13 Absorption coefficient | Absorption generally larger than scattering | Absorption usually small but marked exceptions |

of the difference in identity of the major scattering centres for X-rays and neutrons is that, while the magnitude in electron units of the X-ray form factor for free atoms at zero Q is simply their atomic number Z, the corresponding neutron scattering length b depends on the detailed nuclear energy levels and is an irregular function of Z as shown later in fig. 12. Thus atoms of low atomic number may have comparable or larger neutron scattering lengths than those much higher up the periodic table and atoms of neighbouring atomic number may have quite different scattering lengths. This gives rise to two of the important applications of neutron scattering, namely the location of light atoms and the distinguishing of atoms of neighbouring, relatively high atomic number. An example of the first application is given in section 5.4. The fact that neutrons are scattered by the nucleus means that the neutron scattering length is different for the various isotopes of a given element, which is the basis of the isotopic substitution technique described in section 5.4.1, and also for different nuclear spin states of the same isotope.

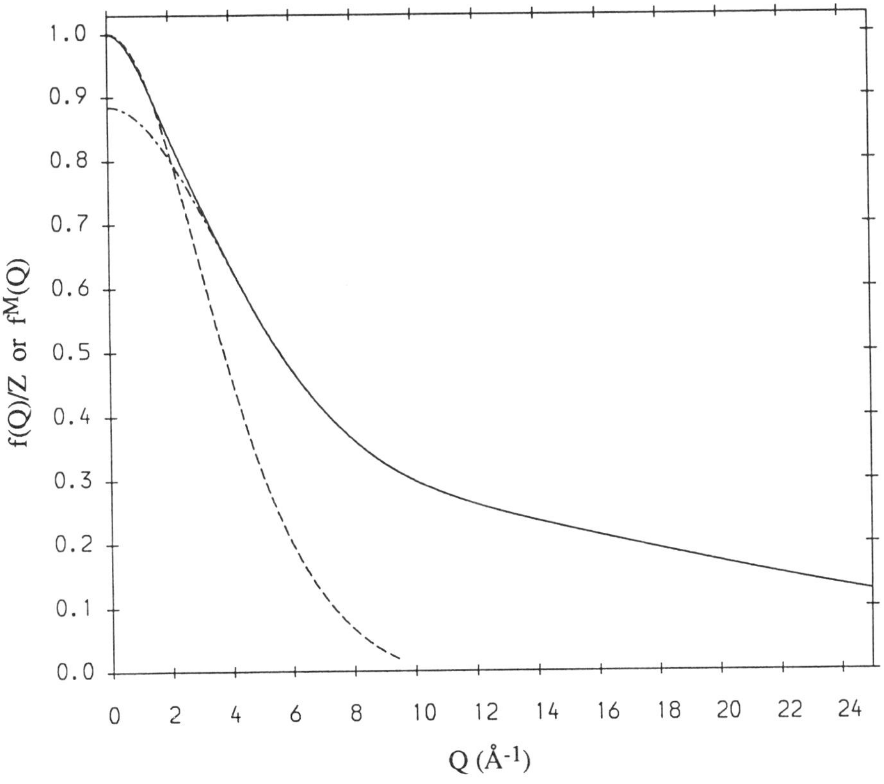


**Figure 11** f(Q)/Z for the Fe Free atom (——) and the $Fe^{3+}$ free ion (-·-·), together with the neutron magnetic form factor $f^M(Q)$ for $Fe^{3+}$ (----).

Unlike X-ray photons the neutron has a magnetic moment and hence is sensitive to the arrangement of the magnetic spins arising from unpaired electrons within a sample. Neutron magnetic scattering can therefore be used to probe the magnetic structure of amorphous solids as reviewed in ref. [49]. The magnetic structure of glasses is discussed in Chapter 13 but the use of neutron magnetic diffraction to yield information about their atomic structure will be briefly mentioned in section 5.4.3.

Both X-rays and neutrons can be scattered inelastically and in each case the inelastic scattering can be used to extract details of the dynamics of the scattering centres. For X-rays the Compton scattering profile yields the electron momentum density distribution within the constituent atoms. There is also a contribution to the inelastic scattering from the atomic thermal vibrations but the energy transfers involved are very small compared to the incident X-ray energy (c.f. table 4). On the other hand the energy of thermal neutrons is of a similar magnitude to that of thermal excitations and hence neutrons can be used to study the dynamics of both the atoms and magnetic spins within a sample. Compared to optical spectroscopy, neutron inelastic scattering yields a quantity very much closer to the vibrational density of states, due to the lack of the complex matrix elements which link the optical spectra to $g(\omega)$. Neutron inelastic scattering data therefore provide a more direct test of theoretical calculations of the dynamics of amorphous solids, and eventually of the structural models on which these calculations are based, than do infrared and Raman spectroscopy. The use of neutron inelastic scattering techniques to investigate the dynamics of amorphous solids has been reviewed elsewhere [50] and an account of magnetic inelastic scattering studies of amorphous solids is given in ref. [49].

## 4.1 Scattering by an Assembly of Atoms

Consider an atom at a position P a vector distance **r** from an arbitrary origin O as depicted in fig. 10. The phase of the radiation reaching the detector at R (R>>r), relative to that which would be scattered at the origin is

$$\phi = \frac{2\pi(r\cos\alpha - r\cos\beta)}{\lambda} \tag{11}$$

However, $(2\pi/\lambda)r\cos\alpha$ is equal to $\mathbf{r}\cdot\mathbf{k}_o$ and the corresponding quantity involving $\cos\beta$ is $\mathbf{r}.\mathbf{k}$. Hence

$$\phi = \mathbf{r}.(\mathbf{k}_o - \mathbf{k}) = \mathbf{r}.\mathbf{Q} \tag{12}$$

and the amplitude of the scattered wave becomes

$$A_1(Q) = a(Q)e^{i\mathbf{r}.\mathbf{Q}} \tag{13}$$

in which a(Q) is used to represent either the X-ray form factor f(Q) or the neutron scattering length b as appropriate. For a system of N atoms it is merely necessary to add the amplitudes for each atom, taking account of their relative phase

$$A_N(Q) = \sum_p a_p(Q)e^{i\mathbf{r}_p.\mathbf{Q}} \tag{14}$$

and the scattered intensity is obtained by multiplying $A_N(Q)$ by its complex conjugate

$$I_N(Q) = A_N(Q)A_N^*(Q) = \sum_p \sum_q a_p(Q)a_q(Q)e^{i\mathbf{r}_{pq}\cdot\mathbf{Q}} \tag{15}$$

Both the p and q summations are taken over all the atoms in the sample. Note that $I_N(Q)$ depends only on the interatomic vectors $\mathbf{r}_{pq} = \mathbf{r}_p - \mathbf{r}_q$ and is independent of the initial (arbitrary) choice of origin. Also equ. (15) applies to any sample, crystalline, amorphous or gaseous, since it makes no assumption about the way in which the atoms are arranged in space.

As indicated above, the neutron scattering length for an element may vary from atom to atom due to the presence of different isotopes or nuclear spin states. However, in a normal sample the distribution of isotopes or nuclear spin states between the various atoms of a given element is completely random and thus, for neutrons, the neutron scattering length for each atom in equ. (15) must be replaced by the average value for the element in question. In order to do this it is first necessary to separate the terms in the double summation for which p and q refer to the same atom. Hence

$$I_N(Q) = \sum_p \overline{b_p^2} + \sum\sum_{p \neq q} \bar{b}_p \bar{b}_q \, e^{i\mathbf{r}_{pq}\cdot\mathbf{Q}} \tag{16}$$

The first term on the right hand side of equ. (16) is known as the self scattering and since it involves the square of the scattering length for atom p it is this quantity which must be averaged. For the second term (the distinct scattering), on the other hand, the isotopes/nuclear spins at sites p and q are uncorrelated, even if they are of the same element, which means that independent averages are required for $b_p$ and $b_q$.

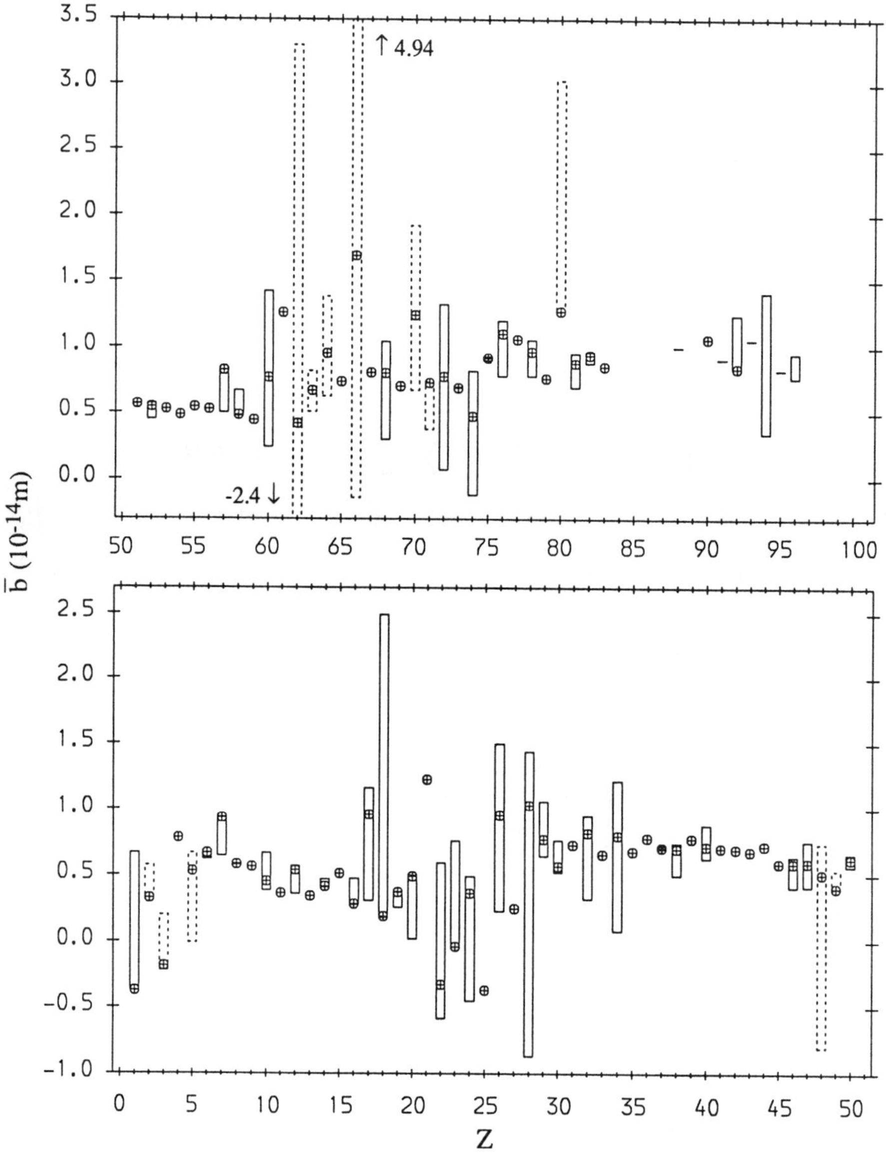


**Figure 12** The neutron scattering length $\bar{b}$ as a function of atomic number Z, showing the value for the natural element (⊕) and the range of isotopic variation (bar). A dashed bar denotes elements with an isotope having an imaginary component to $\bar{b}$ and indicates the variation in the real part of $\bar{b}$ at a wavelength of 1Å. The values of $\bar{b}$ are taken from ref. [51].

The variation of the average neutron scattering length $\overline{b}$ with atomic number is illustrated in fig. 12. The limits of the bars denote the minimum and maximum values of $\overline{b}$ for the individual isotopes of each element and hence any scattering length within the bar can be obtained by a suitable mixture of isotopes. For isotopes where the scattering length is wavelength dependent (see section 5.4.2) the values given are those for a wavelength of 1Å.

## 4.2 Scattering by an Isotropic Sample

In the case of an isotropic sample, such as an amorphous, gaseous or polycrystalline material, for any given interatomic vector $\mathbf{r}_{pq}$ there will be equivalent vectors elsewhere in the sample in all possible orientations with respect to the scattering vector $\mathbf{Q}$. Hence, for such materials, it is possible to perform an isotropic average to yield

$$I_N(Q) = \sum_p \overline{a_p^2(Q)} + \sum_{p \neq q}\sum \overline{a_p(Q)}\;\overline{a_q(Q)}\,\frac{\sin r_{pq}Q}{r_{pq}Q} \tag{17}$$

where, as before, $\overline{a(Q)}$ represents either f(Q) or $\overline{b}$ for X-rays or neutrons respectively. Equation (17) is known as the Debye equation and may be used to calculate the scattered intensity if all the interatomic vectors are either known or can be calculated. This would be the case for a structural model (c.f. section 7) or a molecular gas, in which intermolecular correlations can be neglected and the double summation need only be performed for a single molecule.

For an amorphous sample the interatomic vectors are unknown and hence it is necessary to work in terms of the correlation functions introduced in section 3.6. The double summation over interatomic vectors in equ. (17) may thus be replaced by an integration over the component correlation functions $t_{ij}(r)$ to give

$$I(Q) = \sum_i \overline{a_i^2(Q)} + \sum_i \sum_j \overline{a_i(Q)}\;\overline{a_j(Q)} \int_0^\infty t_{ij}(r)\frac{\sin Qr}{Q}\,dr \tag{18}$$

the intensity I(Q) now being normalised to one composition unit (c.u.). The i summation is taken over the atoms in one composition unit, while that for j is over atom types. The second term on the right hand side of equ. (18) may be split into two parts, the first involving the differential component correlation functions $d_{ij}(r)$ and the second the average composition unit number density $\rho^0$.

$$I(Q) = \sum_i \overline{a_i^2(Q)} + \sum_i \sum_j \overline{a_i(Q)}\,\overline{a_j(Q)} \int_0^\infty d_{ij}(r) \frac{\sin Qr}{Q} dr$$

$$+ \int_0^\infty 4\pi r \rho^\circ \left[ \sum_i \overline{a_i(Q)} \right]^2 \frac{\sin Qr}{Q} dr \qquad (19)$$

The reason for doing this is that the final (average density) term, usually denoted $I^\circ(Q)$, is effectively a $\delta$ function at Q=0 (This is rigorously correct for an infinite sample.) and hence is not recorded in a diffraction experiment since it is included with the unscattered primary beam transmitted by the sample.

The diffraction properties of a sample in reciprocal space may be represented by the interference function

$$Qi(Q) = Q\left[ I(Q) - \sum_i \overline{a_i^2(Q)} - I^\circ(Q) \right] \qquad (20)$$

and it is this reciprocal space function which is usually employed, for example, in the comparison of experimental data with the predictions of structural models (c.f. section 8). From equ. (19) it may be seen that the interference function is a weighted sum of the Fourier sine transforms of the component differential correlation functions $d_{ij}(r)$, the weighting factors being $\overline{a_i(Q)}\,\overline{a_j(Q)}$.

$$Qi(Q) = \sum_i \sum_j \overline{a_i(Q)}\,\overline{a_j(Q)} \int_0^\infty d_{ij}(r) \sin Qr\, dr \qquad (21)$$

## 4.3 Experimental Real Space Correlation Functions

In order to extract a real space correlation function from an experimental interference function it is necessary to make use of the Fourier inversion theorem and define the experimental differential correlation function D(r) as the Fourier sine transform of Qi(Q). The relationship between D(r) and the component differential correlation functions defined in section 3.6 is complicated by the fact that experimental data are limited to a finite range of Q ($Q \leq Q_{max}$). Hence a modification function, M(Q), is introduced into the Fourier transformation such that [52]

$$D(r) = \frac{2}{\pi} \int_0^\infty Qi(Q)\, M(Q) \sin rQ\, dQ \tag{22}$$

M(Q) being zero for $Q \geq Q_{max}$. The experimental total correlation function T(r) is obtained by adding an average density term T°(r) which is the Fourier transform of QI°(Q)

$$T(r) = D(r) + T^{\circ}(r) = \sum_i \sum_j t'_{ij}(r) \tag{23}$$

The prime in equ. (23) denotes either N or X, for neutrons or X-rays respectively and indicates that $t'_{ij}(r)$ is not the same function as $t_{ij}(r)$, the relationship between them being one of convolution.

$$t'_{ij}(r) = \int_0^\infty t_{ij}(\mathbf{r'})[P'_{ij}(r-r') - P'_{ij}(r+r')]\, dr' \tag{24}$$

r' is a dummy convolution variable and the component peak functions $P'_{ij}(r)$ define the experimental resolution in real space caused by the finite upper limit in Q. For neutrons $\bar{b}$ is independent of Q and $P^N_{ij}(r)$ is simply proportional to the Fourier cosine transform of M(Q)

$$P^N_{ij}(r) = \frac{\bar{b}_i \bar{b}_j}{\pi} \int_0^\infty M(Q) \cos rQ\, dQ \tag{25}$$

[M(Q) is an even function and therefore a cosine transform is involved rather than the sine transform required for odd functions such as Qi(Q) or $d_{ij}(r)$.] The corresponding average density term T° (r) becomes

$$T^{\circ}(r) = 4\pi r \rho^{\circ}\left(\sum_i \bar{b}_i\right)^2 \tag{26}$$

A simple truncation of the data at $Q=Q_{max}$

$$\begin{aligned} M(Q) &= 1 \qquad Q \leq Q_{max} \\ M(Q) &= 0 \qquad Q > Q_{max} \end{aligned} \tag{27}$$

leads to a component peak function $P_{ij}^{N}(r)$ which has pronounced "termination ripples" on either side of the central maximum, which might be identified as real features in an experimental correlation function, and indeed it is not that rare in the literature to find cases where this has actually happened. Hence it is more common to use the more gradual cutoff provided by the Lorch function [53]

$$\begin{aligned} M(Q) &= \frac{\sin \Delta r Q}{\Delta r Q} \qquad Q \leq Q_{max} \\ M(Q) &= 0 \qquad Q > Q_{max} \end{aligned} \tag{28}$$

the resolution width $\Delta r$ being equal to $\pi/Q_{max}$. Figure 13 shows the reduced

**Table 5. Peak Heights and Widths for Three Common Reduced Neutron Peak Functions**

| $M(Q)$ $(Q \leq Q_{max})$ | Height | Width (fwhm) |
|---|---|---|
| 1 | $0.3183Q_{max}$ | $3.791/Q_{max}$ |
| $\exp(-BQ^2)$ $(B = \ln 10/Q^2_{max})$ | $0.1800Q_{max}$ | $5.389/Q_{max}$ |
| $\frac{\sin \Delta r Q}{\Delta r Q}$ $(\Delta r = \pi/Q_{max})$ | $0.1876Q_{max}$ | $5.437/Q_{max}$ |

neutron peak function for both the step function [equ. (27)] and the Lorch function together with that for the earlier truncated exponential modification function [52] and it can be seen that the latter two modification functions greatly reduce the satellite oscillations on either side of the central maximum but at the expense of a slight loss in real space resolution. The heights and widths (full width at half maximum, fwhm) for each peak function are given in table 5, from which it may be seen that the experimental resolution is inversely proportional to $Q_{max}$.

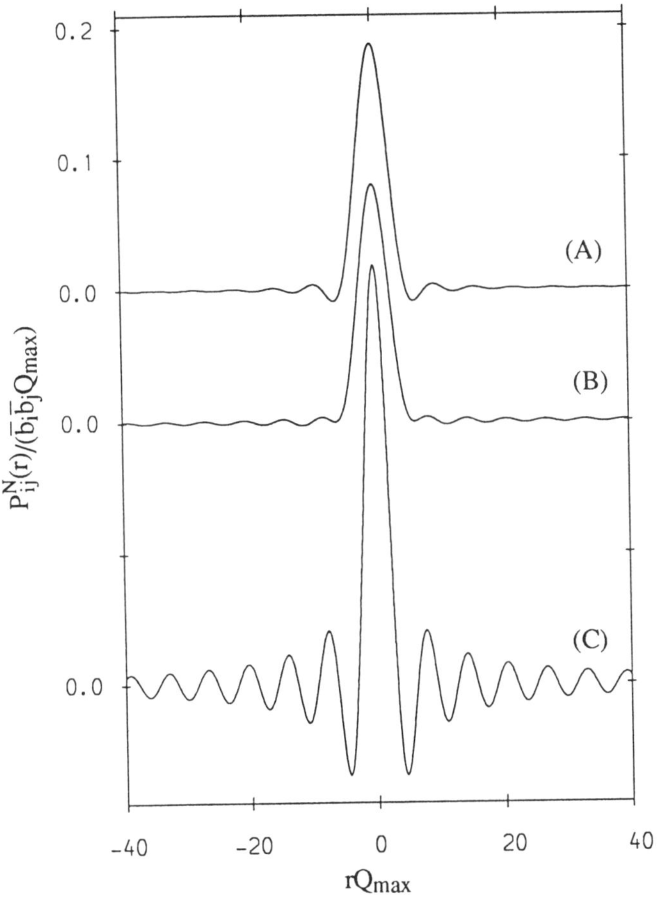


**Figure 13** Reduced neutron peak functions:
(A) $M(Q) = \sin(\Delta rQ)/(\Delta rQ)$, $\Delta r = \pi/Q_{max}$;
(B) $M(Q) = \exp(-BQ^2)$, $B = \ln 10/Q^2_{max}$; (C) $M(Q) = 1$.

The situation for X-rays is more complicated than that for neutrons since, as noted at the beginning of this section, X-rays are primarily scattered by electrons and thus the immediate result of an X-ray diffraction experiment is an electron-electron correlation function. A simple Fourier sine transformation of the X-ray interference function along the lines of equ. (22), therefore, leads to the electronic correlation function of Finbak [54], which suffers from poor real space resolution since the peaks reflect the finite size of the electron clouds surrounding each atom/ion. For this reason it is conventional to divide $Qi^X(Q)$ by a "sharpening function" before Fourier transformation to give a pseudo "atomic" correlation function. For systems containing more than one atomic species the usual

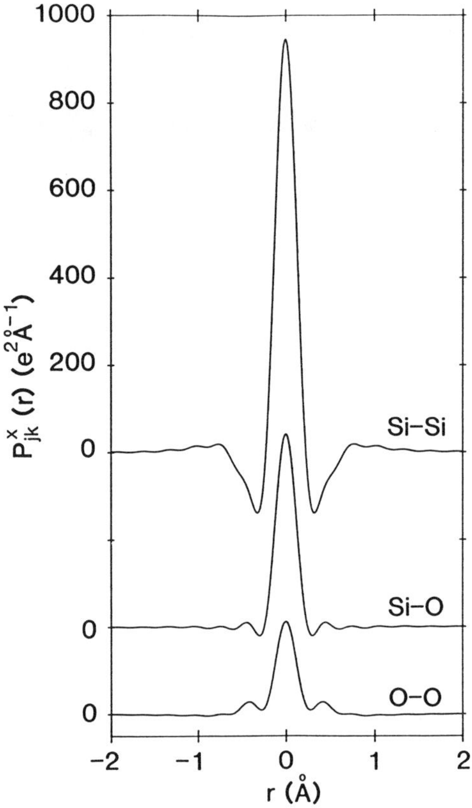


**Figure 14** X-ray component peak functions for vitreous silica appropriate to the X-ray study of Mozzi and Warren [55].

sharpening function is $f_e^2(Q)$,

$$f_e(Q) = \frac{\sum_i |f_i(Q)|}{\sum_i Z_i} \qquad (29)$$

being the average scattering factor per electron. This leads to X-ray component peak functions of the form

$$P_{ij}^X(r) = \frac{1}{\pi}\int_0^\infty \frac{f_i(Q)f_j(Q)}{f_e^2(Q)} M(Q) \cos rQ \, dQ \qquad (30)$$

Note that, since the scattering factor f(Q) is a different function of Q for each element, the X-ray component peak functions have a different shape for each component, as illustrated in fig. 14 for vitreous silica. [Compare with the case for neutrons, in equ. (25), where the scattering lengths do not depend on Q and can be taken outside the integral sign to yield component peak functions which have the same shape and differ only in amplitude.] The corresponding average density term is similar to equ. (26) except that the atomic number, Z, replaces the neutron scattering length, $\bar{b}$.

## 5. EXPERIMENTAL TECHNIQUES

An X-ray or neutron diffraction experiment involves a measurement of the scattered intensity as a function of the magnitude of the scattering vector, Q. In order to achieve the required variation in Q it is possible to scan the scattering angle, $2\theta$, at fixed wavelength, $\lambda$, (conventional technique) or to make measurements as a function of $\lambda$ at constant $2\theta$ (dispersion or time-of-flight technique). The technique employed for any given experiment depends primarily on the nature of the radiation source but may also be influenced by the need for a special sample environment (e.g. cryostats, furnaces, high pressure cells, etc.). For example, it is much easier to use the variable wavelength technique at high pressures or temperatures since the constant scattering angle only requires relatively small fixed windows in the sample cell for the incident and scattered beams. Similarly, all values of Q are examined simultaneously making the variable wavelength technique ideally suited for following phase changes or studying kinetic effects.

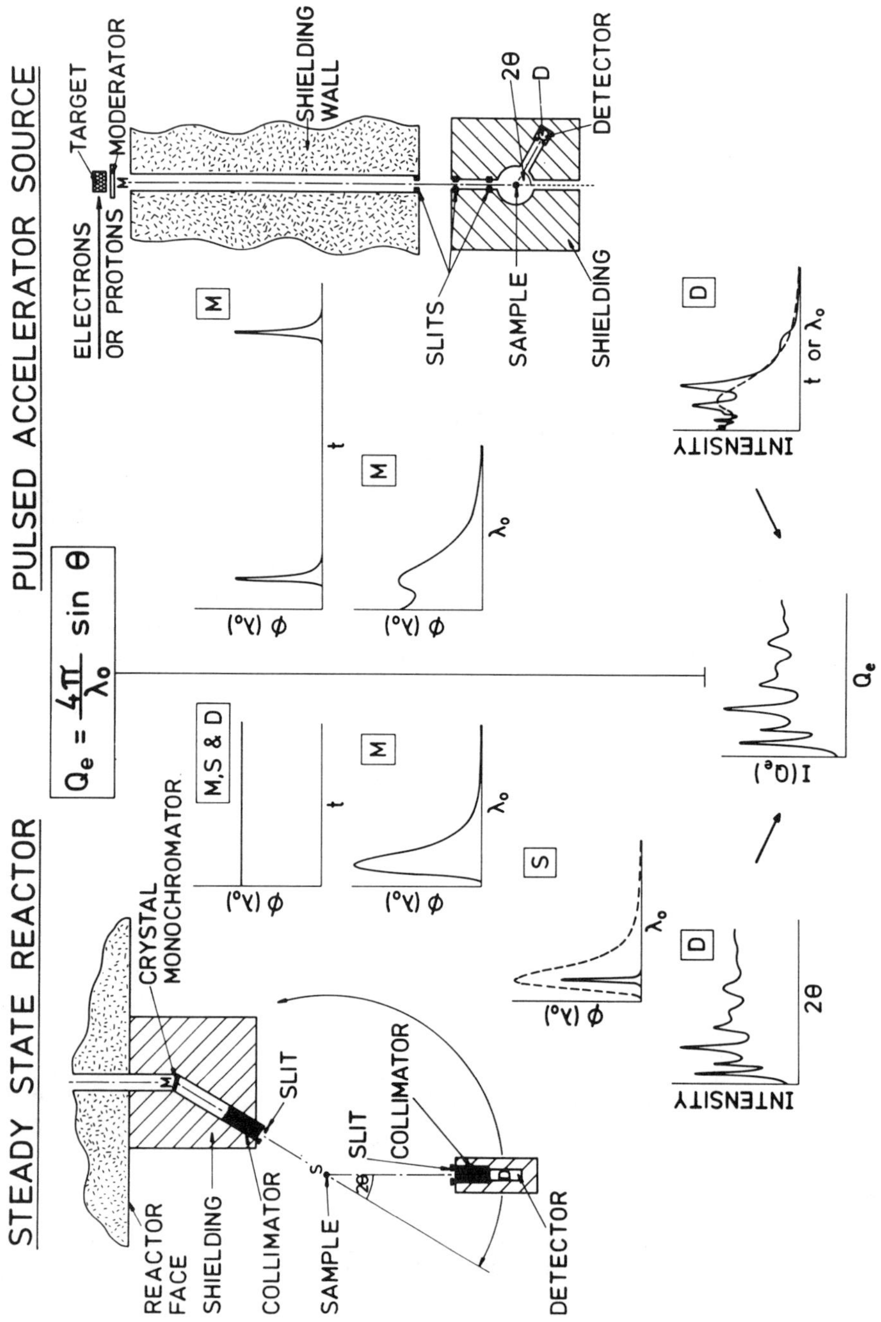
STEADY STATE REACTOR
PULSED ACCELERATOR SOURCE
$Q_e = \frac{4\pi}{\lambda_o} \sin \theta$
REACTOR FACE
SHIELDING
COLLIMATOR
SAMPLE
DETECTOR
CRYSTAL MONOCHROMATOR
SLIT
COLLIMATOR
M,S & D
M
S
D
INTENSITY
2θ
I(Qₑ)
Qₑ
ELECTRONS OR PROTONS
TARGET
MODERATOR
SHIELDING WALL
SLITS
2θ
DETECTOR
t or λₒ

## 5.1 Neutron Diffraction

The two techniques for neutrons are compared in fig. 15. In the conventional steady state reactor twin-axis experiment the flux $\phi(\lambda)$ of thermal neutrons extracted from the moderator/reflector of a steady-state reactor has a Maxwellian distribution of velocities and is time independent. A beam of wavelength $\lambda$, selected by the monochromator crystal, is incident on the sample and scattered into the detector through a variable angle $2\theta$. The variable $\lambda$ time-of-flight technique is usually employed with a pulsed accelerator source. Electrons or protons strike a heavy metal target producing pulses of fast neutrons which are partially moderated before being incident on the sample and scattered into a detector situated at a fixed angle $2\theta$. The detector records the scattered intensity as a function of the time-of-flight for the distance from the moderator via the sample to

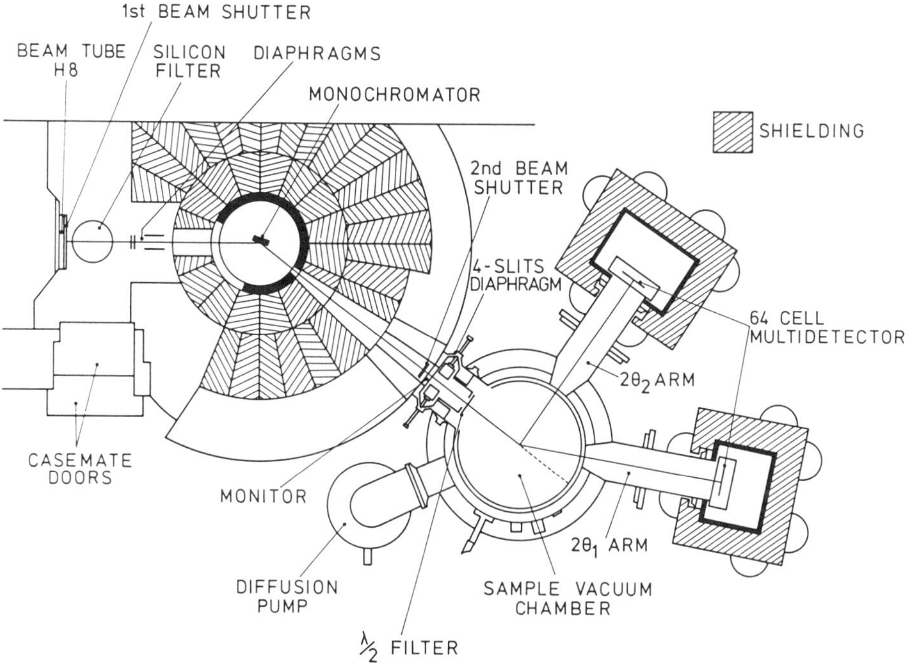


**Figure 16** The D4b diffractometer at the Institut Laue-Langevin [56].

← **Figure 15** A Schematic comparison of reactor and pulsed source techniques for measuring the neutron diffraction pattern from an amorphous sample [47].

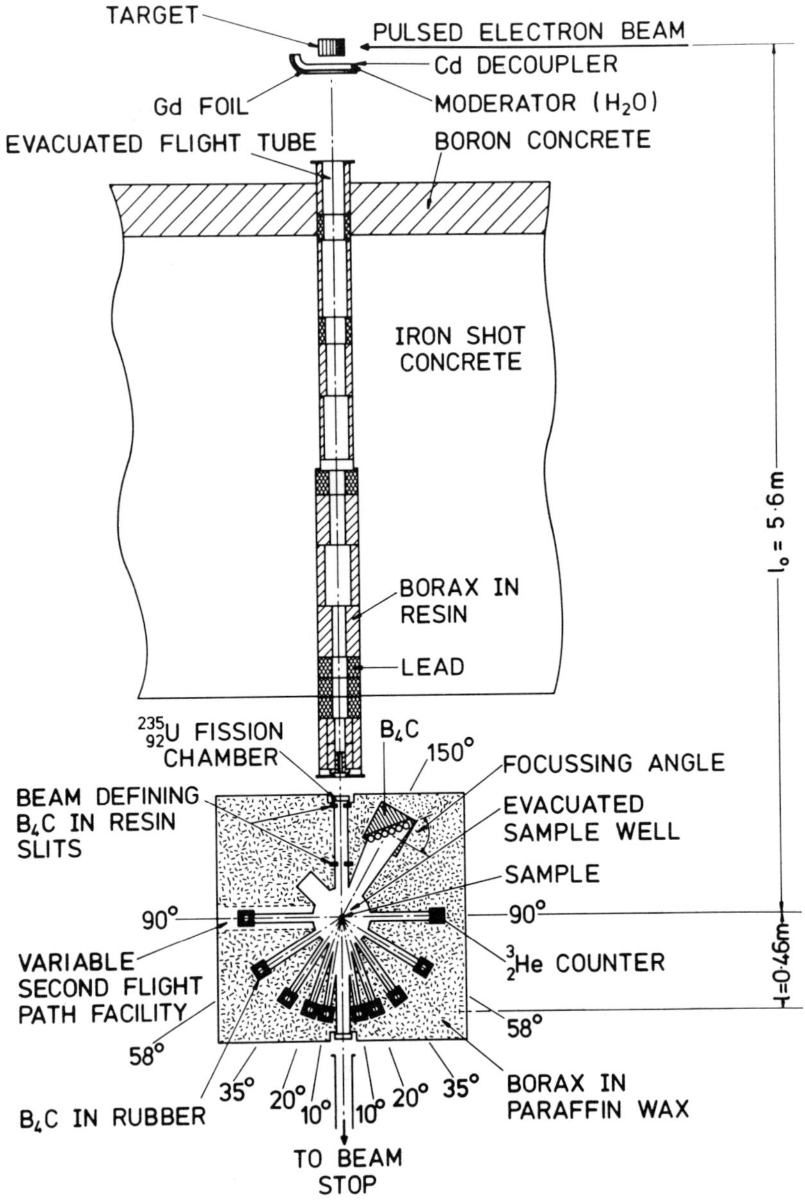


**Figure 17** The total scattering spectrometer (TSS) at the HELIOS pulsed neutron source [47].

the detector. For any neutron the time-of-flight is simply related to $\lambda$ through the velocity and De Broglie's relationship and the diffraction pattern I(Q) may be extracted by dividing the measured intensity by the incident neutron spectrum shape.

Examples of the type of conventional and time-of-flight diffractometers used for amorphous solids are given in figs. 16 and 17 respectively. The main difference between the real instruments and the schematic diagrams of fig. 15 is that the former have multiple counters to increase the count rate and, for the time-of-flight spectrometer, to provide coverage of the full range of Q. Typical corrected diffraction patterns from the two instruments are shown in figs. 18 and 19. The great advantage of a pulsed accelerator source over a steady-state reactor is that the former is under-moderated which gives rise to a strong epithermal component in $\phi(\lambda)$ [c.f. the plots of $\phi(\lambda)$ in fig. 15]. These short wavelength neutrons allow data to be obtained to much higher values of Q, resulting in a corresponding increase in real space resolution, and as an example the interference function for vitreous silica from the Harwell Total Scattering Spectrometer (TSS) is shown in fig. 20. In order to illustrate the improvement in real space resolution gained by the use of a pulsed source time-of-flight diffractometer an idealised version of the correlation function for vitreous silica is shown in fig. 21 together with the correlation functions obtained with and without a Lorch modification function and $Q_{max}$ equal to $22Å^{-1}$ (D4) and $40Å^{-1}$ (pulsed source diffractometer). Older time-of-flight spectrometers such as the TSS employ large time-focussed banks of detectors at scattering angles in excess of 90° in order to increase count rates. However, theoretical studies of the form of the Placzek corrections for departures from the static approximation (c.f. section 5.3), for both conventional twin axis and time-of-flight diffractometers, indicate that to avoid distortion of $Qi^N(Q)$ measurements should be performed at as low a scattering angle as possible, consistent with adequate Q resolution, and for this reason the latest pulsed source instruments, such as the Glass, Liquids and Amorphous Diffractometer (GLAD) at the Argonne Intense Pulsed Neutron Source (IPNS), are designed to measure a complete diffraction pattern at scattering angles of less than ~45°.

## 5.2 X-ray diffraction

Accurate quantitative laboratory X-ray diffraction measurements on glasses usually employ a modified powder counter diffractometer. The situation for X-rays is complicated by the presence of incoherent Compton scattering which at high values of Q can be of much higher intensity than the required coherent contribution. Since neither the wavelength distribution nor the integrated intensity of the Compton scattering from a given sample can be satisfactorily calculated it is necessary either to remove this contribution by means of a diffracted beam monochromator or to measure the energy distribution of the scattered intensity at each angle and then extract the desired coherent scattering using peak fitting

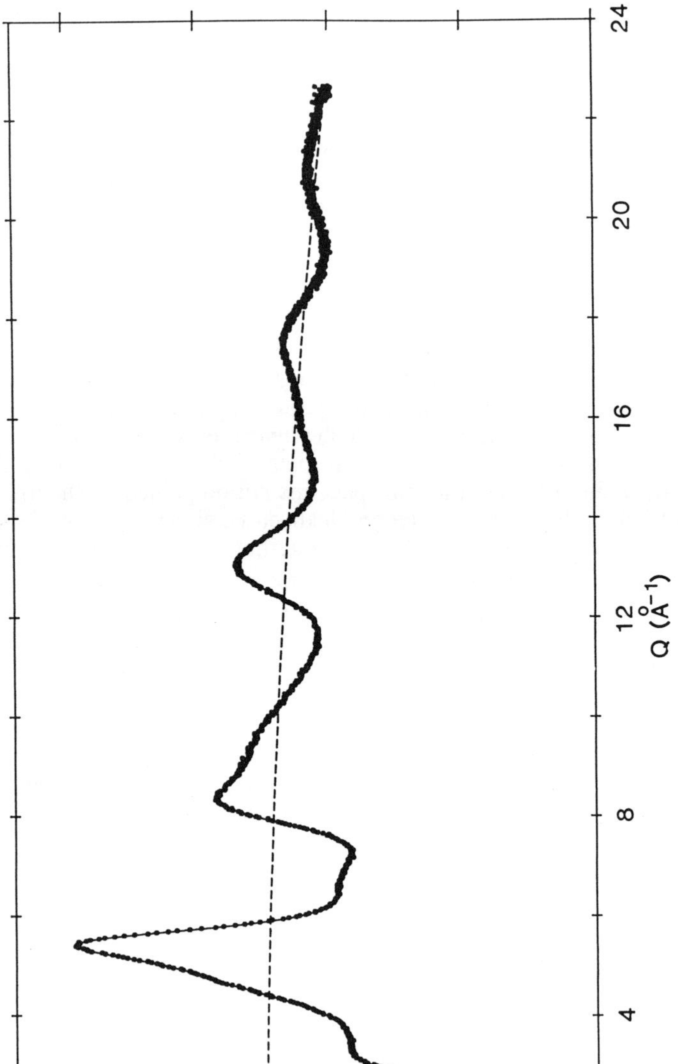
I(Q) (barns)
Q (Å$^{-1}$)
0.0
0.5
1.0
1.5
2.0
0
4
8
12
16
20
24

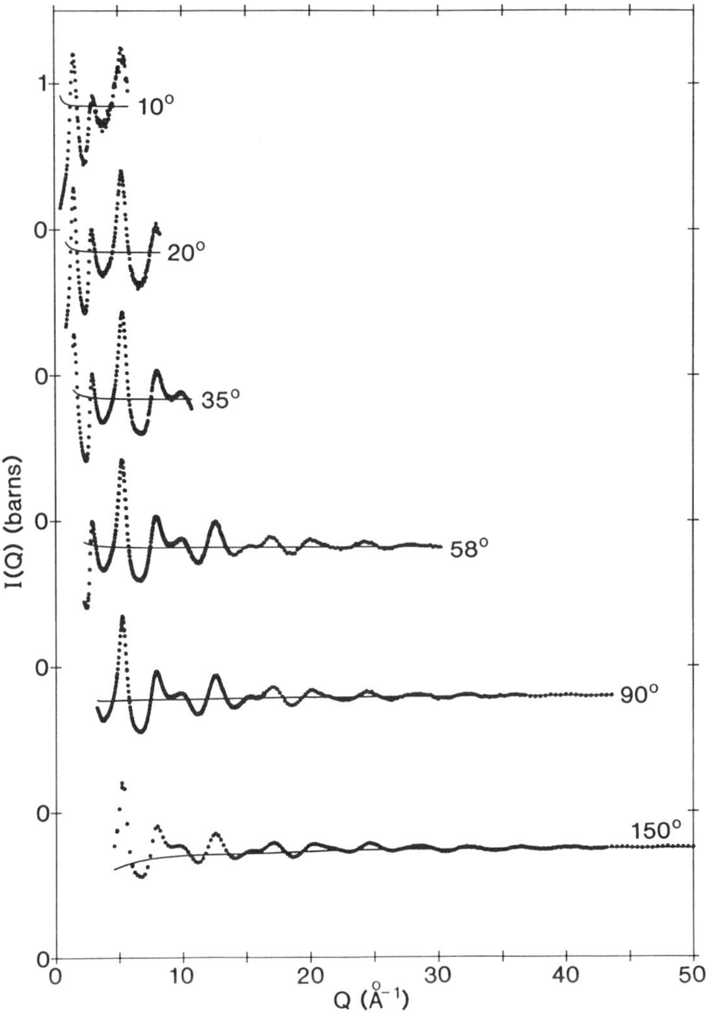


**Figure 19** The corrected intensity I(Q) for vitreous silica from the Harwell TSS [47] (•) normalized to the self scattering (——).

← **Figure 18** The corrected diffraction pattern I(Q) for vitreous $BeF_2$ obtained with the D4b diffractometer [56]. •, experimental points; ——, cubic spline fit; -----, self scattering.

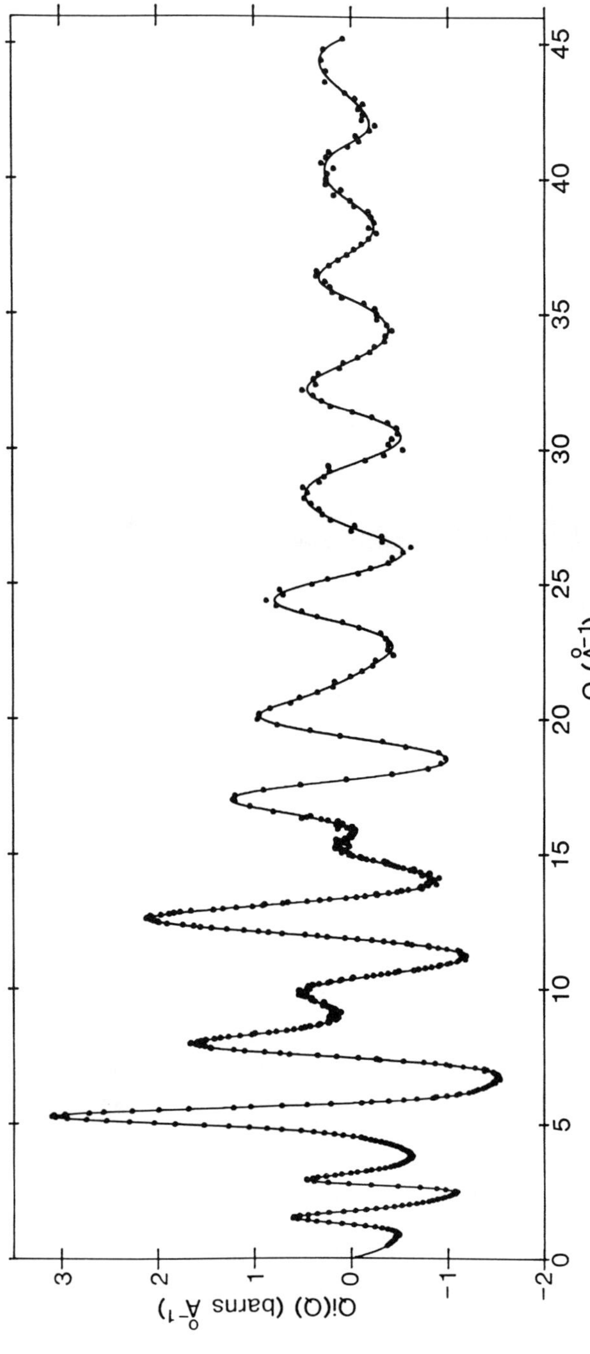

Qi(Q) (barns Å$^{-1}$)
Q (Å$^{-1}$)

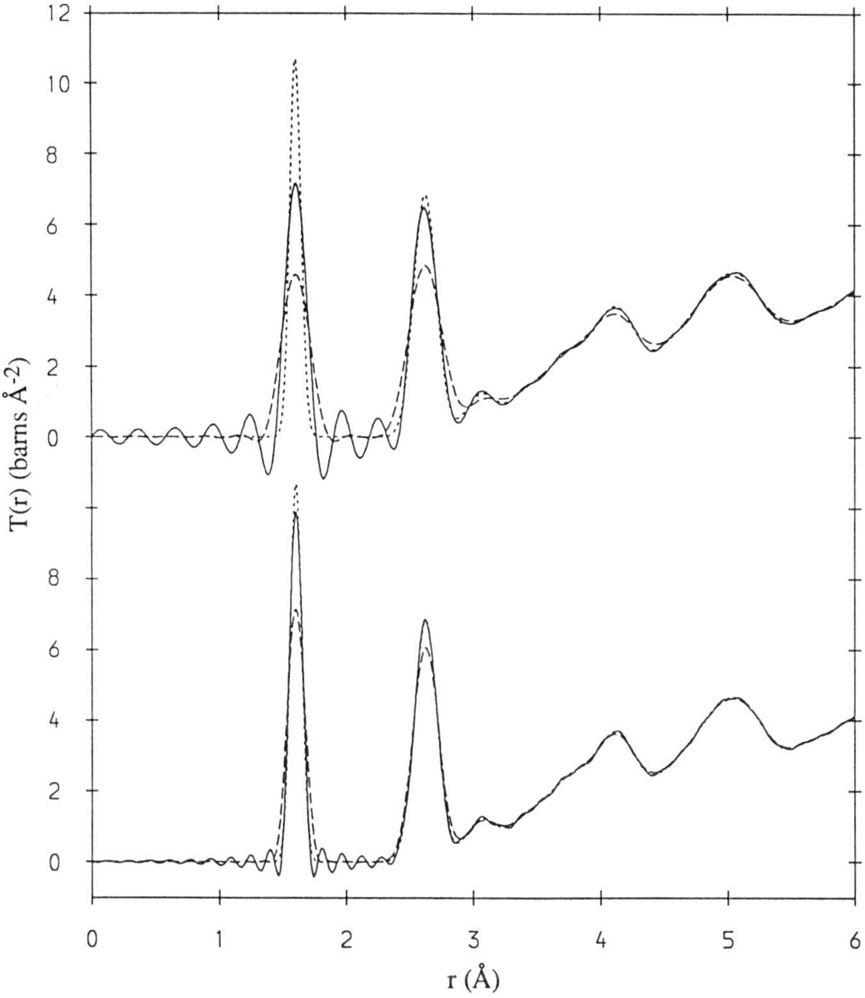


**Figure 21** A ***best estimate*** of the neutron correlation function for vitreous silica (·····) [57], showing the broadening introduced by Fourier transforming the corresponding interference function with $Q_{max} = 22Å^{-1}$ (upper curves) and $40Å^{-1}$ (lower curves) with (----) and without (——) a Lorch modification function.

← **Figure 20** The neutron interference function for vitreous silica [47]. •, experimental points; ——, cubic spline fit.

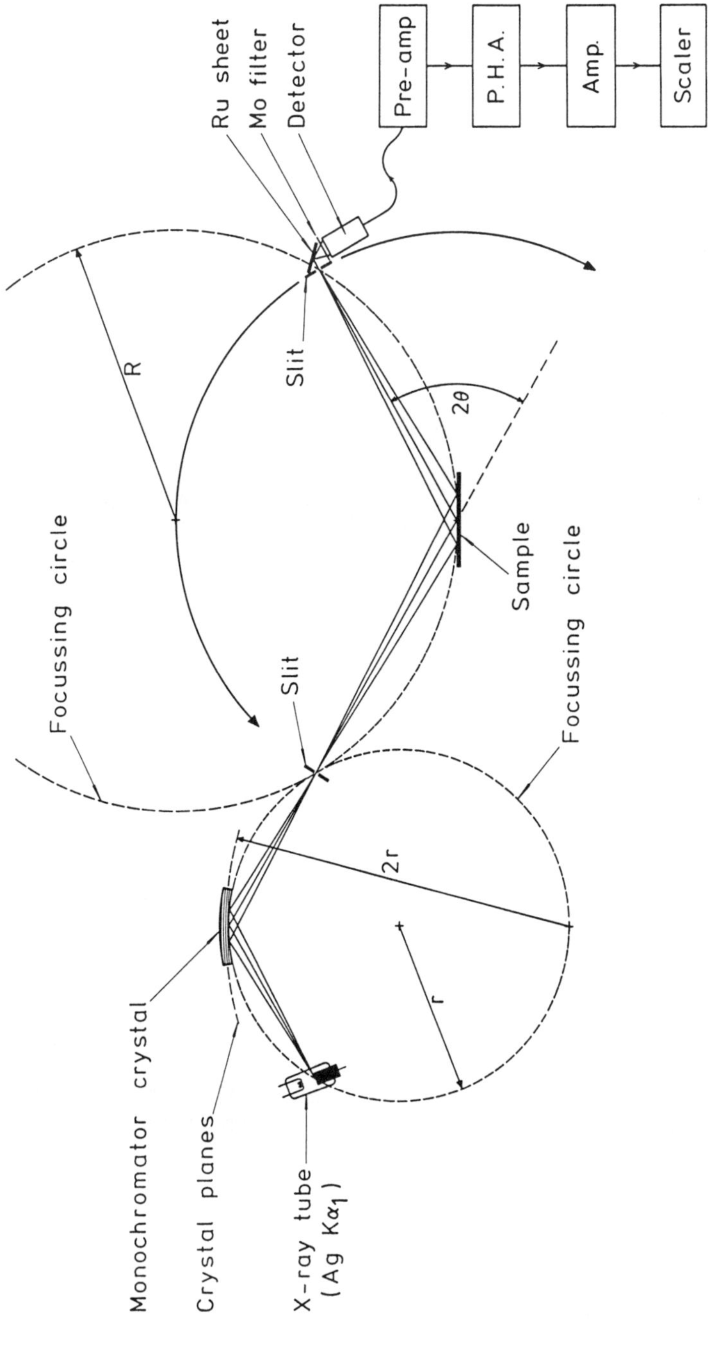

Ru sheet
Mo filter
Detector
Pre-amp
P.H.A.
Amp.
Scaler
R
Slit
2θ
Focussing circle
Sample
Focussing circle
Slit
2r
r
Monochromator crystal
Crystal planes
X-ray tube
(Ag Kα1)

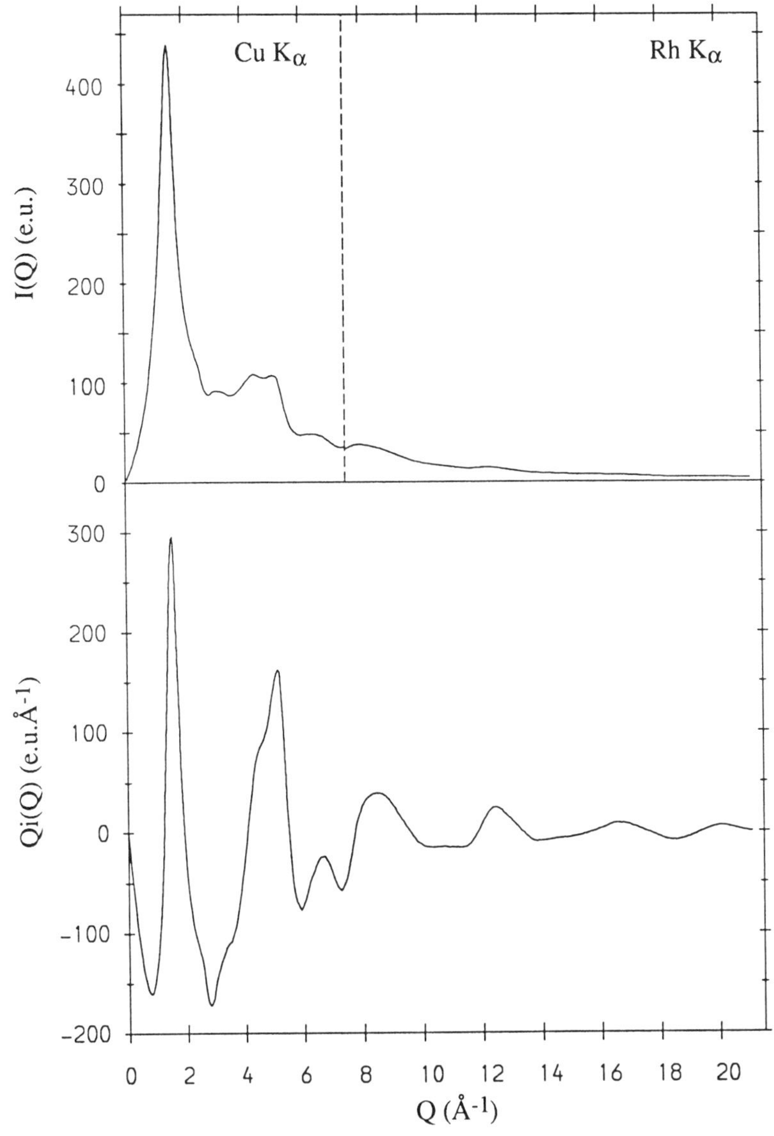


**Figure 23** Mozzi and Warren's [55] X-ray data for vitreous silica. The vertical dashed line indicates the change from Cu $K_\alpha$ to Rh $K_\alpha$ radiation.

← **Figure 22** The fluorescent foil X-ray technique [58].

techniques. Many authors have used a single fixed monochromator in the diffracted beam but the difficulty with this arrangement is that white background radiation from the X-ray tube can be Compton scattered into the monochromator envelope. A much better arrangement is that due to Warren and Mavel [58], shown in fig. 22, which employs a monochromator in both the incident and diffracted beams. A conventional curved crystal monochromator is used in the incident beam (Ag $K_{\alpha 1}$: $\lambda$=0.5594Å), but that in the diffracted beam comprises a foil with an absorption edge at a wavelength slightly longer than the characteristic line of the X-ray tube (Ru K edge: $\lambda$=0.5605Å) such that the coherent intensity will excite fluorescence whereas the Compton scattering will not. The fluorescent radiation is then recorded by the detector. Despite the obvious superiority of this technique, however, it is very rarely used with the result that X-ray data quality is frequently not up to that of its neutron counterpart. Mozzi and Warren's [55] X-ray data for vitreous silica are shown in fig. 23.

The dispersion technique is also possible with X-rays using a semiconductor detector. The advantage of this method for complex sample environments and for studying time dependent phenomena has already been mentioned. Outside such special applications, however, X-ray variable wavelength experiments are unlikely to prove important for accurate work on amorphous materials owing to the difficulty of accurately determining the incident spectrum and making corrections for absorption and Compton scattering.

The most important recent development in X-ray diffraction studies of glasses is the increasing use of synchrotron radiation sources, which produce an intense beam of well-collimated X-rays and normally include a wiggler to increase the flux at short wavelengths [59,60]. This type of source has several advantages over a standard laboratory X-ray generator and with an appropriate diffractometer, optimised for amorphous materials, is capable of yielding high quality data. The continuous spectrum of a synchrotron radiation source allows the selection of the optimum wavelengths for an anomalous dispersion experiment (c.f. section 5.4.2), rather than the few discrete values afforded by the characteristic line spectra of conventional X-ray tubes, and has resulted in a much greater use of this technique. Similarly, the highly collimated synchrotron X-ray beams have made it possible to perform grazing incidence scattering experiments to investigate very thin films or surface layers. (See for example refs [59] and [60].)

## 5.3 Data reduction

Any measurement involves a certain number of corrections to the raw data but the good experiment minimises these corrections or puts them in a form in which they are easily handled. Normally corrections are included for:

(i) Counter paralysis time. This is becoming increasingly important with modern high intensity X-ray and neutron sources.

(ii) Instrumental background and the scattering from any sample container used.

(iii) Absorption, self-shielding and multiple scattering. These corrections are closely coupled and can only be treated independently if one or other of them is small.

(iv) X-ray effects including polarisation and residual Compton scattering not removed by the diffracted beam monochromator.

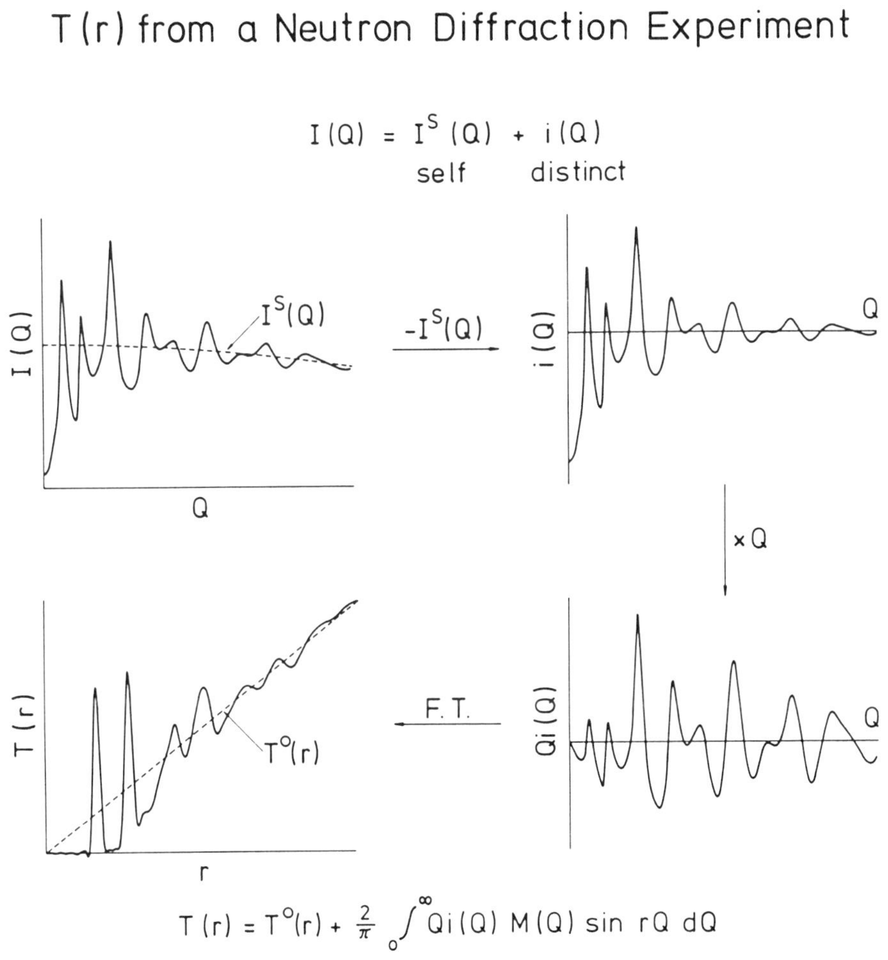


**Figure 24** The relationship between the corrected, normalised diffraction pattern I(Q) and the real space correlation function T(r). The neutron diffraction data illustrated are for vitreous silica.

Following correction it is necessary to normalise the data to absolute units. This may be achieved either by the use of self consistent (Krogh-Moe [61] - Norman [62]) integration techniques or, for neutrons, by measuring the (incoherent) scattering from a standard vanadium sample. Full details of the experimental corrections and normalisation procedures can be found in the standard X-ray and neutron scattering texts and in refs [7] and [44-47].

The procedure for extracting the real space correlation function T(r) from the corrected, normalised diffraction pattern I(Q) is illustrated pictorially for neutrons in fig.24. The distinct contribution i(Q) is obtained by subtracting the self scattering from I(Q) and in the case of neutrons it is necessary at this stage to allow for departures from the static approximation by making so-called Placzek corrections [7,44,45,47,63]. These arise because the detector in a total diffraction experiment performs its integration over $\omega$ at constant $2\theta$ rather than at constant Q as required by the static approximation which leads to G(r,0). As a result the self scattering is Q dependent and not a constant as would be inferred from equ. (20). The form of the Placzek corrections for conventional and time-of-flight measurements is given in refs. [45] and [47] respectively.

A controversial aspect of the analysis of X-ray and neutron diffraction data for glasses is the Fourier transformation of the interference function to give the correlation function T(r). This may be achieved using either numerical integration techniques, such as Filon's quadrature [64], or a fast Fourier transformation algorithm [65], the latter being more economical on computer time but much less flexible. The data may also be smoothed before Fourier transformation but this has very little effect on the resulting transform since the frequency range of the noise removed mainly corresponds to distances in excess of those of interest [65].

The accuracy of experimental data can most easily be gauged by the behaviour of the transform at low r, below the first true peak. If the transform is well behaved in this region then the data are of reasonable quality. Great care is needed in making this assessment, however, as some authors go to great lengths to disguise the fact that their data are of poor quality, as illustrated in fig. 25. One method is to use figures with grossly distorted axes (figs. 25B and C) and/or to plot the radial distribution function rT(r) (fig. 25D), in which the extra factor of r has the effect of reducing the amplitude of the error ripples at low r relative to the structure at higher r. It is also possible to plot figures which are very small (Some journals are remarkably co-operative in this respect, especially if the original figure is of poor quality, since small figures save space.). The most blatant, and frequently used, approach is simply not to plot the correlation function below the point where the low r side of the first true peak crosses the r axis (fig. 25E). A particularly devious technique, however, is to use the false oscillations below the first peak as a criterion for cleaning up the data before publication and then not to include the original unadulterated transform (fig. 25F). In the absence of other information, such data must be treated with the utmost suspicion, since the methods normally employed ***only*** remove the oscillations below the first peak and do ***not*** in any way

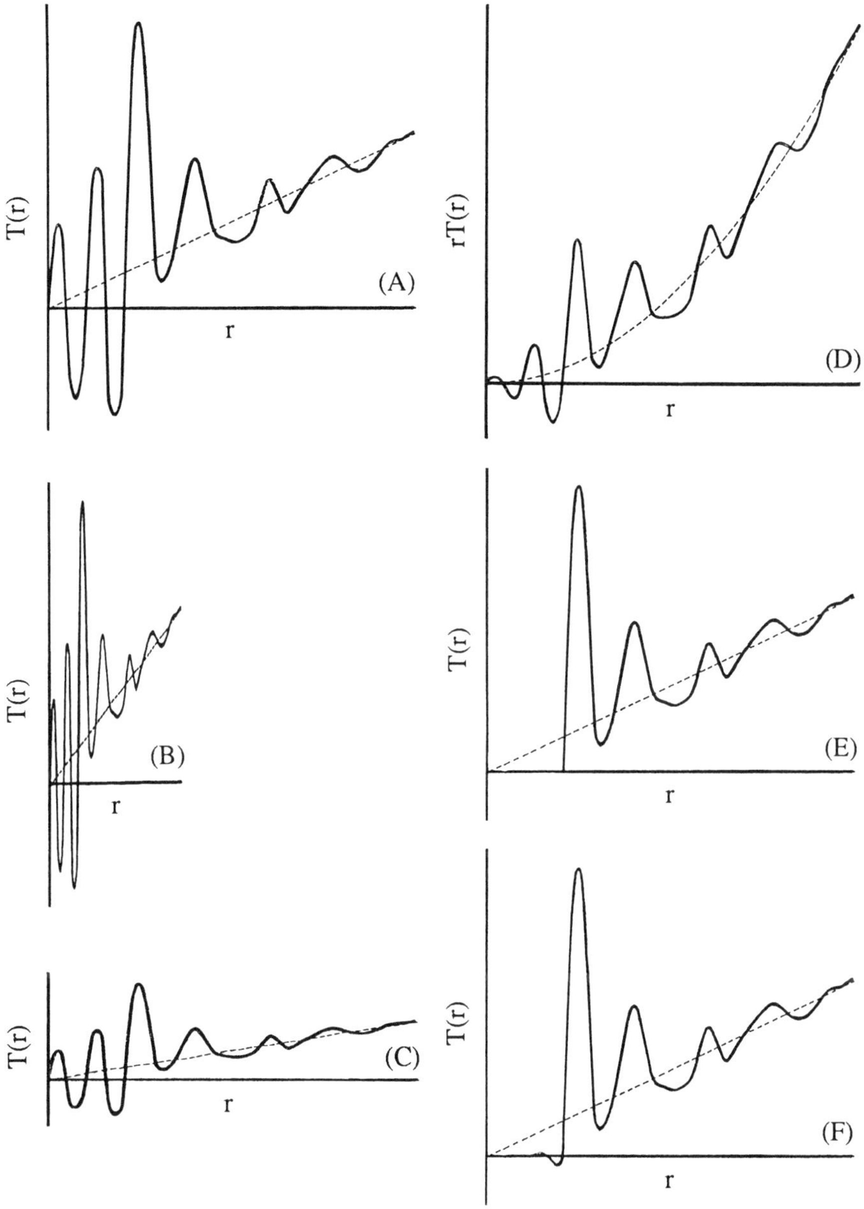


**Figure 25** Methods used by authors to disguise poor data quality: (A) original data; (B) and (C) grossly distort the axes; (D) plot the radial distribution function rT(r); (E) omit data below the first real peak and (F) use iterative techniques to "massage" the data before publication and remove the unwanted noise.

improve the accuracy at higher r. Similar methods are sometimes used to "remove" the effects of terminating the data at finite $Q_{max}$. Information theory, however, indicates that it is impossible to replace the unmeasured data without making some assumption about the material under investigation so that the resulting correlation function merely becomes one possible model which fits the results rather than an unbiased Fourier transform. Similar objections can be raised to the maximum entropy technique [66], which also has the additional problem that it is impossible to know what to do to the correlation function from a structural model in order to be able to directly compare it with the experimental maximum entropy version.

## 5.4 Separation of Component Correlation Functions

A completely unambiguous assignment of the peaks in T(r) can only be made by separating the correlation function into its individual components, especially in regions where adjacent peaks overlap. Such a separation of individual component correlation functions, or alternatively of linear combinations of subsets of these components, may be achieved by varying the neutron scattering length or X-ray form factor for one or more of the constituent elements. If two experiments are performed in which the scattering length of element A is varied, the difference between the experimental correlation functions is of the form [67]

$$\Delta T(r) = T(r) - T'(r)$$
$$= m_A\left[\overline{a_A(Q)^2} - \overline{a_{A'}(Q)^2}\right]t'_{AA}(r) + 2m_A\left[\overline{a_A(Q)} - \overline{a_{A'}(Q)}\right]\sum_{j\neq A}\overline{a_j(Q)t'_{Aj}(r)} \quad (31)$$

$m_A$ being the number of A atoms in the composition unit. This difference correlation function is essentially equivalent to the result of an EXAFS measurement at the absorption edge of element A [68] and involves a linear combination of the A-A and A-X component correlation functions, where X is any element other than A. As with EXAFS spectroscopy, a single variation for element A may thus be used to study its environment in a complex multicomponent sample. Alternatively, the same data may be combined to yield linear combinations of the A-A and X-X or of the A-X and X-X components. A further variation for element A allows two more first-order difference correlation functions to be obtained and the separation of the A-A, A-X and X-X correlation functions [67,69]. This represents a complete separation into the individual component correlation functions $t'_{ij}(r)$ for a binary system, while to achieve the same result for an n element system it is necessary to perform a minimum of n(n + 1)/2 experiments in which the scattering amplitudes of at least n - 1 elements are varied.

The methods of varying $\bar{b}$ or f(Q) are discussed below (sections 5.4.1 to 5.4.3) together with isomorphous substitution (5.4.4) and the difference technique

(5.4.5), which although not rigorous may nevertheless prove extremely useful in favourable cases provided they are used with the necessary caution. Unfortunately for many amorphous solids none of these techniques is feasible, but even in such cases valuable extra information can usually be obtained from a combination of neutron and X-ray diffraction, for which the component weighting factors are different ($\bar{b}_i\bar{b}_j$ and $\sim Z_iZ_j$ respectively; c.f. section 4.3). A particularly striking example of the complementarity of X-ray and neutron diffraction is afforded by fig. 26, which compares the X-ray [70] and neutron [71] correlation functions for vitreous $As_2O_3$ ($Q_{max}$=17.04Å$^{-1}$ ) and illustrates the use of neutron diffraction to

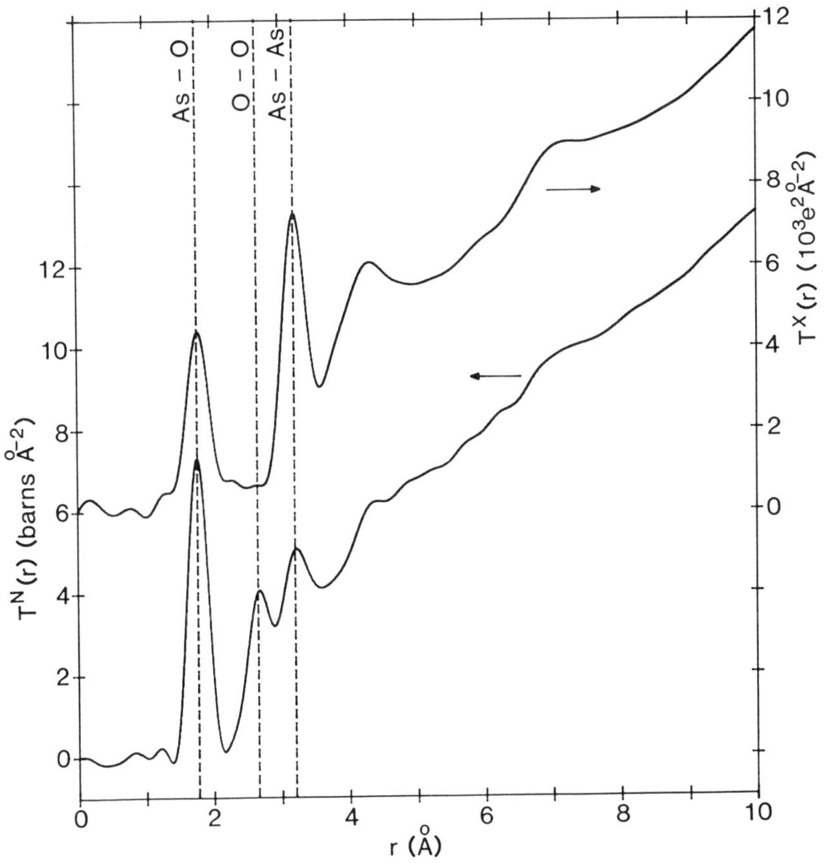


**Figure 26** A comparison of the X-ray [70] and neutron [71] correlation functions for vitreous $As_2O_3$ [71]. The vertical dashed lines indicate the positions of the first As-O, O-O and As-As peaks.

investigate light atoms in the presence of atoms of much higher atomic number as mentioned in the introduction to section 4. Note that with X-rays there is no evidence of the O-O distance within the basic $AsO_3$ pyramidal structural units whereas the first As-As distance is much more prominent than in the neutron case. The O-O component correlation function is particularly important since it is primarily the oxygen atoms which determine the overall packing of the $AsO_3$ pyramids in the vitreous network [71].

**5.4.1** ***Isotopic substitution.*** The scattering length of a given element A, with more than one isotope, may be varied by altering the relative isotopic abundances. Samples may thus be prepared with different values of $\bar{b}_A$ as may be seen from fig. 12 which shows the range of $\bar{b}$ available for each element. For a few elements (H, Li, Ti, Cr, Ni, Sm, Dy and W) isotopes exist with both positive and negative scattering lengths, the latter corresponding to scattering without the normal phase change of $\pi$. This gives rise to a special form of isotopic substitution known as the null technique [69] in which a combination of isotopes is used such that element A has a zero scattering length and hence does not contribute to the measured interference or correlation functions. A measurement of the scattering from such a sample yields the X-X component directly. Examples of the use of the isotropic substitution technique are given later in figs. 41 and 57.

**5.4.2** ***Anomalous dispersion.*** The anomalous dispersion technique was first proposed for X-rays by Krogh-Moe [72] and is based on the fact that near an absorption edge the X-ray form factor has both real and imaginary parts which are wavelength dependent.

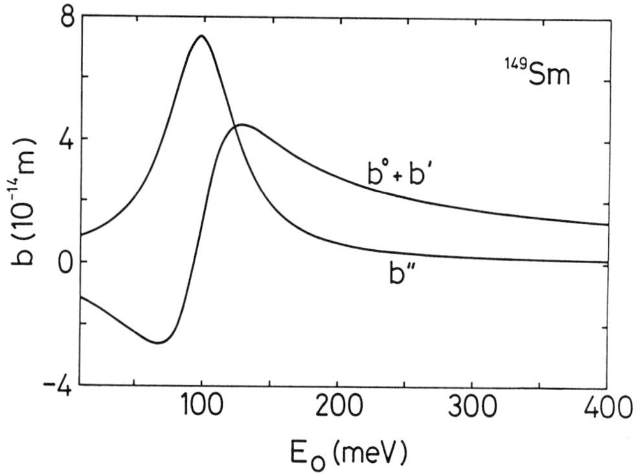


**Figure 27** The real ($b^o$+b') and imaginary (b") parts of the neutron scattering length for $^{149}$Sm [73].

$$f(Q) = f^{\circ}(Q) + \Delta f'(\lambda) + i\Delta f''(\lambda) \quad (32)$$

The utilisation of anomalous dispersion to separate component correlation functions has, however, only really become feasible with the advent of suitable diffractometers at synchrotron X-ray sources which allow the wavelength to be continuously varied to obtain the optimum values of $\Delta f'(\lambda)$ and $\Delta f''(\lambda)$.

Anomalous dispersion experiments are also possible with neutrons [73] for which a variation in scattering length occurs close to an absorption resonance and is much larger for the isotope in question than is the case for X-ray scattering. As an example the energy dependence of the real and imaginary parts of the scattering length for $^{149}Sm$ is shown in fig. 27. At present the neutron anomalous dispersion technique is limited to very few elements but this number should increase with the development of the next generation of amorphous diffractometers at pulsed neutron sources which make measurements at smaller scattering angles and much shorter neutron wavelengths as discussed in section 5.1.

**5.4.3** ***Neutron magnetic diffraction.*** The spins of the magnetic atoms/ions in paramagnetic glasses (i.e. many transition metal and rare earth atoms/ions) can frequently be aligned either by cooling to temperatures below a magnetic ordering transition or by applying a high magnetic field at low temperatures. In either case the difference between the diffraction patterns of the ordered and paramagnetic states yields the magnetic interference function $Qi^M(Q)$ [49]. For a system in which there is no correlation between the magnetic moment directions and the interatomic vectors, $Qi^M(Q)$ may be Fourier transformed to give a magnetic correlation function which depends on both the relative positions and the magnetic moments of the magnetic ions [49] and hence it is possible to extract information concerning both the atomic and magnetic structure. The former application is particularly useful when studying complex systems containing a single magnetic species A, since it yields the A-A distances alone, as may be seen from fig. 28 which shows the magnetic correlation function for vitreous $0.79Fe_2O_3{\cdot}P_2O_5$ [74]. This glass exhibits antiferromagnetic short range ordering at low temperatures and hence the negative peaks in fig. 28 correspond to neighbours with spins antiparallel to that at the origin. The vertical bars indicate the Fe-Fe distances in crystalline α-quartz $FePO_4$. Measurements on magnetic systems are in general limited to $Q \lesssim 10Å^{-1}$ due to the fall-off in the magnetic scattering with increasing Q (c.f. fig. 11), which limits real space resolution, but on the other hand the correlation functions obtained are simpler since they refer only to the magnetic ions.

**5.4.4** ***Isomorphous substitution.*** Isomorphous substitution involves the replacement of one element with another of different scattering amplitude along with the assumption that the structure is otherwise unchanged. In order for this to be the case the chemistry of the two elements must be identical for the system in question and also the bonding to the other elements present, particularly in respect of bond lengths and angles. In general isomorphous substitution, like doping, is not nearly so useful for amorphous solids as for crystals since there is no longer the constraint of a lattice and hence an atom/ion has a much greater tendency to modify its immediate environment to give the optimum (minimum energy) local structure. The most likely candidates for isomorphous substitution are the lanthanide rare earth ions, since their chemistry and ionic radius are relatively insensitive to changes in the 4f electron configuration. Other possibilities include transition metal ions such as the substitution of Hf for Zr in fluorozirconate glasses [75].

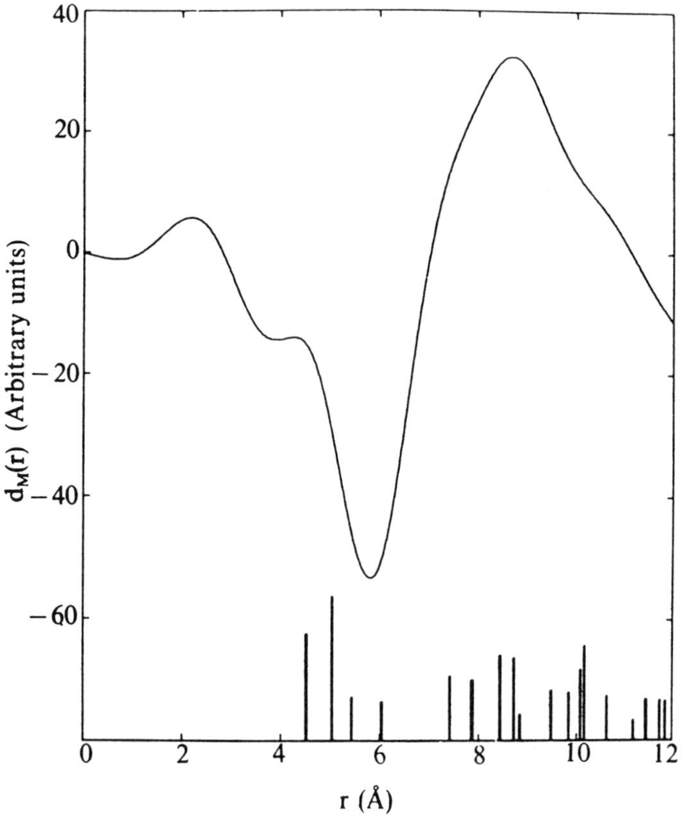


**Figure 28** The neutron magnetic correlation function for vitreous $0.79Fe_2O_3 \cdot P_2O_5$ [74] ($Q_{max} = 4$ Å$^{-1}$). The histogram gives the Fe-Fe distances in crystalline $FePO_4$.

**5.4.5** ***The difference technique.*** In any multicomponent system the information obtainable from diffraction experiments is greatly increased if a systematic study is carried out as a function of composition. One approach in such cases is to use the difference technique in which the correlation function for the base glass is subtracted from those of the more complex glasses and the assumption made that the resulting difference correlation functions apply only to the added component(s). Unless the concentration of the added component(s) is small, however, this assumption is unlikely to be valid except perhaps for the first coordination shell (c.f. ref. [76]).

## 6. METHODS OF INTERPRETATION

The final real space data obtained from a diffraction investigation of an amorphous solid may range from a single correlation function T(r) to a complete determination of all the individual components $t'_{ij}(r)$ extracted using the techniques just described. There is an unfortunate tendency in the literature to regard the correlation function, or its individual components, as the final result of a structural study and to merely describe its general form and list a few peak positions. As may be inferred from section 3.6, however, the correlation function is closely related to the crystallographer's Patterson function and no crystallographer would ever end a paper with just a Patterson function. The experimental interference and correlation functions should thus be considered the starting point for the real science, once the data reduction is complete, the object being to extract the maximum information concerning the structure of the material in question. The methods for doing this vary for the different ranges of order. In range I the parameters for the basic structural unit / co-ordination polyhedron are best extracted using peak fitting techniques whereas in ranges II and III the non-unique, one-dimensional nature of the data leads to an extensive use of modelling techniques. Before discussing each range in detail, however, it is first appropriate to make a few general comments about the form of the data themselves.

The neutron correlation function for vitreous silica at ambient temperature, obtained from a combination of the data in fig. 20 and those obtained with various twin-axis spectrometers [45], is shown in fig. 29 marked with the approximate extent of ranges I, II and III. The high real space resolution ($Q_{max}$=45.2Å$^{-1}$) means that the characteristics of the three ranges of order can be clearly seen. The two peaks from the intrastructural unit order (range I) are extremely sharp and, even at this value of $Q_{max}$, there is still a significant contribution to their width from the peak function $P^N_{ij}(r)$ (c.f. section 5.1 fig. 21). Ranges II and III overlap but the main features between 3.0 and 5.5Å are due to distances between atoms in adjacent tetrahedra (range II). Their increased breadth compared to the peaks in range I arises from the wide distribution of intertetrahedral Si-O-Si and bond torsion angles. Finally range III is characterised by the broadest features in the correlation function which extend out to ~10-20Å, the limit of the intermediate range order.

The amount of structure in the experimental interference and correlation functions, for a given material, and hence their information content, varies considerably but tends to be greater for amorphous network solids than for their random packing based counterparts as may be seen from fig. 30 which compares the correlation functions for amorphous and liquid Ge. The correlation function for liquid Ge is obtained from the data of Salmon [77] and the amorphous solid data are those of ref. [78] Fourier transformed with the same value of $Q_{max}$ (16.1Å$^{-1}$) so that the curves are directly comparable.

The general form of the diffraction pattern, or interference function, for an amorphous solid is also worthy of note since many structural models, particularly those derived from crystalline structures, fail to get this correct. The absence of long range order leads to a correlation function in which the peaks tend to increase in width and decrease in amplitude with increasing r, as may be seen from fig. 29. This means that in reciprocal space the highest frequency Fourier components decay

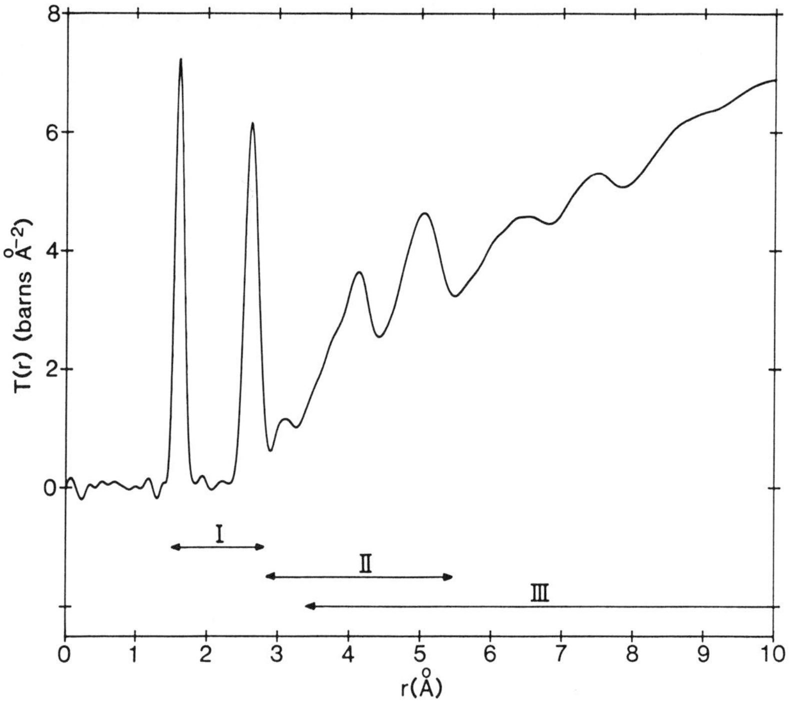


**Figure 29** The correlation function for vitreous silica [57]. The roman numerals indicate the extent of the ranges of order defined in table 2.

the most rapidly and hence the interference function for an amorphous solid is in general characterised by a succession of peaks which get broader with increasing Q until only the Fourier component arising from the sharpest real space peak, usually due to the first interatomic distance, remains (c.f. fig. 20). Thus it is the first few peaks in reciprocal space which are particularly sensitive to the range III order and hence a disagreement between a structural model and experiment in this region ($Q \lesssim 5Å^{-1}$) is predominantly indicative of incorrect intermediate range order. Similarly the rapid decay of the highest frequency Fourier components means that the first peak in the diffraction pattern from an amorphous material is almost invariably the sharpest.

## 6.1 Range I Order

A complete specification of the order within range I includes not only an identification of the basic structural unit(s) or co-ordination polyhedra present but

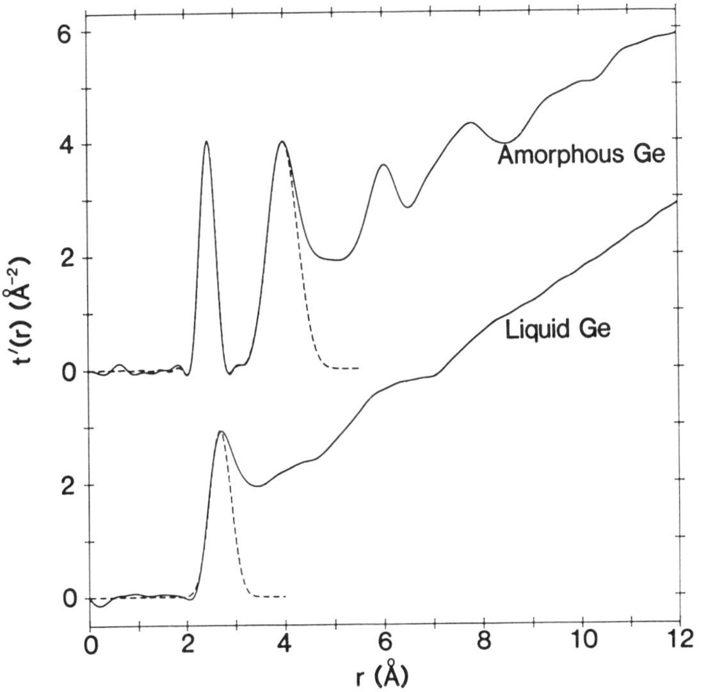


**Figure 30** The correlation function for amorphous Ge at ambient temperature [78] compared to that for the liquid [77] at 1000°C.

also quantitative internal parameters such as co-ordination numbers and the detailed distribution of bond lengths and angles (c.f. table 2 and fig. 2). Too often interpretation is limited to the derivation of bond lengths from peak positions and co-ordination numbers from peak areas, the latter being obtained by assuming a peak is symmetric about its maximum in the function rT(r). This procedure can, however, lead to considerable errors, particularly if a peak is asymmetric and/or only partially resolved as is frequently the case in multicomponent systems. Note that the experimental broadening is asymmetric in rT(r) and also that to obtain the true area under any peak it is necessary to include the area under all the termination ripples on either side of the central maximum.

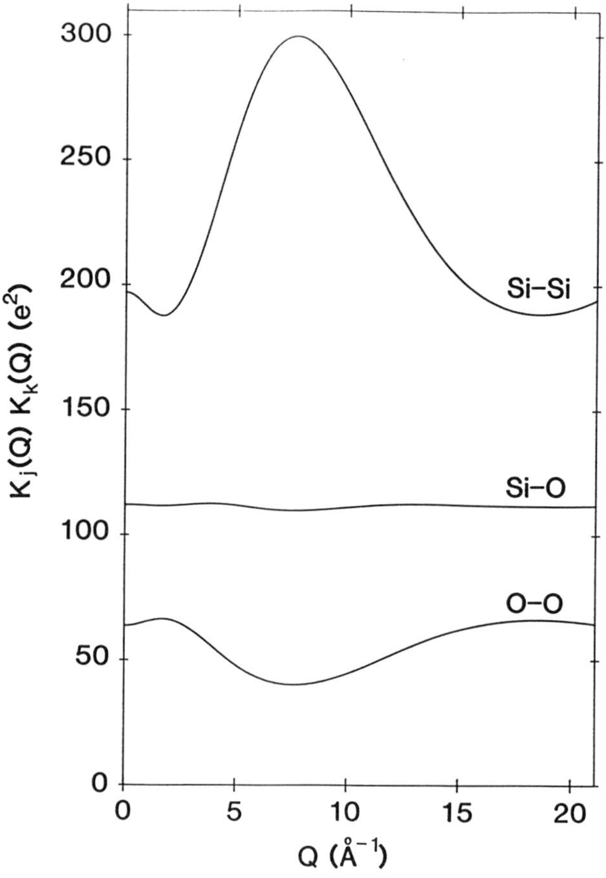


**Figure 31** Effective electron products, $K_i(Q)K_j(Q)$, for vitreous silica.

In order to extract co-ordination numbers from peak areas it is necessary to know $A'_{ij}$, the area under the appropriate peak function $P'_{ij}(r)$. For neutrons $A^N_{ij} = \bar{b}_i\bar{b}_j$, but there is considerable confusion in the literature as to the correct value for X-rays. Many authors following the traditional K approximation [79]

$$f_i(Q) = K_i(Q)f_e(Q) \approx \bar{K}_i f_e(Q) \tag{33}$$

assume that the area under $P^X_{ij}(r)$ is $\bar{K}_i\bar{K}_j$, where $\bar{K}$ is taken as the average value over the Q range of the experiment, but this is incorrect as may be seen by Fourier transforming equ. (30) and taking the limit as $Q \to 0$ which yields

$$A^X_{ij} = 2\int_0^\infty P^X_{ij}(r)dr = K_i(0)K_j(0) \tag{34}$$

The area under $P^X_{ij}(r)$ is thus given by the effective electron product at zero Q, $K_i(0)K_j(0)$, but there is still the problem in a real system of knowing the effective number of electrons associated with each atomic species. For most amorphous solids the K approximation is extremely poor, as may be seen from fig. 31 which plots the effective electron products $K_i(Q)K_j(Q)$ for vitreous silica, and hence should not be used.

It is important to realise that, while structure is conventionally discussed in terms of "atoms", in practice neither neutron nor X-ray diffraction experiments give information about these entities. The neutron correlation function refers to nuclei which may not be in the centre of their associated electron cloud if there is appreciable covalency or polarisation effects in the chemical bonding. The situation for X-rays is complicated by the use of a sharpening function which means that the accuracy of the interpretation of an X-ray correlation function is governed by that to which the electron densities used to calculate the relevant form factors reflect those in the real sample. This is particularly important for elements of low atomic number where a significant fraction of the electrons are involved in the bonding process. The end result of an X-ray diffraction experiment is that (radial) distribution of atoms/ions, with the electron distribution used to calculate the form factors, which would be required to reproduce the measured electron correlation function (before sharpening). Thus not only may co-ordination numbers be wrong, due to an incorrect assumption concerning the effective number of electrons associated with a given "atom", but also the distribution of interatomic distances extracted from the data, particularly in systems with aspherical electron distributions surrounding each nucleus. Similarly $T^X(r)$ will not necessarily be zero below the first peak but will reflect the modification of the free atom/ion electron density to account for any bonding/polarisation. This totally invalidates the use of the procedures described in section 5.3 to "massage" X-ray data before publication, which constrain $T^X(r)$ to be zero in this region.

The effect of bonding electrons on $T^X(r)$ below the first peak and on the determination of peak areas and co-ordination numbers is illustrated in fig. 32 which presents a simulation of the contribution to $T^X(r)$ for vitreous $BeF_2$ from the first Be-F, F-F and Be-Be interactions using peak parameters obtained from a recent neutron diffraction study [56]. The two curves were calculated using spherical free atom ($A^X_{BeF} = 36e^2$) and spherical free ion ($A^X_{BeF} = 20e^2$) form factors with $Q_{max}$=20Å$^{-1}$. Note in particular that, although there is a (relatively small)

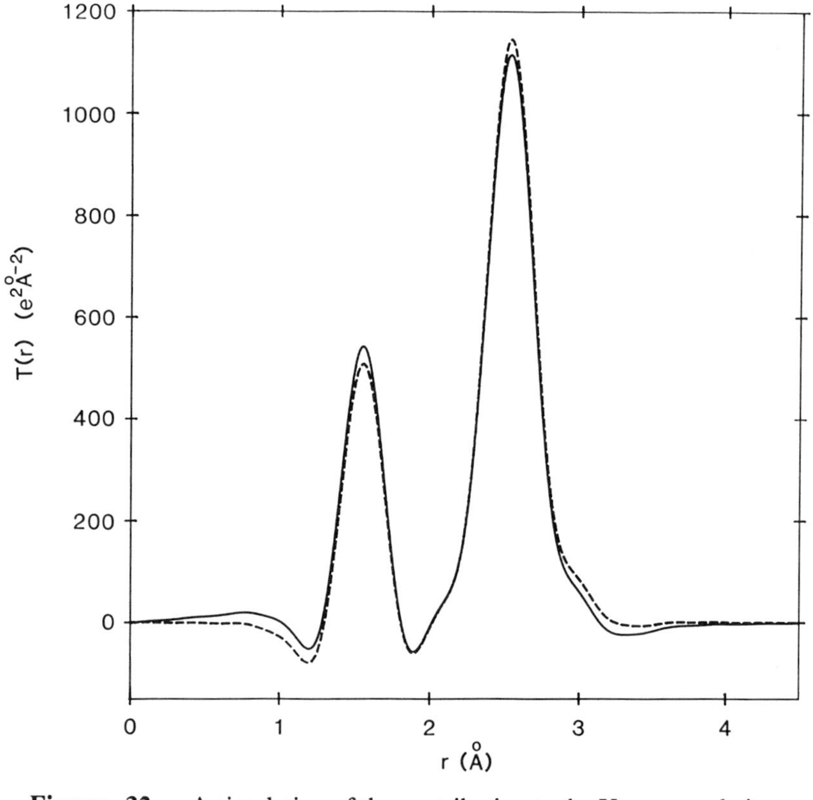


**Figure 32** A simulation of the contribution to the X-ray correlation function of vitreous $BeF_2$ from the first Be-F, F-F and Be-Be interactions assuming free atom (——) and free ion (---) form factors [56].

change in the height of the first Be-F peak, the difference of a factor of 1.8 in area mainly occurs away from the central maximum and is a reflection of the fact that it is the electrons in outer orbitals which are involved in the transition from free atoms to free ions, or in any other type of bonding. Great care is thus required in the interpretation of peak areas under X-ray correlation functions, particularly in systems involving light atoms. As demonstrated by fig. 32 large errors can occur if

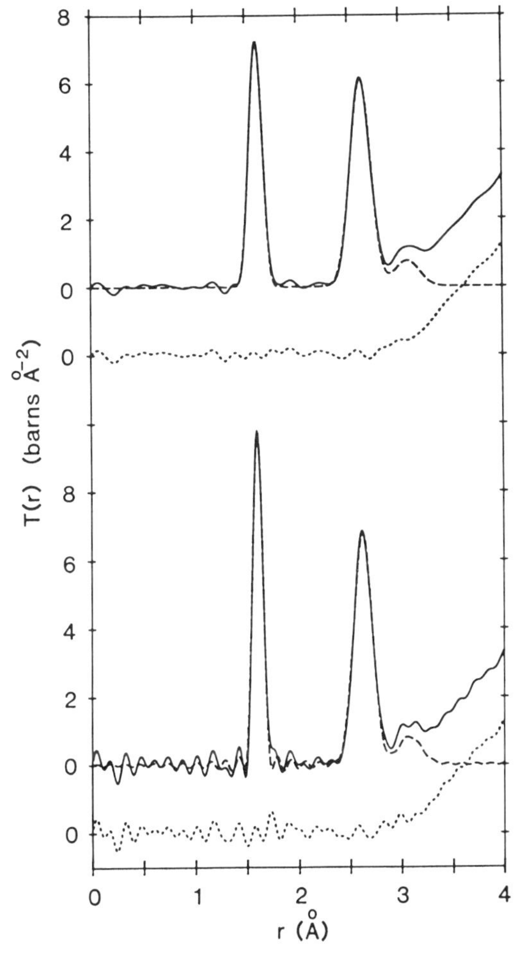


**Figure 33** A fit to the first two peaks in T(r) for vitreous silica with (upper curves) and without (lower curves) a modification function [47]. ——, experiment; ----, fit; ·····, residual.

the peak area is taken as simply that beneath the central maximum or if the outer features of the first peak are suppressed by data "massaging" techniques.

A much better method of obtaining structural parameters in range I is by means of fitting techniques. Usually it is assumed that the peaks involved have a Gaussian distribution in $t_{ij}(r)$ centred at $r_{ij}$, with co-ordination number $n_{i(j)}$ and rms deviation $(\overline{u_{ij}^2})^{1/2}$, and these parameters are extracted by fitting either the relevant peaks in T(r), which involves convolving the Gaussian with the correct component peak function $P'_{ij}(r)$, or else Qi(Q) at high Q

$$Q\, i_{ij}(Q) = n_{i(j)}\overline{a_i(Q)}\;\overline{a_j(Q)}\,\frac{\sin r_{ij}Q}{r_{ij}}\,e^{-\overline{u_{ij}^2}Q^2/2} \tag{35}$$

The advantage of fitting in real space is that the Fourier transform localises each peak and has the effect of reducing the coupling between the parameters for adjacent

**Table 6. Peak Parameters for Vitreous Silica**

| Radiation | Modification function | $R_\chi$ (%) | i-j | $n_{i(j)}$ | $r_{ij}$ (Å) | $(\overline{u^2_{ij}})^{1/2}$ (Å) |
|---|---|---|---|---|---|---|
| Neutrons | Yes | 3.8 | Si-O | 3.85±0.16 | 1.608±0.004 | 0.047±0.004 |
| | | | O-O | 5.94±0.23 | 2.626±0.006 | 0.091±0.005 |
| | | | Si-Si* | 4.0 | 3.077 | 0.111 |
| Neutrons | No | 6.9 | Si-O | 3.79±0.14 | 1.605±0.003 | 0.045±0.002 |
| | | | O-O | 5.92±0.22 | 2.625±0.005 | 0.090±0.004 |
| | | | Si-Si* | 4.0 | 3.077 | 0.111 |
| X-rays | Yes | 7.3 | Si-O | 3.77±0.12 | 1.626±0.004 | 0.053±0.009 |
| | | | O-O | 6.15±0.38 | 2.657±0.012 | 0.102±0.016 |
| | | | Si-Si* | 4.0 | 3.077 | 0.111 |
| X-rays | No | 9.0 | Si-O | 3.58±0.09 | 1.619±0.002 | 0.036±0.004 |
| | | | O-O | 5.37±0.32 | 2.643±0.006 | 0.078±0.007 |
| | | | Si-Si† | 4.0 | 3.060 | 0.135 |

* From ref. [80]
† Fixed

peaks. It is also possible to use other peak shapes such as a Gaussian in the component radial distribution function $rt_{ij}(r)$. Fits to the first two peaks in the neutron correlation function for vitreous silica, obtained with and without a modification function, are shown in fig. 33 and the resulting peak parameters are summarised in table 6. Parameters for the overlapping first Si-Si peak are taken

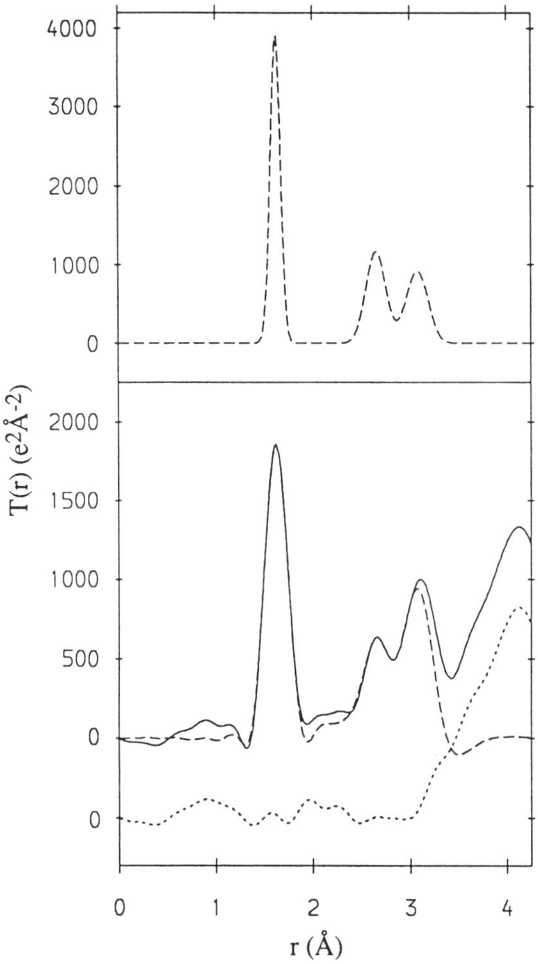


**Figure 34** A peak fit to Mozzi and Warren's [55] X-ray data for vitreous silica (lower curves), together with the unbroadened peaks (upper curve) weighted according to the product $Z_iZ_j$. ———, experiment; ----, fit; ·····, residual.

from X-ray data [80]. The fits were performed in real space using a $\chi^2$ minimisation procedure [81] and the quality of fit is expressed in terms of the $R_\chi$ factor [82]

$$R_\chi = \left( \frac{\sum_i \left[T_{exp}(r_i) - T_{fit}(r_i)\right]^2}{\sum_i T_{exp}^2(r_i)} \right)^{1/2} \qquad (36)$$

calculated over the fitting range 1.0-2.9 Å. The errors quoted in table 6 relate only to the $\chi^2$ fit and do not include any systematic errors in the original correlation function.

A corresponding fit to the first three peaks in Mozzi and Warren's [55] X-ray correlation function for vitreous silica is illustrated in fig. 34, together with the unbroadened peaks weighted according to the atomic numbers of Si and O. The parameters for this fit are also given in table 6. Note that the relative heights of the O-O and Si-Si peaks are not what would be expected from the atomic numbers of Si and O due to the satellite features on either side of the central maximum of the appropriate component peak functions (fig. 14). The areas under the O-O and Si-Si peaks are $768e^2$ and $784e^2$ respectively. It is also interesting that there appears to be a significant difference between the Si-O bond lengths as determined from X-ray and neutron diffraction, which may be due to the fact that the electron distributions around the silicon and oxygen atoms are not spherical.

As pointed out by Mozzi and Warren, the most probable bond angle is not simply calculated from the mean bond length and equivalent second neighbour distance. To investigate the O-Si-O bond angle variation for vitreous silica a Monte Carlo approach has been used [47], with the neutron data, in which pairs of Si-O bond lengths and an O-Si-O bond angle were chosen using a Gaussian random number generator and the resulting O-O distance distribution compared with experiment. This procedure, which assumes that the two bond lengths and the bond angle are uncorrelated, yields a mean bond angle of 109.7±0.6° and an rms bond angle variation of 4.5 ± 0.4°.

## 6.2 Range II Order

In addition to interatomic distances and bond angles, the quantification of order in range II involves the distribution of bond torsion angles $A(\alpha)$ which cannot be obtained directly from diffraction data without assuming some model. For materials such as $SiO_2$ the difference between the structures of the crystal and glass are that in the latter there is a broad distribution of Si-O-Si and bond torsion angles whereas in general the former is characterised by one, or at most a few, discrete values. One simple model for the interconnection of structural units in a random network assumes a uniform distribution of torsion angles except where excluded by steric effects. This model has been used by Mozzi and Warren and later by Galeener [83] to predict Mozzi and Warren's Si-O-Si bond angle distribution, which is given in fig. 35. The contribution to Mozzi and Warren's X-ray correlation function from range I and II order, according to their model, is shown in fig. 36.

The structural models discussed in the next section on range III order must also by definition, contain some inherent bond and torsion angle distribution and indeed often fail in this respect, making a comparison of range III order with experiment pointless. An early example of a structural model is that due to Bell and Dean [84] for vitreous silica which has subsequently been relaxed by Gaskell and Tarrant [85] and in its relaxed form is compared with both neutron and X-ray data in fig. 37. The range II order gives rise to the peaks in $T^N(r)$at 4.1Å (mainly Si-O interactions) and 5.0Å (O-O) and it can be seen that the shoulder on the low r side of the 4.1Å peak is much too pronounced compared to experiment, indicating that the range II order in the model is incorrect. This is a common failing of models of vitreous silica and a similar discrepancy between model and experiment can be seen for most of the other published random network models of vitreous silica, such as those discussed by Evans et al. [86].

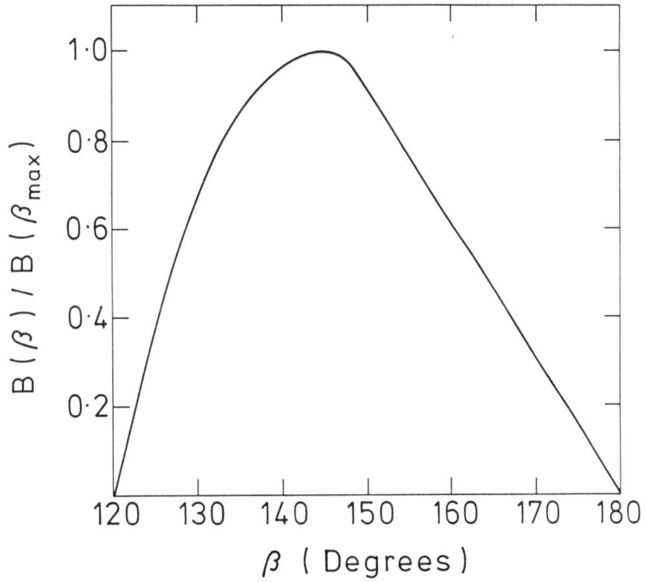


**Figure 35** The Si-O-Si bond angle distribution for vitreous silica as determined by Mozzi and Warren [55].

## 6.3 Range III Order

As already noted the normal method of investigating range III order is via modelling studies in which an attempt is made to build a "typical" region of the structure which can be used both to compare to diffraction data and in the calculation of other properties. Models may be built either by hand or using a computer and may be in the form of a cluster or, in the case of computer models, be generated with periodic boundary conditions. At present the main difficulty in producing models of amorphous solids, using either technique, is to make them large enough to predict the various distributions of structural parameters with sufficient statistical accuracy and in the case of cluster models to ensure that they are capable of effectively infinite extension. The advantage of hand building is the feel it gives for a structure, the disadvantage is that it is tedious and difficult to vary parameters in a systematic way. The use of computer relaxation techniques, however, reduces the accuracy with which it is necessary to measure the atomic co-ordinates and also removes any anisotropy which might result from the effects of gravity on a large model.

Structural transformation techniques may also be used to generate new models, the simplest being relaxation to a new set of equilibrium bond lengths and angles such as used by Davis et al. [87] to produce models of amorphous antimony and phosphorus from an existing model of amorphous arsenic [88]. This type of relaxation may be likened to the displacive transformations which occur for the

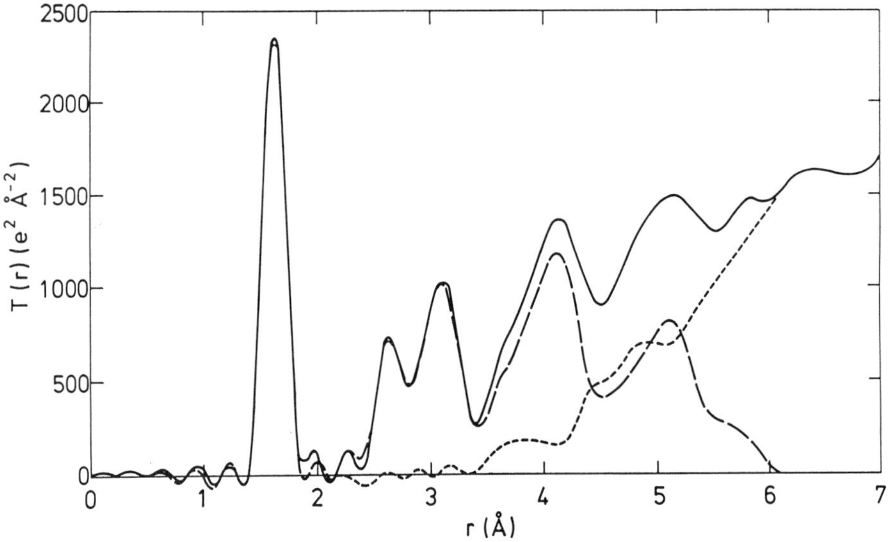


**Figure 36** The contribution to the X-ray correlation function of vitreous silica from range I and II order [55]. ——, experiment; -----, fit; ·····, residual.

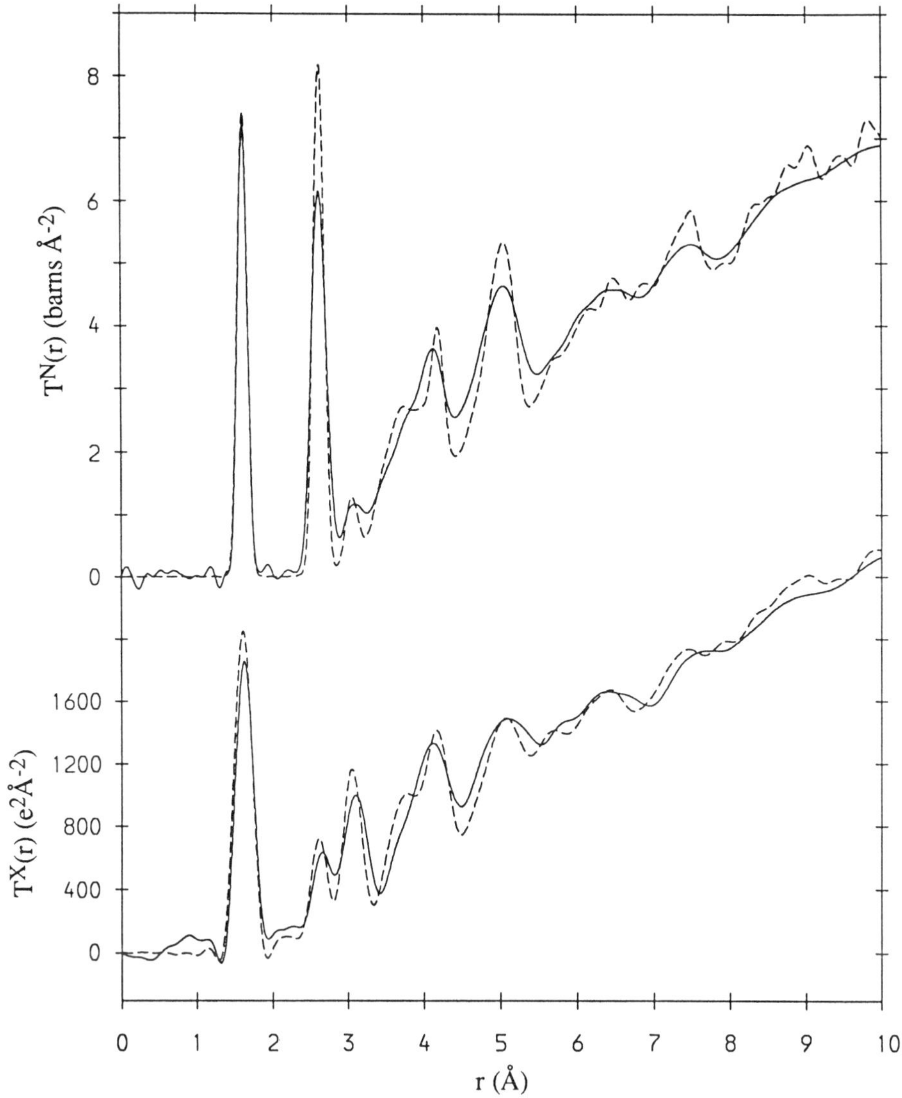


**Figure 37** Neutron (upper curves) and X-ray (lower curves) correlation functions for the Bell and Dean [84] model of vitreous silica, as relaxed by Gaskell and Tarrant [85], (-----) compared to experiment (——) [47,55].

crystalline polymorphs of $SiO_2$, such as the $\alpha \rightleftharpoons \beta$ transition in quartz and cristobalite, since in both cases the network topology is unchanged. The relationship between the structures in figs 6B and 6C has been discussed in section 3.3. Given one of them it is clearly possible to generate the other, by the addition or removal of the X atoms followed by energy relaxation. Such a transformation has been termed a decoration transformation [38,89] and is also topologically invariant. Examples of decoration transformations in the literature include the As to $B_2O_3$ transformation of Elliott [90] and the conversion of the Evans and King [91] model of $SiO_2$ to amorphous Si [92,93].

Topological transformations are of two basic types. In a purely reconstructive transformation only the connectivity matrix $c_{pq}$ is affected whereas a connectivity transformation involves a change in the connectivity of at least some of the topological units present. Reconstructive transformations have been used to generate models of amorphous Si and Ge, the most successful being the "Sillium" models of Wooten and Weaire [94-96] which start from the crystal structure. An example of a connectivity transformation is the conversion of the Bell and Dean [84] model of $SiO_2$ to $B_2O_3$-$SiO_2$ [97] and $B_2O_3$ [98].

## 7. STRUCTURAL MODELS

A wide range of structural models has been presented in the literature and the most important of these models will now be considered in more detail. First, however, it is necessary to make some general comments concerning the comparison of models with experiment since this is frequently grossly inadequate. Indeed, for most of the models so far reported, a rigorous comparison reveals serious discrepancies between the model and diffraction data. It should also be stressed that any valid structural model must not only be consistent with diffraction data but also with the available data from all the other techniques described elsewhere in this book. For example a Monte Carlo simulation may yield a structure which is consistent with diffraction data but has to be rejected due to the presence of too many over or under co-ordinated atoms as determined by magnetic resonance measurements.

The average number density $\rho^\circ$ is a very simple structural parameter, but nevertheless an extremely important one since any model must be capable of predicting the correct density. In the words of Finney [99], "Models are wrong whose density is wrong." A difficulty in this respect is that the equilibrium bond length for a given material is frequently not known to better than 1% which is equivalent to a 3% uncertainty in the density. Uniformity of density may also be a problem for cluster models, particularly for those which are computer generated, in that the average density $\rho^\circ$ is not maintained as extra units are added to an initial nucleus. A useful check [100] is to determine $\rho^\circ$ as a function of model size from the number of atoms inside a sphere of increasing radius R, centred at the average co-ordinate $\bar{x}$, $\bar{y}$, $\bar{z}$. As R increases $\rho^{\circ}$ should approach a constant value and then fall off at the edge of the model. The point just before the density begins to fall off determines the largest possible spherical model for use in calculating the model component correlation functions. A spherical cluster model is required to correct for finite model size, the model correlation functions being reduced relative to those for an infinite model by the factor [100]

$$F(r) = \frac{1}{16}\left(\frac{r}{R} - 2\right)^2\left(\frac{r}{R} + 4\right) \qquad r \leq 2R$$
$$F(r) = 0 \qquad r \geq 2R \tag{37}$$

The functions Qi(Q) and T(r) emphasise different aspects of a given structure and hence it is important when comparing models with experiment to make such comparisons in both real and reciprocal space. Many authors term good agreement with experiment as getting the peaks in either Qi(Q) or T(r) in the right place. Only very poor models fail to achieve this. What is necessary is also to get peak shapes and areas correct, which involves including the effects of thermal vibration for static models (e.g. computer relaxed random networks) and in real space folding the model component correlation functions with the correct component peak functions $P'_{ij}(r)$ for the experiment in question. Otherwise the comparison between model and experiment is meaningless. Note that $P'_{ij}(r)$ has satellite features on either side of the central maximum which depend on the exact form of M(Q) and themselves lead to features in T(r). Thus a Gaussian approximation to $P'_{ij}(r)$ is simply not adequate, especially when the width is arbitrarily chosen to give the best agreement with experiment.

Resolution effects are also important in reciprocal space and may involve both model and experimental interference functions. In calculating the interference function for a real-space model it is necessary to truncate the model correlation function at some value $r_{max}$. This leads to a broadening of the model interference function and the introduction of satellite features in reciprocal space in exactly the same way as for an experimental correlation function in real space. Thus in order to compare model and experiment in reciprocal space it is necessary to fold the ***experimental*** interference function with the reciprocal space peak function P(Q) appropriate to the model calculation. For a spherical model of radius R, the reciprocal space peak function is the Fourier cosine transform of F(r) [equ. (37)] and takes the form [101]

$$P(Q) = \frac{3}{4\pi RQ^2}\left[1 - \frac{2\sin(2RQ)}{2RQ} + \frac{\sin^2(RQ)}{(RQ)^2}\right] \tag{38}$$

Similarly the model interference function must be broadened by the reciprocal space resolution function for the instrument used in the experiment.

## 7.1 Random Network Cluster Models

A large number of random network cluster models has been reported in the literature, particularly for amorphous Ge and vitreous $SiO_2$. The main problem in building such models by hand is to incorporate the correct distribution of structural parameters such as bond lengths and angles. As an example consider the Polk and Boudreaux model of amorphous Si and Ge [102] which is compared to neutron data [78] in fig. 38. This model was constructed from plastic tetrahedra joined together with rigid aluminium sleeves, like those in fig. 39, which had the effect of constraining the Ge-Ge-Ge bond angle distribution to be much too narrow as may be seen from the second peak in the correlation function. As with most cluster models, the Polk model has been relaxed to the minimum energy configuration using a Keating potential [103,104] and hence represents equilibrium atomic co-ordinates. Thus in comparing with experiment it is not only necessary to include the correct peak function $P^N_{GeGe}(r)$ but also appropriate thermal factors, the rms thermal displacement, $\langle u^2_{Ge}\rangle^{1/2}$ used in the present comparison being 0.070Å.

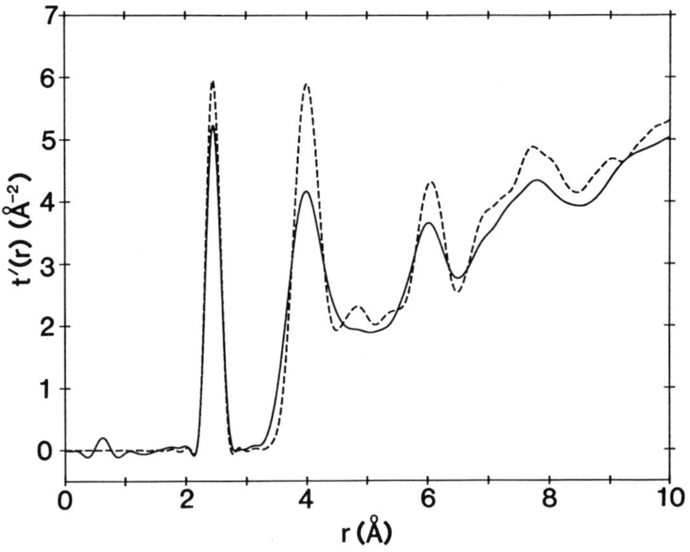


**Figure 38** The correlation function for the Polk and Boudreaux [102,103] model of amorphous Ge compared to the neutron data of ref. [78]. -----, model; ——, experiment [57].

The Bell and Dean [84] model of vitreous silica, as relaxed by Gaskell and Tarrant [85], has already been mentioned in section 6.2 (fig. 37) where it was pointed out that the range I and II order is incorrect. There is also excess structure in the region of T(r) above 6Å which characterises the range III order. This is a consequence of poor statistical accuracy due to the small size of the model (614 atoms) which also means that too much strain is released on relaxation as demonstrated by the fact that the strain energy is too small [85] and the $SiO_4$ structural units too regular.

Recently a new computer algorithm has been developed by Gladden [105] which mimics the hand building process and allows the various structural parameters to be varied in a systematic way. This algorithm has been used to generate a number of random network models of vitreous silica and the optimum equilibrium Si-O-Si bond angle and minimum ring size have been extracted by comparison with earlier neutron diffraction data by the present author and co-workers [45,46]. A model generated using these optimum parameters is compared with neutron data in fig. 40. The very much higher real space resolution reveals that further parameter refinement is still required in the light of these new data, probably in respect of the relative strength of the force constants used for structural relaxation. The comparison between the model and the lower resolution neutron data in fig. 5(a) of ref. [105], on the other hand, demonstrates the potential of the algorithm to yield very much better agreement with diffraction data than hitherto,

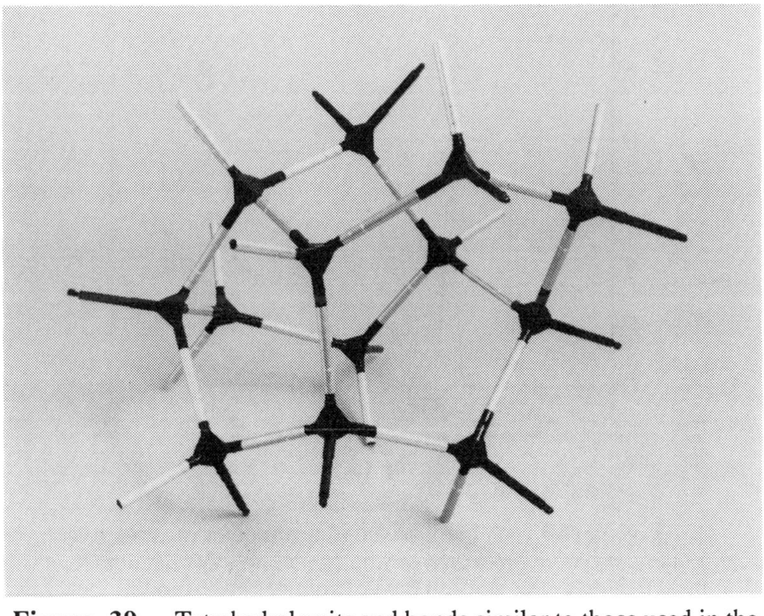

**Figure 39** Tetrahedral units and bonds similar to those used in the Polk and Boudreaux [102] model of amorphous Ge [57].

particularly in respect of the range II order. The most important aspect of fig. 40, however, is that it clearly emphasises the need for the highest possible real space resolution when using diffraction data to test structural models. As with any model fit, the reliability of the parameters extracted is critically dependent on the quality of the original experimental data.

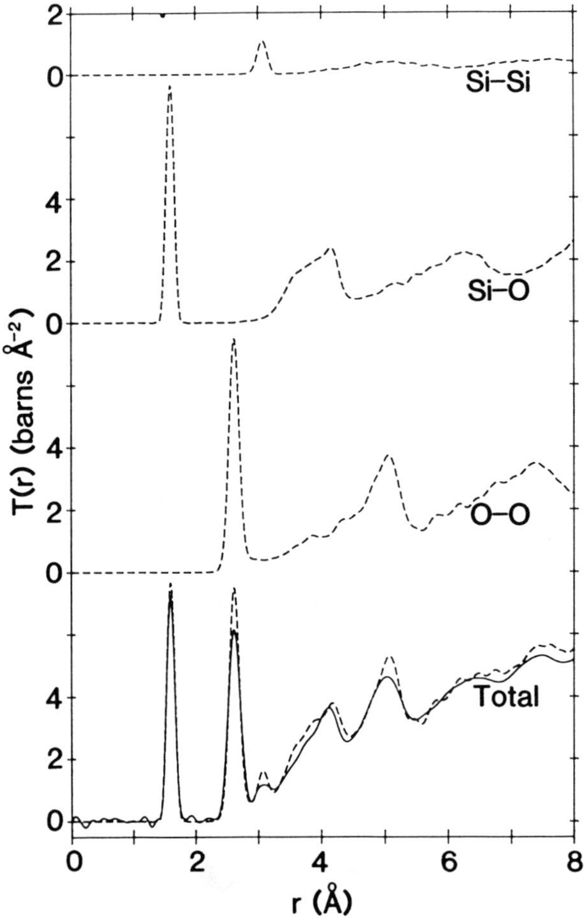


**Figure 40** A comparison of a computer random network model of vitreous silica generated by Gladden [105] with the neutron data of fig. 29. -----, model; ——, experiment [57].

## 7.2 Random Sphere Packing Models

The earliest random sphere packing models comprised hand constructed random close packings of single sized spheres, the most well known example being that due to Finney [14,106]. Sphere packings can also be computer generated for both hard and soft spheres with a wide range of packing densities and in the case of binary systems with varying radius ratio. An alternative approach is to use the Percus-Yevick hard sphere analytical expressions for either single sized spheres [107] or a binary system [108].

For a simple binary hard sphere system $A_{1-x}X_x$ there should be all three types of interatomic contact (A-A, A-X and X-X) and the A-X first neighbour distance should be the mean of the A-A and X-X distances. For most metallic glasses, however, there appear to be no mutual contacts between the smaller atoms (i.e. the metalloid atoms in transition metal — metalloid glasses and the transition metal atoms in rare earth - transition metal glasses) as may be seen from fig. 41 [109] which compares the three component correlation functions for melt spun $Dy_7Ni_3$, obtained using the double null isotopic substitution technique [69], with those calculated for a binary hard sphere liquid using the Percus-Yevick equation. The partial interference functions $Qi_{ij}(Q)$ were multiplied by a Debye-Waller factor with a value of 0.147Å for the rms bond length variation so as to include the effects of thermal vibrations. Fourier transforms were then performed, using the same $Q_{max}$ and modification function as for the experimental data. The hard sphere diameters used in the calculation were adjusted so as to obtain coincidence between the calculated and experimental first peaks of $t^N_{DyDy}(r)$ and $t^N_{DyNi}(r)$, yielding final values for the Dy and Ni atom diameters of 3.347 and 1.938Å respectively. The composition and number density ($x_{Ni} = 0.3$; $\rho° = 0.03938$ atoms Å$^{-3}$) were not adjusted. If the atoms are treated as hard spheres then it must be concluded that the Ni atoms are close to each other but not touching. Thus a more ordered structural model than a random close packing of hard spheres is required as indicated in section 2.2.

An example of a predominantly ionic system in which the anions are approximately randomly close packed is vitreous zinc chloride. In crystalline δ-$ZnCl_2$ the $Cl^-$ ions form a hexagonal close packed array and the $Cl^-$ ion packing density for the glass (0.656 [110]) is very close to that for a random close packing. Chlorine has a higher neutron scattering length ($0.9577 \times 10^{-14}$m) than zinc ($0.568 \times 10^{-14}$m) [51] and hence the Cl-Cl contribution plays a dominant role in $T^N(r)$. The Cl-Cl contribution calculated from the Finney [14, 106] random close packing is compared with experiment in fig. 42. The upper histogram shows t(r) for the original packing scaled to a hard sphere diameter of 3.70Å and the lower dashed curve the same function including the appropriate weighting factors and broadened by the experimental resolution and a Gaussian of rms deviation 0.23Å to approximately account for thermal effects and the softer ions. It can be seen that the model Cl-Cl contribution accounts for all the major features in the correlation function above 3Å. The $Zn^{2+}$ ions occupy tetrahedral holes in such a way that the resulting $ZnCl_4^{2-}$ tetrahedra form a corner linked $AX_2$ random network. An alternative way of viewing the same structure is that it is the result of reducing the

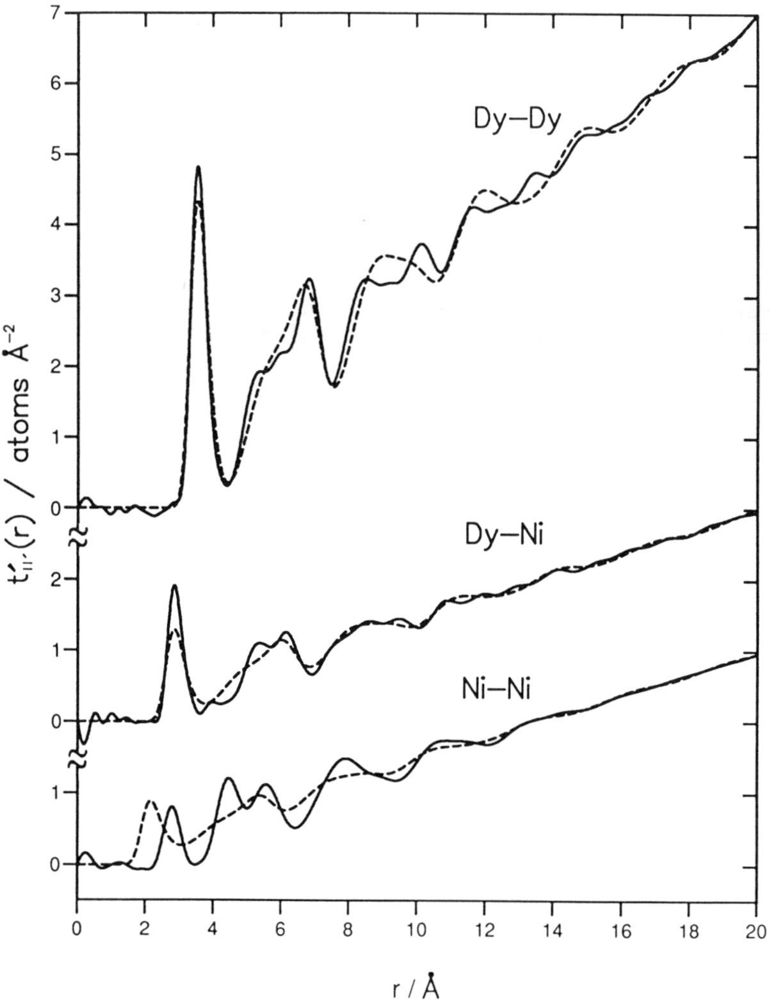


**Figure 41** The three component correlation functions for melt-spun $Dy_7Ni_3$ (——), compared to the results of a Percus-Yevick calculation for a binary hard sphere liquid (-----) [109].

A-X-A angles in a silica-like random network, allowing bond rotations, until the anions become approximately randomly close packed. Vitreous $ZnCl_2$ is not, however, purely ionic as revealed by molecular dynamics simulations using simple ionic potentials which fail to reproduce the experimental correlation function [110].

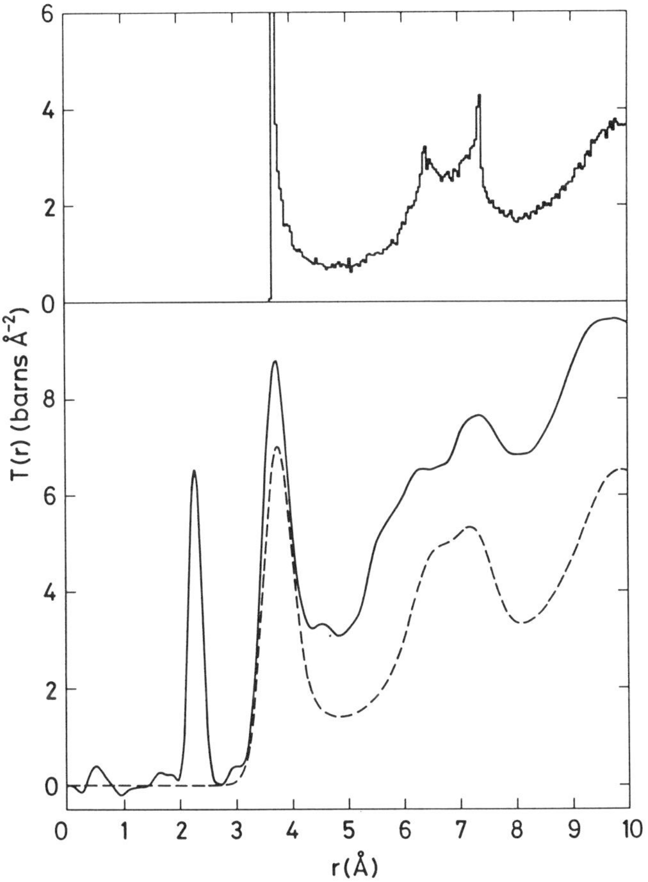


**Figure 42** The component correlation function, $t'_{ClCl}(r)$, for a random close packed array of $Cl^-$ ions (-----) compared to the experimental $T^N(r)$ ($Q_{max}$=23.6Å$^{-1}$) for vitreous $ZnCl_2$ [110]. The histogram gives the unbroadened $t_{ClCl}(r)$.

## 7.3 Crystal Based Models

Useful information on the structure of amorphous solids can often be obtained from a simple comparison with the corresponding crystalline polymorphs and for this reason the crystalline state is frequently a starting point for structural models. This does not necessarily imply an acceptance of the crystallite theory, merely the fact that the short range order may resemble that of the associated crystals. The great value of such models lies not in obtaining a perfect fit but in their use as a vehicle for comparing intermediate range order in the crystalline and amorphous solid states.

A convenient formalism is provided by the quasi-crystalline* model [111] in which the component differential correlation functions for the crystal, $d_{ij}^{xtal}(r)$ are multiplied by F(r) from equ. (37) to restrict the longer range order to interatomic spacings less than some characteristic correlation length L=2R. The model is equivalent to embedding a crystallite of diameter L in a structureless matrix of the same average density (to eliminate small Q scattering) and then averaging over all relative orientations of the crystallite and the scattering vector Q. Thus

$$Qi(Q) = \sum_i \sum_j \overline{a_i(Q)}\, \overline{a_j(Q)} \int_0^L d_{ij}^{xtal}(r)\, F(r) \sin Qr\, dr \qquad (39)$$

The model correlation functon D(r) can be obtained by the usual Fourier transformation as in fig. 43 which compares a β-cristobalite quasi-crystalline model with a correlation length of 10Å to the experimental neutron interference and correlation functions for vitreous silica. Multiplying the component differential correlation functions $d_{ij}^{xtal}(r)$ by F(r) in real space is equivalent to folding the interference function for the crystal with the reciprocal space peak function P(Q) in equ. (38). The fact that the width of P(Q) is independent of the positions of the Bragg peaks, $Q_{hkl}$, means that the model interference function exhibits far too much structure at high Q. However, simple models of this type are useful for comparing range III order. For example the β-cristobalite model in fig. 43 yields three peaks in the region below 5Å$^{-1}$ whereas the real material has only two indicating that the structure of vitreous silica is not like that of β-cristobalite despite several claims in the literature to the contrary [112,113].

---

* Unfortunately, since details of this model were published the term "quasi-crystal" has been used to denote a completely different class of materials which have long range order without periodicity.

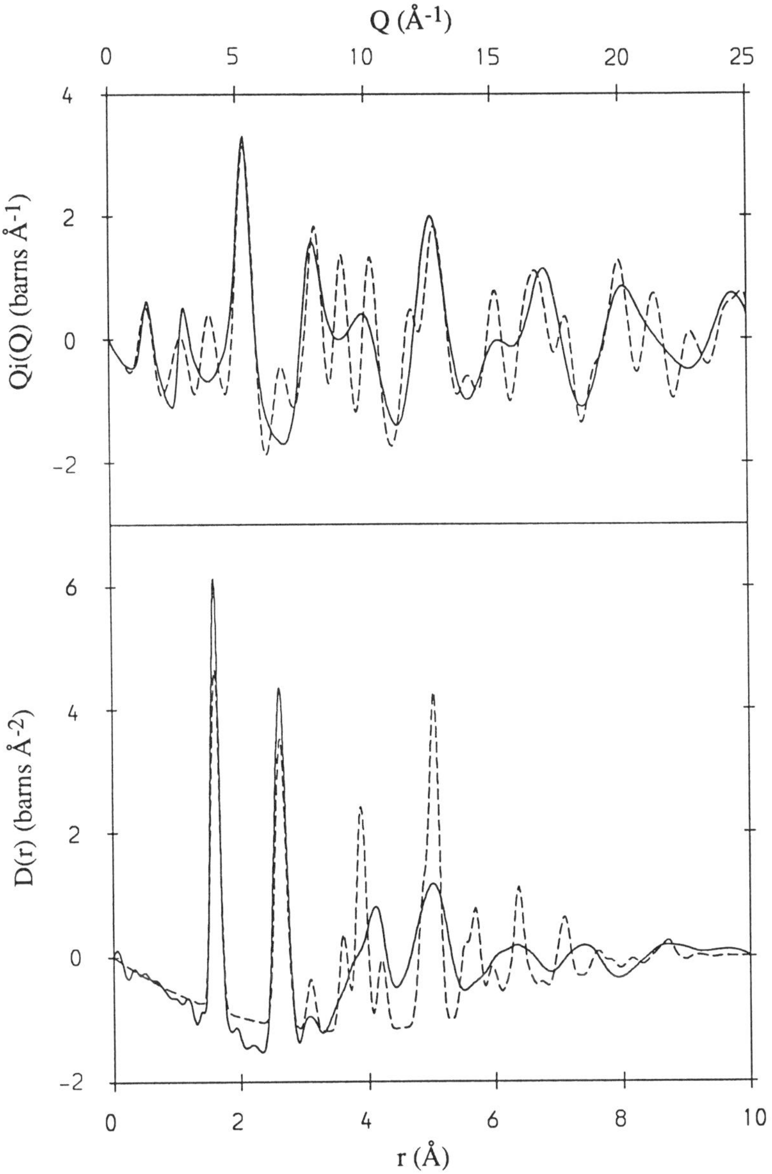


**Figure 43** The interference and correlation functions for a quasi-crystalline model [111] of vitreous silica, based on the β-cristobalite structure (-----), compared to experiment (——).

In order to obtain the correct form of diffraction pattern from a crystal based model it is necessary to include disorder such that the width of the peaks in the model correlation function increases with increasing r. This may be achieved by the introduction of strain but the problem is that simple models for the introduction of strain also lead to an unacceptable distortion of the basic structural units and hence the model becomes incompatible with experiment in respect of the range I order as may be seen for the Hosemann paracrystalline model for vitreous silica [114,115] in fig. 44. The paracrystalline model involves the introduction of strain at constant network topology and for the model in fig. 44 leads to intratetrahedral O-Si-O bond angles of 106 and 117° and a splitting of the intratetrahedral O-O peak.

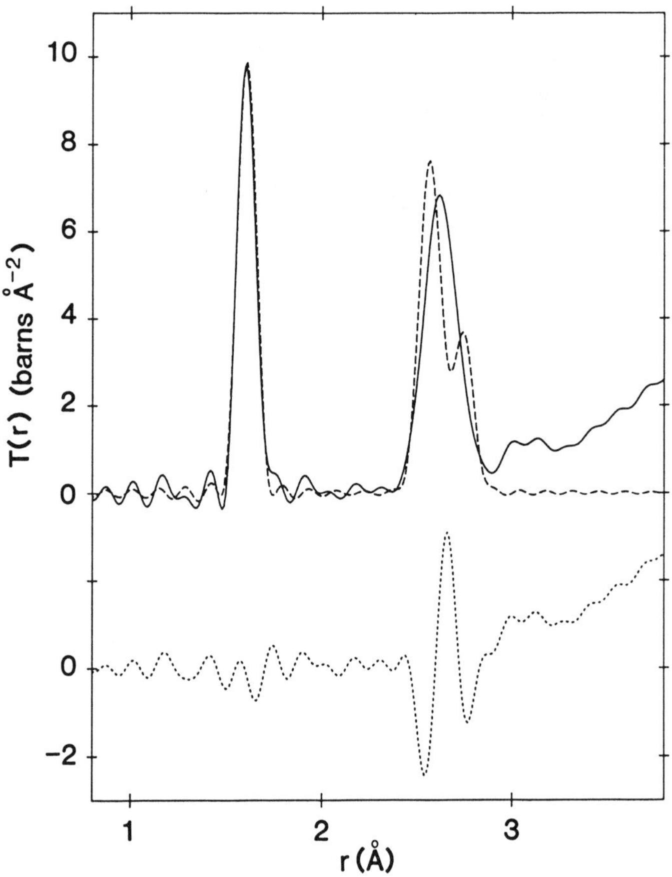


**Figure 44** The first three peaks in $T^N(r)$ as predicted by the Hosemann [114,115] paracrystalline model of vitreous silica [57]. -----, model; ·····, residual; ——, experiment.

Various mathematical models have been employed in the theory of strained crystals (e.g. see Warren [116]), both for the functional form of the strain and for the way in which the strain increases with the spacing between equivalent (hkl) planes. Three such models are summarised in table 7. In each case the introduction of strain leads to a progressive broadening of the Bragg peaks in reciprocal space with increasing Q but not a shift in their position. The consequences of model 2 in reciprocal space are very similar to those of the paracrystalline model due to Hosemann [114,115], although in the latter case the theoretical approach is rather different in that it involves multiple convolutions. It should also be noted that in a real strained polycrystalline material there may be additional reciprocal space broadening due to crystallite size effects and hence the overall peak shape will be the convolution of those due to strain and finite crystallite size. This is particularly relevant when using this type of formalism as a basis for models of amorphous materials.

**Table 7 Strained Crystal Models**

| | Real Space | | Reciprocal Space | |
|---|---|---|---|---|
| Model | Functional Form | Width | Functional form | Width |
| 1 | Gaussian | $\propto r$ | Gaussian | $\propto Q$ |
| 2 | Gaussian | $\propto \sqrt{r}$ | Lorentzian | $\propto Q^2$ |
| 3 | Lorentzian | $\propto r$ | Lorentzian | $\propto Q$ |

## 7.4 Amorphous Cluster Models

The concept of an amorphous cluster has already been outlined in section 2.3. Such models are more appropriate for materials such as amorphous Ge, which exhibit small angle scattering, rather than simple melt-quenched glasses which do not. The amorphous cluster polytetrahedral model of amorphous Ge due to Gaskell [117] was constructed using 14-atom structural units of tetrahedral shape and diamond cubic symmetry. The method of joining the tetrahedral modules was such that the resulting model is not space filling and exhibits a significant amount of distortion, particularly near its surface. In order to show the effect of strain, correlation functions for two spherical sections are given in fig. 45, one at the model centre containing 982 atoms and the other at the model surface containing 250 atoms. (The model comprises a total of 3094 atoms). It is obvious that t'(r) for the central section exhibits far too much structure above the first peak. The agreement between the edge section and experiment is much better, but there are still serious discrepancies in the region of the second peak in t'(r), particularly in respect of the high r shoulder at 4.6Å. The correlation function for the total model would be intermediate between the two extremes shown and thus it can be concluded that the structure of amorphous Ge is much closer to the best random network models of Wooten and Weaire [94-96] (c.f. section 7.7.1) than the polytetrahedral model.

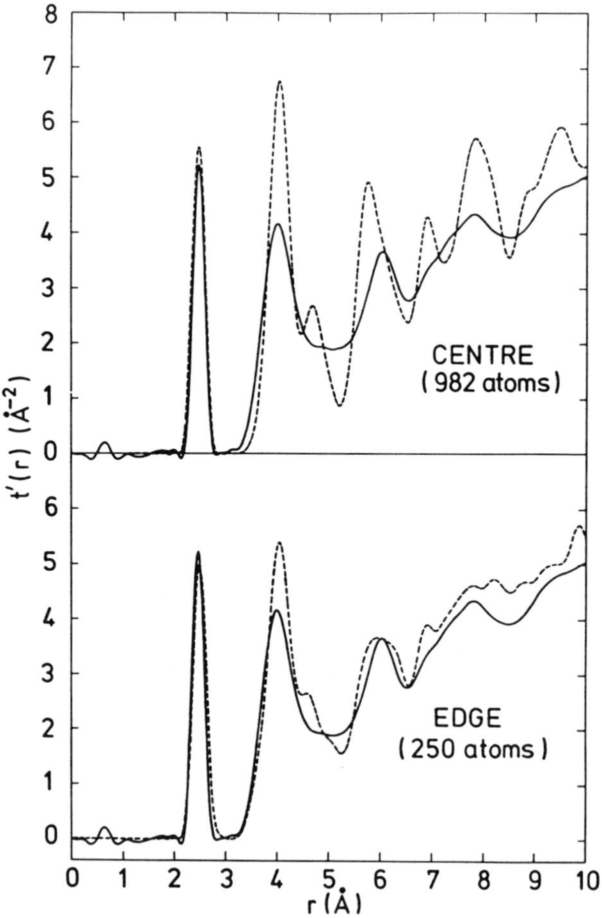


**Figure 45** Correlation functions for two spherical sections from the Gaskell [117] polytetrahedral model of amorphous Ge situated at the model centre and close to its edge [89]. ——, experiment; -----, model.

## 7.5 Layer Models

Grigorovici [27] has associated threefold co-ordination with layer formation. If layers are present, with an interlayer spacing $d_L$, then a relatively sharp pseudo-Bragg peak will occur in the diffraction pattern at

$$Q_L = \frac{2\pi}{d_L} \tag{40}$$

Many workers in interpreting diffraction data from amorphous chalcogenides have placed great emphasis on the first diffraction peak which it is claimed is anomalous in that it occurs at very low Q and is very sharp. In the case of vitreous arsenic sulphide (fig. 46) this peak is close to the 020 reflection for the mineral orpiment which has a layer structure, the 020 reflection corresponding to the interlayer spacing [118]. A similar situation exists for other chalcogenides (e.g. $As_2Se_3$, $GeS_2$ and $GeSe_2$) and has consequently led to the suggestion that the glasses also have layer structures. It is, however, extremely instructive to consider whether the

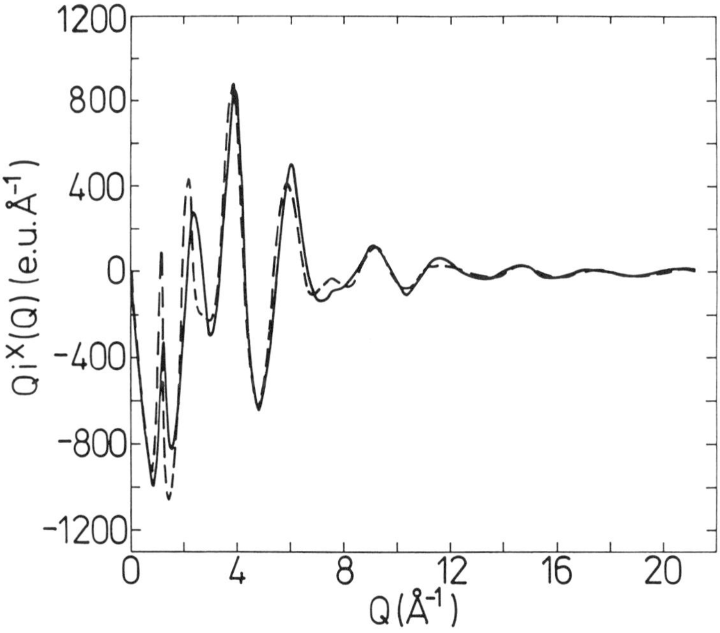


**Figure 46** X-ray interference functions for amorphous thin film $AsS_{1.38}$ (-----) and bulk vitreous $AsS_{1.60}$ (——) [119].

first diffraction peak for chalcogenide glasses is indeed anomalous. Two chalcogenide glasses will be taken as examples, $GeSe_2$ and $AsS_{1.66}$, which represent 4:2 and 3:2 connected systems respectively. Consider first the case of vitreous $GeSe_2$ which has a $GeSe_4$ tetrahedral structural unit analagous to the $SiO_4$ tetrahedron in vitreous $SiO_2$. The first neutron diffraction peak for vitreous $GeSe_2$ is at a lower value of Q than that for vitreous $SiO_2$, but this would be expected since the Ge-Se bond is much longer than the Si-O bond (2.370Å as opposed to 1.608Å). If the vitreous $SiO_2$ structure were expanded to give the Ge-Se bond length the Q scale would contract by the same factor. Thus in order to compare neutron diffraction patterns for the two materials it is necessary to plot the structure factor S(Q) against $r_{A\text{-}X}Q$ where $r_{A\text{-}X}$ is the A-X bond length (A = Si or Ge; X = O or Se). This is done in fig. 47 [119], for the four glasses $SiO_2$, $GeSe_2$, $AsS_{1.66}$ and $B_2O_3$. It can be seen that the first three neutron diffraction peaks for $GeSe_2$ are almost exactly coincident with those of vitreous $SiO_2$. Similarly the first peak for $AsS_{1.66}$ is at a higher value of $r_{A\text{-}X}Q$ than for $SiO_2$ and $GeSe_2$ while that for vitreous $B_2O_3$ is actually at the lowest $r_{A\text{-}X}Q$ of all four glasses.

Not only is the first diffraction peak for vitreous silica coincident with that for vitreous $GeSe_2$, when the data are plotted vs $r_{A\text{-}X}Q$, but so is the 101 reflection for $\alpha$-cristobalite, which certainly does not have a layer structure. Thus both layered and non-layered structures can give a peak at the same value of $r_{A\text{-}X}Q$. The situation for vitreous $AsS_{1.66}$ is also ambiguous in that its first diffraction peak is approximately co-incident with both the 020 orpiment interlayer reflection and Bragg peaks for all three $As_4S_4$ molecular crystals. There is in fact a wide variety of structural features which can give rise to a sharp first diffraction peak, including layers, molecules (c.f. section 7.6), ionicity, chemical ordering etc. and thus it is impossible to associate the presence of such a peak with any one of them in particular. To the author's knowledge layer formation has not been satisfactorily established in any amorphous solid and is usually suggested merely because the corresponding crystalline compound has a layer structure. Nevertheless it must be emphasised that a layer structure has not been shown to be incorrect. It remains a possible explanation of the experimental diffraction data but not the only one.

A variation on the layer theme is the "outrigger raft" model of vitreous $GeSe_2$ due to Phillips [120]. In this model a segment of the ß-$GeSe_2$ crystalline layer structure is terminated by Se-Se bonds which Phillips claims gives rise to a peak at 219 $cm^{-1}$ in the Raman spectrum. However, this peak is almost certainly associated with edge sharing $GeSe_4$ tetrahedra (2-membered rings) which are also present in both ß-$GeSe_2$ and the Phillips model [36]. The crystal is chemically ordered and there is evidence that the same is true for the glass so that at stoichiometry it seems unnecessary to invoke the presence of Se-Se bonds which must be balanced by energetically unfavourable Ge-Ge bonds. A much more detailed discussion of the Phillips model, and of layer models in general, can be found in refs [119] and [121] and a review of the theories of the first diffraction peak for amorphous materials is given by Moss and Price [122].

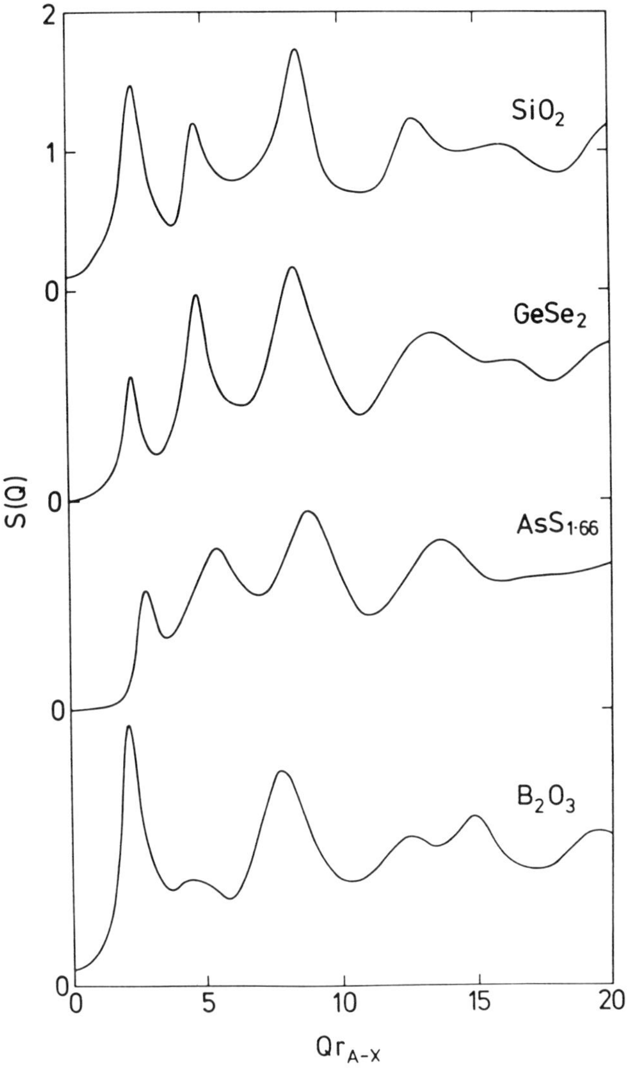


**Figure 47** Reduced neutron diffraction patterns for vitreous $SiO_2$, $GeSe_2$, $AsS_{1.66}$ and $B_2O_3$ [119].

## 7.6 Molecular Models

As indicated in section 2.3 a simple model for molecular amorphous solids is a random packing of approximately spherical molecules with no orientational correlation between adjacent molecules. In this case the scattered intensity may be

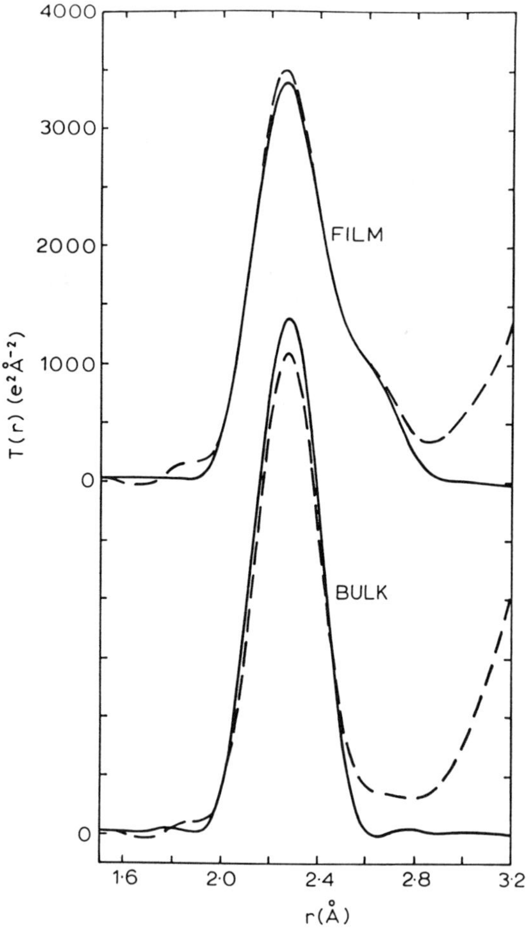


(A)

**Figure 48** X-ray (A) and neutron (B→) peak fits for $AsS_x$ bulk and thin film samples and also for the film after annealing for one week at ~120°C [12]. ——, model; -----, experiment.

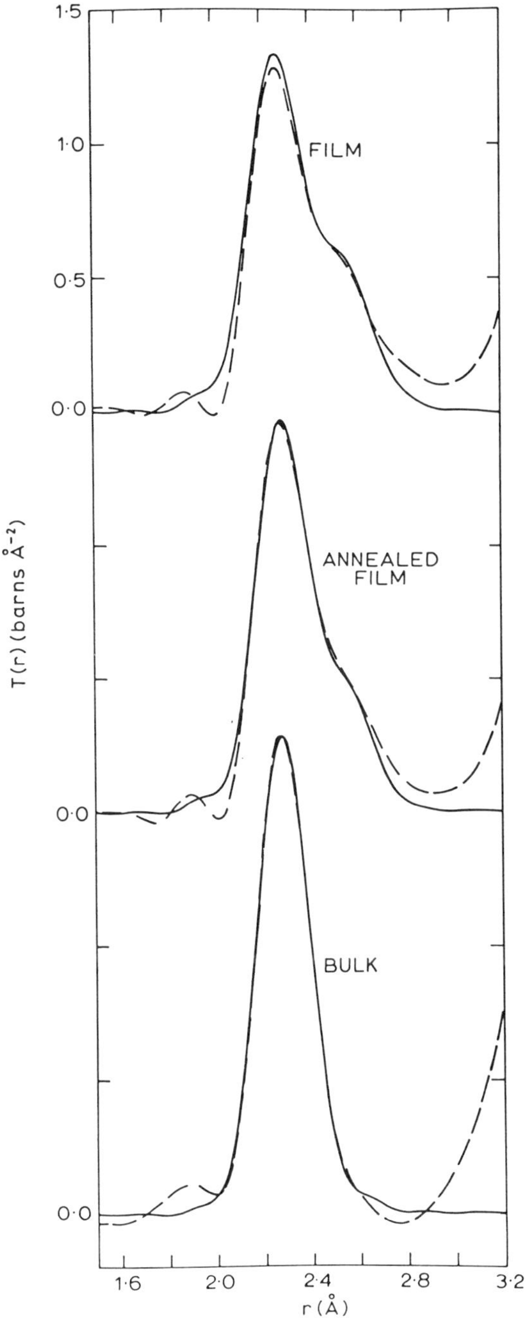

1·5
1·0
0·5
0·0
FILM
ANNEALED
FILM
0·0
BULK
0·0
T(r) (barns Å⁻²)
1·6
2·0
2·4
2·8
3·2
r (Å)


(B)

written [12,123]

$$I(Q) = \overline{F^2(Q)} + \overline{F(Q)}^2 i_c(Q) \tag{41}$$

where F(Q) is the molecular form factor

$$F(Q) = \sum_i \overline{a_i(Q)}\, e^{iQ.r_i} \tag{42}$$

and $i_c(Q)$ is the reduced distinct intensity for the molecular centres. The i summation is taken over the atoms in a molecule and $r_i$ is the distance of the $i^{th}$ atom from the molecular centre. A convenient source of $i_c(Q)$ is the analytical expression for a Percus-Yevick hard sphere fluid (c.f. section 7.2) which is available [107] as a function of packing density, $\eta$. The packing density, inter-molecular distance, d, and the molecular density, $\rho_M^o$, are related by

$$\eta = \frac{\pi d^3 \rho_M^o}{6} \tag{43}$$

Examples of vapour deposited amorphous solids have already been given in section 2.3, including arsenic sulphide [12] where the molecules have undergone at least partial polymerisation. The first peaks in T(r) for vitreous ($As_{0.38}S_{0.62}$) and vapour-deposited ($As_{0.43}S_{0.57}$) arsenic sulphide are compared in fig. 48. The peak fits indicate that, while the bulk glass is chemically ordered, the vapour deposited film contains far more As-As bonds than the minimum required by stoichiometry as revealed by the high r shoulder. Moreover, the mean As-As distance (2.57Å) exceeds the normal covalent bond length (2.49Å) and is much nearer that (2.59Å) found in the $As_4S_4$ realgar molecule (fig. 49) which is known from mass spectrometry studies to be present in the vapour phase. Hence it can be concluded that vapour deposited arsenic sulphide contains a significant fraction of $As_4S_4$ molecules / molecular fragments. A random orientation molecular model based on the realgar-I molecule is compared to X-ray data in fig. 50. At high Q, Qi(Q) is dominated by intra-molecular contributions and in this region the model is in best agreement with experiment. The differences between model and experiment at low Q are the result of polymerisation and orientational effects arising from the fact that the $As_4S_4$-I molecule is in reality not spherical. The first diffraction peak for the molecular model arises mainly from the first peak in $i_c(Q)$ although its shape is considerably modified by the simultaneous rapid fall off in F(Q).

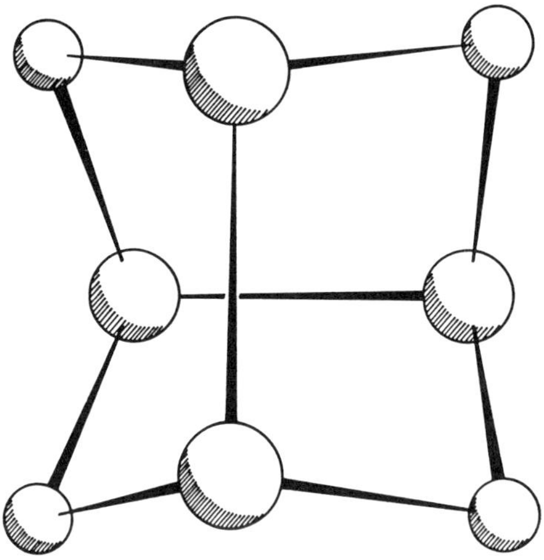

**Figure 49** The $As_4S_4$-I realgar molecule [119].

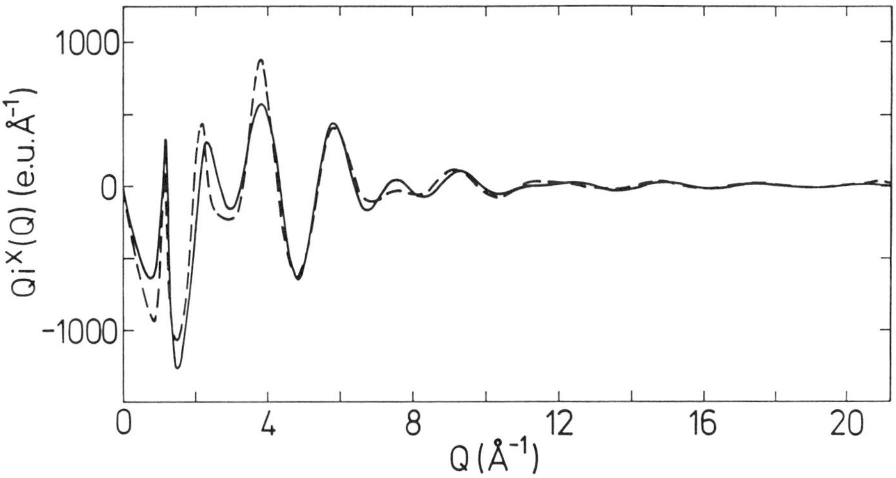


**Figure 50** The X-ray interference function for an $As_4S_4$-I molecular model (——) compared with experimental data for thin film $AsS_{1.38}$ (-----) [119].

## 7.7 Monte Carlo and Related Techniques

A number of different modelling methods may be grouped under the general heading of Monte Carlo techniques. The essential features of the Monte Carlo approach is that a starting configuration is progressively modified using random numbers, each modification being tested against some criterion (either potential

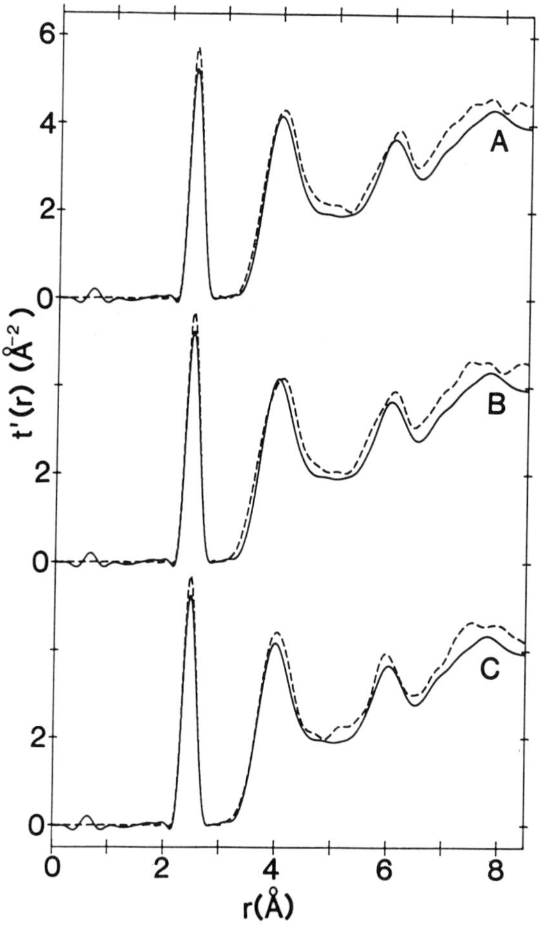


**Figure 51** A series of periodic boundary models of amorphous Ge generated by Wooten et al. [94-96] (-----) compared to experiment [78] (——) [57]. (A) model relaxed with Keating potential, (B) as model (A) but relaxed with Weber potential and (C) model containing 4-membered rings.

energy or agreement with experiment) to see whether it is favourable. Favourable modifications are accepted whereas, to prevent the system becoming stuck in a local minimum, unfavourable modifications are usually tested against an exponential probability to see whether they should be accepted or rejected. Monte Carlo and related methods can be divided into two classes depending on whether it is the potential energy or $\chi^2$, the discrepancy between model and experiment, which is being minimised.

**7.7.1** ***Potential energy minimisation.*** The original Monte Carlo technique involves moving individual atoms at random within a periodic boundary and has been successfully employed for liquids using Van der Waals and ionic potentials. Metallic systems have also been simulated with hard sphere and Lennard-Jones potentials. Brawer has performed a Monte Carlo simulation of

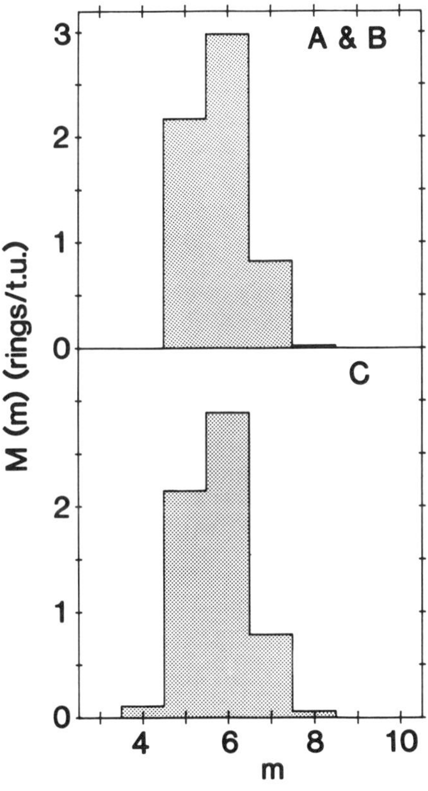


**Figure 52** Ring statistics for the amorphous Ge models in fig. 51 [57].

vitreous $BeF_2$ using a simple 2-body ionic potential [124] but as with the molecular dynamics simulations discussed in section 7.8 accurate studies of amorphous network solids require 3-body forces.

A novel Monte Carlo technique for generating random network models of amorphous Si and Ge ("Sillium" models) has been evolved by Wooten and Weaire [94-96] which starts with the ambient crystalline network and then introduces disorder via a series of topological transformations which are tested against a Maxwell-Boltzmann factor as in the case of conventional Monte Carlo simulations. Correlation functions for three models due to Wooten and Weaire are compared to the data of Etherington et al [78] in fig. 51. Models A and B are topologically equivalent but have been relaxed with, A, a Keating potential [104] and, B, a generalisation of Weber's adiabatic bond-charge model [125]. Model C, on the other hand, contains 4-membered rings which were specifically excluded from models A and B. The complete shortest path ring statistics for both model topologies are given in fig. 52. It can be seen that all three models have a bond angle distribution very much closer to that of the real material than the Polk model and in general are in very good agreement with experiment, except that the average model number densities are too high. The additional structure in t'(r) at higher r is almost certainly statistical in nature due to the small model size (216 atoms).

**7.7.2 *$\chi^2$ minimisation.*** As an alternative to minimising the potential energy in Monte Carlo simulations it is possible to maximise the agreement between model and experiment as indicated by a reliability factor $R_\chi$ such as that given in equ. (36). This technique was first employed by Averbach and co-workers [126,127] to investigate the structure of a series of chalcogenide glasses and thin films, although the number of atoms in the cluster models thus generated was severely limited (~100-200) by the then available computer power.

When used with a periodic boundary the $\chi^2$ minimisation technique is essentially equivalent to the Rietveld analysis method used to refine crystal structures from polycrystalline powder diffraction data. A modified form of the Rietveld program, which includes strain broadening using models 1 and 2 in table 7, has been used by Le Bail et al. [128] to obtain an excellent fit to the neutron diffraction patterns for a series of heavy metal fluoride glasses as may be seen from fig. 53.

The most recent advances in this type of modelling have been made by McGreevy and co-workers [129-131] who have introduced an exponential probability for accepting unfavourable moves, along the lines of the Metropolis condition used in conventional Monte Carlo simulations [132], and used very much larger systems. As an example the neutron and X-ray distinct scattering functions, F(Q), from a recent simulation of vitreous silica [130] are illustrated in fig. 54. Clearly, given adequate computer time, this type of simulation is capable of yielding excellent agreement with experiment but this is to be expected since performing a simulation with N atoms is equivalent to fitting the experimental data with 3N adjustable parameters. (Typically N is of the order of 3000.) By comparison, a cubic spline fit along the lines of that in fig. 18 requires only a few hundred parameters and hence the number of excess parameters in the simulation is extremely large. For this reason it is vital when using this type of technique to carefully examine the resulting configuration to ensure that it is chemically

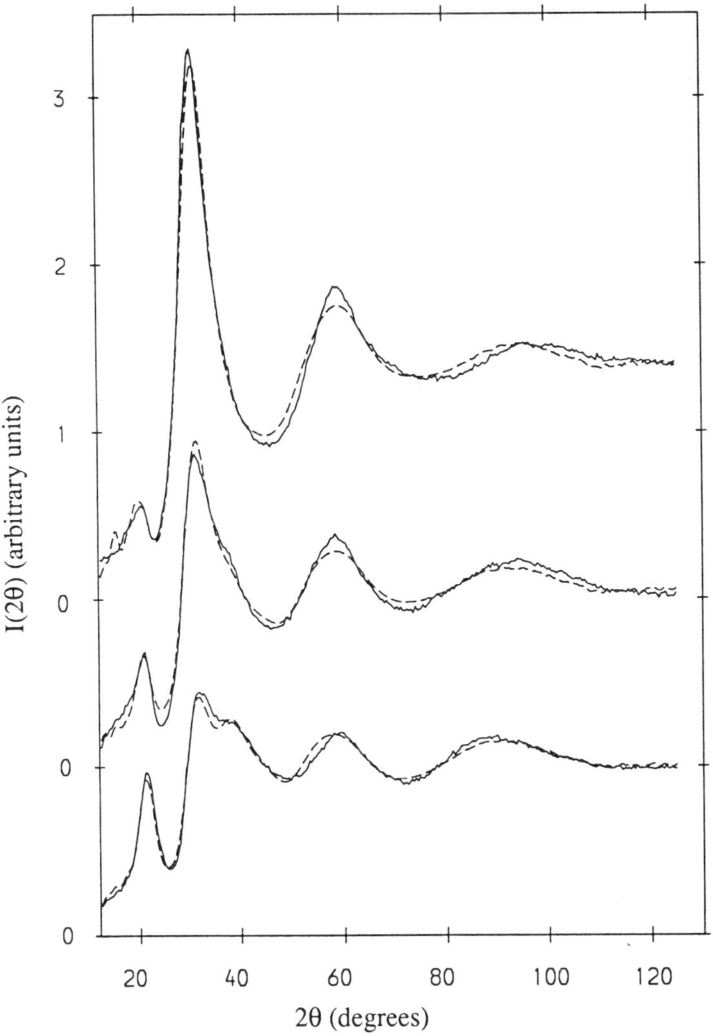


**Figure 53** Modified Rietveld fit to $2PbF_2{\cdot}TF_2{\cdot}MF_3$ glasses [128]. Top, $2PbF_2{\cdot}MnF_2{\cdot}VF_3$; middle, $2PbF_2{\cdot}MnF_2{\cdot}FeF_3$ and bottom, $2PbF_2{\cdot}ZnF_2{\cdot}FeF_3$. ——, experiment; -----, fit.

reasonable and is in accord with all the other available data for the material in question. Thus the simulation in fig. 54 must be rejected in that it contains far too many silicon atoms with co-ordination numbers other than 4. Additional constraints, such as the maximum percentage of incorrectly co-ordinated atoms, may of course be included in the original simulation at the expense of considerable extra computational time.

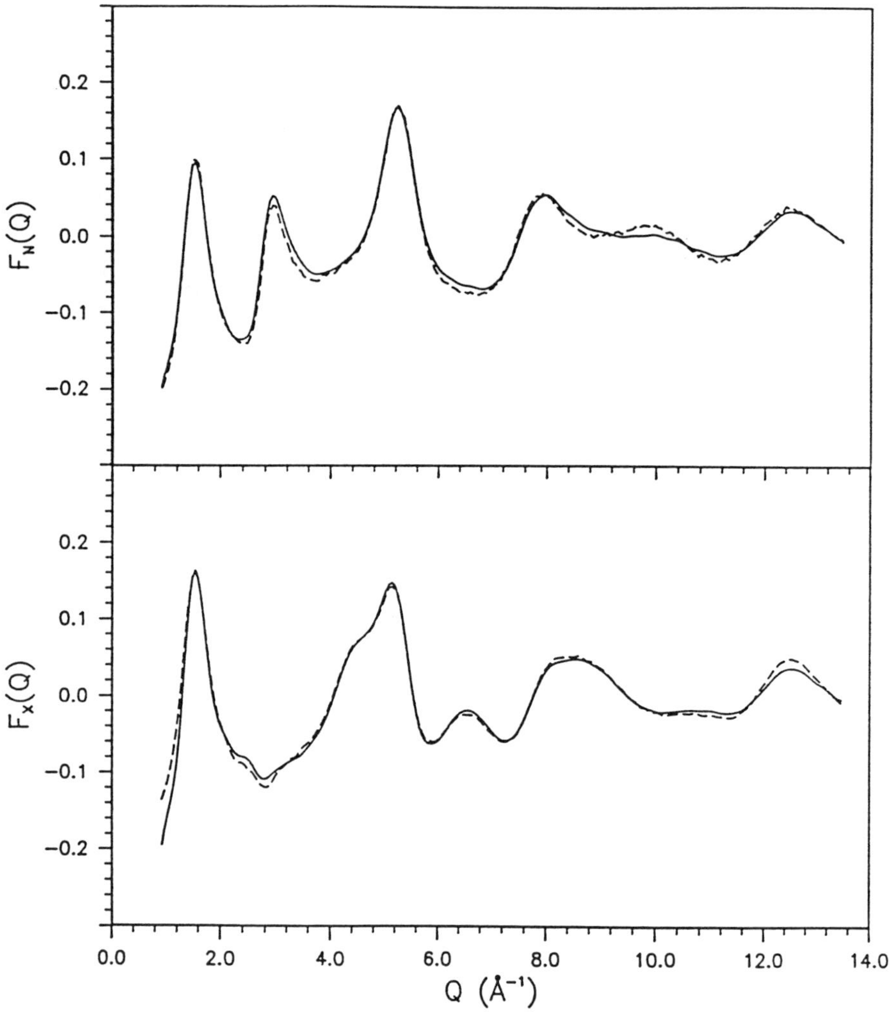


**Figure 54** Reverse Monte Carlo fit to the neutron and X-ray distinct scattering functions, F(Q), for vitreous silica [130].

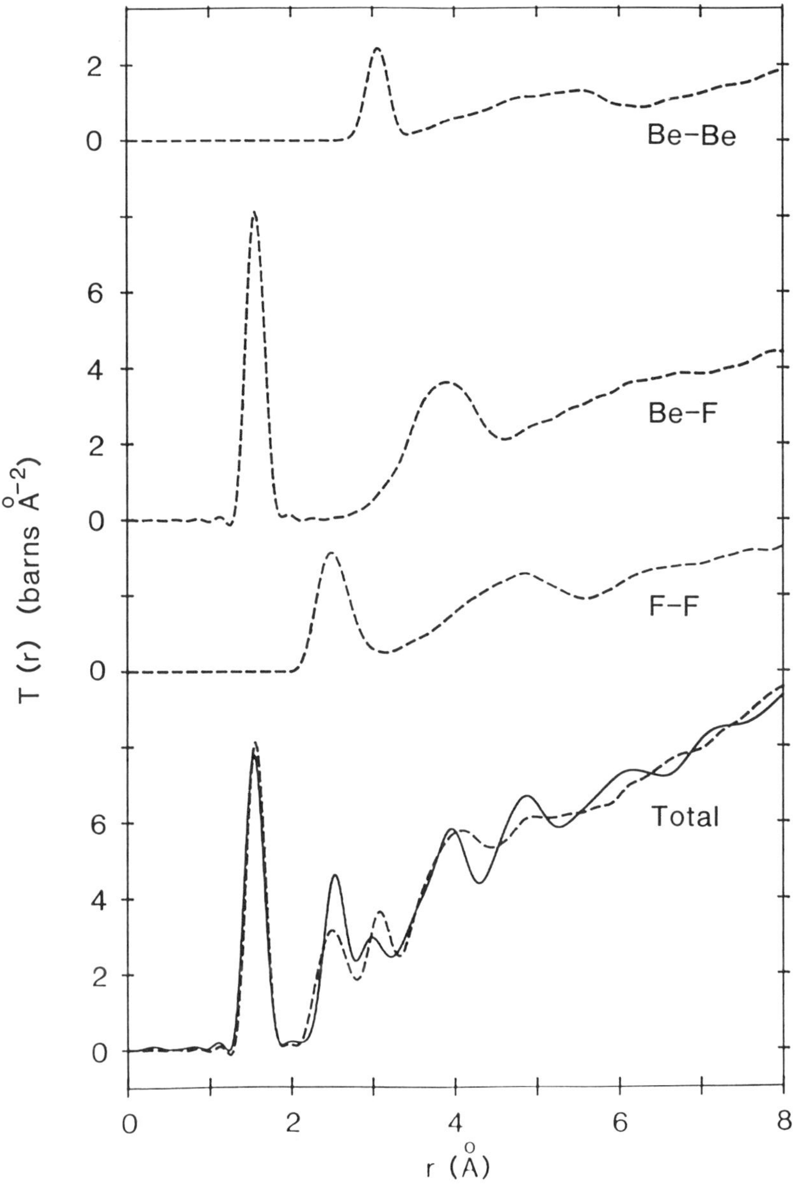


**Figure 55** The neutron correlation function for a molecular dynamics simulation (-----) of vitreous $BeF_2$, using 2-body potentials, compared to experiment (——) [56].

## 7.8 Molecular dynamics

The molecular dynamics technique is described in detail in Chapter 9 of this book and hence discussion here will be limited to the comparison of the results of molecular dynamics simulations with experiment. Molecular dynamics simulations have been performed for a number of glasses, mainly using simple ionic potentials, an example being the simulation of vitreous $BeF_2$ by Brawer compared to neutron data in fig. 55 [56,76]. The great majority of these simulations published in the literature claim to be in (good) agreement with diffraction data in that they yield the appropriate structural unit. However, when the resulting correlation functions are correctly compared to experiment it is usually found that the structural units are much too distorted as demonstrated in fig. 55 by the broad distribution of intratetrahedral F-F distances and hence F-Be-F bond angles. (The width of the F-Be-F bond angle distribution [133] is ~30°.) It is imperative when making such

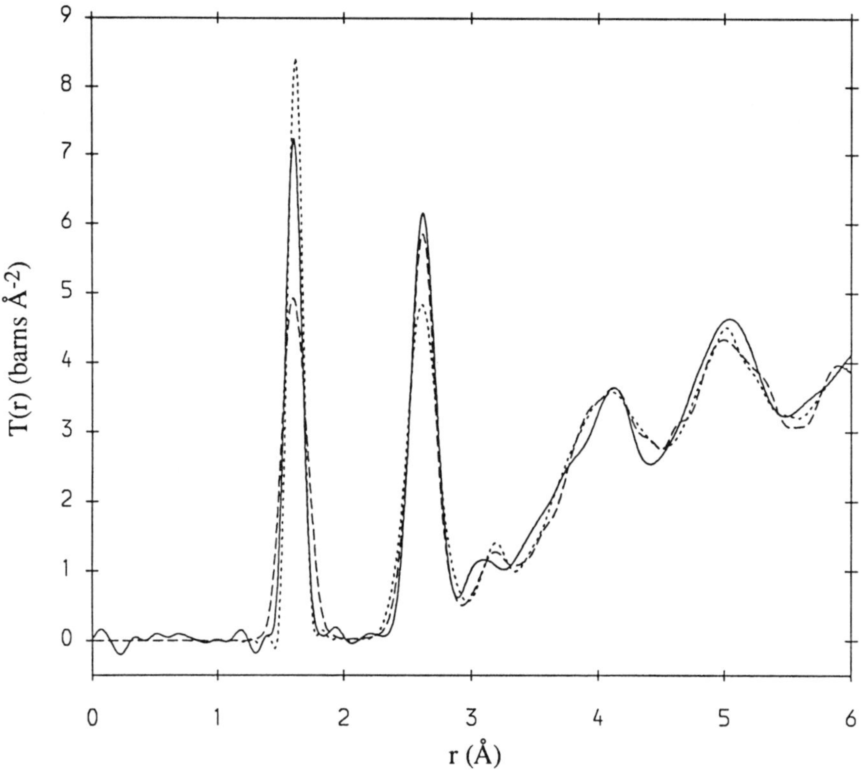


**Figure 56** Neutron correlation functions for molecular dynamics simulations of vitreous silica, using 2- (·····) and 3-body (-----) potentials [135], compared to experiment (——).

comparisons that the interval $\Delta r$ (histogram column width), used in calculating the component correlation functions for the simulation, is small compared to the experimental resolution in real space, $\Delta r_{1/2}$, in order to avoid extra peak broadening. A useful criterion is given by $\Delta r \lesssim 0.1\Delta r_{1/2}$. Note that the statistical

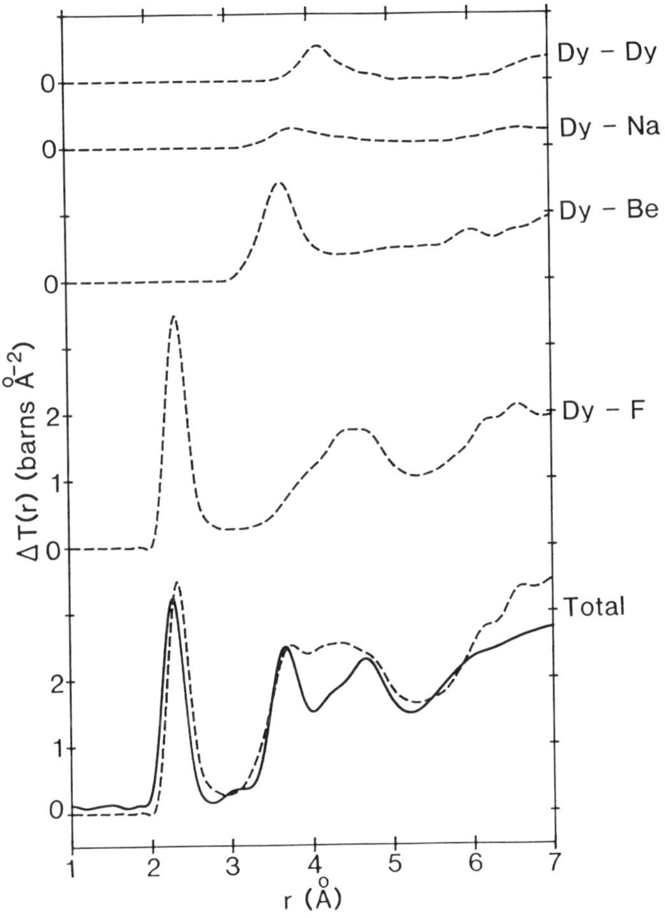


**Figure 57** The Dy-Dy + Dy-X (X = Na, Be or F) isotopic difference correlation function for vitreous NaF-$DyF_3$-$BeF_2$ compared to the predictions of a molecular dynamics simulation [76].

noise in the simulation correlation function arising from the small value of $\Delta r$ is smoothed by the convolution with $P'_{ij}(r)$.

The main reason for the tetrahedral distortion is the absence of any covalent character in the potential used for the simulation and the corresponding lack of any bond angle restoring forces. A comparison of the third peak in the correlation function shows that the Be-F-Be bond angle distribution is also incorrect while the structure in the simulation correlation function at higher r is much less than that for the real material. The increase in the mean Be-F-Be bond angle and the narrower distribution is probably the result of $Be^{2+}$ ion repulsion shifting the distribution towards 180°. Indeed a similar simulation [134] of vitreous NaF-$BeF_2$ shows the Be-F-Be bond angle to have a most probable value of 180°. The necessity for 3-body potentials in molecular dynamics simulations of network glasses can be seen from fig. 56 which compares preliminary simulations of vitreous silica using 2- and 3-body potentials [135] with neutron data. The 3-body potentials lead to a much narrower distribution of O-Si-O bond angles as may be seen from the width of the intratetrahedral O-O peak at 2.63Å, but this occurs at the expense of the Si-O bond length distribution, which becomes too broad. The average first Si-Si distance also remains too large.

One application of molecular dynamics simulations in the analysis of diffraction data, even those using 2-body potentials, is in the interpretation of data for complex systems and an example of this is given in fig. 57 which shows the Dy-Dy + Dy-X (X=Na, Be or F) difference correlation function for vitreous NaF-$DyF_3$-$BeF_2$, obtained by isotopic substitution (c.f. section 5.4.1), together with the results of a molecular dynamics simulation [76]. Due to slightly too large an ionic radius for $Dy^{3+}$, the first neighbour Dy-F peak for the simulation is displaced to higher r but its width is in agreement with experiment. The structure for the simulation between 3.3 and 5.0Å is in qualitative agreement with experiment but is more smeared out. The simulation does, however, identify the experimental peak at 3.7 Å as being due to the first Dy-Be coordination shell, while the feature between 4 and 5 Å arises from higher order Dy-F interactions with a small contribution from Dy-Dy correlations.

## 8. GENERAL DISCUSSION

Diffraction data are frequently held to be insensitive to the exact nature of the structure of an amorphous solid but, if used correctly as above, modern diffraction experiments do in fact eliminate most of the structural models proposed in the literature. The main problem with the analysis of diffraction data from glasses is one of uniqueness in that even if a perfect fit is obtained with experiment this is no guarantee that there are not other models which would fit equally well. Hence it is essential to consider, and if possible eliminate, all reasonable alternative models, particularly those already proposed in the literature for the material under investigation. Too often diffraction data are published along with the author's favoured model, totally ignoring other published models and/or the results of other complementary techniques which would invalidate the model in question. Conversely, diffraction data have been ignored, for the reasons just outlined, which has resulted in structural models being proposed which are totally incompatible with diffraction data, an example being the Phillips model for vitreous $SiO_2$ [112,113]. Agreement with diffraction data is thus a necessary but not sufficient criterion for any valid structural model. What can be stated with absolute certainty however, both for diffraction and for the results of any other experimental technique, is that ***any model which is not consistent with the data is wrong and must be rejected.***

A controversial aspect of the modelling of amorphous solids is the use of periodic boundary conditions, as in Monte Carlo and molecular dynamics simulations, and hence it is important both to appreciate the effect these have on the overall model structure and to understand their influence on comparisons between models and experiment, especially in reciprocal space. A periodic boundary model is in reality a crystal. The lack of internal symmetry within the unit cell means that the structure belongs to the triclinic P1 space group, although the unit cell itself is frequently pseudo-cubic (pseudo-orthorhombic in the better models which use different repeat distances $a_o$, $b_o$ and $c_o$ along the x, y and z axes). Any calculation which uses periodic boundary conditions, therefore, is appropriate to a P1 crystal and not an amorphous solid. The diffraction pattern for such a structure, after averaging over all orientations of the unit cell with respect to the scattering vector **Q**, comprises a series of δ-function Bragg peaks characteristic of a Debye-Scherrer polycrystalline powder diffraction pattern and not the continuous distribution typical of an amorphous solid. Thus with the exception of the thermal diffuse scattering, the diffraction pattern is only sampled at discrete values of Q ($Q_{hkl}$, the Bragg peaks) and there is no scattering below the Bragg cut-off at $Q = 2\pi/a_o$, $a_o$ being the largest unit cell dimension (assuming orthogonal axes). This should be compared to the case of a cluster model where the finite model size leads to a broadening of the interference function relative to that for an infinite network (c.f. particle size broadening for polycrystalline materials). Note that any calculation which truncates the structure (correlation function) at some distance less than $a_o$ is simply incapable of distinguishing between the crystalline and amorphous states and the limitation of the range of order is a property of the calculation, not the underlying structural model. The correct method of calculating the interference function for a periodic boundary model is to use the formalism appropriate to crystals such that

$$Qi(Q) = \frac{2\pi^2\rho^o}{n_u^2} \sum_h \sum_k \sum_l \frac{|F_{hkl}|^2}{Q_{hkl}} \delta(Q-Q_{hkl}) - Q \sum_i \overline{a_i(Q)}^2 e^{-\overline{u_i^2}Q^2/3} \qquad (44)$$

in which $|F_{hkl}|^2$ is the isotropically averaged square of the structure factor

$$|F_{hkl}|^2 = \sum_i \sum_j \overline{a_i(Q)}\,\overline{a_j(Q)} \cos 2\pi[h(x_i-x_j) + k(y_i-y_j) + 1(z_i-z_j)]e^{(-\overline{u_i^2}+\overline{u_j^2})Q_{hkl}^2/6} \tag{45}$$

The i and j summations are taken over the atoms in the unit cell, with fractional co-ordinates $x_i$, $y_i$ and $z_i$, $n_u$ is the number of composition units in the unit cell and h, k and l are the conventional Miller indices. Equations (44) and (45) assume that the atomic rms thermal displacements $(\overline{u_i^2})^{1/2}$ are spherically symmetric.

The spacing of the Bragg peaks at low Q for model C in fig. 51 relative to the experimental neutron diffraction pattern for amorphous Ge is shown in fig. 58. Each Bragg peak is represented by a vertical line in the appropriate position, the height of which is proportional to the peak intensity for the static model

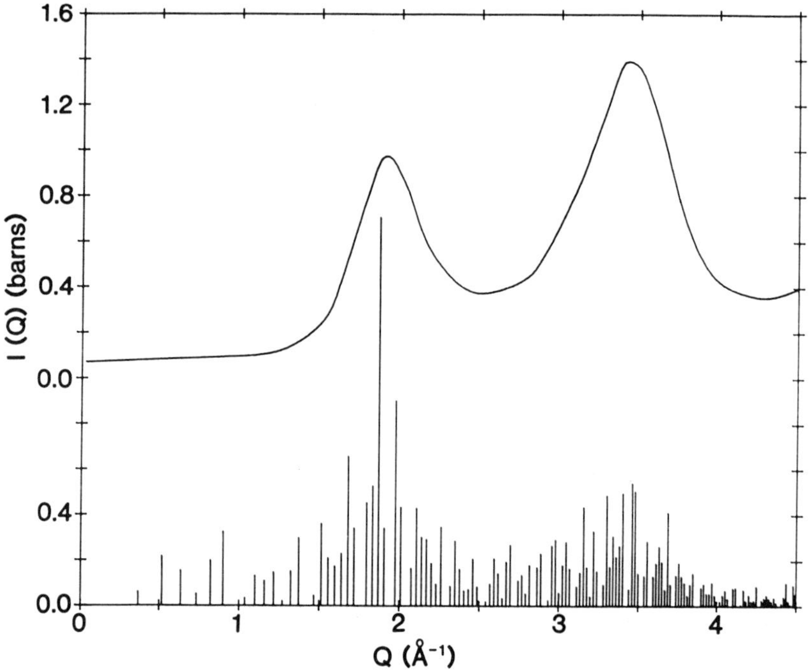


**Figure 58** The positions and intensities of the Bragg peaks from model (C) in fig. 51 (vertical bars) relative to the experimental neutron diffraction pattern for amorphous Ge (solid line) [57].

$\left[\overline{(u^2_{Ge})}^{1/2} = 0\right]$. The prominent Bragg peak at 1.87Å$^{-1}$ is close to the 111 Bragg reflection for crystalline Ge, suggesting that the model may contain a small amount of residual crystallinity. A similar prominent Bragg peak is also observed for Model A. Conventionally such a diffraction pattern is artificially broadened, e.g. by summing the δ-function Bragg peaks into relatively broad histogram columns, in which case it is imperative that before intercomparison the experimental data are treated similarly. However, such a broad spacing of Bragg peaks is unlikely to lead to an accurate prediction of the shape of the first diffraction peak for an amorphous network solid and hence it is important to try to estimate the minimum size of periodic boundary model required for this purpose.

Assuming a pseudo-cubic unit cell, the positions of the Bragg peaks are given by

$$Q_{hkl} = \frac{2\pi}{a_o}(h^2+k^2+l^2)^{1/2} = \frac{2\pi}{a_o} m^{1/2} \tag{46}$$

where m is an integer. One possible criterion is that the calculated diffraction pattern should be continuous over the first peak within the experimental resolution so that the only necessary broadening of the model interference function would be by the experimental reciprocal space resolution function P(Q). Differentiating and taking finite elements the minimum lattice parameter $a_{min}$ becomes

$$a_{min} = \frac{\pi}{Q_1}\left[\frac{2Dm}{\left[\frac{DQ}{Q}\right]}\right]^{1/2} \tag{47}$$

Since m is an integer the minimum Δm is unity and the best experimental resolution in the region of the first diffraction peak at $Q_1$ is ~1% (ΔQ/Q = 0.01). Hence

$$a_{min} \simeq \frac{44.4}{Q_1} \tag{48}$$

This yields values for $a_{min}$ of 23Å and 29Å respectively for amorphous Ge ($Q_1$ = 1.91Å$^{-1}$) and vitreous silica ($Q_1$ = 1.53Å$^{-1}$). It is interesting to compare this method of calculating $a_{min}$ with a criteriton frequently used with periodic boundary models that they should only be used for calculations out to a maximum interatomic separation of $a_o/2$. For vitreous silica Konnert and Karle [80] report measurable structure out to ~20Å and thus for a periodic boundary model to be capable of predicting this structure it should have a minimum lattice parameter of ~40Å.

Similar minimum model size criteria can also be formulated in both real and reciprocal space for cluster models. In reciprocal space the finite model size leads to a broadening of the interference function. The peak function, P(Q), for a spherical model is given by equ. (38), the full width at half maximum height being equal to 3.477/R. A frequently used criterion for experimental broadening to be neglected is that it should not exceed one tenth of the true width of the peak in question. Applying this criterion to the finite model size broadening of the first diffraction peak leads to a minimum radius

$$R_{min} \simeq \frac{34.8}{\Delta Q_1} \tag{49}$$

The width of the first diffraction peak, $\Delta Q_1$, for vitreous silica is 0.63Å$^{-1}$ yielding a minimum model radius of 55Å.

In real space the finite size of a cluster model limits the statistical accuracy with which T(r) can be determined at high r since the multiplicative correction factor (the reciprocal of F(r) in equ. (37) for a spherical model) increases rapidly for $r>R$ ($1/F(R) = 3.2$) and hence the model radius should be greater than the correlation length for the real material. For vitreous silica this yields $R_{min}$ = 20Å which is much smaller than the value obtained from the reciprocal space criteria. Hence for models intermediate in size between the real and reciprocal space criteria the correct broadening of the experimental reciprocal space data will be required before comparison with the model. Note that for good amorphous diffractometers the experimental broadening of the first diffraction peak can frequently be neglected and hence it is not usually necessary to correct model interference functions for experimental reciprocal space resolution.

## 9. CONCLUDING REMARKS

As demonstrated in this chapter, modern neutron and X-ray diffraction experiments are capable of yielding accurate data with high real space resolution which if used correctly provide a stringent test of the various structural theories and models of amorphous solids. Much of the confusion in the literature over the validity of such models arises because many workers have not fully comprehended the finer points of the diffraction method and the comparison between models and experiment has been at a very superficial level. For this reason an emphasis has been placed on the need to understand the exact nature of the information provided by diffraction experiments on amorphous solids and on procedures for the accurate, quantitative comparison of models with experiment. Most models fail due to incorrect range I and II order and there are few, if any, materials for which there is a ***comprehensive*** structural model that can predict the best diffraction data within their known experimental uncertainty. The greatest barrier to progress in understanding the structure of amorphous solids lies not so much with diffraction techniques themselves but in the development of modelling procedures in which model parameters can be varied in a systematic way until agreement is obtained with experiment in both real and reciprocal space. This situation is unique to amorphous materials and has no parallel in corresponding studies of the crystalline state.

## ACKNOWLEDGEMENTS

This chapter was completed while the author was Visiting Professor of Glass Science at the New York State College of Ceramics and the final camera ready copy was produced during time spent as a Visiting Scientist at the Argonne National Laboratory supported by the U.S. Department of Energy, Basic Energy Sciences - Materials Sciences, under contract number W-31-109-ENG-38. The support of both institutions is gratefully acknowledged.

## REFERENCES

[1] "Compilation of ASTM Standard Definitions," 7th Edn (ASTM, Philadelphia, 1990), 212.
[2] W.H. Zachariasen, "The Atomic Arrangement in Glass," J. Amer. Chem. Soc. 54, 3841-3851 (1932).
[3] G.N. Greaves, A. Fontaine, P. Lagarde, D. Raoux and S.J. Gurman, "Local Structure of Silicate Glasses," Nature 293, 611-616 (1981).
[4] P.H. Gaskell, M.C. Eckersley and J. Zhao, "The Structure of Oxide Glasses - A New Insight from Neutron Scattering Studies," Rivista della Staz. Sper. Vetro (5), 21-28 (1990).
[5] P.J. Bray and J.G. O'Keefe, "Nuclear Magnetic Resonance Investigations of the Structure of Alkali Borate Glasses," Phys. Chem. Glasses 4, 37-46 (1963).
[6] M. Ueno, M. Misawa and K. Suzuki, "On the Change in Coordination of Ge Atoms in $Na_2O$-$GeO_2$ Glasses," Physica 120B, 347-351 (1983).
[7] A.C. Wright, "The Structure of Amorphous Solids by X-ray and Neutron Diffraction," Adv. Struct. Res. Diffr. Meth. 5, 1-84 (1974).
[8] A.C. Wright and A.J. Leadbetter, "Diffraction Studies of Glass Structure," Phys. Chem. Glasses 17, 122-145 (1976).
[9] H.J.L. Trap and J.M. Stevels, "Physical Properties of Invert Glasses," Glastech. Ber. 32K, VI/31-52 (1959).
[10] F. Betts, A. Bienenstock and S.R. Ovshinsky, "Radial Distribution Studies of Amorphous $Ge_xTe_{1-x}$ Alloys," J. Non-Cryst. Solids 4, 554-563 (1970).
[11] F. Betts, A. Bienenstock, D.T. Keating and J.P. de Neufville, "Neutron and X-ray Diffraction Radial Distribution Studies of Amorphous $Ge_{0.17}Te_{0.83}$," J. Non-Cryst. Solids 7, 417-432 (1972).

[12] M.F. Daniel, A.J. Leadbetter, A.C. Wright and R.N. Sinclair, "The Structure of Vapour-Deposited Arsenic Sulphides," J. Non-Cryst. Solids 32, 271-293 (1979).
[13] G.O. Brunner, "An Unconventional View of the 'Closest Sphere Packings'," Acta Crystallogr. A27, 388-390 (1971).
[14] J.L. Finney, "Random Packings and the Structure of Simple Liquids I. The Geometry of Random Close Packing," Proc. Roy. Soc. Lond. A319, 479-493 (1970).
[15] W.D. Kingery, H.K. Bowen and D.R. Uhlmann, "Introduction to Ceramics," 2nd Edn. (Wiley, New York, 1976).
[16] J.D. Bernal, "The Geometry of the Structure of Liquids," In: "Liquids: Structure, Properties, Solid Interactions" Ed. T.J. Hughel (Elsevier, Amsterdam, 1965), 25-47.
[17] J.L. Finney and J. Wallace, "Interstice Correlation Functions; A New, Sensitive Characterisation of Non-Crystalline Packed Structures," J. Non-Cryst. Solids 43, 165-187 (1981).
[18] D.E. Polk, "Structural Model for Amorphous Metallic Alloys," Scripta Metall. 4, 117-122 (1970).
[19] D.E. Polk, "The Structure of Glassy Metallic Alloys," Acta Metall. 20 485-491 (1972).
[20] P.H. Gaskell, "A New Structural Model for Amorphous Transition Metal Silicides, Borides, Phosphides and Carbides," J. Non-Cryst. Solids 32, 207-224 (1979).
[21] P.H. Gaskell, "On the Structure of Simple Inorganic Amorphous Solids," J. Phys. C 12, 4337-4368 (1979).
[22] J.P. Laval, B. Frit and J. Lucas, "Crystal Chemistry of Zirconium in Glass-Forming Fluorozircorates," Mater. Sci. Forum 6, 457-464 (1985).
[23] A.A. Lebedev, "The Polymorphism and Annealing of Glass," Trans. State Optical Inst. 2(10), 1-20 (1921).
[24] B.E. Warren, "X-Ray Diffraction Study of the Structure of Glass," Chem. Rev. 26, 237-255 (1940).
[25] N. Valenkov and E.A. Porai-Koshits, "X-Ray Investigation of the Glassy State," Z. Kristallogr. 95, 195-229 (1936).
[26] E.A. Porai-Koshits, "The Possibilities and Results of X-Ray Methods for Investigation of Glassy Substances," In: "The Structure of Glass" (Consultants Bureau, New York, 1958), 25-35.
[27] R. Grigorovici, "The Structure of Amorphous Semiconductors," In: "Electronic and Structural Properties of Amorphous Semiconductors," Eds. P.G. Le Comber and J. Mort (Academic Press, London, 1973), 191-241.
[28] R. Zallen, "The Physics of Amorphous Solids," (Wiley, New York, 1983).
[29] L.H. Sperling, "Introduction to Physical Polymer Science" (Wiley-Interscience, New York, 1986).
[30] R.W. Fawcett, C.N.J. Wagner and G.S. Cargill III, "Radial Distribution Studies of Amorphous $Ge_xSe_{1-x}$ Alloy Films," J. Non-Cryst. Solids 8-10, 369-375 (1972).

[31] M.R. Chowdhury and J.C. Dore, "The Structure of Amorphous Carbon Tetrachloride by Neutron Diffraction," J. Non-Cryst. Solids 46, 343-360 (1981).
[32] J.R. Granada and J.C. Dore, "Neutron Diffraction Studies of Phosphorous II. The Super-Cooled Liquid and Plastic Crystal Phases," Mol. Phys. 46, 757-768 (1982).
[33] M.R. Chowdhury, J.C. Dore and D.G. Montague, "Neutron Diffraction Studies and CRN Model of Amorphous Ice," J. Phys. Chem. 87, 4037-4039 (1983).
[34] B.J. Gellatly and J.L. Finney, "Characterisation of Models of Multicomponent Amorphous Metals: the Radical Alternative to the Voronoi Polyhedron," J. Non-Cryst. Solids 50, 313-329 (1982).
[35] L.F. Gladden and S.R. Elliott, "Computer-Generated Models of a-$SiSe_2$ II. Structural Studies," J. Non-Cryst. Solids 109, 223-236 (1989).
[36] R.J. Nemanich, F.L. Galeener, J.C. Mikkelsen Jr., G.A.N. Connell, G. Etherington, A.C. Wright and R.N. Sinclair, "Configurations of a Chemically Ordered Continuous Random Network to Describe the Structure of $GeSe_2$ Glass," Physica 117B & 118B, 959-961 (1983).
[37] J.H. Simmons, C.J. Simmons, R. Ochoa and A.C. Wright, "Fluoride Glass Structure," In: "Fluoride Glass Fiber Optics," Eds. I.D. Aggarwal and G. Lu (Academic Press, San Diego, 1991), 37-84.
[38] A.C. Wright, G.A.N. Connell and J.W. Allen, "Amorphography and the Modelling of Amorphous Solid Structures by Geometric Transformations," J. Non-Cryst. Solids 42, 69-86 (1980).
[39] S.V. King, "Ring Configurations in a Random Network Model of Vitreous Silica," Nature 213, 1112-1113 (1967).
[40] C.S. Marians and L.W. Hobbs, "A Language for the Study of Network Silica Glasses" Diffusion Defect Data 53-54, 31-36 (1987).
[41] J.F. Shackelford and B.D. Brown, "The Lognormal Distribution in the Random Network Structure," J. Non-Cryst. Solids 44, 379-382 (1981).
[42] A.F. Wright, "Structural Studies of Amorphous Solids by Small-Angle Neutron Scattering," J. Non-Cryst. Solids 76, 43-60 (1985).
[43] L. Van Hove, "Correlations in Space and Time and Born Approximation Scattering in Systems of Interacting Particles" Phys. Rev. 95, 249-262 (1954).
[44] R.N. Sinclair and A.C. Wright, "Neutron Scattering from Vitreous Silica I. The Total Cross-Section," J. Non-Cryst. Solids 57, 447-464 (1983).
[45] P.A.V. Johnson, A.C. Wright and R.N. Sinclair, "Neutron Scattering from Vitreous Silica II. Twin-Axis Diffraction Experiments," J. Non-Cryst. Solids 58, 109-130 (1983).
[46] A.C. Wright and R.N. Sinclair, "Neutron Scattering from Vitreous Silica III. Elastic Diffraction," J. Non-Cryst. Solids 76, 351-368 (1985).
[47] D.I. Grimley, A.C. Wright and R.N. Sinclair, "Neutron Scattering from Vitreous Silica IV. Time-of-Flight Diffraction," J. Non-Cryst. Solids 119, 49-64 (1990).
[48] S. Urnes, "X-Ray Diffraction Studies of Glass," In: "Modern Aspects of the Vitreous State" Ed. J.D. Mackenzie (Butterworths, London, 1960) 1, 10-37.

[49] A.C. Wright, "Neutron Magnetic Scattering Studies of Amorphous Solids," J. Non-Cryst. Solids 40, 325-346 (1980).

[50] R.N. Sinclair, "Inelastic Neutron Scattering from Amorphous Solids," J. Non-Cryst. Solids 76, 61-78 (1985).

[51] V.F. Sears, "Appendix. Neutron Scattering Lengths and Cross Sections," In: Methods of Experimental Physics Vol. 23, Part A, "Neutron Scattering," Eds. K. Sköld and D.L. Price (Academic Press, Orlando, 1986), 521-550.

[52] J. Waser and V. Schomaker,"The Fourier Inversion of Diffraction Data," Rev. Mod. Phys. 25, 671-690 (1953).

[53] E.A. Lorch, "Neutron Diffraction by Germania Silica and Radiation-Damaged Silica Glasses," J. Phys. C 2, 229-237 (1969).

[54] C. Finbak, "The Structure of Liquids I," Acta Chem. Scand. 3, 1279-1292(1949).

[55] R.L. Mozzi and B.E. Warren, "The Structure of Vitreous Silica," J. Appl. Crystallogr. 2, 164-172 (1969).

[56] A.C. Wright, A.G. Clare, G.Etherington, R.N. Sinclair, S.A. Brawer and M.J. Weber, "A Neutron Diffraction and Molecular Dynamics Investigation of the Structure of Vitreous Beryllium Fluoride," J. Non-Cryst. Solids 111, 139-152 (1989).

[57] A.C. Wright, R.A. Hulme, D.I. Grimley, R.N. Sinclair, S.W. Martin, D.L. Price and F.L. Galeener, "The Structure of Some Simple Amorphous Network Solids Revisited," J. Non-Cryst. Solids 129, 213-232 (1991).

[58] B.E. Warren and G. Mavel, "Elimination of the Compton Component in Amorphous Scattering," Rev. Sci. Instrum. 36, 196-197 (1965).

[59] A Bienenstock, A. Fischer-Colbrie, R.D. Lorentz, K.F. Ludwig Jr. and L. Wilson, "Synchrotron Radiation and the Determination of Atomic Arrangements in Amorphous Materials," In: "Physics of Disordered Materials," Eds. D. Adler, H. Fritzsche and S.R. Ovshinsky (Plenum, New York, 1985), 151-158.

[60] A. Bienenstock, S. Brennan, A. Fischer-Colbrie, P.H. Fuoss, K.F. Ludwig Jr., R.D. Lorentz, W.K. Warburton and L. Wilson, "X-Ray Synchrotron Radiation Structural Studies of Amorphous Materials," Diffusion Defect Data 53-54, 245-253 (1987).

[61] J. Krogh-Moe, "A Method for Converting Experimental X-ray Intensities to an Absolute Scale," Acta Crystallogr. 9, 951-953 (1956).

[62] N. Norman, "The Fourier Transform Method for Normalizing Intensities," Acta Crystallogr. 10, 370-373 (1957).

[63] G. Placzek, "The Scattering of Neutrons by Systems of Heavy Nuclei," Phys. Rev. 86, 377-388 (1952).

[64] L.N.G. Filon, "On a Quadrature Formula for Trigonometric Integrals," Proc. Roy. Soc. (Edinburgh) 49, 38-47 (1929).

[65] M. Dixon, A.C. Wright and P. Hutchinson, "The Smoothing and Fast Fourier Transformation of Experimental X-Ray and Neutron Diffraction Data from Amorphous Materials," Nucl. Instrum. Meth. 143, 379-383 (1977).

[66] W. Wei, "A New Approach to High Resolution RDF Analysis," J. Non-Cryst. Solids 81, 239-250 (1986).

[67] C.A. Yarker, P.A.V. Johnson, A.C. Wright, J. Wong, R.B. Greegor, F.W. Lytle and R.N. Sinclair, "Neutron Diffraction and EXAFS Evidence for $TiO_5$ Units in Vitreous $K_2O{\cdot}TiO_2{\cdot}2SiO_2$," J. Non-Cryst. Solids 79, 117-136 (1986).

[68] T.M. Hayes and A.C. Wright, "Diffraction and EXAFS Spectroscopy as Structural Probes of Amorphous Solids," In: "The Structure of Non-Crystalline Materials 1982," Eds. P.H. Gaskell, J.M. Parker and E.A. Davis (Taylor and Francis, London, 1983), 108-119.

[69] A.C. Wright, A.C. Hannon, R.N. Sinclair, W.L. Johnson and M. Atzmon, "The Neutron Diffraction Double-Null Isotopic Substitution Technique," J. Phys. F. 14, L201-L205 (1984).

[70] M. Imaoka and H. Hasegawa, "X-ray Diffraction Study of the Structure of Vitreous Arsenic Oxide," Phys. Chem. Glasses 21, 67-73 (1980).

[71] A.G. Clare, A.C. Wright, R.N. Sinclair, F.L. Galeener and A.E. Geissberger, "A Neutron Diffraction Investigation of the Structure of Vitreous $As_2O_3$," J. Non-Cryst. Solids 111, 123-138 (1989).

[72] J. Krogh-Moe, "A Method for the Resolution of Composite Radial Pair Distribution Functions," Acta. Chem. Scand. 20, 2890-2891 (1966).

[73] A.C. Wright, G. Etherington, J.A.E. Desa and R.N. Sinclair, "Neutron Diffraction Studies of Rare Earth Ions in Glasses," J. Phys. Coll. C9, 31-34 (1982).

[74] F.A. Wedgwood and A.C. Wright, "Short Range Antiferromagnetic Ordering in Vitreous $Fe_2O_3$-$P_2O_5$," J. Non-Cryst. Solids 21, 95-105 (1976).

[75] G. Etherington, L. Keller, A. Lee, C.N.J. Wagner and R.M. Almeida, "An X-ray Diffraction Study of the Structure of Barium Fluorozirconate and Fluorohafnate Glasses," J. Non-Cryst. Solids 69, 69-80 (1984).

[76] A.G. Clare, G. Etherington, A.C. Wright, M.J. Weber, S.A. Brawer, D.D. Kingman and R.N. Sinclair, "A Neutron Diffraction and Molecular Dynamics Investigation of the Environment of $Dy^{3+}$ Ions in a Fluoroberyllate Glass," J. Chem. Phys. 91, 6380-6392 (1989).

[77] P.S. Salmon, "A Neutron Diffraction Study on the Structure of Liquid Germanium," J. Phys. F 18, 2345-2352 (1988).

[78] G. Etherington, A.C. Wright, J.T. Wenzel, J.C. Dore, J.H. Clarke and R.N. Sinclair, "A Neutron Diffraction Study of the Structure of Evaporated Amorphous Germanium," J. Non-Cryst. Solids 48, 265-289 (1982).

[79] B.E. Warren, H. Krutter and O. Morningstar, "Fourier Analysis of X-Ray Patterns of Vitreous $SiO_2$ and $B_2O_3$," J. Amer. Ceram. Soc. 19, 202-206 (1936).

[80] J.H. Konnert and J. Karle, "The Computation of Radial Distribution Functions for Glassy Materials," Acta Crystallogr. A29, 702-710 (1973).

[81] P.R. Bevington, "Data Reduction and Error Analysis for the Physical Sciences" (McGraw Hill, New York, 1969).

[82] G.B. Carpenter, "Principles of Crystal Structure Determination" (Benjamin, New York, 1969).

[83] F.L. Galeener, "A Model for the Distribution of Bond Angles in Vitreous $SiO_2$," Philos. Mag. B51, L1-L6 (1985).

[84] R.J. Bell and P. Dean "The Structure of Vitreous Silica: Validity of the Random Network Theory," Philos. Mag. 25, 1381-1398 (1972).

[85] P.H. Gaskell and I.D. Tarrant, "Refinement of a Random Network Model for Vitreous Silicon Dioxide," Philos. Mag. B42, 265-286 (1980).

[86] K.M. Evans, P.H. Gaskell and C.M.M. Nex, "Modelling the Medium-Range Structure of $SiO_2$," In: "The Structure of Non-Crystalline Materials 1982," Eds. P.H. Gaskell, J.M. Parker and E.A. Davis (Taylor and Francis, London, 1983), 426-438.

[87] E.A. Davis, S.R. Elliott, G.N. Greaves and D.P. Jones, "Three-Fold Coordinated Random Networks," In: "The Structure of Non-Crystalline Materials," Ed. P.H. Gaskell (Taylor and Francis, London, 1977), 205-209.

[88] G.N. Greaves and E.A. Davis, "A Continuous Random Network Model with Three-Fold Co-ordination," Philos. Mag. 29, 1201-1206 (1974).

[89] A.C. Wright, G.Etherington, J.A.E. Desa, R.N. Sinclair, G.A.N. Connell and J.C. Mikkelsen Jr., "Neutron Amorphography," J. Non-Cryst. Solids 49, 63-102 (1982).

[90] S.R. Elliott, "A Continuous Random Network Approach to the Structure of Vitreous Boron Trioxide," Philos. Mag. B37, 435-446 (1978).

[91] D.L. Evans and S.V. King, "Random Network Model of Vitreous Silica," Nature 212, 1353-1354 (1966).

[92] D. L. Evans, M.P. Teter and N.F. Borrelli, "Vitreous Silica Minus Oxygen→Amorphous Silicon:- A Model Study," A.I.P. Conf. Proc. 20, 218-222 (1974).

[93] D.L. Evans, M.P. Teter and N.F. Borrelli, "The Range and Kind of Order in Random Tetrahedral Structures," J. Non-Cryst. Solids 17, 245-258 (1975).

[94] F. Wooten and D. Weaire, "Generation of Random Network Models with Periodic Boundary Conditions," J. Non-Cryst. Solids 64, 325-334 (1984).

[95] F. Wooten, K. Winer and D. Weaire, "Computer Generation of Structural Models of Amorphous Si and Ge," Phys. Rev. Lett. 54, 1392-1395 (1985).

[96] F. Wooten and D. Weaire, "Modeling Tetrahedrally Bonded Random Networks by Computer," Solid State Phys. 40, 1-42 (1987).

[97] R.J. Bell, A. Carnevale, C.R. Kurkjian and G.E. Peterson, "Structure and Phonon Spectra of $SiO_2$, $B_2O_3$ and Mixed $SiO_2$-$B_2O_3$ Glasses," J. Non-Cryst. Solids 35 & 36, 1185-1190 (1980).

[98] R.J. Bell and A. Carnevale, "A Structural Model for $B_2O_3$ Glass," Philos. Mag. B43, 389-399 (1981).

[99] J.L. Finney, "Amorphous Polymorphism: Structural Variability and Characterisation in Amorphous Systems," In: "Diffraction Studies of Non-Crystalline Substances," Eds. I. Hargittai and W.J. Orville-Thomas (Elsevier, Amsterdam, 1981), 440-490.
[100] G. Mason, "Radial Distribution Functions from Small Packings of Spheres," Nature 217, 733-735 (1968).
[101] B.E. Warren, "Calculation of Powder-Pattern Intensity Distributions," J. Appl. Crystallogr. 11, 695-698 (1978).
[102] D.E. Polk and D.S. Boudreaux, "Tetrahedrally Coordinated Random-Network Structure," Phys. Rev. Lett. 31, 92-95 (1973).
[103] P. Steinhardt, R. Alben, M.G. Duffy and D.E. Polk, "Relaxed Continuous-Random-Network Models," Phys. Rev. B8, 6021-6023 (1973).
[104] P.N. Keating, "Effect of Invariance Requirements on the Elastic Strain Energy of Crystals with Application to the Diamond Structure," Phys. Rev. 145, 637-645 (1966).
[105] L.F. Gladden, "Medium-Range Order in v-$SiO_2$," J. Non-Cryst. Solids 119, 318-330 (1990).
[106] J.D. Bernal, I.A. Cherry, J.L. Finney and K.R. Knight, "An Optical Machine for Measuring Sphere Coordinates in Random Packings," J. Phys. E 3, 388-390 (1970).
[107] A.C. Wright, Disc. Farad. Soc. 50, 111-113 (1970).
[108] J.E. Enderby and D.M. North, "Percus-Yevick Structure Factors for Liquid Alloys," Phys. Chem. Liq. 1, 1-11 (1968).
[109] A.C. Hannon, A.C. Wright and R.N. Sinclair, "The Atomic and Magnetic Structure of Melt-Spun Amorphous $Dy_7Ni_3$," Mater. Sci. Eng. A134, 883-887 (1991).
[110] J.A.E. Desa, A.C. Wright, J. Wong and R.N. Sinclair, "A Neutron Diffraction Investigation of the Structure of Vitreous Zinc Chloride," J. Non-Cryst. Solids 51, 57-86 (1982).
[111] A.J. Leadbetter and A.C. Wright, "Diffraction Studies of Glass Structure I. Theory and Quasi-Crystalline Model," J. Non-Cryst. Solids 7, 23-36 (1972).
[112] J.C. Phillips, "Spectroscopic and Morphological Structure of Tetrahedral Oxide Glasses" Solid State Phys. 37, 93-171 (1982).
[113] F.L. Galeener and A.C. Wright, "The J.C. Phillips Model for Vitreous $SiO_2$: A Critical Appraisal," Solid State Commun. 57, 677-682 (1986).
[114] R. Hosemann, M. Hentschel, A. Lange, B. Uther and R. Brückner,"Dreidimensionale Analyse der Parakristallinen Struktur von Kieselglas," Z. Kristallogr. 169, 13-33 (1984).
[115] R. Hosemann, "Microparacrystals - The Building Material in Nature," J. Phys. Coll. C8, 379-383 (1985).
[116] B.E. Warren, "X-ray Diffraction" (Addison-Wesley, Reading, 1969).
[117] P.H. Gaskell, "Construction of a Model for Amorphous Tetrahedral Materials Using Ordered Units," Philos. Mag. 32, 211-229 (1975).
[118] A.J. Leadbetter and A.J. Apling, "Diffraction Studies of Glass Structure (V). The Structure of Some Arsenic Chalcogenide Glasses," J. Non-Cryst. Solids 15, 250-268 (1974).

[119] A.C. Wright, R.N. Sinclair and A.J. Leadbetter, "Effect of Preparation Method on the Structure of Amorphous Solids in the System As-S," J. Non-Cryst. Solids 71, 295-302 (1985).

[120] J.C. Phillips, "Topology of Covalent Non-Crystalline Solids II: Medium-Range Order in Chalcogenide Alloys and A-Si(Ge)," J. Non-Cryst. Solids 43, 37-77 (1981).

[121] A.C. Wright, D.L. Price, A.G. Clare, G. Etherington and R.N. Sinclair, "Diffraction by Glasses with a Layer Structure," Diffusion and Defect Data 53-54, 255-263 (1987).

[122] S.C. Moss and D.L. Price, "Random Packing of Structural Units and the First Sharp Diffraction Peak in Glasses," In: "Physics of Disordered Materials," Eds. D. Adler, H. Fritzsche and S.R. Ovshinsky (Plenum, New York, 1985), 77-95.

[123] A. Guinier, "X-Ray Diffraction" (Freeman, San Francisco, 1963).

[124] S.A. Brawer, "The Glass Transition of $BeF_2$: A Monte Carlo Study," J. Chem. Phys. 72, 4264-4277 (1980).

[125] W. Weber, "Adiabatic Bond Charge Model for the Phonons in Diamond, Si, Ge, and $\alpha$-Sn," Phys. Rev. B15, 4789-4803 (1977).

[126] M.D. Rechtin, A.L. Renninger and B.L. Averbach, "Monte Carlo Models of Amorphous Materials," J. Non-Cryst. Solids 15, 74-82 (1974).

[127] A.L. Renninger, M.D. Rechtin and B.L. Averbach, "Monte Carlo Models of Atomic Arrangements in Arsenic-Selenium Glasses," J. Non-Cryst. Solids 16, 1-14 (1974).

[128] A. Le Bail, C. Jacoboni and R. de Pape, "Refining Structural Models for Glasses: Is it Possible? The Case of $Pb_2M_2F_9$," J. Phys. Coll. C8, 163-168 (1985).

[129] R.L. McGreevy and L. Pusztai, "Reverse Monte Carlo Simulation: A New Technique for the Determination of Disordered Structures," Mol. Simulation 1, 359-367 (1988).

[130] D.A. Keen and R.L. McGreevy, "Structural Modelling of Glasses Using Reverse Monte Carlo Simulation," Nature 344, 423-425 (1990).

[131] L. Pusztai and R.L. McGreevy, "The Structure of Glassy Zinc Chloride: A Reverse Monte Carlo Study," J. Non-Cryst. Solids 117/118, 627-630 (1990).

[132] N. Metropolis, A.W. Rosenbluth, M.N. Rosenbluth, A.H. Teller and E. Teller, "Equation of State Calculations by Fast Computing Machines," J. Chem. Phys. 21, 1087-1092 (1953).

[133] W.J. Dell, R.V. Mulkern, P.J. Bray, M.J. Weber and S.A. Brawer, "Nuclear-Magnetic-Resonance Studies and Molecular-Dynamics Simulations of the Structure of Sodium Fluoroberyllate Glasses: Evidence of Non-Tetrahedrally-Coordinated Beryllium," Phys. Rev. B31, 2624-2633 (1985).

[134] K. Furuhashi, J. Habasaki and I. Okada, "A Molecular Dynamics Study of the Structures and Dynamic Properties of Molten $NaBeF_3$ and $Na_2BeF_4$," Mol. Phys. 59, 1329-1344 (1986).

[135] T.F. Soules, private communication.

# CHAPTER 9

# ATOMISTIC MODELING AND SIMULATION OF THE STRUCTURE AND TRANSPORT PROPERTIES OF GLASS

**Joseph H. Simmons, Romulo Ochoa and Tod M. Neidt**
**University of Florida**
**Gainesville, FL 32611**

## INTRODUCTION

Mathematical modeling of the structure and dynamics of materials at the molecular level holds great promise for the development of an understanding of the behavior of materials and, in particular, the microscopic mechanisms which determine their characteristics. Coupled with an accurate and relevant data-base, mathematical modeling studies can reveal many of the fundamental microscopic processes which control the behavior of materials and, consequently, lead to the elucidation of the relationship between microscopic processes and observed macroscopic behavior. Through this understanding, the causative relationship between local structure/composition and observed physical properties can be established, thus forming the basis for the design of new materials and composites or modifications of existing components to meet the severe and varied constraints imposed by applications in today's high technology society.

Simulating the behavior of materials requires the foresight to select approaches appropriate to the study of the characteristics of interest and a sound experimental or technological base to suggest, provide and verify the requisite assumptions. Without these guidelines, mathematical simulations all too often turn into an academic exercise with little technological value. The major strengths of mathematical simulations are numerous and can be distilled to: (1) the ability to reveal behavior at the molecular level and describe atomistic dynamic processes, (2) the ability to determine structural make up and relate it to macroscopic properties or measurable physical characteristics, (3) the ability to reduce complex processes to their essential components and study them separately, and (4) the ability to extend to large ensembles the behavior of a precisely defined ideal interaction model. The major weakness of mathematical simulation and its greatest challenge has been the difficulty of identifying and separating intrinsic characteristics resulting from the fundamental physical mechanisms apart from artifacts inadvertently developed through the many assumptions and simplifications made during the simulation and the formation of underlying models of the physical system. This is not to say that mathematical modeling must always involve complex interactions. On the contrary, many macroscopic properties can result from simple geometrical constraints which transcend the more complex interactions of atoms. However, this strongly suggests that the modeling and simulation studies must always be well anchored in experimental evidence and must always seek experimental confirmation.

Jim Haile, in an excellent book on the elementary methods of molecular dynamics [1], follows an argument by Casti [2] which establishes the hierarchy of model, physical reality and simulation. He points out that a model or theory is an attempt to simplify physical reality by selecting only those interactions which have the most relevance to and/or greatest influence on the physical process of interest thus reducing the problem to a tractable analysis. Therefore, a model is a simpler subset of the physical system. A simulation, by contrast, begins from such a general precept that it must be severely restricted to apply to physical reality. Thus, the simulation is a more complex superset of the physical system. A key step in the development of a realistic theory or model is the identification of processes which can be neglected and those sufficiently relevant to be transformed to mathematical form, whether tractable and solvable in closed form or by numerical approximation. A key step in the development of a realistic simulation is the selection of appropriate restrictions to a calculation algorithm which define a subset equivalent to the physical system or a subset that is contained within the realm of the physical system (e.g., a simplification of the physical system). That is to say that all simulations need not resemble physical systems to be useful, since much knowledge can be and has been obtained from simulations of simpler systems (e.g., hard spheres and various soft sphere potentials). In fact, as pointed out by Haile [1], the simulation is not of the physical system but rather it is a simulation of a **model** of the physical system. Therefore, no simulation can claim that it exactly reproduces the physical system. In the sections to follow, therefore, we will refer to the development of mathematical forms for the interaction between particles as the **model** and to the calculation of the behavior of the ensemble as the **simulation**.

There are several methods for simulating materials behavior and structure, each with its own advantages and limitations. In general, Monte Carlo studies can be thought to address global thermodynamic states; quantum mechanical models can be thought to address local structural arrangements determined by electronic states, and molecular dynamics can be thought to address transport processes. In most studies, however, much more information is available to these approaches, but in all cases, the assumptions made in forming the systems must be examined carefully to determine whether the kind of analysis sought is appropriate for the desired study. For example, quantum mechanical models can reveal the chemical reactivity and thermodynamic stability of various local structures and molecular dynamics simulations can be used to model structure, and thermodynamic behavior of large ensembles of particles. There are numerous reviews of all these methods available and the reader is referred to them for an in depth description of the various techniques (quantum-mechanical [3], Monte Carlo [4-6] and molecular dynamics [1,7-13]).

This chapter will describe molecular scale simulations of non-metallic, inorganic solids, with special emphasis on Molecular Dynamics (MD). We will address the major approaches to modeling structure and dynamics of solids and give several examples of MD studies to demonstrate the range of capabilities of the method and highlight the kinds of results which may be obtained from simulations. Because the question of static structure and thermodynamics are of primary interest, and because the question of appropriate force functions always hangs above simulations of physical systems, we will begin by briefly reviewing stochastic and quantum mechanical modeling approaches. We will end by sketching future directions in simulation which bring together the advantages of quantum mechanical structural determinations and the dynamical capabilities of molecular dynamics simulations. An in-depth description of the methodology of molecular dynamics is left to excellent recent reviews [1,10,13-15].

## MOLECULAR MODELS AND SIMULATIONS

In a sense, microscopic or molecular modeling studies (including simulations) work in the opposite direction from experimental investigations. The latter are generally well anchored in accurate measurements which describe the characteristics of a material through an observation of the average behavior of a large ensemble of particles. The collection of these measurements is then used to **infer** the causative atomic processes. This inference is often strengthened by measurements of the static microscopic nature of the material which include microscopy (real space structure) and scattering spectroscopy (reciprocal space structure). Dynamic processes may also be studied (relaxation processes, line broadening, spectroscopy). On the other hand, molecular modeling and simulation studies build from individual atomic processes to **deduce** collective or ensemble averages. When viewed in this context, molecular-level studies begin from approximations (models) of real particle interactions and build a system through a set of self-consistent rules to predict the properties and behavior of the particle ensemble. Two different goals may be sought. One is to use simple models and investigate the results of the simulation in term of the characteristics of the simple model. This allows the investigator to view the consequences of various assumptions in forming the model or conducting the simulation. This approach may not lead to a description of a physical system. However, it may help establish relationships between individual atomic processes and ensemble averages which are related to physical properties of the system. The second goal is to simulate a physical system. In this case, when the study is conducted properly, the assumptions and approximations of the simulation technique and the original models disappear in the ensemble averages. If the averages represent adequately the properties of the physical system, the difficult question becomes the following: If the ensemble averaged properties (density, elastic constants, phonon vibration spectrum, transport properties, etc.) of the simulated system agree with those of the real system, then how far can the approach be taken back toward the individual atomic dynamics (which are generally available) and still represent the physical system? Often, this effort is supported by comparison of calculated microscopic quantities, such as radial distribution functions, to physical measurements, such as neutron and x-ray scattering, EXAFS, Raman scattering spectroscopy, XPS or NMR studies of chemical shifts, etc. However, the approach varies drastically between investigators. Many use a multitude of empirical higher order corrections to the model to justify simulations of the details of individual particle behavior. Others limit the models to those simpler interactions which have direct physical basis and only study overall tendencies of the system.

Below are brief examples of several simulation methods to elucidate the structure and behavior of solids. They vary from stochastic models of random close packed hard spheres to quantum mechanical ab-initio calculations of the electronic states of interatomic interactions. Each approach has advantages and limitations, and each can contribute to the understanding of the constraints and characteristics which must be considered in attempts to understand the formation of amorphous and crystalline structures and their dynamic response to external disturbances. In most cases, in structure studies, primary attention will be paid to amorphous structures. Single crystals are trivial and polycrystalline structures have generally received far less attention due to the greater complexity of modeling the grain boundary. We will briefly discuss the most notable benefits of each method in order to focus on the approach of molecular dynamics in the context of what can be done with the other approaches.

## Stochastic Simulations

Stochastic simulation has been applied to the development of glass structures by a variety of methods, the earliest of which consists of placing atoms in close proximity to each other to form complex 3-dimensional structures in quasi mechanical equilibrium. Of particular note is the work of Bernal [16] and Finney [17] with random close packing of hard spheres and the work of Bell and Dean [18] with a ball-and-stick structure of amorphous silica. The former shows similarities to metallic glasses while the latter gives one of the best calculated radial distribution functions of silica, with a relatively accurate phonon density of states.

The stochastic simulations allow a local analysis of the results of extending to a large ensemble of particles a set of strict geometrical constraints applied to forming bonds between atoms. These studies are highly dependent on the rules chosen to construct the geometry (interaction model). However, when they succeed well, as in the case of the structure of silica, they reveal the importance of the geometrical considerations adopted.

One of the most fruitful geometrical considerations in the determination of structure is the cation/anion radius ratio. The coordination polyhedra of systems in which the bonding is at least partially ionic in character are to a large extent determined by the relative sizes of their constituent ions, usually expressed as the radius ratio, $r_{cation}/r_{anion}$. (Anions are in general larger than cations, hence $r_{cation}/r_{anion} < 1$.) For stability, the cation at the center of an $AX_n$ polyhedron should fill the cavity defined by the anions and there exists a critical radius ratio at which the criterion is just fulfilled, and below which the cation cannot remain in contact with all the anions, thus destabilizing the $AX_n$ polyhedron. Above this value, the $AX_n$ polyhedron will remain stable until the point is reached where extra X atoms can be incorporated. In general, small radius ratios (n=3 or 4) tend to result in well defined structural units whereas larger radius ratios (larger values of n) allow a greater latitude for distortion and frequently lead to the presence of more than one type of polyhedron. The effective radius ratio for a given pair of ions is a function of the degree of ionicity of their bond, and small values of n often are associated with a considerable covalent bonding character, because of the compatibility of the corresponding polyhedral shapes with common bond hybridization geometries. Conversely, at high values of n, there is an increased degree of ionicity in the bonds and the different polyhedra are determined by ionic size (radius ratio) and geometrical packing efficiency. There is also an increased probability of both edge and face sharing between adjacent polyhedra since the increase in strain energy required is much less.

For example, it is apparent that first order structural differences between oxide and fluoroberyllate glasses with small coordination polyhedra (due to small radius ratios), on the one hand, and heavy-metal fluoride glasses with much larger coordination polyhedra, on the other hand, arise from the different radius ratios of the glass formers involved. In oxide and fluoroberyllate glasses the coordination polyhedra are generally well defined, therefore, the unit cell in the amorphous phase closely resembles the crystal forms and the amorphous structure results from the packing of crystal-like coordination polyhedra. In heavy-metal fluoride (HMF) glasses, the larger radius ratio leads to less well defined coordination polyhedra and a range of polyhedral types. The variations in polyhedral types in HMF glasses lead to average coordination numbers which are not integers and to structures which exhibit broad peaks in the real space correlation function, making stochastic simulations less useful for their study than for silicates where the tetra-hedral unit and its packing configurations dominate the local structure (Simmons [19]).

## Relaxation Simulations

Relaxation simulations have been used by a variety of investigators (Fletcher and Reeves [20], Yasui and Inoue [21]). The approach varies between investigator. It generally consists of the formation of a structure using either stochastic methods or the random stacking of distributions of crystalline coordination polyhedra followed by a minimization of the total potential energy of the system. The minimization can be realized by finite difference or steepest descent methods or allowing relaxation of the system by molecular dynamics or possibly Monte Carlo calculations. The initial packings generally exhibit sharp peaks in the radial distribution function which subsequently broaden during the relaxation process. If one imagines a 3N dimensional potential energy surface for the ensemble of particles, then the stochastic or trial configuration begins at some non-equilibrium position and the subsequent relaxation approach effectively seeks out a local minimum. But it is difficult to guarantee that the relaxed structure in the local minimum has reached the non-crystalline global or any of several other deep minima of the system. Due to the sluggishness of amorphous systems, it is clear that a search for such global or deep minima is very difficult. In fact, their discovery may not be useful in representing the actual glass structure since the real material in cooling slowly from an equilibrium liquid state reaches a metastable equilibrium state which is not necessarily, if it exists at all, the non-crystalline global minimum. The greatest failing of relaxation simulations, is that since they begin near some sort of crystalline state, they do not follow the same path to the solid structure as the real material and therefore they do not necessarily access the same local minima. The difference in success is therefore determined by the degree of similarity between the structure of the glass and the crystalline forms. The approaches which use a well developed stochastic model may be more successful than those which relax crystalline structures.

An improvement may be had by increasing the temperature of the ensemble, thus raising the probability for transitions out of local minima and increasing the sampled region of phase space.

**Monte-Carlo Simulations**

Monte Carlo simulations use random displacements of each particle in a canonical ensemble to calculate resulting changes in the total internal energy ($\Delta E$) of the ensemble. Statistical effects of fluctuations in local thermal energy are simulated by accepting, in addition to all moves which yield reductions in total energy, those moves which increase the total energy if they yield a Boltzmann factor, $\exp(-\Delta E/kT)$, greater than a random number between 0 and 1. This technique adjusts the structure of the ensemble until a free energy minimum is achieved. It avoids getting stuck in local minima by use of energy fluctuations if the excess energy of the fluctuations is large enough to activate the system out of these minima. If the system appears too sluggish, the search for the global minimum can be quickened by raising the temperature, but the change in temperature may also displace the minima in configurational space. When energy fluctuations are possible, deeper minima are generally achieved. The lowest non-crystalline minimum is most likely neither achieved by this technique nor by the real glass. Excellent structural studies of glasses have been achieved with this approach (Abraham for Lennard Jones particles [22], and Brawer for fluoroberyllates [23]).

**Definition of Models**

The approaches discussed above require the definition of several factors or models, the most important of which are: (1) interparticle interactions, (2) core repulsion and atomic size, (3) size of the simulated cluster and (4) effect of termination of the simulated cluster.

The questions of interatomic forces, core repulsion and ionic size have occupied many investigations. The ball-and-stick appraoch for example only works well when flexible sticks are used which can support a moderate amount of bending. In the computer simulations, a variety of interatomic potentials are used, including: hard core with no attraction (hard sphere potential), softer core with some attraction (Lennard-Jones potentials), soft core with strong electrostatic interactions (Born-Mayer potentials), soft core with electrostatic and directional forces (2 and 3-body potentials), and an innumerable list of empirically defined potentials. Some of these potential functions will be discussed later. Great differences in structure can result from the choice of potentials. For example, hard cores lead inevitably to non-crystalline structures. The system has no flexibility and cannot reach any global minimum without very careful a-priori construction to attain the global minimum on the first attempt. Softening the core repulsion allows crystallization to occur. A soft core with a weak attraction (Lennard Jones) exhibits a greatly enhanced ability to probe various local minima in the configurational energy surface of the ensemble. The ease of probing can lead readily to the absolute minimum (the crystal). A strong attraction or electrostatic interaction makes the system more sluggish and can avoid falling into the absolute minimum. Systems with two or more different components add sufficient entropy to greatly reduce the probability for crystallization, even with Lennard Jones interactions.

The forces discussed above are all approximate classical expressions representing the real quantum mechanical interactions of the system. Many features of solids have been well modeled by these simple classical forces because many of the microscopic and macroscopic properties of materials only depend on the general features of the interatomic forces. Consequently, the details of the quantum mechanical forces are eliminated either by the scale of the interactions, by averaging over a large particle ensemble, or by the insensitivity of the properties of interest. For example, radial distribution functions [24], phonon vibration spectra [25], rheological behavior of fluids [26], elastic constants and brittle fracture of solids [27], have been well modeled with classical force functions. Very recent studies of 2 and 3-body interactions to model partially covalent bonding have allowed several investigators to extend classical studies to surface behavior [28] and some chemical polycondensation processes [29].

None-the-less, quantum mechanical forces are more appropriate for some studies such as those involving electronic and optical properties, structural defects and chemical reactions and one must always be careful to remember that classical methods use numerous empirically adjustable parameters to develop the force functions. Therefore, in the examination of results, an analysis of the effect of the selected classical forces is imperative before interpretation of the results.

Questions pertaining to the size of the simulated clusters and the termination effects on the clusters (e.g. treatment of the particles at the boundary of the cluster) are handled differently by different approaches. Stochastic simulations have used more than 1,000 particles with either free space or some container at the edge of the cluster. Computer calculations (relaxation and Monte Carlo methods) use between 100 and several thousand particles with no termination through application of a cyclic reproduction of the cluster (periodic boundary conditions) in all spatial directions. By contrast, quantum mechanical calculations, because they account for the *electronic states* of each particle can only treat clusters with a few to a few tens of atoms.

## Quantum Mechanical Calculations

Clusters can be simulated using quantum mechanical electronic states to model the interatomic interactions. The level of accuracy is only limited by the assumptions and approximations invoked in the calculation of the atomic orbitals for the components of the cluster. The total molecular wavefunction for the cluster, $\Psi_T$ is obtained by solving the time-independent, non-relativistic Schrodinger Equation:

$$H\,\Psi_T = E_T\,\Psi_T \tag{1}$$

where **H** is the cluster Hamiltonian operator and $\mathbf{E_T}$ is the total cluster energy. The Hamiltonian may be written in terms of its component kinetic energy operators for the nuclei and the electrons and its component potential energy operators for the interactions between nuclei and nuclei, electrons and electrons, and nuclei and electrons.

$$\mathbf{H = KE_n + KE_e + V_{nn} + V_{ne} + V_{ee}} \tag{2}$$

The solution of this equation is a formidable undertaking which is simplified by various approximations.

*Born-Oppenheimer Approximation*

First, the Born-Oppenheimer (adiabatic) approximation separates the nuclear coordinate variable, **R**, by removing the nuclear kinetic energy operator, $\mathbf{KE_n}$ from the Hamiltonian. This allows a separation of the nuclear and electronic wave functions:

$$\Psi_T(\mathbf{R},\mathbf{r}) = \Psi_N(\mathbf{R},\mathbf{r})\ \Psi_e(\mathbf{R},\mathbf{r}) \tag{3}$$

thus leading to a solution of the electronic wave function, $\Psi_e(\mathbf{R},\mathbf{r})$, for a fixed value of the nuclear coordinate wavefunction: $\Psi_N(\mathbf{R},\mathbf{r})$. Since the nuclei move relatively slowly in comparison to the motion of the electrons, then the fields generated by the nuclei appear stationary from the standpoint of the electrons. This is a reasonable approximation since the nuclear masses for most elements are more than 10,000 times the electronic rest mass ($m_e = 9.1\text{x}10^{-31}$ Kg and $m_p = 1.67\text{x}10^{-27}$ kg). The total Schrodinger Equation can then be solved by adding the nuclear kinetic energy to the electronic wavefunction solution to produce a total wavefunction consisting of the products of the electronic and nuclear components. A minimum energy structure is found by successive iterative solutions in which the nuclei are repositioned with respect to each other. This approach, since it calculates the local structure of the cluster, can produce a phonon density of states, however, the Born-Oppenheimer approximation and all its extensions (discussed below), do not naturally lead to studies of dynamics associated with the thermal motions of the nuclei, and a variety of 'repairs' are necessary to handle temperatures above absolute zero.

*Ab-initio Calculations*

Ab-initio calculations use various approximations to calculate the now separate total electronic wave function. In this approach, all approximations are well defined so that with enough computer time, they may be removed and an exact calculation may be obtained. This is rarely done, however, so that although the calculation method is labeled ab-initio, the results are not necessarily accurate.

The next approximations are invoked to solve the electronic wave function equation. The latter cannot be separated into wavefunctions for each individual electron because of the electron-electron interaction energies. There are several major approaches to the solution: (1) the Hartree-Fock (HF) calculations, (2) the Hartree-Fock plus configuration interactions (CI), (3) the modified Hartree-Fock calculations and (4) the density functional (DF) calculations.

The Hartree-Fock Self-Consistent Field approximation [30] assumes that the motion of an electron is only subject to the average potential from all the other electrons. This neglects the correlation between electronic motions and models electronic motions as those of quasi-independent particles. The electronic wavefunction can then be written as a Slater determinant of one-electron molecular orbitals (MO), (e.g., an anti-symmetrized product of one-electron wavefunctions). Anti-symmetrization is used to keep apart electrons of the same spin [31]. This is expressed as a wave function of form:

$$\boldsymbol{\Psi}_e = \mathbf{A}\,[\mathbf{O}_a(1)\mathbf{O}_b(2)....\mathbf{O}_z(n)] \tag{4}$$

where **A** is the electron antisymmetrization operator, and $\mathbf{O}_a$ **(i)** are the **a**'th spin-orbitals of the **i**'th electron, which may be written as an orbital expansion in terms of basis functions $\{\mathbf{X}_u\}$ centered on the $u^{th}$ nuclei:

$$\mathbf{O}_a(\mathbf{i}) = \sum [\mathbf{X}_u(\mathbf{i})\,\mathbf{C}_{ua}] \tag{5}$$

leading to the N dimensional matrix eigenvalue problem

$$\mathbf{F\,C = S\,C\,E} \tag{6}$$

where **F** is the Fock or energy matrix and **S** is the overlap matrix, with the **a**'th column of the matrix **C** representing the orbital coefficients of the orbital $\mathbf{O}_a$.

The one electron molecular orbitals are obtained by minimizing the total electronic energy through an iterative application of the variational principle to trial wavefunctions. Generally the basis functions are written as a linear combination of Gaussian orbitals. The molecular orbitals can be either delocalized (e.g. spread over the entire molecule), or localized (e.g. the electrons are distributed in atom-like and valence bond orbitals). The localized description has successfully calculated heats of formation by assuming that there is a specific energy associated with the individual bonds. Delocalized orbitals provide a more consistent approach to calculations of spectroscopic properties [31] since they concern removal or redistribution of electrons.

The Hartree-Fock plus configuration interaction method reverses the neglect of electron correlation effects (HF+CI). Here, the many electron wavefunction is expanded as a linear combination of anti-symmetrized products of molecular orbitals chosen to cover all possible occupied and excited states (e.g., the wavefunction is fabricated as a superposition of the Slater determinant and all possible determinants obtained by substituting an increasing number of occupied

MO with virtual orbitals). This approach gives a far improved evaluation of the electronic wave function, however, the time taken for calculations is so large that it is used mainly for electronic state calculations of atoms with less than 10 electrons [30]. For example, calculations of the Si-O bond energy must neglect the CI approach.

When calculating HF wave functions for multi-electron atoms, numerous orthonormal basis sets are used [32]. Molecular systems for example can be modeled by linear combinations of atomic orbitals (LCAO) which have the advantage of centering the orbitals on the nuclear coordinates (atomic-like orbitals) [33]. Atomic-like Slater orbitals (STO) used as basis sets can be improved by the addition of normally unoccupied orbitals of the atom in the expansion. STO-3G indicates calculation procedures which use an STO approximated by 3 Gaussian functions to represent each atomic orbital in the MO expansion [34]. STO-3G* adds 3d orbitals to the STO-3G basis of second row atoms [35]. Programs are available for the HF calculations of the electronic states of various elements using a variety of basis sets.

Gibbs [34] and Lasaga and Gibbs [36,37] use HF calculations with atomic orbitals and Gaussian functions (STO-3G* basis set - e.g., Slater-type atomic orbitals represented by a linear combination of 3 Gaussians, including 3d orbitals on Si) to calculate potential energy surfaces for the Si-O bond in a variety of silicate molecules. The calculations on a $H_6Si_2O_7$ molecule reveal a shallow energy minimum at a value of 144° for the Si-O-Si bond angle with an Si-O bond length of 1.60 Å. However, the angle may span from 130 to 155° at the cost of only 1 kcal/mol (see Fig. 1, reprinted from [36]). The result is in agreement with experimental observations on a wide variety of crystalline silica structures.

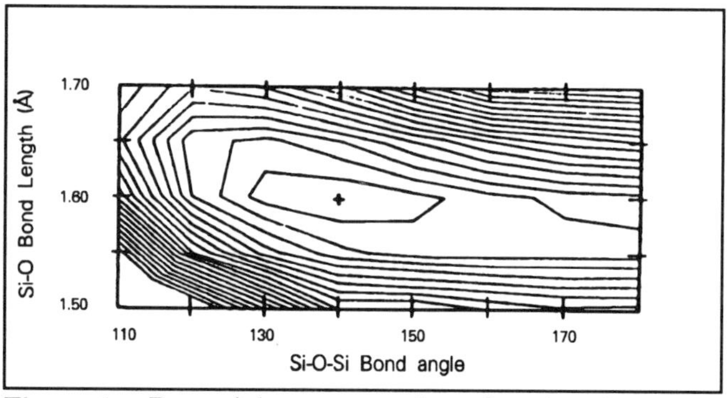


**Figure 1** - Potential energy surface for the Si-O bond length in terms of the Si-O-Si angle from Ref. 36.

*Semi-Empirical Methods*

The formation of the Fock, or energy matrix, **F** is an $N^4$ step (N= number of electrons), limiting in general the size of the system that can be investigated using the Hartree-Fock Self-Consistent Field approach just outlined [31]. Semi-empirical methods attempt to form the Fock matrix in $N^3$ or $N^2$ steps in such a fashion that the rate-determining process in this procedure is the repeating of $N^3$ steps required to solve the matrix eigenvalue problem [38,39]. These methods focus on the degree of differential overlap between the electronic wavefunctions of the basis set. This is represented by the off-diagonal terms in the overlap matrix **S** in the above equation [40,41]. CNDO, MNDO and AM1 [42] use complete neglect of differential overlap. In the development of cluster structures, MNDO and AM1 yield generally closer results to ab-initio calculations. AM1 is optimized for hydrogen bonding. INDO [40,41] and MINDO use intermediate and modified intermediate neglect respectively. They give fairly accurate bond lengths and angles, dipole moments and very accurate ionization energies [42]. INDO emphasizes electronic interactions. Programs developed at the Quantum Theory Project at the University of Florida (ACES, UHF, and ZINDO), programs developed by Dewar (AMPAX) [43], James Stewart (MOPAX) [39] and Pople and his collaborators (Gaussian90) [40] use gradient optimization methods for searching potential energy surfaces for stable structures and for transition state structures [44-46]. (Gaussian90 is an ab-initio method.) Generally, CNDO and INDO methods are directed to the calculation of charge distribution, dipole moments, polarizability, electronic spectra [30]. These methods give reasonably good values of molecular geometry but rather unreliable ionization and dissociation energies. MINDO and its revisions are designed to give accurate molecular geometries and dissociation energies. All neglect overlap methods are parametrized to certain physical properties, such as spectroscopic structure, dissociation energies, etc., therefore the parametrized quantities will always be accurately represented.

West and Wallace [47], using INDO and AM1 methods have modeled the adsorption of an $H_2O$ molecule onto a hydroxylated tetrasiloxane ring structure. They show the formation of a hydrogen bond between the water molecule and a stretched O-H bond in the tetrasiloxane molecule, down shifting the IR $\nu_3$ vibration peak to 2.82 $\mu$m. This calculation was supported by experimental observations presented in the same paper. Further studies by the same authors [48] have revealed the formation of penta-coordinated Si transition states in the tetrasiloxane ring ($D_1$ Raman feature) with one adsorbed water molecule, and have shown how the first water bonded to a *trisiloxane* ring facilitates the breakage of one of the Si-O bonds in agreement with studies of the formation and annihilation of $D_2$ Raman peaks upon drying and rehydration of silicate surfaces.

The last semi-empirical approaches discussed here are the free electron MO and the simple LCAO-MO (linear combination of atomic orbitals) method which includes the Huckel MO (HMO) and the extended Huckel MO (EHMO) [42]. The quantitative predictions of these methods are generally poor, and access to high speed computers has made the great simplifications necessitated by these methods unwarranted. However they do reveal interesting qualitative tendencies to those unable to use the more powerful approaches because of lack of computer power, or to those willing to sacrifice quantitative bonding energy calculations to study large clusters.

*Local Density Approximation*

Several modifications of the HF calculations have also been used. The density functional method simplifies modeling efforts by assuming that the total electronic energy depends on the electronic density, thus the valence or bonding electrons move in the potential defined by the local electron density and the exchange operator likewise is replaced by a function of the electron density [49]. The approach maps exactly the problem of an interacting electron gas onto a system of non-interacting particles in some effective external potential. The electron-ion interaction is modeled by an effective core approximation (also known as Pseudopotential approach), in which the inner shell electrons are assumed relatively inert (e.g., frozen in place) and replaced by an effective core potential, thus significantly reducing the order of the calculation. This method is useful for elements above the second row of the periodic table and for problems involving studies of surfaces and structures involving point defects. DF pseudopotentials have been calculated by a number of investigators and are available from H to Pu [50].

The Local Density Approximation is not considered to be an ab-initio calculation because it is limited in the ultimate accuracy of the technique. The LDA introduces an effective exchange-correlation potential to replace the calculations of the HF approach. This function is empirically determined and not directly related to approximations of the real exchange and correlation processes. The $X\alpha$ scattered wave method in which the effective exchange-correlation potential is modeled by a spherically averaged muffin-tin potential is widely practiced [51].

Carr and Parrinello [52] modified the density functional approach to use molecular dynamics as a method of finite differences to vary both the electronic wave function and the nuclear configuration (structure of the material) simultaneously. As the kinetic energy is withdrawn from the system the method quickly finds the total local energy minimum by finding at the same time the proper total electronic wave function and the correct structure corresponding to the minimum energy configuration subject to the determined electronic configuration. This method leads to the zero-temperature structure and saves time if both the nuclear configuration and the electronic wave function converge during the procedure. In the semi-empirical methods discussed above, the minimum electronic wave function is calculated for a variety of nuclear configurations and these are step-wise adjusted to lower the calculated minimum energy. Modifications of the Carr-Parrinello approach alternate MD calculations of the dynamics of the nuclei using LD force models to simulate higher temperature behavior.

Joannopoulos and co-workers [53,54] produced interesting results in the studies of electronic defects in semiconductors (Si, Ge) and in the electronic states of E' centers in silica and germania. For example, pseudopotential studies show a 140° minimum in the Si-O-Si bond angle energy with only a difference of 0.15 eV between 140 and 180°. Examination of the structure of oxygen vacancy defects in alpha-cristobalite (Joannopoulos) [54] and alpha-quartz (Allen) [55] revealed that the removal of an oxygen atom allows an Si-Si interaction and the formation of a dimer thanks to the weak back-bond interactions from the remnant 3 oxygens on each Si. Such reactions are generally strong enough to resist dimer formation. However, in silica, the resistance is weak due to the strong ionic component of the Si-O bond, and the Si-Si dimer bond distance can reach a 2.65 Å equilibrium distance for a neutral vacancy (the dimer bond distance in silicon is 2.35 Å). The negatively charged vacancy strengthens the bond and reduces the equilibrium distance to 2.43 Å, while the bond associated with the positive vacancy is 3.02 Å in cristobalite and 3.10 Å in quartz. This result supports the concept that the bonding in silica is predominantly ionic, because in purely covalent silicon, adding electrons to the anti-bonding states weaken the bond, whereas in silica the other oxygens ionically remove the electrons from the Si-Si bond, and adding one or two extra electrons strengthens the bond. Furthermore, the energy balance favors a neutral vacancy with a symmetric structure, and asymmetric structures cost a significant increase in energy. Allen [56] finds that the +1 charge state has another metastable structure which involves a puckering distortion in the local structure. The puckered Si atom regains 4-coordination while the oxygen atom in the stretched bond beside the vacancy acquires 3-coordination. Rudra and Fowler [57] attribute the E' spin signal to the unpaired electron which is associated with the remaining dangling bond of the other Si atom.

In general these techniques require the use of periodic boundary conditions on a relatively small set of atoms. Supercomputers may extend cluster sizes to hundreds of atoms. While this approach bridges well the gap between quantum mechanical cluster calculations and molecular dynamics, it does not give dynamical information only available from MD approaches and does not give reaction path chemistry available from the MO approaches.

*Limitations of Quantum Mechanical Methods*

In looking at limitations of simulation methods, one rule always holds: *More advanced computers allow studies of larger ensembles with more complex interactions*. Due to the $N^x$ dependence of quantum chemical procedures, ab-initio techniques ($N^4$ and $N^3$) are seldom used for more than about 15 atoms and 60 electrons, although larger systems have been recently reported in the literature. Semi-empirical calculation ($N^2$) can be much larger. (An SCF - configuration interaction calculation was conducted on the photosynthetic reaction center, with over 600 atoms and 1500 electrons [58,59].)

Once methods are limited to small clusters, the problems of cluster size and termination become important. Investigators differ in their approach to these problems and use either periodic conditions to extend their clusters or precise termination procedures such as the use of univalent atoms which end the cluster. Such termination methods suffer from the interaction between the termination atoms if the clusters are too small. For example, hydrogen or fluorine are often used as termination atoms. Tetrasiloxane molecules consisting of a 4-membered ring of silicon oxide tetrahedra terminated with hydrogen, yield Si-O bond lengths of about 1.715Å instead of the experimentally determined value of 1.62Å for silica glass and crystals [47]. The calculation is not incorrect and the deviations do not result from the various assumptions made to find the electronic wave functions. Instead, the longer bond lengths result from the repulsive forces between the terminating hydrogen atoms which place the molecule under a non-negligible tensile pressure. If the cluster is made larger in size or a moderate compressive pressure is applied, then the bond length may be reduced. However, the choice of appropriate pressure to obtain the expected 1.62Å bond length is empirical. A question then arises regarding the calculations of chemical reactions occurring on a cluster whose bond lengths are incorrect. What is found is that the model, even with incorrect bond lengths predicts relatively accurately energetics of reaction paths and activation energies. This result indicates that not everything must be perfect to obtain useful data and insight into modeled processes. What is necessary, however, is an understanding of the source of the discrepancies and a measure of the sensitivity of the properties of interest to the processes either strongly or weakly associated with the measured discrepancies.

Periodic boundary conditions avoid the problem of the effect of termination atoms, but replace them with problems related to the applied periodicity of the cluster. This periodicity can drastically affect the structure of the minimum energy configuration and may favor crystallization or the repeat geometry may strongly restrict the local structure. Periodic boundaries are only of interest in systems with a very large number of molecules ($>1,000$ particles).

*Media Effects*

Media effects have recently come under investigation (Zerner) [60,61]. This approach consists of subjecting the simulated cluster to underlying fields (i.e., polarization) representing the collective effect of the matrix atoms.

The first order correction treats a quantum chemical cluster in a dielectric continuum containing charge and dipole fields that simulates the environment. Some degree of success has been achieved in structure and thermodynamic studies [60,62] and even in spectroscopy [61] and dynamics [63]. In this model the one-electron part of the Hamiltonian is corrected for the interaction of the ground state cluster by the dipole-induced-dipole interaction with the medium. This is an iterative procedure as charge separation is stabilized by the polarizable medium. At present, this theory in its simplest form, assumes a spherical cavity [61-66]. The dielectric continuum model assumes enough thermal motion of the atoms in the medium such that its properties can be modeled in classical terms of a bulk dielectric and index of refraction. Clearly, the dielectric continuum model has limitations in that it averages the medium effect and neglects higher order polarizations.

## MOLECULAR DYNAMICS SIMULATIONS

One of the great strengths of the MD approach is that it simulates the random constant motion of particles from thermal excitation. The resulting ever changing position and velocity coordinates of all atoms are calculated via Newton's Equations, under the sum of forces of all contributing neighbors. Collisions between atoms are handled by taking into consideration strong interatomic repulsions in the potential or force functions. The form of this repulsion component provides for a 'hard' or 'soft' collision between nearest neighbors and defines the effective size of the atoms. The back-and-forth motion of simulated atoms in MD calculations can then be analyzed to reveal the phonon density spectrum of the material under study. Because thermal excitations of the atoms are inherent in the MD simulation this approach gives MD studies the ability to describe and analyze dynamic processes in materials, which include, not just the motion of isolated atoms in a frozen landscape, but rather collective motions of large groups of particles. Often, it is found that these motions can provide the essential features of complex processes even though the interatomic forces used in the modeling are a clear oversimplification of the local quantum mechanical interactions. Examples of such results abound. Several examples are listed below and some are described in more detail at the end of this chapter. Of great interest to the glass industry, because of its impact on glass forming behavior, is the simulation of shear thinning processes in viscous liquids and glasses at high strain rates whose essential behavior was accurately described by a unary Lennard Jones system [26,67]. Comparison of the results of the MD study with measurements on real glasses showed exact simulation of the stress overshoot and viscosity decrease with shear rate. The MD study contributed structural information about the process which was not obtainable from experimental measurements. Another example is the use of Embedded-Atom Model (EAM) and Lennard Jones interactions to obtain the essential features of interfaces between FCC grains [68] of silicon. Most recently, we have made profound discoveries in studies of the fundamental mechanisms of brittle failure in glass using MD simulations with pair-wise ionic functions [27]. These have elucidated the effect of stress rate on failure mode, the role of structural rearrangements to form free surfaces on the fracture process and the effect of pre-existing flaws in the amorphous structure on the predisposition of the material for crack formation and propagation. Using 3-body potentials, Garofalini has investigated free surface formation [28] and the polycondensation reaction in silica sols [29]. In both instances, results of the MD simulation match closely experimental findings and provide experimentally inaccessible structural information. Brawer and Weber with simpler 2-body potentials have shown remarkably accurate descriptions of the structure of rare-earth doped $BeF_2$ glasses [69], and the luminescence behavior of rare-earth dopant atoms in fluoroberyllate glasses [70].

## Methodology

Computer programs for conducting molecular dynamics are available from a variety of sources and the methodology used is reviewed in several excellent papers [1,7-13]. Here, we will discuss the methodology in general terms only and attempt to explain to the reader the differences between the various approaches.

Molecular dynamics simulations consist of solving numerically the equations of motion of interacting particles in an ensemble. Therefore the spatial position and momentum of each member of the ensemble (particle) is described in time. Generally, simulations are conducted from picosecond to nanosecond durations. The differential equations for the interacting ensemble of particles cannot be solved in closed form, therefore several methods are implemented to solve the problem numerically using finite difference methods. While there are many approaches which may be used with increasingly accurate estimates, calculation speed is very important, and reduces the number of commonly used algorithms to a handful. Below we describe only two: the Verlet algorithm [1,71-74] and the Predictor-Corrector (PC) methods [1,75-76]. Both approaches calculate the positions and velocities of the ensemble of particles at fixed time intervals. Without correction, the Verlet algorithm assumes that the particles move under constant acceleration within the time intervals. The Predictor-Corrector methods use a higher order derivative. If the time interval is short enough, the constant acceleration or constant higher order derivative assumption does not significantly affect the results. However, short time intervals limit the duration of the total simulation because of computer calculation time. How investigators compromise between intervals short enough to yield a good approximation of the motion of the particles and long enough to yield a simulation duration which can access mechanisms of interest is discussed below.

Whichever method is used for calculating the particle trajectories, the stability of the system has to be checked by performing runs of several thousand time steps and calculating the total energy fluctuation and the temperature variation. The total energy consists of the sum of kinetic energies of all the particles and the potential energies calculated from the force function selected. The temperature is calculated from kinetic gas theory as follows:

$$\frac{3}{2}NkT = \sum_i \left[\frac{1}{2} m_i v_i^2\right] \tag{7}$$

Usually, if a time step is adequate, the temperature will oscillate about the set or desired value which is obtained as described below.

After the initial positions and velocities of all particles have been assigned there are several paths that can be followed to perform studies on solids. The path selected will depend on the experiments to be performed and on the properties desired. Algorithms have been devised to so that the averages of the trajectories correspond to microcanonical, isoenthalpic-isobaric, canonical or isothermal-isobaric ensembles. Since the initial conditions are given at high temperatures in order to allow a rapid equilibration of the initial system, either a constant volume or a constant pressure is assumed in cooling to the solid phase. In the constant volume case the Hamiltonian is the sum of the kinetic energy and the potential energy of the system and this method is preferred when control over the density is desired and in experiments where strain is going to be applied. To simulate experimental conditions of constant pressure, Andersen [77] modified the Lagrangian of the MD system by coupling it to a piston with an adjustable mass. This method was extended by Parrinello and Rahman [78] to allow for the shape of the cell to change and respond to the full pressure tensor. In either case, after a simulation is started, the system is allowed to relax at the preset temperature for several thousand time steps. At this point the investigator is presented again with several options to control the temperature of the system. Velocity scaling (Woodcock [79]) to maintain a reference temperature has been widely used and found to have negligible effects on the structure and dynamical properties of the system [80]. Andersen suggested a Maxwellian rethermalization by stochastic collisions [77] and stochastic coupling to a heat bath has also been applied [80,81]. Some methods add a dissipative or velocity dependent term to the force (see Hoover et al. [82] and Evans [83]). Berendsen et al. [84] proposed coupling the MD system to an external bath with constant temperature with adjustable time constants. The method chosen is applied to the system until the system reaches

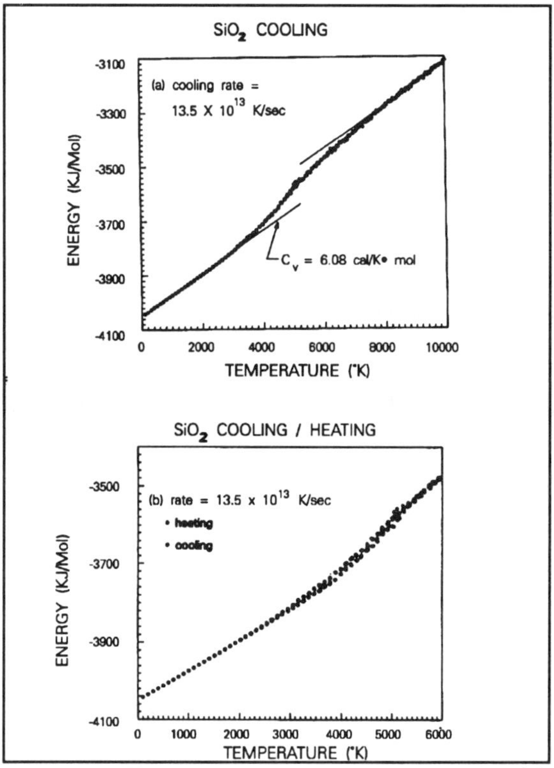


**Figure 2** - Energy vs temperature curves for (a) cooling and (b) cooling and heating showing the hysteresis expected from a silica simulation of Soules [13].

the desired temperature and equilibrates and then the system is usually allowed to continue at constant energy before measuring its properties.

In the case of glasses the final temperature is reached by lowering the temperature of the system in steps of a few hundred degrees kelvin at a time while adjusting the volume of the system to yield the desired density, then allowing the system to thermodynamically equilibrate at each stage. Even in the case of small temperature steps the final glass structure will have been quenched orders of magnitude more rapidly than is achievable in laboratory conditions, therefore, it will have a very high fictive temperature. None-the-less, hysteresis in cooling and heating through the transformation range has been reproduced (Fig. 2) (Soules [24]). Several cooling and heating paths used on NaCl simulations essentially represent the glass transition and crystal melting transition (Fig. 3) (Kumta [85]).

At any stage of the simulation, thermodynamic, structural, transport and mechanical properties of the system can be easily calculated since positions and velocities of all particles are known at all times. The most commonly calculated thermodynamic quantities are the pressure tensor, temperature, total energy and free energy of the system. The pair radial and bond angle distribution functions characterize the structure of the sample and can be compared, through appropriate transformation, to X-ray and neutron diffraction data [86]. Transport properties like self diffusion can be calculated from mean square-displacement calculations. Time correlations of velocities by themselves give information about the memory of the system or if a Fourier transform is applied they can be used to calculate the vibrational spectra of the material. The elastic constants such as Young's modulus and bulk modulus can be calculated by straining the system a small percentage and measuring the change in stress. Other properties that have been calculated include bulk and shear viscosity, free volume, local density fluctuations, local elastic constants, stress tensor, etc.

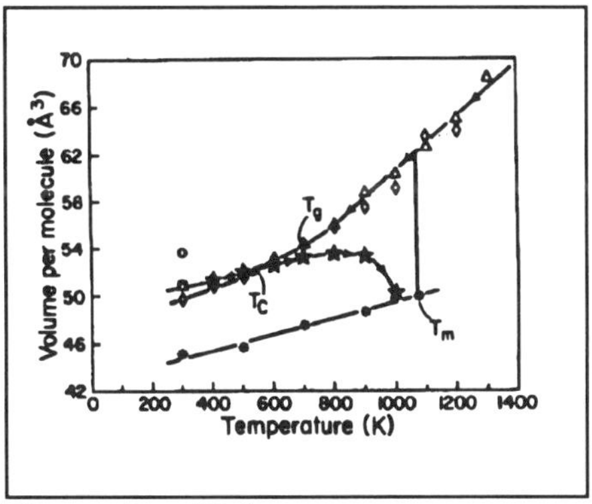


**Figure 3** - Volume vs temperature curves for cooling of NaCl in constant pressure MD simuations of Kumta [85].

Returning to the calculation algorithm, let us examine the various steps and assumptions of the simulation, in order to gain insight into the capabilities and limitations of the method. The Verlet algorithm is probably the most popular approach for calculating particle trajectories. Although it is generally not as accurate as the Predictor-Corrector (PC) methods, it is much faster. The Verlet algorithm uses a combination of Taylor series expansions to calculate the position of a particle in terms of its two previous position calculations, as follows:

$$x(t+\Delta t) = 2x(t) - x(t-\Delta t) + [\frac{d^2x(t)}{dt^2}](\Delta t)^2 + O[(\Delta t)^4] \tag{8}$$

The expansion truncation error varies as $(\Delta t)^4$ where $\Delta t$ is the time interval. The velocity is generally calculated for the middle of the interval:

$$v(t+\frac{\Delta t}{2}) = \frac{x(t+\Delta t) - x(t)}{\Delta t} \tag{9}$$

or

$$v(t) = \frac{x(t+\Delta t) - x(t-\Delta t)}{2\Delta t} \tag{10}$$

This algorithm is simple and stable for moderately large time steps, [72-74] and it requires the storage of only 3 vectors per particle. Typical time steps, $\Delta t$, are about 0.1 to 10 femtoseconds.

The Predictor-Corrector methods use a variety of higher order derivatives to calculate the Taylor series approximations for the particle positions and their derivatives. Then, the particle positions and velocities are corrected by the use of corrector expressions associated with the specific predictor chosen. This method is potentially more accurate than the Verlet Algorithm, but it generally requires more time to perform, and generally requires the storage of 6 or more vectors per particle. The various equations used for the PC methods are available in numerous papers [75,76], including Haile [1].

Both techniques have been used extensively. We find that investigators using Lennard-Jones force functions tend toward the PC method while those using Born-Mayer functions tend toward the Verlet algorithm. This is generally due to the size of the ensembles. Lennard-Jones ensembles are smaller due to the shorter range of the force. We refer the reader to Haile's book [1] for an excellent discourse of both approaches.

The second derivative of the particle position is the acceleration calculated from the sum of forces on each particle. We will first examine the summation over all particles and then the force function itself.

The summation covers all the particles in the solid and is infinite in principle. However, all force functions eventually become negligible with separation distance, therefore only those particles within a specified range from each particle must be considered. In most studies, even this distance is too large and force truncation terms are used to drive the force function to zero at some reasonable separation. If this is done sufficiently smoothly, then fluctuations in energy resulting from the impulse received by particles entering the effective force range are minimized.

Several truncation techniques have been employed, each adding one or two adjustable parameters to the force function. Among the most popular are those of Ewald and Soules. The Ewald sum uses a cubic crystalline lattice to calculate the truncation term based on the fact that the sum of forces from regularly alternating charges on an ionic lattice exhibit a short range characteristic (i.e. the lattice far away averages to a negligible contribution). The Ewald sum truncation [87], as modified by Woodcock and Singer [88], and adapted to non-cubic geometry by Garofalini [89], adds an error function to the potential energy term of the form:

$$T_E = erfc\left(\frac{r}{\beta_{ij}}\right) \tag{11}$$

For a non-cubic crystal, there are two empirically adjustable parameters, $\beta_{ij}$. The Soules approach [24] uses an empirical truncation term on the force function. It also has two adjustable parameters, $\mathbf{R}_{\mathbf{max}}$ and the exponent **n**:

$$T_S = \left[1 - \left(\frac{r}{R_{max}}\right)^n\right] \tag{12}$$

The Soules truncation term has a microscopic basis, as it results from a sum of spherical muffin-tin potentials from isotropically distributed point charges. It is quite acceptable for amorphous structures, but may yield isolated charge clusters in crystals [24].

In principle, if the range of the truncation function is sufficiently large, its form will have little effect on the calculations. However, this is not generally the case. Studies of fluorozirconate glasses using the ionic Born-Mayer force with a large Zr-Zr repulsion term showed a strong sensitivity to the range of truncation.

Short truncation ranges (5.5 Å) exhibited an accumulation of Zr atoms about each other between 4.5 and 5 Å. A range of 9Å was required to eliminate the effect of the truncation process on the Zr-Zr pair correlation function [19,90]. However, using the same range on silicates yielded drastically altered radial distribution functions and a range of 5.5Å was found more suitable (see Fig. 4) [27].

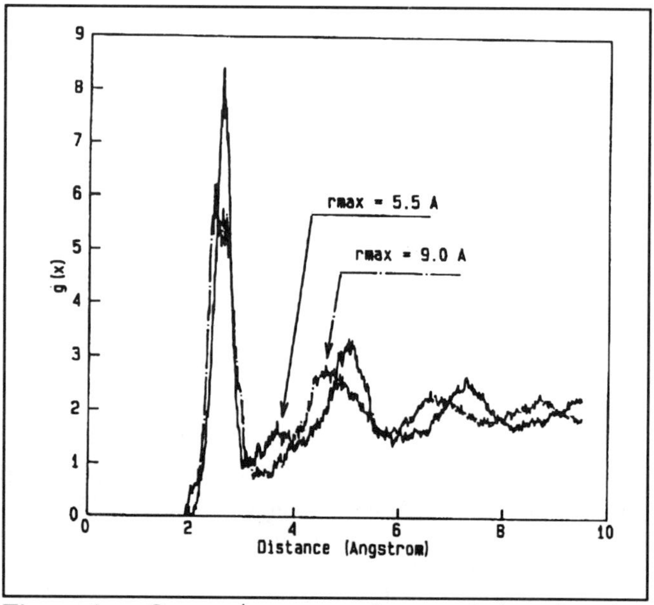


**Figure 4** - Oxygen/oxygen pair correlation function showing the difference between truncation ranges in MD studies on silica glass [27]

These results show that the truncation process cannot be developed independently from the force function and must be an integral part of the force optimization process.

An added benefit of truncating the force function is that by limiting the number of particles with which an atom interacts, the computational time is greatly reduced. Also, this allows the formation of neighbor lists about each particle which only require updating every 10 or 20 time steps.

A multitude of force functions has been used in MD studies. The most critical step in an MD simulation is the adequate choice of a function representing the interparticle force or potential energy. The potential form chosen will depend on the material to be simulated and to some degree on the properties to be measured for the system to be studied. An increased complexity in the potential function will require more empirically adjustable parameters which in turn increases the difficulty of interpretation in the analysis of the simulation output. This is especially true in case of simulations involving the measurement of mechanical and dynamical properties of a system. The selection of the force or potential function also depends on the main objective of the simulation. For example, an interesting objective is the calculation of structures resulting from the random stacking of hard spheres [91,92]. The appropriate potential function is simply:

$$\begin{aligned} U_i(r) &= \infty \qquad \text{for } r < \sigma_i \\ U_i(r) &= 0 \qquad \text{for } r > \sigma_i \end{aligned} \tag{13}$$

where $\sigma_i$ is the diameter of the ith particle. (Because of the form of the potential, hard spheres are treated by a different algorithm from that presented here. The interested reader is referred to several publications on the subject [1,91,92].) A more interesting question might relate to the properties of systems with softer repulsion potentials with or without attraction components.

A simple yet useful potential function is the Lennard Jones (LJ) potential [93]:

$$U_i(r) = kC_i[(\frac{\sigma_i}{r})^n - (\frac{\sigma_i}{r})^m] \tag{14}$$

where the first term is the repulsion component of the potential, avoiding particle overlap and the second term is the attraction component:

$$\mathbf{k} = [n/(n-m)][n/m]^{m/(n-m)}.$$

The exponents are generally large making both terms of the potential short range. The most popular form of this potential sets the attraction term exponent, **m**, to 6, following the dispersion of the London force (Van Der Waals fluids) [94], and the repulsion term empirically to 12 (e.g., the repulsion term exponent for a Xenon core [$5s^2\,5p^6$] in the Born Potential). In the model, **k** = 4, $C_i$ is the energy at the minimum in the potential and $\sigma_i$ is the spatial position of the minimum. Both unary and binary systems have been studied and have revealed interesting phase diagrams. The popularity of the Lennard Jones (LJ) potential function stems from its simplicity and the fact that its form offers a first order representation of the essential components of more realistic potentials. In general it displays characteristics of materials whose behavior is dominated by geometric considerations without much effect from the details of the force function. This clearly applies to noble gases such as argon. However, the LJ potential has also accurately described numerous processes, notably, rheological properties of liquids [26], glass formation [95], shear thinning effects in molten glasses [67] and surface melting effects in crystals [68]. The LJ potential also benefits from analytical tractability so that closed form solutions can be compared to the results of simulations [71,96].

A slightly more complex potential form is found in the Born-Mayer-Huggins

(BMH) potential [97], for which the attraction term is the Coulomb potential and the repulsion term has the form of a Yukawa potential.

$$U_{ij}(r) = A_{ij}\, e^{-r/\rho} + [z_i z_j e^2/r] \tag{15}$$

where $z_i$ is the ionic valence, and $\rho$ is a softness parameter in the core repulsion representing the effect of core overlap. The parameter $\rho$ has a strong effect on the system behavior and is commonly assumed to be 0.29 Å for oxide glasses. The parameter $A_{ij}$ is written according to Huggins-Mayer as:

$$A_{ij} = b(1 + z_i/m_i + z_j/m_j)\exp[(\sigma_i + \sigma_j)/\rho] \tag{16}$$

where **b** is a constant, typically $0.19 \times 10^{-12}$ ergs, $m_i$ is the number of valence electrons and $\sigma_i$ is the ionic size parameter of the ith species. This potential has a somewhat harder core than LJ, and a much larger interaction range. It has been very successfully used to simulate the structure and dynamics of ionic glasses. With the truncation discussed above, examples of the fits of radial distribution functions from silica glass and from zirconium-barium fluoride glass are given in Figs. 5 and 6. The forces used in these figures were derived from the BMH potential for a single component ionic system, which means one cation and one anion, therefore, 6 adjustable parameters were employed, including those needed for truncation. The silica simulation was obtained by Soules [24] and the zirconium fluoride simulation by Simmons [19] using the following force function:

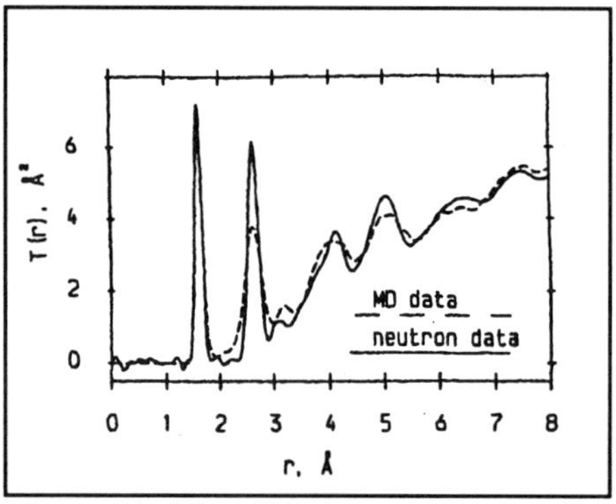


**Figure 5** - Fit of MD generated radial distribution function [24] to neutron scattering data [86] using the interatomic force function of Equ. 26.

$$F_{ij}(r) = (A_{ij}/\rho)\exp\left(-\frac{r}{\rho}\right) + \frac{z_i z_j}{r^2} \tag{17}$$

where $A_{OO} = 2.3 \times 10^{-9}$ ergs, $A_{OSi} = 3.0 \times 10^{-9}$ ergs, $A_{SiSi} = 3.3 \times 10^{-9}$ ergs, $R_{max} = 5.5$ Å $\rho = 0.29$ Å, $n = 3$.

Garofalini uses a BMH potential with the Ewald truncation which requires an additional adjustable parameter:

where $\mathbf{A}_{OO}$ = 0.72x10$^{-9}$ ergs, $\mathbf{A}_{OSi}$ = 2.96x10$^{-9}$ ergs, $\mathbf{A}_{SiSi}$ = 1.88x10$^{-9}$ ergs, $\rho$ = 0.29 Å, $\beta_{OO}$ = 2.55 Å, $\beta_{OSi}$ = 2.60 Å, $\beta_{SiSi}$ = 2.53 Å.

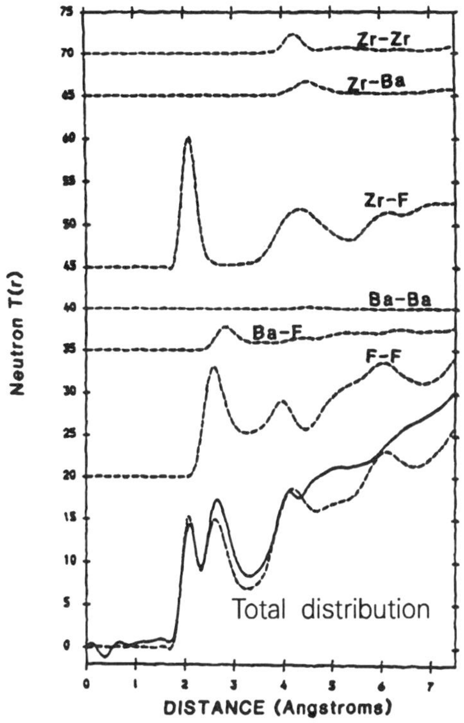


**Figure 6** - Fit of MD generated radial distribution function for $Zr_2BaF_{10}$, reprinted from [19].

$$U_{ij}(r) = A_{ij}\exp(-\frac{r}{\rho}) + \frac{z_i z_j e^2}{r}\, erfc(\frac{r}{\beta_{ij}}) \qquad (18)$$

More complex potentials have been developed to represent 3-body interactions, as would be expected from system with some degree of covalent bonding. Structures like diamond, silicon, germanium, etc. require more than the 2-body potentials shown above. Keating and others have used combinations of 2-body and 3-body potentials to describe the interactions of partially covalent tetrahedral structures.

Keating Potential [98]:

$$U_{ij}(r) = \frac{3}{16}\frac{\alpha}{d^2}(r^2-d^2) \tag{19}$$

$$U_{ijk} = \frac{3}{8}\frac{\beta}{d^2}(r_{ij}r_{jk}\cos\theta_{ijk} + \frac{1}{3}d^2)^2 \tag{20}$$

Stillinger-Weber Potential as adapted by Garofalini uses the BMH 2-body potential plus a 3-body term [99,100]:

$$u_{ijk} = h(r_{ij},r_{ik},\theta_{jik}) + h(r_{jk},r_{ji},\theta_{kji}) + h(r_{ki},r_{kj},\theta_{ijk}) \tag{21}$$

$$h(r_{ij},r_{ik},\theta_{jik}) = \lambda_i(\cos\theta_{jik} - \cos\theta_{jik}^c)^2 \exp[\frac{\gamma_i}{(r_{ij}-r_i^c)} + \frac{\gamma_i}{(r_{ik}-r_i^c)}] \tag{22}$$

for $\mathbf{r}_{ij} < \mathbf{r}_i^c$ and $\mathbf{r}_{ik} < \mathbf{r}_i^c$ only. The function $\mathbf{h}(\mathbf{r}_{ij},\mathbf{r}_{ik},\boldsymbol{\theta}_{jik}) = \mathbf{0}$ otherwise. The additional constants are $\lambda_{Si} = 18\times10^{-11}$ ergs, $\lambda_O = 0.3\times10^{-11}$ ergs, $\gamma_{Si} = 2.6$ Å, $\gamma_O = 2.0$ Å, $\mathbf{r}_{Si}^c = 3.0$ Å, $\mathbf{r}_O^c = 2.6$ Å, $\rho = 0.29$ Å, $\cos\theta_{O\text{-}Si\text{-}O} = -1/3$, $\cos\theta_{Si\text{-}O\text{-}Si} = -1/3$. The latter is based on the tetrahedral structure of ice, resulting from the effect of the 2 lone pair electrons on the oxygen.

The use of combinations of 2 and 3 body potentials for a unary ionic system requires 15 parameters for $SiO_2$. Garofalini and Soules have fitted the radial distributions of fused silica with combinations of 2 and 3 body potentials and the results are shown in Fig. 7. Garofalini [100] finds that by forcing the O-Si-O and Si-O-Si angle terms to 109.471°, the oxygen-oxygen pair correlation function was significantly narrowed as required by the experimental results (see Fig. 7), and the Si-O-Si bond angle distribution was broadened and shifted to smaller angles from the 2-body only simulation (Fig. 8). Of greater significance is the elimination of the 2-body bond angle peak at 105° presumably due to 5-coordinated silicon ions.

Traditionally, simulations of metals have used the pair-wise Morse potential with occasional modifications.

Morse potential: $\mathbf{U_{ij}(r) = U_0 \{exp[-2a(r-r_0)] - 2exp[-a(r-r_0)]\}}$ **(23)**

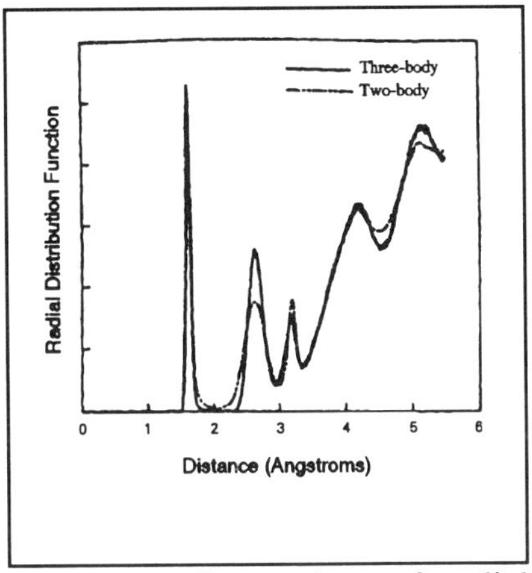


**Figure 7** - Comparison of radial distribution functions calculated with 2-body (Equ.18) and 3-body (Equs. 18,21,22) potentials on 648 atoms. Reprinted from [100].

Lennard-Jones and the Johnson potential (originally used for simulating iron) have also received some use.

Johnson Potential: $\mathbf{U_{ij}(r) = - A_{ij}(r-B_{ij})^3 + C_{ij}r - D_{ij}}$ **(24)**

However, most efforts rely on the addition of artificial terms, generally ill-defined, to represent the material's cohesion through what could be called *volume-dependent* terms. Daw and Baskes [101] recently developed a many-body description of the cohesion energy of metals using density functional theory and produced the embedded-atom method, later adapted to treat a wide range of alloys. The approach is semi-empirical in that the method uses parametrized functions to describe metallic bonding whose forms were obtained from fundamental theory. The parameters are defined from experimental data. The cohesive energy term is the sum of the embedding energies calculated from the free atomic electron densities of the surrounding atoms.

$$\Phi_{ij}(r) = \sum_i F_i(\rho_i) + 0.5 \sum_i \sum_j U_{ij}(r) \qquad (25)$$

where **F** is the embedding energy, $\boldsymbol{\rho}$ is the spherically averaged atomic density and $\mathbf{U_{ij}(r)}$ is the electrostatic two-body interaction. The background density for each atom is determined by evaluating at its nucleus the superposition of atomic density tails from its neighbors. The method has been successfully applied to structural defect and surface studies of FCC metals and alloys [102,103].

Needless to say that this list is not nearly exhausted. For example, Vashista [104] has used a modified 2/3-body potential to simulate the crystalline forms of silica in the melt and Vessal et al.[105] have modeled vitreous silica with a sum of power series.

With the large number of adjustable parameters and the infinite variety of acceptable forms for the potential or force function any reasonable reader would quickly abandon this subject. However, this is where common sense and experimental verification play a critical role. Anyone can write a molecular dynamics program, or get a working program from a colleague. What distinguishes successful simulations are: the nature of the questions asked of the computation, the selection of parameters for the computation and the experimental support used for interpretation of the results.

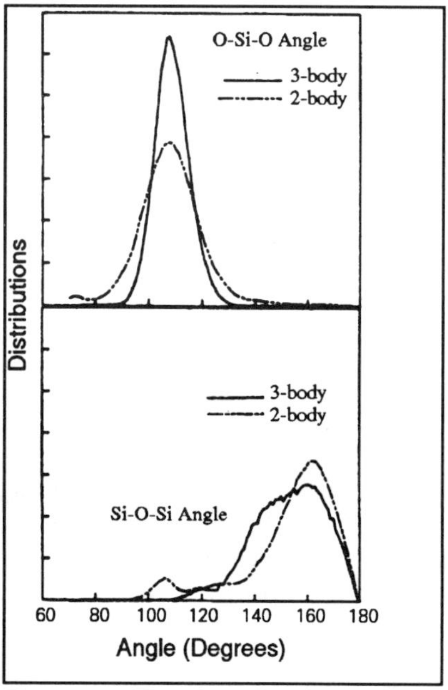


**Figure 8** - Bond angle distributions for the (a) O-Si-O and (b) Si-O-Si angles. Reprinted from [100].

*Tests of MD Methodology for Realistic Simulations of Materials Behavior*

The first test of the simulation includes comparing the various macroscopic characteristics of the system with results from experiments on the targeted physical system. Good agreement spurs an improvement in the modeling, while disagreement with data can lead to a re-examination of the self-consistent rules or parameters selected. In such studies, however, great care must be taken to compare simulations and experimental results, especially, diffraction data, in the correct manner.

The main problem with the analysis of diffraction data from amorphous solids is one of uniqueness in that, even if a perfect fit is obtain with experiment, there is no guarantee that there are no other models or simulations that would fit equally well. For this reason, it is essential to consider all reasonable alternative approaches, particularly, those already proposed in the literature for the material under investigation.

Agreement with diffraction data is therefore a necessary but not sufficient criterion for any valid structural model. Comparisons of the simulations with other experimental results are essential to establishing credibility for the proposed simulation if the goal is to represent the behavior of a specific material.

For example, the density must equal the known material density in order to obtain a reasonable facsimile of the material's structure. The elastic constants are not as critical unless the mechanical properties of the material are sought. Accurate reproduction of the elastic constants, however, is very difficult since it involves adjusting the overlapping core softness parameter in the force function. Generally, reasonable models will reproduce reasonably well the vibrational frequency spectrum of the material.

The vibrational density of states **D(ω)** of an MD system can be obtained from a Fourier Transform of the velocity autocorrelation function, **γ(t)** [106]:

$$\gamma(t) = \frac{\langle \sum_i v_i(t_0) \cdot v_i(t_0 + t) \rangle}{\langle \sum_i v_i(t_0) \cdot v_i(t_0) \rangle} \tag{26}$$

with the following relationship:

$$D(\omega) = \int_0^\infty \gamma(t) \cos(\omega t)\, dt \tag{27}$$

The IR absorption spectrum is obtained from a Fourier Transform of the time correlation of the dipole moment.

In order to minimize termination errors in the numerical integration, a variety of windowing and smoothing functions are applied. However, there are serious problems with box size and investigators must generally ignore the low frequency (long wavelength) end of the spectrum. Recent studies of beta cristobalite by Kieffer [107] show the presence of the 1200 and 800 $cm^{-1}$ modes, but the lower frequency modes are strongly box-size dependent. Also if the modulus is too high, then the low frequency modes can be severely upshifted. Garofalini [25] has conducted studies of vitreous silica and reported vibrational modes at 400,800 and 1100 $cm^{-1}$, in comparison with 450 ($\upsilon_1$ TO), 495 ($\upsilon_1$ LO, $D_1$), 606 ($D_2$), 800 ($\upsilon_3$ LO & TO), 1050 ($\upsilon_4$ TO) and 1200 ($\upsilon_4$ LO) $cm^{-1}$ from Raman [108], 465, 800 and 1100 $cm^{-1}$ from IR absorption data, 350, 800, 1080 and 1200 $cm^{-1}$ from neutron diffraction, and 410, 730 and 1040 $cm^{-1}$ from the stochastic simulation of Bell and Dean [18].

The next step involves a comparison of the radial distribution function and/or values of the interatomic distances. By altering the repulsion term in the force function (more specifically, the size parameters, $\sigma_i$, the density and the core softness factor, $\rho$), the interparticle distances are adjusted to match those of the real material. This is done by a variety of methods. Wright [86, 19] recommends matching not only the peak positions in the real space correlation function, but also peak shapes and peak areas. This requires accounting for the effects of thermal vibrations for static simulations (ie. computer relaxed random networks) and, in real space, folding the model component correlation functions, $\mathbf{t_{jk}(r)}$ with the correct component peak functions, $\mathbf{P_{jk}(r)}$, which define the resolution in real space for the experiment in question. Otherwise the comparison between model and experiment is not useful. Note that the peak functions $\mathbf{P_{jk}(r)}$ have satellite features on each side of the central maximum which depend on the exact form of the modification function, **M(Q)**, and these can lead to features in the real space correlation function **T(r)**. For neutron diffraction studies, the correlation function, $\mathbf{T^N(r)}$ is written as follows [see A. Wright article in this publication]:

$$T^N(r) = \sum_j \sum_k \int_0^\infty t_{jk}(r') \, [P_{jk}(r-r') - P_{jk}(r+r')] \, dr'$$

$$T^N(r) = 4\pi r \rho^0 \sum_j b_j + \frac{2}{\pi} \int_0^\infty Q I^N(Q) M(Q) \sin(rQ) \, dQ \qquad (28)$$

The **j** summation is taken over the atoms in one composition unit and that for **k** over all atoms. $\rho^0$ is the average number density of the sample; $\mathbf{b_j}$ is the isotopically averaged neutron scattering length of atom **j**; $Q = (4\pi/\lambda) \sin\theta$, with $\lambda$ as the incident radiation wavelength, and $2\theta$ as the scattering angle; $\mathbf{Q\ I^N(Q)}$ is the interference function. This kind of analysis was used in comparing the correlation functions from the simulations and the neutron diffraction data in Figs. 5 and 6.

Resolution effects are also important in reciprocal space and typically involve both simulated and experimental interference functions. In calculating the interference function for a real-space model without periodic boundary conditions, it is usually necessary to truncate the simulated correlation function at some value $\mathbf{r_{max}}$. This leads to a broadening of the model interference function and the introduction of satellite features in reciprocal space in the same way as for an experimental correlation function in real space. Thus, in order to compare simulation and experiment in reciprocal space, it is necessary to fold the **experimental** interference function with the reciprocal space peak function **P(Q)** appropriate to the model calculation. Similarly, the simulated interference function must be broadened by the reciprocal space resolution function for the instrument used in the experiment.

Finally, it is important to remember that the real and reciprocal space representations of the data, **T(r)** and **Q I(Q)**, emphasize different aspects of any given structure and it is necessary to examine any deficiencies between simulation and experiment in both spaces (for example, a layered structure will give a well-defined peak in reciprocal space, but shows less obvious features in **T(r)**).

MD calculations are constrained by two characteristic times: the time step and the duration of the run. The time step must be small because of the assumption of constant acceleration on each particle for the duration of each time step (Verlet algorithm). This is not a problem when the particles are well separated, but can cause enormous difficulties during interparticle collisions. A great amount of energy can be arbitrarily pumped into the system if the time step is too long. In general, most investigators choose a time step such that thermal vibrations last 50-200 time steps (i.e. $\Delta t \sim 0.1$ - 10 fs). The difficulty with choosing too small a time step is that it may take too many steps to describe a relaxation process in the materials during a reasonable computer run. In general, most investigators with small minicomputers can conduct 1000-10,000 time steps in a reasonable time. Therefore, only processes with characteristic times much less than 1000 $\Delta t$ - 10,000 $\Delta t$ can be studied.

The question of box size or number of atoms is often ignored and can cause serious misinterpretation of behavior. For example, it is well known that too low a number of atoms with periodic boundary conditions can enhance crystallization kinetics. The box should be large enough so that a particle's image in another box is too far away to interact with itself. Therefore, the minimum box size is dependent on the range of the interatomic forces. Simmons et al. (19) using a screened Born-Mayer potential showed that a screening range of 9 Å was required to adequately model the Zr-Zr repulsion in fluorozirconate glasses. If one requires images to be at least two 9-angstrom screening distances apart in order to avoid sharing neighbors, then the minimum number of atoms in the box with periodic boundaries should be about 480 atoms.

The rate and the means of cooling from the equilibrium liquid state to form the glass can have a great influence on the structure developed. Because of the requirements on the size of the time step, it is not possible to simulate the cooling rate of real materials with MD calculations. MD glasses are cooled very fast and have correspondingly high fictive temperatures. The most often used cooling technique involves both a reduction in temperature and in volume to achieve a reasonable density at room temperature. Also possible, as discussed above, are methods of cooling at constant pressure [77,78], which yield a noticeable glass transition hysteresis [24,77,85]. However, due to the rapid cooling rates employed in all these methods, the structure simulated can only resemble that of real glasses if the viscoelastic process scales in time over about 13 orders of magnitude (relaxation times of 10 ps for the MD glass and 100s for the real glass). The interesting aspect of glass science that one uncovers when considering this question is that the viscoelastic behavior of some glasses ($SiO_2$, $GeO_2$) *does scale over 10-14 orders of magnitude* (Angell's strong glasses [109]). Most glasses, however, show non-Arrhenian viscosity with only piecewise scaling over 3-4 orders of magnitude (Angell's fragile glasses [109]).

## EXAMPLES OF APPLICATIONS

There have been so many successful applications of MD studies for modeling the structure and dynamics of solids that a complete list would be a feat in itself and a description of the investigations would easily fill several books. Instead, we have decided to give a few examples of research studies with which we have accumulated much experience.

### High Shear Rate Studies of Viscous Flow in Liquids

The flow behavior of inorganic liquids at high shear rates has long been of interest to the glass processing industry [110]. While viscous liquids have generally exhibited Newtonian behavior (strain rate or stress independent viscosity), organic materials have undergone large deviations both in the thixotropic and dilatant directions. Since a correspondence between structural change and the non-Newtonian behavior has not been determined for these materials, it has generally been surmised that the non-Newtonian behavior resulted from complex chain kinetics, such as unfolding, stretching, cross-linking, etc. in the organic materials. Since most inorganic glasses do not have complex chain-like structures, it was assumed that non-Newtonian behavior should not be expected.

Experimental studies supported by molecular dynamics simulations have profoundly altered this picture and drastically changed the understanding of viscous flow at high shear rates in simple liquids [67]. The experimental studies demonstrated a clear occurrence of non-Newtonian shear flow in inorganic glasses whereby the viscosity of the liquid decreased drastically at high shear rates. The effect appeared to be relatively insensitive of the exact composition of the liquid. MD studies conducted with a *Lennard-Jones* interaction potential demonstrated the same effect. Figure 9 shows a comparison of the results of the MD study and the experimental investigations [67]. The behavior of the shear thinning process as a function of shear rate appears identical in all 3 systems analyzed. While the time scales of the experiments were significantly different from the simulation, the features of the observed process were qualitatively identical, including a stress overshoot which was observed at high shear rates. It is however very clear that the Lennard-Jones potential has no hope of reproducing adequately the structure of any oxide glass in a simulation. Therefore, the occurrence of shear thinning effects in the MD simulation indicate that the shear thinning process is a fundamental characteristic of flow in simple liquids, and does not require complex structural dynamics.

Experimental studies of the shear thinning process in oxide glasses have continued [26]. However, investigators have had to turn to the MD investigation to seek a structural explanation of the process [111]. This was made possible by the ability to examine the atomic structure of the simulated system at any point in the investigation. Figure 10 reveals the results of the MD examination. The material appears to form distinct layers under shear deformation, which layers facilitate flow. While the layering effect has not yet been demonstrated experimentally, the highly simplified MD study has provided both the impetus for the experimental work and has provided the structural atomistic mechanism responsible for the observed macroscopic behavior.

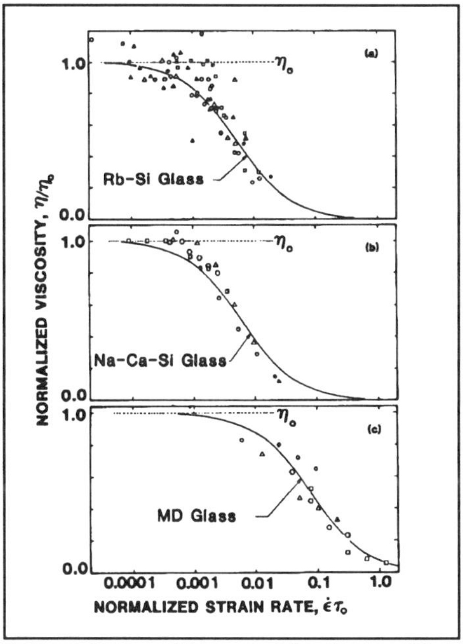


**Figure 9** - Shear thinning effect measured in 2 real glasses and a Lennard-Jones (Equ.14) MD glass. Reprinted from [67].

## Brittle Fracture Processes in Silica

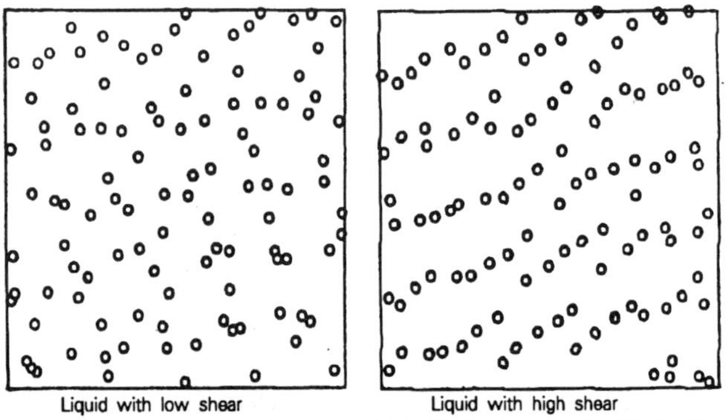

**Figure 10** - At low shear rates, the structure is random. At high shear rates, the structure shows significant layering and pronounced shear thinning [111].

The study of brittle fracture in solids has generally resisted microscopic scrutiny, because of the difficulty with accessing the atomic scale during a fracture event. Furthermore, the process is so fast that even if adequate experimental tools were available, the recording of the event would be difficult. Today, the discovery of the fractal nature of crack surfaces has allowed the extension of macroscopic structural features to the microscopic scale through self-similarity [112].

Investigations of brittle fracture in silica by molecular dynamics have been conducted by several investigators [27,113,114]. The results begin to form a coherent description of the dynamic processes associated with the accumulation of strains at the atomic scale which eventually lead to bond breakage, and their collective growth to cause failure or fracture of the solid. MD studies have shown that during the period of stress build-up, there occurs a significant strain concentration at regions of reduced local density (pre-existing flaws) by means of atomic rearrangements from thermal motions of the atoms associated with phonon propagation, defect diffusion or any rapid mass transport process. These structural rearrangements precede failure and act to weaken the loaded structure. These effects are seen to occur as fast as the speed of sound and create failure zones ahead of the propagating crack. The act of bond breakage has also been investigated and appears to result from sudden atomic motions associated with the reconstruction of the free surface [115]. Fracto-emission has also been observed to occur in the MD studies [116].

The effect of adsorbed water during slow crack growth in silica has also been analyzed with quasi-empirical quantum-mechanical studies. A strained tetrasiloxane ring is first modeled and then brought into the proximity of a water molecule. The water molecule is seen to polarize along the strained bond while the ring rearranges its structure to increase the coordination of the Si atom in the strained bond and then break the bond to form two hydroxyl groups. The energetics of the reaction can be calculated and show a significantly lowered activation energy for bond breakage [47].

## Modeling of Surfaces and Interfaces

Recent studies with modified BMH 2/3-body potentials have led to an elucidation of the formation of surfaces in silica, with a predominance of small rings of tetrahedra [28], and the development of dynamics of sol-gel polycondensation reactions [29]. Further extensions of these studies have investigated the formation of thin films and the adsorption of water molecules on glass surfaces [117].

Studies with LJ and EAM potentials have looked at interface formation between grains of a bicrystal of Cu and Si, the behavior of twist boundaries near the melting point, melting effects in small Si clusters, and variations in elastic constants across grain boundaries [118,119]. Grain boundary equilibration was successfully demonstrated by heating the MD structure with 2-dimensional periodic boundaries (in directions parallel to the boundary surface), using the constant stress tensor method of Parrinello and Rahman [78] which allows for a distortion of the cell. Due to the unique equilibration scheme used, the interface was simulated at high temperatures without instability problems found when the periodic boundary conditions were applied in 3 dimensions, since the latter can cause the boundaries to interact with their own images. Below the melting point, grain boundary migration was observed to occur by the transition to an amorphous state (transient melting) of the plane adjacent to the boundary and subsequent recrystallization in the crystal symmetry of the other grain.

## Hidden Structure in Liquids

Molecular Dynamics studies can also reveal important thermodynamic quantities, following an analysis first developed by Stillinger and Weber [120]. Following this method, the particle positions derived in an MD simulation at a finite temperature are referred, essentially uniquely, by means of a method of steepest descent, to relative minima on a potential energy surface in 3N-dimensional configurational space. This methods allows separation of the thermal excitations from the temperature independent packing statistics which characterize the simulated solid. The result of these analytical quenches reveal that an unexpected degree of polycrystalline order resides under the cloak of disorder introduced by the thermal fluctuations in atomic positions. This approach makes possible comparisons of various developed structures from simulations, to determine relative configurational energies, in order to seek the relationship between various local minima representing the amorphous structures. Thus, simulated configurations corresponding to local and global minima may be distinguished.

Similar studies of the melting of an FCC crystal using a truncated Lennard Jones potential with 5-12 exponents, showed that significant portions of the potential energy surface exhibited a negative curvature at temperatures as low as 1/4 the melting point. Furthermore, by examining the quenched configurations at the melting point, it was discovered that the dominating defects produced in melting were vacancy-interstitial pairs [121].

## FURTHER STUDIES

Molecular dynamics studies have only begun to make their impact on the development of understanding of the relationship between atomic structure and macroscopic behavior. The greatest strength of the MD approach which is its ability to yield *dynamic atomistic behavior* has received very little attention to date. When investigators do study dynamic behavior, they most often only analyze ensemble averages and correlation functions and overlook the most exciting result of the MD simulations which is information about the motions of individual atoms and groups of atoms. Analyses of collective atomic dynamics are, however, very difficult and often require visualization of the moving system. The human eye seems best able to integrate the visual field and discern the presence of collective motions. Therefore, a powerful analysis tool is the seldom used visualization system.

The major objection to MD studies has been the use of classical forces when either electronic or covalent bonding characteristics play a major role in defining the cohesive energy of the material. Furthermore, the use of 2-body potentials greatly simplifies the complex interactions which occur in real materials. A variety of investigators have developed force functions which produce potential energy surfaces obtained from quantum mechanical (QM) calculations [36,122,123]. These clearly offer an improvement in the formulation of the interaction, especially in studies of local structural defects and small ensembles of atoms. However, even the QM approaches have limitations and the reader is reminded of the profound advances in understanding obtained by use of very simplified force functions, when the characteristics of interest are not as sensitive to the form of the individual interatomic interaction.

Finally, computer speed is not a panacea. Clearly, the faster machines allow more complex systems to be studies in larger ensembles. However, when the calculation of interest can be done faster than the analysis and interpretation of the results, then little is gained by a faster machine.

## BIBLIOGRAPHY


[1] J.M. Haile, *Molecular Dynamics Simulation*, John Wiley, New York NY 1991.

[2] J. L. Casti, *Alternate Realities*, Wiley-Interscience, New York, NY 1989.

[3] J. Sauer, "Molecular Models in Ab Initio Studies of Solids and Surfaces: from ionic crystals and semiconductors to catalysts," *Chem. Rev.* **89**, 199-255 (1989).

[4] M.H. Kalos and P.A. Whitlock, *Monte Carlo Methods. 1: Basics*, Wiley and Sons, New York, 1986.

[5] K. Binder, ed., *Monte Carlo Methods in Statistical Physics*, 2nd ed., Springer-Verlag, Berlin, 1986.

[6] K. Binder, ed., *Applications of the Monte Carlo Method in Statistical Physics*, Springer-Verlag, Berlin, 1984.

[7] F. Vesely, *Computerexperimente an Flussigkeitmodellen*, Physik Verlag, Weinheim, 1978.

[8] W.G. Hoover, *Molecular Dynamics*, Lecture Notes in Physics, **258**, Springer-Verlag, Berlin, 1986.

[9] D.J. Evans and G. Morriss, *Statistical Mechanics of Nonequilibrium Liquids*, Academic Press, New York, 1990.

[10] M.P. Allen and D.J. Tildesley, *Computer Simulation of Liquids*, Clarendon Press, Oxford, 1987.

[11] H. Gould and J. Tobochnik, *An Introduction to Computer Simulation Methods: Applications to Physical Systems*, Addison Wesley, Reading MA, 1988.

[12] G.T. Evans, "Liquid State Dynamics of Alkaline Chains," in *Molecular-Based Study of Fluids*, J.M. Haile and G.A. Mansoori, eds. ACS Adv. Chem. Ser., **204**, 423 (1983).

[13] T.F. Soules, "Stochastic and Molecular Dynamic Models of Glass Structure"; in *Glass Science and Technology* **4A**. Edited by Kreidl and

Uhlmann, Academic Press, New York (1990).
[14] S. Brawer, *Relaxation in Viscous Liquids and Glasses*, The American Ceramic Society, Columbus, OH, 1985.
[15] G. Ciccotti and W.G. Hoover, eds., *Molecular-Dynamics Simulation of Statistical-Mechanical Systems*, North-Holland, Amsterdam, 1986.
[16] J.D. Bernal, *Proc. Roy. Soc. (London)* **A280**, 299 (1964).
[17] J.L. Finney, "Random Packings and Structure of Simple Liquids. 1. Geometry of Random Close Packings," *Proc. Roy. Soc. (London)* **A319**, 479, (1970).
[18] R.J. Bell and P. Dean, "The Structure of Vitreous Silica: Validity of the Random Network Theory," *Phil. Mag.* **25**, 1381 (1972).
[19] J.H. Simmons, C.J. Simmons, R. Ochoa and A.C. Wright, "Fluoride Glass Structure," in *Fluoride Glass Fiber Optics*. Edited by I.D. Aggarwal and G. Lu. Academic Press, New York, 1991.
[20] R. Fletcher and C.M. Reeves, "Function Minimization by Conjugate Gradients," *Computer J.* **7**, 149-54 (1964).
[21] I. Yasui and H. Inoue, "Molecular Dynamics Simulations of Changes in Structure in $ZrF_4$-Based Glasses," *J. Non-Cryst. Sol.* **71**, 39-47 (1985); *Mat. Sci. Forum* **19**, 103 (1987).
[22] F.F. Abraham, "Isothermal-Isobaric Computer Simulation of the Supercooled-Liquid-Glass Transition Region in The Amorphous Solid FCC," *J. Chem. Phys.* **72**, 359-65 (1980).
[23] S.A. Brawer and M.J. Weber, "Monte Carlo Simulation of $Eu^{+3}$-doped $BeF_2$ glass," *Phys. Rev. Lett.* **45**, 460 (1980).
[24] T.F. Soules, "Computer Simulations of Glass Structures," *J. Non-Cryst. Sol.* **123**, 48-70 (1990).
[25] S. H. Garofalini, "Molecular Dynamics Simulation of the Frequency Spectrum of Amorphous Silica," *J. Chem. Phys.* **76**, 3189-92 (1982).
[26] D. Fincham and D. M. Heyes, "Comparison Between Experimental Argon and Lennard-Jones 12:6 Shear Viscosities," *Chem. Phys.* **78**, 425-441 (1983).
[27] R. Ochoa, T.P. Swiler and J.H. Simmons, "Molecular Dynamics Studies of Brittle Failure in Silica: Effect of Thermal Vibrations," *J. Non-Cryst. Sol.* **128**, 57-68 (1991).
[28] S.H. Garofalini, "Molecular Dynamics Computer Simulation of Silica Surface Structure and Adsorption of Water Molecules," *J. Non-Cryst. Sol.* **120**, 1-12 (1990).
[29] S.H. Garofalini, "Molecular Simulations of Sol-Gel Polymerization," *Bull. Am. Ceram. Soc.* **69**, 1549 (1990).
[30] G.M. Zhidomirov and V.B. Kazansky, "Quantum-Chemical Cluster Models of Acid-Base Sites of Oxide Catalysts"; pp. 131-202 in *Advances*

*in Catalysis*, **Vol 34**, Edited by D.D. Eley, H. Pines and P.B. Weisz, Academic Press, Inc., New York, 1986; also A. Szabo and N.S. Ostlund, *Modern Quantum Chemistry*, McGraw Hill, New York, 1989.

[31] J.A. Pople, "The Molecular-Orbital and Equivalent-Orbital Approach to Molecular Structure," *Quarterly Reviews* **11**, 273-290 (1957).

[32] E. R. Davidson and D. Feller, "Basis Set Selection for Molecular Calculations," *Chem. Rev.* **86**, 681-696 (1986).

[33] K.A.R. Mitchell, "The Use of Outer d Orbitals in Bonding," *Chem. Rev.* **69**, 157-178 (1969).

[34] G.V. Gibbs, "Molecules as Models for Bonding in Silicates," *Am. Mineral.* **67**, 421-450 (1982).

[35] R. Meier and T.K. Ha, "A Theoretical Study of the Electronic Structure of Disiloxane [$(SiH_3)_2O$] and its Relation to Silicates," *Phys. Chem. Miner.* **6**, 37-46 (1980).

[36] A.C. Lasaga and G.V. Gibbs, "Applications of Quantum Mechanical Potential Surfaces to Mineral Physics Calculations," *Phys. Chem. Minerals* **14**, 107-117 (1987).

[37] A.C. Lasaga and G.V. Gibbs, "Quantum-Mechanical Potential Surfaces and Calculations on Minerals and Molecular Clusters. 1. STO-3G and 6-31 G* Results," *Phys. Chem. Miner.* **16**, 29-41 (1988).

[38] M.C. Zerner, in *Reviews in Computational Chemistry*, **Vol. 2**. Edited by K.B. Lipkowitz and D.B. Boyd, VCH, New York (1991).

[39] J. Stewart, in *Reviews in Computational Chemistry*, **Vol. 1**. Edited by K.B. Lipkowitz and D.B. Boyd, VCH, New York (1990).

[40] J. Pople and D.L. Beveridge, *Approximate Molecular Orbital Theory*, McGraw Hill Co., New York, 1970.

[41] A.D. Bacon and M.C. Zerner, "Intermediate Neglect of Differential Overlap Theory for Transition-Metal Complexes - Fe, Co and Cu Chlorides," *Theoret. Chimica Acta* **53**, 21-54 (1979); also M.C. Zerner, G.H. Loew, R.F. Kirchner and U.T. Mueller-Westerhoff, "Intermediate Neglect of Differential Overlap Technique for Spectroscopy of Transition-Metal Complexes - Ferrocene," *J. Am. Chem. Soc.* **102**, 589-99 (1980).

[42] I.R. Levine, *Physical Chemistry*, McGraw Hill Co., New York, 1983.

[43] M.J.S. Dewar, E.G. Zoebisch, E.F. Healy and J.J.P. Stewart, "The Development and Use of Quantum-Mechanical Molecular Models. AM1 - A New General Purpose Quantum-Mechanical Molecular Model," *J. Am. Chem. Soc.*, **107**, 3902-09 (1985). M.J.S. Dewar and C. Jie, "AM1 Calculations for Compounds Containing Silicon," *J. Am. Chem. Soc.* **6**, 1486-90 (1987).

[44] J.D. Head and M.C. Zerner, *Adv. Quantum Chem.* **20**, 241 (1990).

[45] H.B. Schlegel, *Adv. Chem. Physics* **67**, 249 (1987).

[46] P. Jorgensen and J. Simons, *Geometric Derivatives of Energy Surfaces and Molecular Properties*, NATO ASI Series C, **Vol. 166**, Reidel, Boston.

[47] J.K. West and S. Wallace, "Molecular Orbital Modeling of Water Adsorption on a Tetrasiloxane Ring"; in *Better Ceramics Through Chemistry IV.* Edited by Brinker, Scherer, Zelinski and Ulrich, San Francisco, California.

[48] J.K. West and S. Wallace, "AM1 Molecular Orbital Calculations and NIR Spectroscopy on Water Adsorption onto a Tetrasiloxane Ring," *Trans. Am. Ceram. Soc. Symposium on Composites*, Orlando, Florida, 1990.

[49] W. Kohn and L.J. Sham, "Self-Consistent Equations Including Exchange and Correlation Effects," *Phys. Rev.* **140**, A1133-38 (1965).

[50] G.B. Bachelet, D.R. Hamann and M. Schluter, "Pseudopotentials That Work From H to Pu," *Phys. Rev. B*, **26**, 4199-228 (1982).

[51] J.A. Tossel and G.V. Gibbs, "Molecular Orbital Studies of Geometries and Spectra of Minerals and Inorganic Compounds," *Phys. Chem. Minerals* **2**, 21-57 (1977).

[52] R. Carr and M. Parrinello, "Unified Approach for Molecular Dynamics and Density Functional Theory," *Phys. Rev. Lett.* **55**, 2471 (1985).

[53] J.D. Joannopoulos, "Predicting the Structure of Solids," in *Physics of Disordered Materials*, Plenum Press, (1985).

[54] Y.Bar-Yam, S.T. Pantelides, J.D. Joannopoulos, D.C. Allan and M.P. Teter, "The Oxygen Vacancy and the $E_1$' Center in $SiO_2$," *Mat. Res. Soc. Symp.* **105**, 223-8 (1988).

[55] D. C. Allen, M. P. Teter, J. D. Joannopoulos, Y. Bar-Yam and S. T. Pantelides, "Defect Studies in Silicon Dioxide By Local Density Approximation Total Energy Methods," *Mat. Res. Soc. Symp. Proc.* **141**, 255-260 (1989); also, D. C. Allen and M.P. Teter, "Nonlocal Pseudopotentials in Molecular Dynamical Density Functional Theory: Application to $SiO_2$ ," *Phys. Rev. Lett.* **59**, 1136-39 (1987).

[56] D.C. Allen, M. P. Teter, "New Developments In Electronic Structure Theory Applied to Oxides," *Annual Meeting of the Am. Ceram. Soc.*, Dallas, TX, paper 44-SI-90 (1990).

[57] J. K. Rudra and W. B. Fowler, "Oxygen Vacancy and The $E_1$' Center in Crystalline $SiO_2$ ," *Phys. Rev. B* **35**, 8223-30 (1987).

[58] M.A. Thompson and M.C. Zerner, "Effect of a Polarizable Medium on the Charge Transfer States of the Photosynthetic Reaction Center from Rhodopseudomonas-Viridis," *J. Am. Chem. Soc.* **112**, 7828-30 (1990).

[59] P.O. J. Scherer and S.F. Fischer, "Quantum Treatment of the Optical Spectra and the Initial Electron Transfer Process with the Reaction Center of Rhodopseudomonas-viridis," *Chem. Phys.* **131**, 115-27 (1989).

[60] A.R. Katrizky, M.C. Zerner and M.M. Karelson, "A Quantitative Assessment of the Mesostabilization Energy of Carbon-Centered Radicals," *J. Amer. Chem. Soc.* **108**, 7213-14 (1986).
[61] M. Karelson and M.C. Zerner, "On the N-Pi Star Blue Shift Accompanying Solvation," *J. Am. Chem. Soc.*, **112**, 9405-6 (1990).
[62] O. Tapia and O. Goscinski, "Self-Consistent Reaction Field Theory of Solvent Effects," *Molec. Phys.* **29**, 1653-61 (1975).
[63] J.H. Kim and J.T. Hynes, *J. Chem. Phys.*, in press; *Intern. J. Quantum Chem.*, in press.
[64] L. Onsager, "Electric Moments of Molecules in Liquids," *J. Am. Chem. Soc.* **58**, 1486-93 (1936).
[65] J. Kirkwood, "The Dielectric Polarization of Polar Liquids," *J. Chem. Phys.* **7**, 911-19 (1939).
[66] C.J.F. Boettcher, *Theory of Electron Polarization*, **Vol. 1**, Elsevier, 2nd ed., Amsterdam (1973).
[67] J.H. Simmons, R.K. Mohr and C.J. Montrose, "Non-Newtonian Viscous Flow in Glass," *J. Appl. Phys.* **53**, 4075-4080 (1982).
[68] S. Phillpot, D. Wolf and S. Yip, "Effects of Atomic-Level Disorder at Solid Interfaces," *MRS Bull.* **16**, 38 (1990).
[69] S.A. Brawer and M.J. Weber, "Microscopic Model of Anomalous Low Frequency Excitations in Inorganic Glasses: Molecular Dynamics Study of $BeF_2$," *Phys. Rev. Lett.* **46**, 778 (1981).
[70] S.A. Brawer and M.J. Weber, "Molecular Dynamics Simulations of the Structure of Rare Earth Doped Beryllium Fluoride Glasses," *J. Chem. Phys.* **75**, 3522 (1981).
[71] L. Verlet, "Computer 'Experiments' on Classical Fluids. I. Thermodynamical Properties of Lennard-Jones Molecules," *Phys. Rev.* **159**, 98 (1967).
[72] D. Beeman, "Some Multistep Methods for Use in Molecular Dynamics Calculations," *J. Comput. Phys.* **20**, 130 (1976).
[73] W. C. Swope, H. C. Andersen, P. H. Berens and K. R. Wilson, "A Computer Simulation Method for the Calculation of Equilibrium Constants for the Formation of Physical Clusters of Molecules: Application to Small Water Clusters," *J. Chem. Phys.* **76**, 637 (1982).
[74] H. J. C. Berendsen and W. F. van Gunsteren, "Practical Algorithms for Dynamic Simulations," in *Molecular Dynamics Simulation of Statistical-Mechanical Systems*, G. Civvotti and W. G. Hoover, eds., North-Holland, Amsterdam, Netherlands (1986).
[75] A. Rahman, "Correlations in the Motion of Atoms in Liquid Argon," *Phys. Rev. A* **136**, 405 (1964).
[76] C.W. Gear, *Numerical Initial Value Problems in Ordinary Differential*

*Equations*, Prentice-Hall, Englewood Cliffs, 1971.

[77] H.C. Andersen, "Molecular Dynamics Simulations at Constant Pressure and/or Temperature," *J. Chem. Phys.* **72**, 2384 (1980).

[78] M. Parrinello and A. Rahman, "Polymorphic Transitions in Single Crystals: A New Molecular Dynamics Method," *J. Appl. Phys.* **52**, 7182 (1981).

[79] L.V. Woodcock, "Isothermal Molecular Dynamics Calculations for Liquid Slats," *Chem. Phys. Lett.* **10**, 257 (1970).

[80] T. Schneider and E. Stoll, "Molecular Dynamics Study of Structural Phase Transitions. 1. One-Component Displacement Models," *Phys. Rev. B* **13**, 1216-37 (1976).

[81] Y. Hiwatari, E. Stoll and T. Schneider, "Molecular Dynamics Investigation of Solid-Liquid Coexistence," *J. Chem. Phys.* **68**, 3401-04 (1978).

[82] W.G. Hoover, A.J.C. Ladd and B. Moran, "High-Strain-Rate Plastic-Flow Studied Via Non-Equilibrium Molecular Dynamics," *Phys. Rev. Lett.* **48**, 1818-20 (1982).

[83] D.J. Evans, "Computer Experiments for Non-Linear Thermodynamics of Couette Flow," *J. Chem Phys.* **78**, 3297-302 (1983).

[84] H.J.C. Berendsen, J.P.M. Postma, W.F. van Gunsteren, A. DiNola and J.R. Haak, "Molecular Dynamics with Coupling to an External Bath," *J. Chem. Phys.* **81**, 3684-3690 (1984).

[85] P. N. Kumta, P. A. Deymier and S. H. Risbud, "Glass Formation in Simple Ionic Systems Via Constant Pressure Molecular Dynamics," *J. Chem. Phys.* **90**, 7384-94 (1989).

[86] A. C. Wright, "Diffraction Studies of Glass Structure," *J. Non-Cryst. Sol.*, **123**, 129 (1990).

[87] P. P. Ewald, "Die Berechnung Optischer Und Elektrostatischer Gitterpotentiale," *Ann. Phys.(Paris)* **21**, 1087 (1921).

[88] L.V. Woodcock and K. Singer, "Thermodynamic and Structural Properties of Liquid Ionic Salts Obtained by Monte Carlo Computation. 1. Potassium Chloride," *Trans. Faraday Soc.* **67**, 12 (1971).

[89] S. H. Garofalini and H. Melman, *Better Ceramics Through Chemistry II, MRS Symp. Proc.* **73**, C. J. Brinker, D. E. Clark and D. R. Ulrich, eds, Mat. Res. Soc. Pittsburgh, PA, p497 (1986); also A. B. Rosenthal and S. H. Garofalini, "Structural Role of Zinc Oxide in Silica and Soda-Silica Glasses," *J. Am. Ceram. Soc.* **70**, 821-26 (1987).

[90] J. H. Simmons, R. Faith and G. O'Rear, "Molecular Dynamics Simulations Of Fluoride Glass Structures," *Mater. Sci. Forum* **19**, 121-126 (1987).

[91] B.J. Alder and T.E. Wainwright, "Studies in Molecular Dynamics. I.

General Method," *J. Chem. Phys.* **31**, 459 (1959).
[92] J.J. Erpenbeck and W.W. Wood, "Molecular Dynamics Technique for Hard-Core Systems," in *Statistical Mechanics, Part B: Time-Dependent Processes*. Edited by B.J. Berne, Plenum Press, New York, 1976.
[93] J.E. Lennard-Jones, "The Determination of Molecular Fields. I. From the Variation of the Viscosity of a Gas with Temperature," *Proc. Roy. Soc. (London)* **106A**, 441 (1924); "The Determination of Molecular Fields. II. From the Equation of State of a Gas," ibid., 463 (1924).
[94] F. London, "Properties and Applications of Molecular Forces," *Zeit. Physik. Chem. B* **11**, 222 (1930).
[95] J. Fox and H. C. Andersen, "Molecular Dynamics Simulations of a Supercooled Monoatomic Liquid and Glass," *J. Phys. Chem.* **88**, 4019 (1984).
[96] J.J. Nicholas, K.E. Gubbins, W.B. Streett and D.J. Tildesley, "Equation of State for the Lennard-Jones Fluid," *Mol. Phys.* **37**, 1429 (1979).
[97] W.R. Busing, "Interpretation of Crystal Structure of $Li_2BeF_4$ in Terms of Born-Mayer-Huggins Model," *J. Chem. Phys.* **57**, 3008 (1972).
[98] P.N. Keating, "Effect of Invariance Requirements on the Elastic Strain Energy of Crystals with Application to the Diamond Structure," *Phys. Rev.* **145**, 637 (1966).
[99] F.H. Stillinger and T.A. Weber, "Computer Simulation of Local Order in Condensed Phases of Silicon," *Phys. Rev. B* **31**, 5262 (1985).
[100] B.P. Feuston and S.H. Garofalini, "Empirical Three-Body Potential for Vitreous Silica," *J. Chem. Phys.* **89**, 5818-24 (1988).
[101] M. S. Daw, "Model of Metallic Cohesion: The Embedded-Atom Method," *Phys. Rev. B* **39**, 7441-52 (1989).
[102] M. I. Baskes, M. S. Daw and S. M. Foiles, "The Embedded Atom Method: Theory and Application," *Mater. Res. Soc. Symp. Proc.* **141**, 31-41 (1989).
[103] S. M. Foiles, "Calculation of the Surface Segregation of Ni-Cu Alloys with the Use of the Embedded-Atom Method," *Phys. Rev. B* **32**, 7685-93 (1985).
[104] P. Vashista, R.K. Kalia, J.P. Rino and I. Ebbsjo, "Interaction Potential for $SiO_2$: A Molecular Dynamics Study of Structural Correlations," *Phys. Rev. B* **41**, 12197 (1990).
[105] B. Vessal, M. Leslie and C.R.A. Catlow, "Effect of Pressure on the Structure of Simulated Silica Glass," *Molec. Simul.* **3**, 123 (1989).
[106] D. M. McQuarrie, *Statistical Mechanics*, Harper and Row, New York, NY (1976).
[107] D. C. Anderson, K. Kremeyer and J. Kieffer, "Computer Simulations of Vibrational Spectra in Oxide Glasses," *Am. Ceram. Soc. Bul.* **70**, 557

(1991).

[108] F. L. Galeener, A. J. Leadbetter and M. W. Stringfellow, "Comparison of the Neutron, Raman, and Infrared Vibrational Spectra of Vitreous $SiO_2$, $GeO_2$ and $BeF_2$," *Phys. Rev. B* **27**, 1052-78 (1983); also S. Wallace, "Porous Silica Gel Monoliths: Structural Evolution and Interactions With Water," *PhD Dissertation*, University of Florida (1991).

[109] C. A. Angell, "Dynamic Processes in Ionic Glasses," *Chem. Rev.* **90**, 523-42 (1990).

[110] J. H. Simmons and C. J. Simmons, "Non-Linear Viscous Flow in Glass Forming," *Am. Ceram. Soc. Bul.* **68**, 1949-55 (1989).

[111] D. M. Heyes, J. J. Kim, C. J. Montrose and T. A. Litovitz, "Time Dependent Nonlinear Shear Stress Effects in Simple Liquids: A Molecular Dynamics Study," *J. Chem. Phys.* **73**, 3987-96 (1980).

[112] J. J. Mecholsky, D. E. Passoja and K. S. Feinberg-Ringel, "Quantitative Analysis of Brittle Fracture Surfaces Using Fractal Geometry," *J. Am. Ceram. Soc.* **72**, (1989).

[113] T. F. Soules and R. F. Busbey, "The Rheological Properties and Fracture of a Molecular Dynamic Simulation of Sodium Silicate Glass," *J. Chem. Phys.* **78**, 6307-16 (1983); also T. F. Soules, "Models of Glass Strength and Relaxation Phenomena Suggested By Molecular Dynamic Simulations," *J. Non-Cryst. Solids* **73**, 315-30 (1985).

[114] R. Ochoa and J. H. Simmons, "High Strain Rate Effects on the Structure of a Simulated Silica Glass," *J. Non-Cryst. Solids* **75**, 413-18 (1985).

[115] J. H. Simmons, T. P. Swiler and R. Ochoa, "Molecular Dynamics Studies of Brittle Failure in Silica: Bond Fracture," *J. Non-Cryst. Solids*, to be published.

[116] T. F. Soules and S. H. Garofalini, "Computer Simulation in Glass Studies," *Am. Ceram. Soc. Bul.* **70**, 557 (1991).

[117] D. Athanasopolous and S. H. Garofalini, "Computer Simulation of Deposition of Model adsorbates on Glass," *Am. Ceram. Soc. Bul.* **70**, 557 (1991).

[118] J. F. Lutsko, D. Wolf and S. Yip, "Molecular Dynamics Calculation of Free Energy," *J. Chem. Phys.* **88**, 6525-28 (1988).

[119] J. F. Lutsko, D. Wolf, S. Yip, S. R. Phillpot and T. Nguyen, "Molecular Dynamics Method for the Simulation of Bulk-Solid Interfaces at High temperatures," *Phys. Rev. B* **38**, 11572-581 (1988).

[120] F. H. Stillinger and T. A. Weber, "Hidden Structure in Liquids," *Phys. Rev. A* **25**, 978-89 (1982).

[121] R. A. LaViolette and F. H. Stillinger, "Multidimensional Geometric Aspects of the Solid-Liquid Transition in Simple Substances," *J. Chem. Phys.* **83**, 4079-85 (1985); also, R. A. LaViolette, "Amorphous Deposits

with Energies Below the Crystal Energy," *Phys. Rev. B* **40**, 9952-54 (1989).

[122] S. Tsuneyuki, M. Tsukada, H. Aoki and Y. Matsui, "First Principles Interatomic Potential of Silica Applied to Molecular Dynamics," *Phys. Rev. Lett.* **61**, 869-872 (1988); also "Molecular Dynamics Study of the $\alpha$ to $\beta$ Structural Phase Transition of Quartz," *Phys. Rev. Lett.* **64**, 776-9 (1990).

[123] B. W. H. van Beest, G. J. Kramer and R. A. van Santen, "Force Fields for Silicas and Aluminophosphates Based on Ab-Initio Calculations," *Phys. Rev. Lett.* **64**, 1955-58 (1990).

# CHAPTER 10

## GAS DIFFUSION MEASUREMENTS IN GLASSES

**J. E. Shelby**
Glass Science Laboratory
NYS College of Ceramics at Alfred University
Alfred, NY 14802

## INTRODUCTION

A large number of chemical and physical processes are controlled by the diffusion and solubility of gases in glasses and melts.[1] These processes include the fining of melts,[2] control of the oxidation state of many ions,[3,4] the formation of hydroxyl[5] and metallic colloids,[6-8] the elimination of radiation and mechanically induced defects,[9,10] and the formation of bubbles during sealing processes[11] or by radioactive decay via alpha particles. In addition, a significant number of commercial products are strongly influenced by gas diffusion. Obviously, the choice of materials for glass vacuum tubes and systems, glass to metal seals for vacuum components, and lamp bulbs are determined by the permeability of these materials to gases from the surrounding atmosphere.[1]

An understanding of these processes requires accurate values for the diffusivity and solubility of many gases. Unfortunately, the large range of gases of interest requires that a correspondingly large number of experimental techniques be used to measure their diffusivity and solubility. In particular, many of these gases undergo reaction with the glass or melt, which results in trapping of the gas and the formation of a new species. As a results, there are a large number of techniques available to measure the diffusivity and solubility of gases in glasses. The following discussion will be limited to the most common techniques and will be subdivided into sections dealing with processes where no reaction occurs and those where a chemical reaction results in the formation or elimination of an observable species in the sample.

## PHYSICAL DIFFUSION

Physical diffusion occurs whenever the conditions are such that no chemical interaction occurs between the diffusing species and the sample. The most obvious examples of such diffusion include the inert gases (helium, neon, argon, krypton, and xenon), relatively inert gases such as nitrogen, and even normally chemically reactive gases such as hydrogen and oxygen at sufficiently low temperatures that reaction rates are negligible as compared to the diffusion rate. The diffusion of these gases can be measured by either a flow technique or by a saturation/outgassing method. In general, flow techniques are more accurate, but are more limited in application.

## Methods based on flow through a membrane

Any discussion of flow techniques for the determination of the diffusivity of gases must begin with a clarification of the terminology used. The permeability of a material determines the flow rate through a specimen once steady-state flow has been attained. The diffusivity, which refers to the actual rate of movement of individual atoms or molecules, determines the time necessary to attain steady-state flow. Finally, the solubility refers to the dissolved gas concentration per unit of applied pressure of the gas under consideration. These parameters are related through the simple expression[1]

$$K = DS, \tag{1}$$

where K is the permeability, D is the diffusivity, and S is the solubility. It is obvious from equation 1 that the determination of any two of these parameters automatically defines the value of the third. Although all three can be measured using flow techniques, most studies only report measured values of the permeability and diffusivity and calculated values of the solubility based on equation 1.

All flow techniques require the formation of a membrane of the material to be studied and the sealing of this membrane into a vacuum system in such a manner that it forms a barrier to the movement of the gas under consideration. Historically, the formation of a leak-free seal, which is essential in order to ensure that all of the gas must diffuse through the sample to reach the detector, limited the study of gas permeation to commercial glasses, where traditional glass-to-glass seals could be made. Examples of early sample configurations are shown in Figure 1. These samples included disks, where the area and thickness of the sample could be easily measured, and blown bulbs, where the area and thickness were difficult to determine. In the latter case, the area to thickness ratio was often determined by filling the bulb with mercury, immersing the bulb in mercury to an equal depth, and measuring the capacitance of the system. Since the dielectric constant of the glass could be measured independently using a more traditional shape, the area to thickness ratio of the bulb could be calculated from the capacitance. (As will be shown later, only the ratio of the area to thickness is needed to determine the permeability.) After the permeation measurement was completed, the bulb was sectioned and the average wall thickness measured using a micrometer.

The complex sample configurations shown in Figure 1 severely limited both the identity and the number of materials that could be studied by flow methods. The formation of the sample usually controlled the nature and scope of permeation

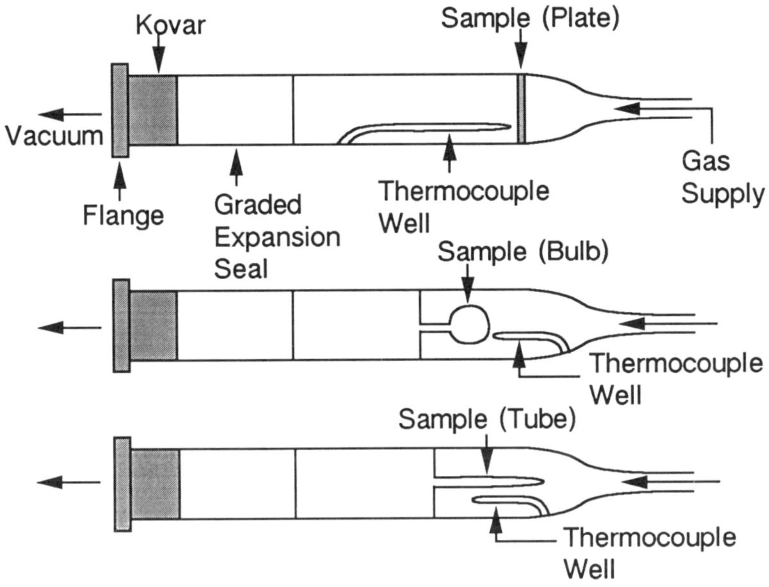


Figure 1. Typical sample configurations for gas permeation measurements.

studies. This problem was largely solved in 1972 by the development by the author and co-workers of a satisfactory sample holder which did not involve a fusion seal.

A typical design for such a sample holder is shown[1] in Figure 2. This sample holder uses a small area (0.25 to 2 $cm^2$), thin (0.01 to 0.05 cm) disk of the material to be studied. This disk is held between concentric silicone O-rings, which can be used to 300°C. The sample holder consists of three principal units: the upper flange, the gas supply line, and the vacuum jacket. The upper flange supplies one sealing surface and provides the connection to the detector system. The gas supply line supplies the other sealing surface and allows the application of the desired gas to one surface of the sample. The permeability of the silicone O-ring is much greater than that of typical glasses. The vacuum jacket is required to prevent the gas from simply diffusing through the lower O-ring and then through the upper O-ring, thus bypassing the sample. A vacuum of $<10^{-2}$ Torr is maintained in the vacuum jacket at all times. Any gas diffusing through the lower O-ring is thus removed from the system before it can diffuse through the upper O-ring. It follows that the gas reaching the detector has diffused through the sample. The background level of the system is easily determined by replacing the sample with an identical sample of a material known to be impermeable to the gas of interest, e.g. stainless steel for measurements of helium permeability.

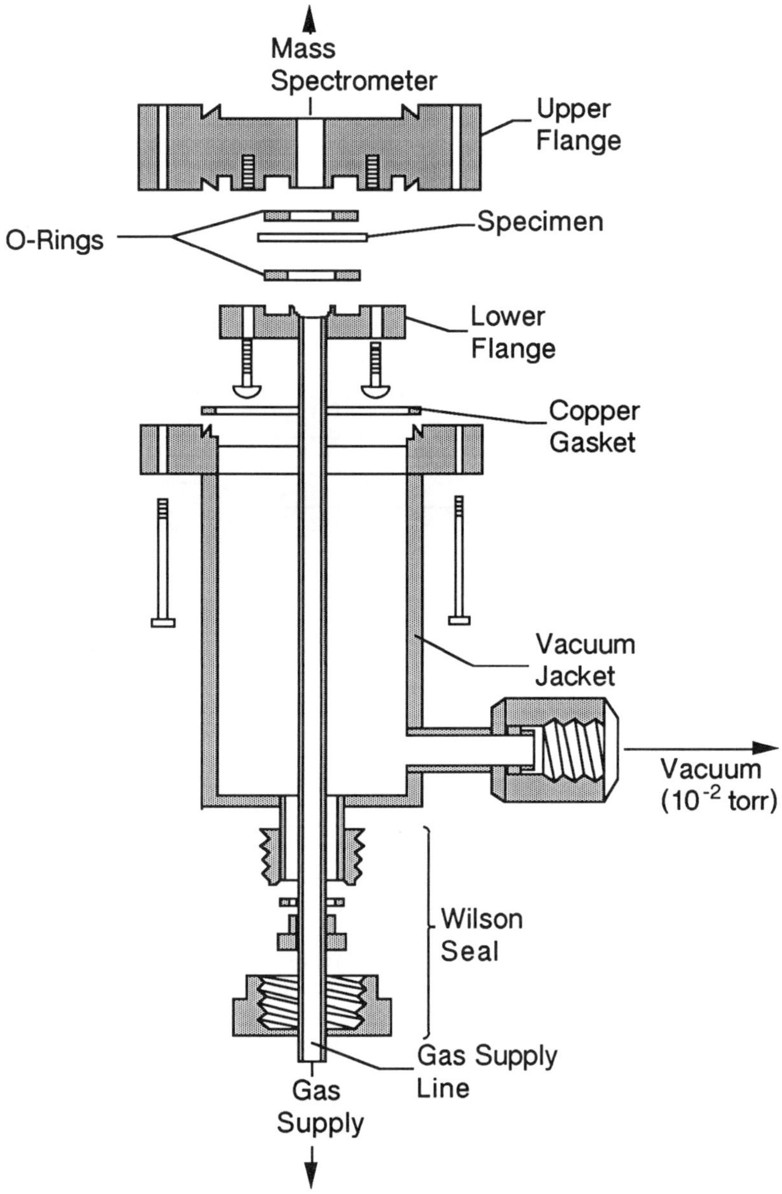


Figure 2. Sample configuration for gas permeation measurements

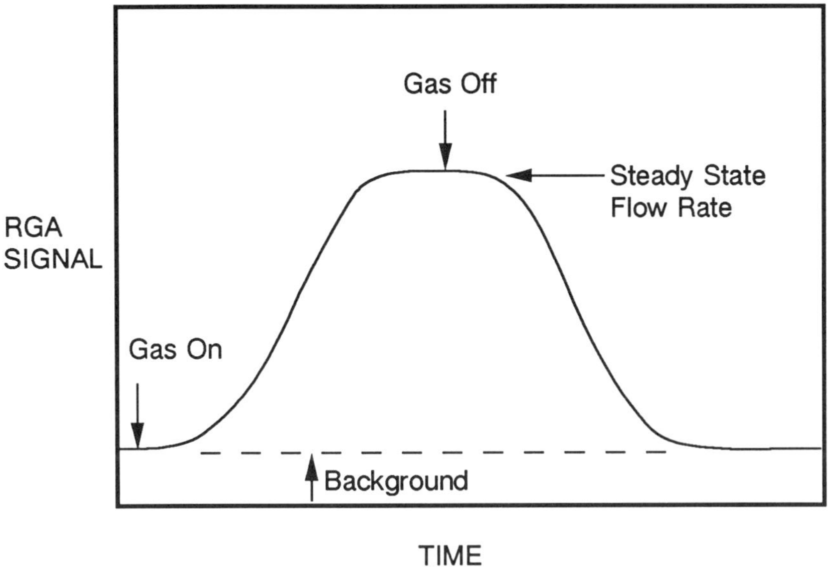


Figure 3. Typical RGA signal as a function of time for gas permeation and diffusion measurements.

Although a number of methods are used to extract the permeability, diffusivity, and solubility from flow techniques, the basic experimental methods are essentially the same. The sample is sealed in place by one of the methods discussed above. The volumes on both sides of the sample are evacuated, the sample is heated to the desired measurement temperature, and the gas to be studied is applied at a known pressure to one side of the sample. The flow of gas through the sample is monitored as a function of time until steady-state flow is attained. The gas is then removed from the gas supply side of the sample. The rate of gas release into the detector system continues to be measured until it decreases to a few percent of the steady-state flow value. A typical curve for the flow rate versus time is shown in Figure 3.

The method used to extract the permeability, diffusivity, and/or solubility from the data depends largely upon the details of the experimental procedure. For example, while the most commonly used method today for these measurements involves the direct determination of the flow rate, older studies simply determined the rate of pressure rise in the detector portion of the system by use of a pressure gauge of the appropriate type for the pressure range actually observed. This method has the advantage of low cost and simplicity of design, but has the drawbacks inherent in the inability to determine if all of the pressure rise is indeed due to the gas of interest, as well as those due to the lower sensitivity of the method. This method is also especially useful for porous materials, where the flow rate is too great for the direct determination method. The permeability is determined in this case via the expression[2]

$$K = (VL/AP)(dP/dt), \quad (2)$$

where V is the volume of the detector system, L and A are the sample thickness and area, respectively, P is the applied gas pressure on the gas supply side of the sample, and dP/dt is the rate of pressure rise in the detector system once steady-state flow is attained. If the total pressure is plotted against time, the diffusivity can also be determined by this technique by extrapolating the steady-state line, i.e. the portion of the data where dP/dt is constant, to its intercept on the time axis. In this case, Barrer[12] has shown that the diffusivity is given by

$$D = L^2/6t_c, \quad (3)$$

where $t_c$ is the intercept (or so-called critical) time. A typical plot of pressure rise versus time is shown in Figure 4.

Other techniques based on total pressure measurements (see Crank[13]) involve analysis of the pre-steady-state data to yield the diffusivity and solubility. The

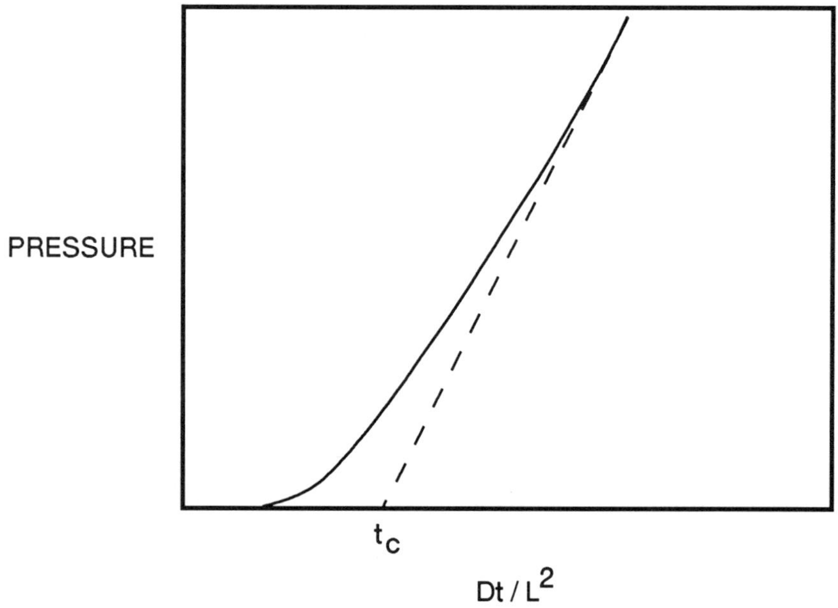


Figure 4. Total pressure as a function of time for the determination of gas diffusivities by the pressure rise method.

permeability is then calculated via equation 1. Since these techniques are rarely used, they will not be discussed further.

The direct measurement of the flow rate is usually used today for the determination of the permeability and diffusivity. In this case, the flow rate is measured using a mass spectrometer tuned for the gas of interest. The permeating gas is constantly removed from the system by continuous pumping. The permeability is determined from the expression

$$K = BL/AP \tag{4}$$

where B is the flow rate, and L, A, and P are defined as above. Since the mass spectrometer does not measure an absolute flow rate, the system must be calibrated for the exact operating conditions by measurement of the flow rate from a standard having a known flow rate. Since the sensitivity of the mass spectrometer can drift with time, the calibration must be made immediately before or after the measurement of the flow from the sample.

The diffusivity can be determined from either the rate of rise to steady-state flow or from the rate of decay of the mass spectrometer signal after removal of the gas from the gas supply side of the sample. Although both techniques have been used, the latter is usually found to yield more reproducible data. It can be shown that the flow rate can be described by a simple exponential decay after a short induction period. The time constant, $t_o$, for this decay is given by[14]

$$t_o = L^2/\pi^2 D. \tag{5}$$

The slope of a plot of ln(flow rate) versus time then yields the value of $t_o$ and hence the diffusivity. The solubility can be calculated via equation 1 or by integrating the flow rate during the decay process to yield the total amount of gas released during decay. Lee et al.[15] have shown that this amount of gas is equal to S/6.

### Saturation/outgassing methods

The techniques discussed above are capable of yielding all three parameters of interest. Other techniques based on the rate of absorption or desorption of gas from powders, rods or fibers, or plates can also be used to determine the diffusivity, using solutions of the diffusion equations in spherical, cylindrical, or plate form. These techniques are frequently used where either the sample is not available in a form suitable for flow measurements or has such a low diffusivity that the flow methods are prohibitively time consuming. If, for example, the particles of a powder are assumed to be spherical, with an average radius *a* (determined by screening or estimated by some other technique), the amount of gas absorbed or desorbed in time t ($M_t$)is given by[13]

$$M_t = M_\infty[1-(6/\pi^2)\Sigma\,(1/\pi^2)\exp(-n^2\pi^2Dt/a^2)], \quad (6)$$

where $M_\infty$ is the total amount of gas absorbed or desorbed. This expression assumes that the concentration of gas is uniform in the specimen at $t = 0$, which is usually easily done for the absorption case, but is frequently more difficult to control for the desorption case. In either case, the value of $M_\infty$ divided by the sample volume and normalized for the pressure yields the solubility of the gas in the material.

If an outgassing measurement is made during continuous pumping of a vacuum system containing the sample, the quantity measured is $dM_t/dt$, which is given by the derivative of equation 6. In either case, the diffusivity must be obtained by finding the value of D which yields a best fit to the data. Unfortunately, the radius of the particle is not identical for every particle in a powder. As a result, the fitting procedure should usually include *a* as a distribution function describing the particle size distribution. Since each additional complication of the analysis naturally leads to an increase in the uncertainity of the diffusivity determined by this technique, it would be best to use a monosized powder for every experiment. This is usually impossible due to other constraints. The use of fibers of constant diameter (and the appropriate modification of equation 6 for cylindrical geometry) can eliminate this problem for many glasses. Unfortunately, many glasses cannot be formed into fibers of the diameter uniformity needed to improve on the powder case. The use of plate samples (and solutions in planar geometry) would be even better, but the greater diffusion path length obtainable for practical geometries limits this method to relatively large diffusivities, as will be discussed later.

A large number of other techniques have also been used for the determination of physical gas diffusivity and solubility in glasses. The reader is referred to Crank[13] and Barrer[12] for the solutions to many special cases of the diffusion equations. In general, it can be stated that the flow techniques yield the best data, while the saturation/outgassing techniques require the least sample preparation and often can be performed with less expensive apparatus. However, the limitation that the gas must diffuse completely through a membrane, combined with the limits on forming such a membrane, have restricted flow techniques to values of $D>10^{-11}$ $cm^2/s$. Saturation/outgassing techniques, especially those using very fine powders, can be used to significantly lower diffusivity values and for materials which cannot be formed into membranes.

## DIFFUSION WITH CHEMICAL REACTION

A great number of processes in glasses and melts are controlled by the diffusion of a reactive molecule either to a reaction site or away from a decomposition site. Although many gases can be involved in such reactions, the most commonly studied processes involve molecular water and hydrogen isotopes, with lesser emphasis on oxygen, carbon dioxide, sulphur oxides, and other gases. These

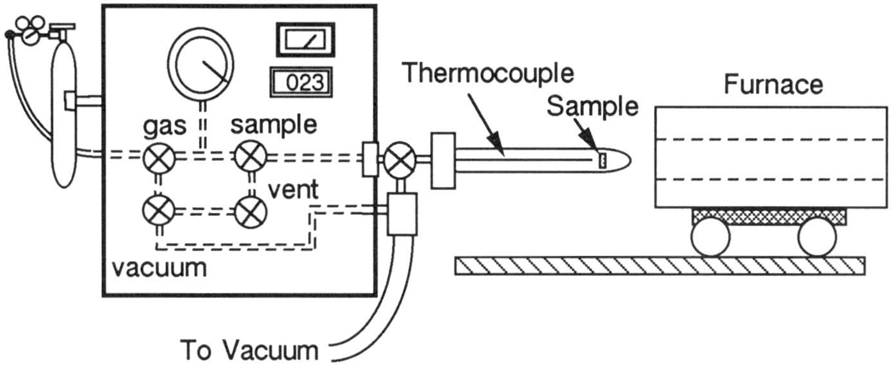


Figure 5. System for heat treatment of glass samples in controlled atmospheres.

processes include the formation[5] and elimination of hydroxyl, the reduction of ions to a lower ionic state[3,4] or even to the atomic state,[6-8,16] which can often lead to the formation of colloids, the elimination of optical defects induced by radiation[9,10] or by fiber drawing, the oxidation of ions to a higer ionic state in melts, and a myriad of others.

## Water diffusion

The diffusion of water is particularly important because it frequently controls the hydroxyl concentration in a glass. Many studies[17] have shown the great influence of hydroxyl on the properties of glasses, including reductions in viscosity and glass transformation temperature.[18,19] and increases in the rates of phase separation and crystallization with increasing hydroxyl concentration. Hydroxyl can be added to or removed from melts by simply altering the partial pressure of water vapor over the melt as a result of diffusion controlled reaction between the water and the oxygen in the melt. Hydroxyl can also be added or removed from solid glasses by treatment in the appropriate atmosphere.

Studies of diffusion controlled reaction processes generally involve the determination of the concentration of some species, e.g. hydroxyl, in a sample as a function of time, temperature, gas pressure, or other variable. A typical apparatus used for such studies is shown schematically in Figure 5. This apparatus allows heat treatments in vacuum or any desired atmosphere for temperatures ranging from ambient to 1000°C at pressures from $10^{-2}$ torr to 1 atmosphere. The limits are primarily imposed by the use of a vitreous silica tube to contain the sample and atmosphere. These studies are often carried out in planar geometry to take advantage of the simplicity of analysis available for plate samples. A large number of solutions to the diffusion equations exist for planar geometry for different boundary conditions. In general, these solutions can be subdivided into two types of measurements, i.e. (1) those where one measures the profile of a species in the sample and (2) those where one measures the average concentration of the species in the sample.

Studies of the dehydroxylation of vitreous silica offer a good example of these studies. The relative hydroxyl content of a glass can be determined by infrared spectroscopy. In many cases, it is not necessary to convert these relative values to absolute values, since only the fractional change in concentration with time is required for the determination of the effective diffusion coefficient. Suppose that one simply measures the initial height of the hydroxyl absorption band at 2.72 micrometers for vitreous silica. The sample is now placed in a furnace at a constant temperature and held for a known time. If the temperature is high enough for reasonable diffusion to occur, the hydroxyl concentration will be reduced. The actual profile of hydroxyl in the sample will be described by the expression[13]

$$\frac{C-C_o}{C_s-C_o} = 1-\frac{4}{\pi}\sum_{n=o}^{\infty} \frac{(-1)^n}{2n+1} \exp\{-D(2n+1)^2\pi^2 t/4L^2\} \cos\frac{(2n+1)\pi x}{2L} \qquad (7)$$

where C is the concentration of hydroxyl at a distance x from the sample surface, $C_o$ is the initial, uniform concentration of hydroxyl in the sample, and $C_s$ is the constant surface concentration of hydroxyl.

It is now possible to determine the effective diffusivity of hydroxyl in the sample by several methods. The most straightforward method requires only a measurement of the average concentration of hydroxyl in the sample. The average concentration is given by the expression[13]

$$\frac{M_t}{M_\infty} = 1-\sum_{n=0}^{\infty} \frac{8}{(2n+1)^2\pi^2} \exp\{-D(2n+1)^2\pi^2 t/4L^2\} \qquad (8)$$

It is possible to solve this expression for the exact value of $M_t/M_\infty$ found in the experiment. However, it is usually much easier to refer to a plot of the average concentration of the measured species as a function of the square root of $Dt/L^2$, which can be found in a number of texts. (A plot for the case of planar geometry is shown in Figure 6.) The value of $Dt/L^2$ for the fractional reduction in concentration is determined, and the value of D evaluated for the known treatment time and sample thickness.

A more accurate value for the diffusivity can be obtained by making a series of measurements for a number of treatment times and samples thickness, plotting the reduced curve, and then calculating the value of D which yields the best fit to all of the data. Again, it would be possible to fit the entire curve to Eq. 8. However, a highly accurate approximation can be used for the value of $t/L^2$ where $M_t/M_\infty$ = 0.5, i.e. where one half of the hydroxyl has been removed. In this case, the value of $t/L^2$ is given by the expression[13]

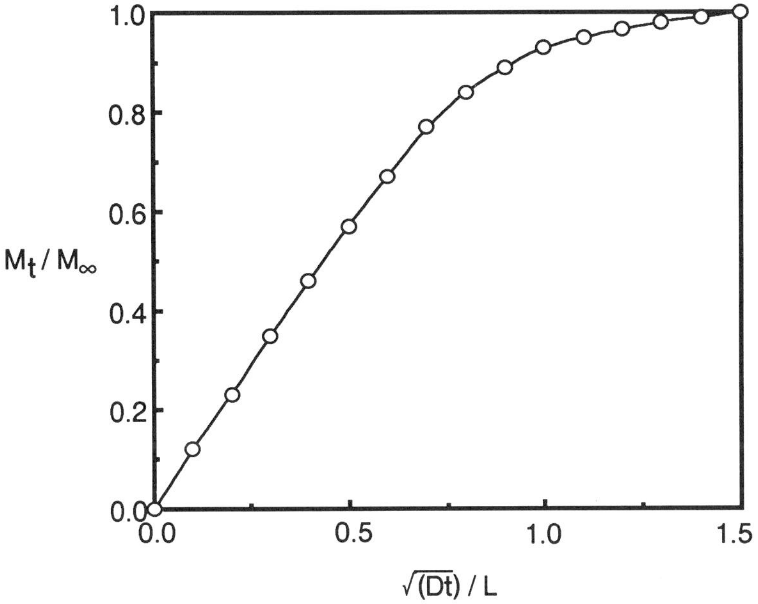


Figure 6. Theoretical curve for the percent of gas removed from a sample as a function of time, sample thickness, and diffusivity.

$$t/L^2 = -(1/\pi^2 D)\ln[(\pi^2/16)-(1/9)(\pi^2/16)^9]. \tag{9}$$

Solving this expression for D yields the expression[13]

$$D = 0.049/(t/L^2), \tag{10}$$

where the value of $t/L^2$ is that for removal of exactly one half of the hydroxyl. Crank[13] indicates that the error in this approximation is of the order of 0.001%. A typical set of data for the dehydroxylation of vitreous silica is shown in Figure 7. The value of $\sqrt{t}/L$ at 50% hydroxyl removal is indicated on the figure.

It is also possible to determine the diffusivity from a plot of $M_t/M_\infty$ versus $t/L^2$ by determing the initial slope of this plot.[13] During the early stages of the process,

$$M_t/M_\infty = (4/\sqrt{\pi})(Dt/L^2)^{1/2} \tag{11}$$

The initial slope of this curve is given by

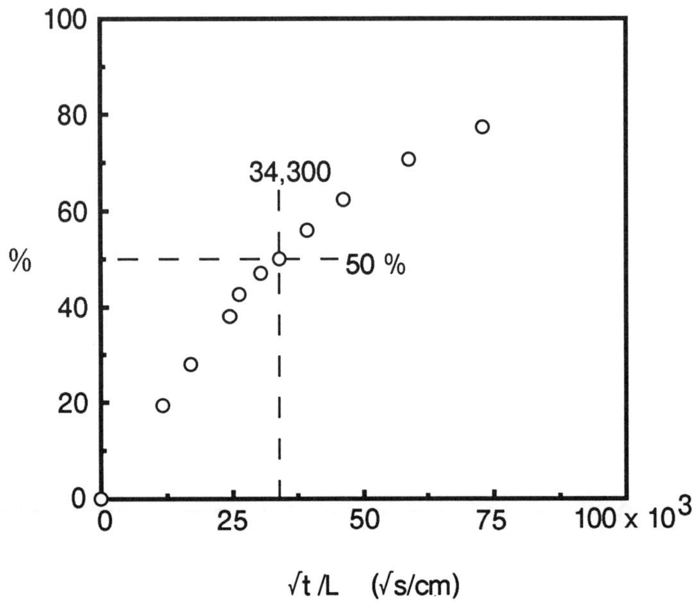


Figure 7. Typical experimental results for the dehydroxylation of vitreous silica at a constant temperature.

$$R = d(M_t/M_\infty)/d(t/L^2)^{1/2}, \tag{12}$$

then

$$D = (\pi/16)R^2. \tag{13}$$

This expression should yield the same value of D as Eq. 10. Since it is usually more accurate to determine a value from a curve rather than to determine the slope, the value obtained from Eq. 10 is often preferred. However, if data cannot be obtained up to a value of $M_t/M_\infty = 0.5$, Eq. 13 still allows the determination of D. Of course, determination of D by both methods will allow a check on the calculation and should be done wherever possible.

The dehydroxylation process can also be studied by determining the concentration profile of hydroxyl after a given time of heat treatment. In this case, the concentration C at any given distance x into the sample is given by Eq. 7. The experimental problem lies in the determination of the concentration at any given depth. This problem is directly analogous to the problem encountered in radioactive tracer studies of ionic diffusion. In this case, however, it is not possible to simply dissolve or grind a layer from the surface and determine the hydroxyl content. Two alternative approaches can be used. First, the sample can be sliced into a series of thin samples and the hydroxyl content of each slice analyzed by infrared spectroscopy. Unfortunately, the slicing process results in the loss of material due to the saw kerf. The thickness of each of these layers must be factored into the profile in order to correctly determine the depth represented by each slice of the sample. (An optical method for making these measurements will be discussed later. This method partially solves this problem.) Alternatively, the surface of the sample can be removed to a known depth by either etching or grinding. The hydroxyl content of the remaining sample can be determined. The reduction in hydroxyl content of the thinned sample can be attributed to the layer removed. A series of consecutive etchings or grindings can then allow the determination of the concentration profile of hydroxyl in the sample. It should be noted that this technique works best when the layer removed either is relatively thick compared to the total sample thickness, which means that only a few points are obtained for constructing the profile, or when the concentration in the layer removed is very different from that of the bulk of the sample. The latter situation occurs for the outermost layers of a dehydroxation sample, but is not true for the layers near the center. As a result, it is frequently best to remove successively thicker layers as the center of the sample is approached. The best fit of the profile obtained can then determined for Eq. 7.

It should be obvious that the measurement of the diffusivity by use of Eqs. 10 and 13 is much easier than the actual determination of the concentration profile. Unfortunately, the simple measurement of the changes in the bulk concentration often yields only the average diffusivity and does not yield as much information about the details of the process as does the determination of the concentration profile. This is especially true if the diffusivity is a relatively small function of the concentration of the hydroxyl. It follows that the experimenter must be cautious when using these techniques to insure that the results of the experiment are truly independent of the hydroxyl concentration and the thickness of the sample.

## Molecular hydrogen

Hydrogen isotope diffusion is often the controlling factor in a number of chemical reactions in glasses. Since hydrogen from the atmosphere can readily diffuse into many glasses, a number of reduction and/or reaction processes can occur in the presence of hydrogen gas. These processes, which include changes in the oxidation state of transition and rare earth ions, which may even lead to the formation of metallic colloids if the ion is reduced to the metallic state, and the removal of defects induced by radiation, manufacturing, or mechanical stress, also produce a form of bound hydrogen, e.g. either hydroxyl or hydride. The availablity of two isotopes of hydrogen has also been shown to be important in controlling the infrared transmission of glasses, since these isotopes can readily exchange at elevated temperatures.[20]

Hydrogen isotopes (the discussion will be limited to hydrogen and deuterium) diffuse as essentially chemically inert molecules in glasses at relatively low temperatures. Unfortunately, the definition of the term "low temperature" is strongly dependent upon the nature of the sample, including the presence of defects and reducible species. For example, hydrogen diffuses through very high purity vitreous silica without interaction up to temperatures of about 500°C. (This case was covered in the first portion of this chapter.) However, if the sample contains radiation–induced defects, reactions which produce hydroxyl and eliminate the defects can occur even at room temperature.

The kinetics of the interaction between a diffusing molecule of hydrogen and a given reaction site in a glass are controlled by both the diffusivity of the hydrogen and the rate of the chemical reaction. In general, the activation energy for chemical reactions is much greater than that for hydrogen molecular diffusion. As a result, the process will be limited by the reaction rate at lower temperatures and by the hydrogen diffusivity at higher temperatures.

If the reaction rate is slow enough, the diffusion process will be effectively independent of the reaction and a hydrogen molecule profile described by Eq. 7 will be present in the sample. If the sample is allowed to equilibrate, a uniform hydrogen profile will exist in the sample. If the reaction is allowed to occur in the presence of a source of hydrogen, the formation of hydroxyl may occur very slowly over a long period of time until all of the reaction sites are consumed. While there will be a slight profile of the hydroxyl produced by the reaction, and an accompanying profile in the concentration of reaction sites, the profile may well be so flat as to be unmeasureable.

If the reaction rate is very fast compared to the diffusion rate of molecules to the reaction sites, a very different result is obtained. In this case, the hydrogen will be consumed as soon as it encounters a reaction site. This process is analogous to the tarnishing of metals, i.e. rust formation, and has been shown to be adequately described by a similar model, which is usually referred to as the tarnishing model.[2] This model, which will be discussed in detail below, has been used to describe the reduction of iron in float glass from the ferric to the ferrous state,[3] the elimination of radiation–induced defects in vitreous silica,[9,10] and the reaction of hydrogen with a manufacturing defect in certain forms of commercial vitreous silica.[5] Other related processes involve the formation of silver,[8] arsenic,[6,7] antimony,[7] and bismuth[7] colloids in glasses upon exposure to hydrogen gas. Colloid formation is often further complicated by the fact that the actual formation of colloids is a two stage process involving first the reduction of the ion to the metallic state by reaction with the hydrogen and then the diffusion of the metallic atom to a growing colloid site. The formation of hydroxyl, which depends only on the first stage of this process, has been found to follow the tarnishing model, while the actual changes in the optical spectrum of the sample, which depend upon the colloid growth, exhibit a somewhat different time dependency.[21]

The tarnishing model has been discussed[5,9,10] in a number of papers. In general, it is assumed that the reaction which leads to the formation of the bound species is very fast compared to the diffusion rate of the mobile species. The number of reaction sites is fixed for a given sample and would be independent of time at the temperature of study in the absence of the mobile species. Under these conditions, it can be shown that the profile of each of the species, i.e. reaction sites and mobile species, are those shown in Figure 8. A sharp interface exists between the reacted and unreacted regions. This interface moves into the sample as the treatment time increases. None of the mobile species can enter the unreacted region until all of the reaction sites have been eliminated in the reacted region, since any remaining unreacted sites will immediately consume any passing mobile species. Under these conditions, Crank[13] has shown that the reacted layer thickness, X, is given by the expression

$$X = (2C_gD_gt/C_x)^{1/2}, \tag{14}$$

where $C_g$ is the concentration of dissolved gas at the outer surface of the sample, $D_g$ is the diffusivity of the gaseous species in the reacted layer (which is assumed to be identical to that in the unreacted layer), t is the heat treatment time, and $C_x$ is the concentration of reaction sites. This expression is based on the assumption that $C_g<C_x$. Since $C_g$ = SP, where S is the solubility of gas in the glass and P is the partial pressure of that gas in the surrounding atmosphere, this expression can be

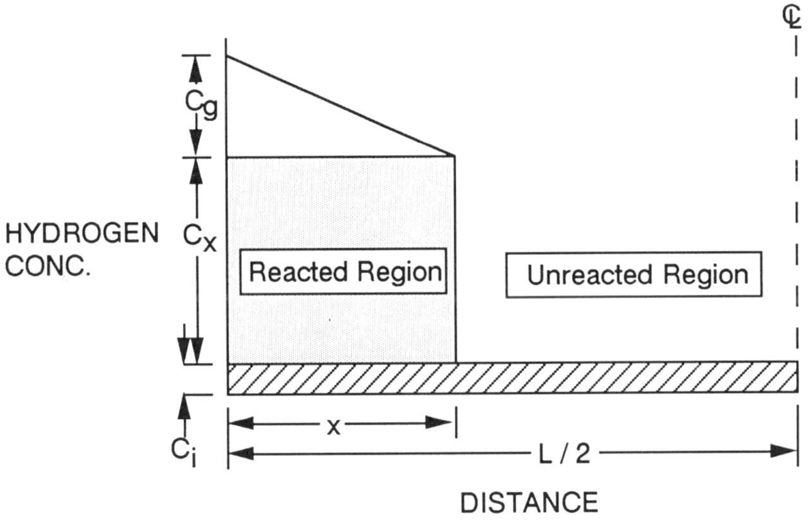


Figure 8. Concentration profiles for hydroxyl and hydrogen gas during a heat treatment based on the tarnishing model for gas diffusion.

rewritten as

$$X = (2SPD_g t/C_x)^{1/2}. \quad (15)$$

Since we also know that $SD_g = K$, where K is the permeability of the gas in the film, the expression can also be written in the form

$$X = (2KPt/C_x)^{1/2}. \quad (16)$$

If the film thickness is measured directly, this equation can be used to determine the value of K if $C_x$ is known, or the value of $C_x$ if K is known. Since $C_x$ is equal to the concentration of reacted sites, it is frequently possible to determine the concentration of the reaction product by other means, so that K can be determined. It should be noted that only the product of S and $D_g$ is determined by this method. Unfortunately, in many cases where this analysis would be correct, data analysis has been incorrectly based on Eq. 7. Under those circumstances, only an "effective diffusivity" is obtained. This approach leads to the unsettling result that the effective diffusivity is a function of pressure.

Eqs. 14–16 are based on the measurement of the thickness of the reacted layer. This method can often be used if the layer can be easily observed in an optical microscope. For example, the reacted layer formed during the reaction of hydrogen with AlOHC (aluminum oxygen hole centers) in irradiated vitreous silica can be easily determined by eye since the unreacted material is purple while the reacted layer is colorless.[9,10] On the other hand, many defects do not produce a colored glass, which means that the elimination of the defect does not lead to a visible change in the sample. For example, the reaction of hydrogen with certain forms of vitreous silica results in the formation of hydroxyl with no visible change in the sample. In this case, the determination of the layer thickness may be quite complicated. The author[5] has shown that in this particular case, it is possible to measure the hydroxyl concentration profile by an optical method. A narrow slit (0.5 mm wide) was placed in the center of the beam of an infrared spectrometer. Alfter removing the front and back faces of the treated sample to eliminate edge contributions, the sample was mounted on a translation stage with a resolution of 0.01 mm. The sample was moved across the beam in small increments, with a spectrum measured for each increment. The data obtained represent the hydroxyl profile as broadened by the slit width. A profile of this type is shown in Figure 9.

It is often more convenient to measure the total concentration of the reaction product (or the decrease in the concentration of the reaction sites) than to measure the reacted layer thickness, as discussed above. Under these conditions, one can measure the total concentration of either species and use the expression

$$(C-C_i)/(C_f-C_i) = 2X/L, \quad (17)$$

where C is the bulk concentration, $C_i$ is the initial concentration of the species measured, and $C_f$ is the final concentration of the species measured after the sample

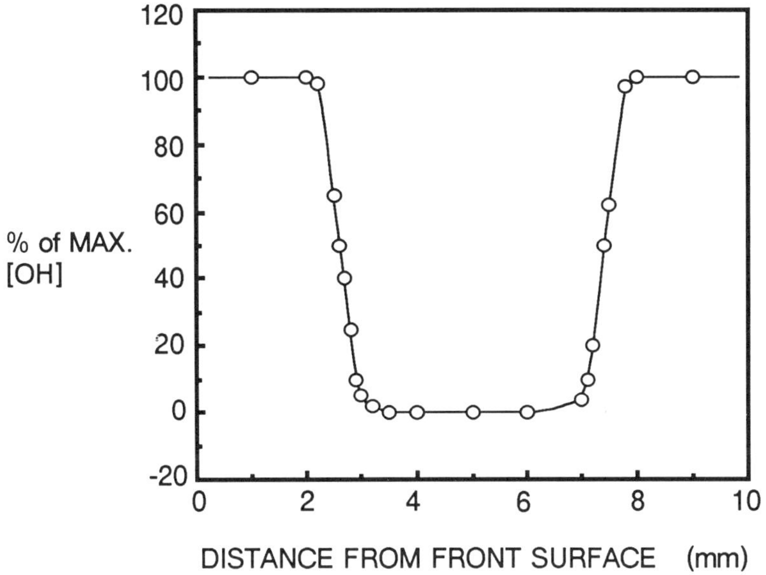


Figure 9. Concentration profile resulting from a tarnishing reaction for hydroxyl formation in a sample of initially hydroxyl-free vitreous silica.

is completely reacted. By combining the last two equations, this equation can be rewritten as

$$(C-C_i)/(C_f-C_i) = (8KPt/C_xL^2)^{1/2}. \tag{18}$$

Since P, t, and L are known, the slope of the fit to the data on a plot of the concentration change expressed as a percentage of the total change against the quantity $(Pt)^{1/2}/L$ will yield the quantity $K/C_x$. If either K or $C_x$ can be independently evaluated, the other can be calculated.

The permeability is the only temperature dependent term in Eq. 18. (It is assumed that $C_x$ is independent of temperature.) It follows that, even if K and $C_x$ cannot be separated, the temperature dependence of the permeability can still be determined by this method. Again, it must be stressed that the temperature dependence is that of the permeability and not that of the diffusivity. The activation energy for permeation is the sum of the activation energy for diffusion and the heat of solution of the diffusing species. Attributing the entire temperature dependence of the reaction rate to diffusion is incorrect.

## CONCLUSIONS

A large number of techniques exist for measuring the diffusion and reaction of gases with glasses. Flow methods yield the best data, but are restricted in application to physical solubility and to relatively large diffusion coefficients. Saturation/outgassing methods are less accurate for systems where no chemical interaction occurs between the diffusing gas and the sample, but can often be extended to lower diffusivities and materials which cannot be prepared in the geometries required for flow measurements. The saturation/outgassing methods serve as the basis for most of the methods used to study systems where reaction between the diffusing species and the glass occur.

## ACKNOWLEDGMENTS

Figures 2 and 5 were prepared by Bradley Lindquist.

## REFERENCES

1. J. E. Shelby, "Molecular Solubility and Diffusion," Chapter 1, pp. 1–40, TREATISE ON MATERIALS SCIENCE AND TECHNOLOGY, VOL. 17: GLASS II, ED. M. Tomozawa and R. H. Doremus, Academic Press, New York, 1979.
2. R. H. Doremus, GLASS SCIENCE, Wiley, New York, 1973.
3. J. E. Shelby and J. Vitko, "Reduction of Iron in Soda Lime Silicate Glasses by Reaction with Hydrogen," J. Non-Cryst. Solids 53(1,2) 155-63 (1982).
4. E. A. Weaver, R. W. Heckman, and E. L. Williams, "Determining Hydrogen Diffusion in Doped Vitreous Silica from the Fluorescence," J. Chem. Phys. 47(12) 4891-95 (1967).
5. J. E. Shelby, "Reaction of Hydrogen with Hydroxyl-Free Vitreous Silica," J. Appl. Phys. 51(5) 2589–93 (1980).
6. J. T. Kohli, B. M. Wright, J. M. Jewell, and J. E. Shelby, "Hydrogen-Induced Coloration of Glasses Containing Arsenic," Phys. Chem. Glasses 31(4) 133–37 (1990).
7. M. R. Tuzzolo, J. T. Kohli, and J. E. Shelby, "Hydrogen-Induced Absorption in Glasses Containing Arsenic and Antimony," Proceedings of the S.P.I.E., San Diego, CA, 1990.
8. J. L. Barton and M. Morain, "Hydrogen Diffusion in Silicate Glasses," J. Non-Cryst. Solids 3 115-26 (1970).
9. J. E. Shelby, "Reaction of Hydrogen with Optical Defects in Glasses, " J. Am. Ceram. Soc. 67(5) C93–94 (1984).
10. J. E. Shelby, "Chemical Annealing of the Aluminum Oxygen Hole Center," J. Appl. Phys. 60(12) 4325–27 (1986).
11. R. J. Eagan and G. J. Jones, "Bubble Formation in Glass by Reaction with Silicon and Silicon-Germanium Alloys," J. Am. Ceram. Soc. 58(7-8) 300-301 (1975).
12. R. M. Barrer, DIFFUSION IN AND THROUGH SOLIDS, Cambridge University Press, London, 1941.

13. J. Crank, THE MATHEMATICS OF DIFFUSION, 2nd Edition, Clarendon Press, Oxford, 1975.
14. D. E. Swets, R. W. Lee, and R. C. Frank, "Diffusion Coefficients of Helium in Fused Quartz," J. Chem. Phys. 34(1) 17-22 (1961).
15. R. W. Lee, R. C. Frank, and D. W. Swets, "Diffusion of Hydrogen and Deuterium in Fused Quartz," J. Chem. Phys. 36(4) 1062-71 (1962).
16. J. E. Shelby and J. Vitko, "Hydrogen Transport in a Machinable Glass-Ceramic," J. Non-Cryst. Solids 45(1) 83-92 (1981).
17. E. N. Boulos and N. J. Kreidl, "Water in Glass: A Review," J. Can. Ceram. Soc. 41 83-90 (1972).
18. J. M. Jewell, M. S. Spess, and J. E. Shelby, "Effect of Water Concentration on the Properties of Commercial Soda-Lime-Silica Glasses, J. Am. Ceram. Soc. 73(1) 379-82 (1990).
19. J. M. Jewell and J. E. Shelby, "Effect of Variations in Water Content on Sub-$T_g$ Aging of Glass," J. Appl. Phys. 68(12) 6159-69 (1990).
20. J. E. Shelby, "Molecular Diffusion and Solubility of Hydrogen Isotopes in Vitreous Silica," J. Appl. Phys. 48(8) 3387–94 (1977).
21. M. R. Tuzzolo, "Hydrogen–Induced Colloid Formation in Glasses," M.S. Thesis, NYS College of Ceramics at Alfred University, Nov. 1990.

# CHAPTER 11

# VISCOUS FLOW AND RELAXATION PROCESSES IN GLASS

**Joseph H. Simmons**
**University of Florida**
**Gainesville, FL 32611**

## INTRODUCTION

Viscous flow and relaxation processes are directly relevant to glass formation because they characterize the sluggishness of the liquid and the transition to the non-crystalline solid state. Without some degree of sluggishness in flow processes, the metastable amorphous state of a glass could not be reached in preference to the crystalline state during cooling from an equilibrium liquid. Therefore, if a glass exhibits a strong increase in viscosity with decreasing temperature near its liquidus point, then its glass-forming tendency is high, while a weak viscosity increase tends to lead to crystallization problems [1]. This effect is primarily responsible for the difference between silicates which are good glassformers and fluorides which are not [2,3].

This chapter will review the fundamental concepts associated with viscous flow, viscoelasticity and structural relaxation in glassforming liquids. Special emphasis will be placed upon concepts which have a microscopic basis and, where feasible, a relationship will be sought between structural characteristics and macroscopic behavior. At the outset, this is a difficult task, because the macroscopic measurements of viscous flow and relaxation processes are generally ensemble averages and the details of the atomistic behavior are lost in the averaging process. Therefore, investigators have had to turn to comparison of empirical models with the results of experiments. This approach does not yield a unique connection between model and experiment and agreement does not even guarantee that the model has any physical basis. Consequently, the literature is full of empirical models which have as their justification only the fitting of ancillary data or behavior.

The form of the chapter will therefore be to review fundamental concepts and see how far they can be extended to explain observations. Then, a variety of models will be examined in light of the experimental evidence and their benefit and limitation identified. At the end, new directions will be suggested. The reader is referred to several excellent articles and books which have also reviewed the subject [4-7].

## GENERAL CONCEPTS OF ELASTICITY AND VISCOELASTICITY

Elastic constants, $C_{ij}$, relate a stress, $\sigma_i$, to a strain, $\gamma_j$:

$$\sigma_i = \frac{F_i}{A_L} = \sum_{j=1}^{6} C_{ij}\gamma_j = \sum_{j=1}^{6} C_{ij}\left(\frac{\Delta \ell}{\ell}\right)_j \qquad (1)$$

In three dimensions, there are 36 possible elastic constants of which 21 are independent. However, isotropic systems, such as glasses, only require 2 elastic constant: the bulk modulus, **B**, and the shear modulus, **G**. There are known relations between the various other forms of elastic constant, such as **E**, Young's modulus, $\upsilon$, Poisson's ratio and $\lambda$, the Lame constant for volume elasticity, as in Table 1 [8].

| **Pairs** | **B,G** | **E,G** | **B,$\upsilon$** | **$\lambda$,G** |
|---|---|---|---|---|
| **B=** | B | EG/[3(3G-E)] | B | $\lambda$+(2G/3) |
| **G=** | G | G | 3B(1-2$\upsilon$)/ [2(1+$\upsilon$)] | G |
| **E=** | 9BG/(3B+G) | E | 3B(1-2$\upsilon$) | [G(3$\lambda$+2G)]/ ($\lambda$+G) |
| **$\upsilon$=** | (3B-2G)/ (6B+2G) | (E/2G)-1 | $\upsilon$ | $\lambda$/[2($\lambda$+G)] |
| **$\lambda$=** | B-(2G/3) | [(E-2G)G]/ (3G-E) | 3$\upsilon$B(1+$\upsilon$) | $\lambda$ |

Viscoelasticity is the time-dependent response of a material to an applied mechanical load. Therefore, viscoelasticity includes the study of stress or strain relaxation. By contrast, structural relaxation is the time-dependent response of the material to a change in temperature. A recent chapter by Rekhson is of particular interest on this subject [6].

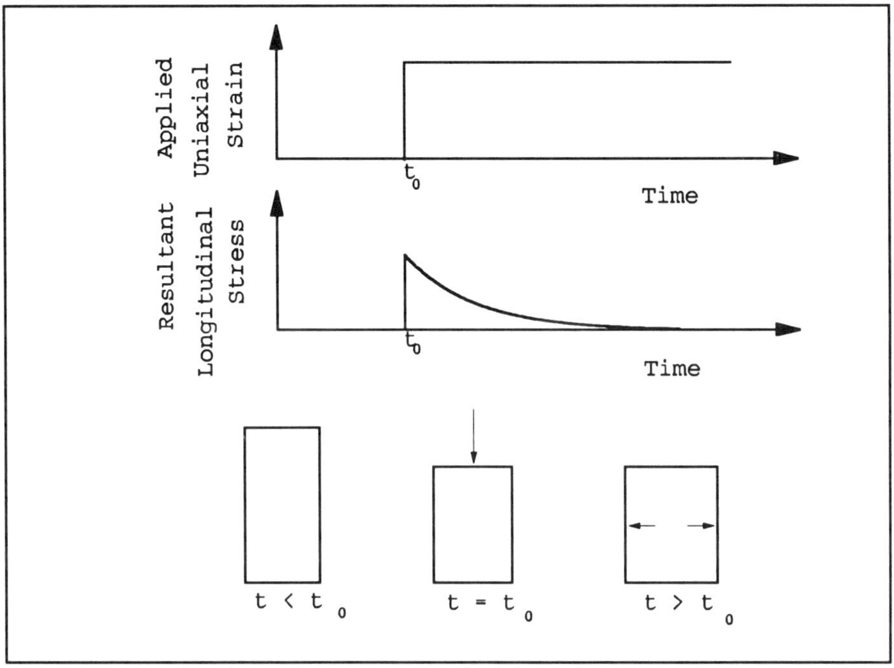


**Figure 1** - Application of a constant uniaxial strain.

Consider as an example, a material under the application of a *uniaxial strain*, $\epsilon_x$. The resulting elastic *stress* is $\sigma_x = E \cdot \epsilon_x$. Figure 1 shows the effect of a constant *strain* applied at $t = t_0$. The instantaneous elastic *stress* is followed by *stress relaxation*, as the material flows to reduce the stress to zero. The illustration shows the associated viscoelastic deformation of a cube. The material's dimensions, normal to the applied strain, increase elastically as defined by Poisson's ratio: $[\epsilon_z = \epsilon_y = -\upsilon\epsilon_x]$, then they increase further due to viscous flow until the longitudinal stress, $\sigma_x$ is completely relieved. In a second example, the response to an applied *stress* (often called creep) is analyzed. Three components of the distortion are identified: $\epsilon_E$ the instantaneous elastic strain, $\epsilon_D$ the delayed elastic strain and $\epsilon_V$ the viscous strain. Figure 2 shows the result of application and subsequent removal of a constant *stress*, $\sigma_0$. Under constant *stress* application, the strain asymptotically approaches a constant slope, defining the viscosity, $\eta$, by the equation:

$$\frac{d\epsilon}{dt} = \frac{\sigma_0}{\alpha\eta} \tag{2}$$

where the shape factor $\alpha$ is **1** for pure shear, and **3** for extensional flow. When the load is removed at a later time, $\mathbf{t_1}$, both the elastic and delayed elastic strains are recovered. The permanent strain, $\epsilon_v$, is produced by viscous flow, and can only be recovered by the application of a *stress* of the same magnitude in the opposite direction for the same time interval.

On an atomic scale, the instantaneous response is produced by

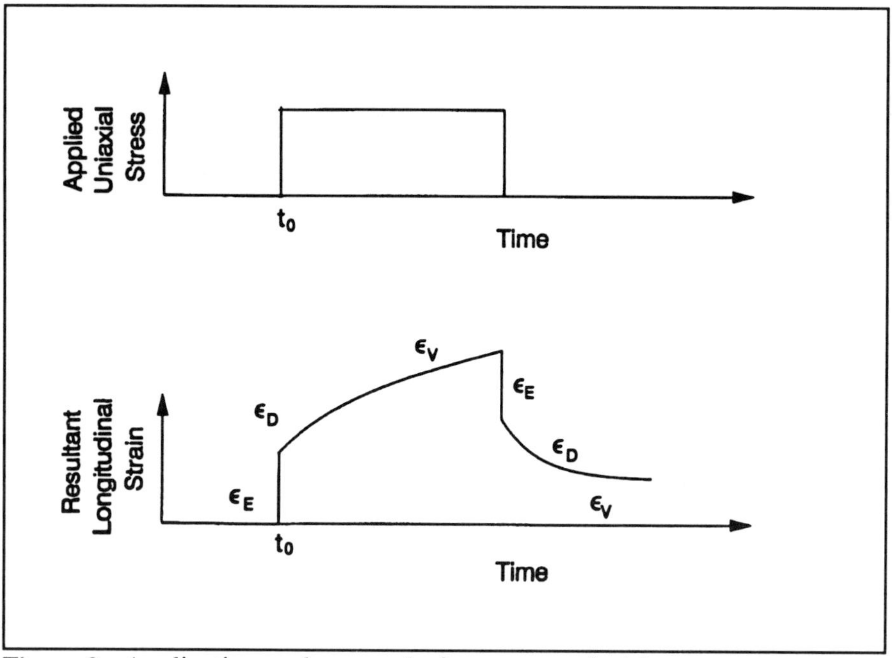


**Figure 2** - Application and removal of constant uniaxial stress.

elastically stretching the interatomic bonds. The delayed elastic response results from a reversible expansion (or compaction) of the structure (bond angle extensions, or disentanglements of structural elements) which raises the internal energy of the system and reduces the entropy. Viscous flow involves physical displacements of groups of atoms or structural elements past each other along quasi-equilibrium paths. Since this does not alter the free energy of the system, there is no thermodynamic recovery force when the stress is removed, therefore the viscous deformation is permanent.

**Models of Shear Response in a Viscoelastic Material**

It is generally helpful to begin by analyzing the behavior of simple combinations of elastic and viscous components (e.g., springs and dashpots).

A. Maxwell model

This mathematical model analyzes the response of a series circuit consisting of a spring and dashpot. The spring element contributes an elastic component, **G**, to the system, while the dashpot contributes a purely viscous component, $\eta$. The total strain is the sum of the elastic and viscous strains:

$$\epsilon(t) = \frac{\sigma(t)}{G} + \frac{1}{\eta}\int_0^t \sigma(t')\,dt' \tag{3}$$

When a constant stress, $\sigma_0$, is applied, the strain is simply written as:

$$\epsilon(t) = \frac{\sigma_0}{G}\left(1 + \frac{t}{\tau_s}\right) \tag{4}$$

where the shear relaxation time is $\tau_s = \eta/G$. This model gives an elastic and a viscous response, but no delayed elastic effect. Under constant strain, $\epsilon_0$, the Maxwell model yields a simple exponential decay of the stress according to the shear stress relaxation time:

$$\sigma(t) = G\epsilon_0 e^{-t/\tau_s} \tag{5}$$

B. Kelvin-Voigt model

This model has the spring and dashpot in parallel, thus the applied stress is distributed between them and the strain is the same for both elements. The dashpot does not permit an instantaneous elastic response. Under an applied stress, the strain is calculated as follows:

$$\epsilon(t) = \frac{1}{\eta}\int_0^t \sigma(t')e^{-(t-t')/\tau_r}\,dt' \tag{6}$$

This expression gives delayed elasticity but no instantaneous elasticity or viscous flow. $\tau_r$ is the retardation time. This expression simplifies under constant stress to:

$$\epsilon(t) = \frac{\sigma_0}{G}(1-e^{-t/\tau_r}) \tag{7}$$

C. Burger model

The essential features of the viscoelastic behavior of glass-forming liquids can be reproduced with a variety of combinations of Maxwell and Voigt elements. The simplest is the Burger model which puts them in series with each other, by using series Maxwell elements, $\mathbf{G_1}$ and $\eta_1$ and Voigt elements, $\mathbf{G_2}$ and $\eta_2$. Under an applied constant stress:

$$\epsilon(t) = \sigma_0[\frac{1}{G_1} + \frac{1}{G_2}(1-e^{-t/\tau_2}) + \frac{t}{\eta_1}] \tag{8}$$

The equation thus produces an instantaneous elastic strain, $\sigma_0/G_1$, a delayed elastic strain, $[1\text{-}\exp(-t/\tau_2)]/G_2$, and a viscous strain, $t/\eta_1$. Materials which do not have an exponential delayed elastic strain can be represented by sums (distributions) of Voigt elements with different characteristic times (e.g., retardation times). Sums of Maxwell elements introduce distributions of relaxation times to the viscosity.

D. Generalized models

Groups of Maxwell elements are best described in terms of the shear modulus: $G(t) = \Sigma G_i \cdot \exp(-t/\tau_i)$, while groups of Voigt elements are best described in terms of the inverse of the shear modulus which is the shear compliance, $J(t) = \Sigma(1/G_i)[1 - \exp(-t/\tau_{r,i})]$. These equations simplify the description of constant applied stress or constant applied strain responses.

| **Applied constant stress** | **Applied constant strain** |
|---|---|
| $\sigma(t) = \sigma_0$ | $\epsilon(t) = \epsilon_0$ |
| $\epsilon(t) = J(t) \cdot \sigma_0$ | $\sigma(t) = G(t) \cdot \epsilon_0$ |
| $\epsilon(t) = \sigma_0\{J_\infty + J_d \cdot [1\text{-}\exp(-t/\tau_r)]\}$ | $\sigma(t) = \epsilon_0 \cdot [G_0 + G_d \cdot \exp(-t/\tau_s)]$ |
| $J(t) = J_\infty + J_d \cdot [1\text{-}\exp(-t/\tau_r)]\}$ | $G(t) = G_0 + G_d \cdot \exp(-t/\tau_s)$ |
| with | with |
| $J(t=0) = J_\infty$ | $G(t=0) = G_\infty = 1/J_\infty$ |
| $J(t=\infty) = J_0$ | $G(t=\infty) = G_0 = 1/J_0$ |
| $J_d = J_0 - J_\infty$ | $G_d = G_\infty - G_0$ |
| $\tau_r$ = retardation time | $\tau_s$ = relaxation time = $G_0\tau_r/G_\infty$ |

The tradition of using subscripts to represent frequency is followed throughout this chapter, therefore, a slowly varying process is indicated by (0) while an instantaneous or rapid process is represented by ($\infty$). The above equations are a special solution of the operator differential equation for the mechanical response of a viscoelastic system:

$$\left[\frac{d}{dt} + \frac{1}{\tau}\right] \epsilon(t) = \left[J_\infty \frac{d}{dt} + \frac{J_0}{\tau}\right] \sigma(t) \tag{9}$$

Continuous spectra can be generated in a similar fashion to yield equations for the time-dependent shear modulus and compliance in terms of distributions of relaxation times **g($\tau$)** and **$\ell(\tau)$**, respectively:

$$\begin{aligned} G(t) &= G_0 + G_d \int_{-\infty}^{\infty} g(\ln\tau)\, e^{-t/\tau}\, d\ln\tau \\ J(t) &= J_\infty + \frac{t}{\eta} + J_d \int_{-\infty}^{\infty} \ell(\ln\tau)\left[1 - e^{-t/\tau}\right] d\ln\tau \end{aligned} \tag{10}$$

where the integrals are taken over **ln$\tau$** because the distribution functions are best represented as functions of **ln$\tau$** instead of $\tau$. (In liquids, the static shear modulus, $\mathbf{G_0}$ is zero.) Longitudinal or bulk moduli and compliances are written in a similar fashion [9,10].

The relaxation functions **g($\tau$)** and **$\ell(\tau)$** are deduced from experiments. Most experiments use an oscillatory stress or strain (e.g., acoustic, ultrasonic, optical vibrations) which are represented by complex functions:

$$\sigma^* = \sigma_0 e^{i\omega t} \quad \text{and} \quad \epsilon^* = \epsilon_0 e^{i(\omega t - \delta)} \tag{12}$$

These complex representations subsequently lead to complex moduli and compliances:

$$\begin{aligned} G^* &= G' + iG'' = G_0 + G_d \int_{-\infty}^{\infty} g(\ln\tau)\left[\frac{i\omega\tau}{1 + i\omega\tau}\right] d\ln\tau \\ J^* &= J' + iJ'' = J_\infty + \frac{i}{\omega\eta} + J_d \int_{-\infty}^{\infty} \ell(\ln\tau)\left[\frac{1}{1 - i\omega\tau}\right] d\ln\tau \end{aligned} \tag{13}$$

The real and imaginary components of the modulus, **G'** and **G"**, and the compliance, **J'** and **J"**, can be calculated directly from the oscillatory experimental measurements. They can then be fitted through the above equations by generating various distributions of relaxation times. The loss tangent **tan$\delta$** is calculated from the ratio of the imaginary to the real components of the modulus or compliance functions:

$$\tan\delta = G''/G' = J''/J' \qquad (14)$$

The viscosity can be expressed as a dynamic quantity with a complex form:

$$\eta^* = \eta' + i\eta'' = G^*/(i\omega) \qquad (15)$$

The steady-state or 'static' shear viscosity commonly measured at low frequencies by methods discussed in the next section can be shown to reduce to the following expression:

$$\eta_0 = \lim_{\omega \to 0} \eta'(\omega) = G_\infty \int_{-\infty}^{\infty} g(\tau)\,\tau\, d\ln\tau = G_\infty \bar{\tau} \qquad (16)$$

The Boltzmann superposition principle allows stress or strain application to be treated as additive processes, however, one must be careful to handle the summations properly [9].

## MEASUREMENTS

### Static Viscosity

There are numerous methods for measuring viscosity. Only the most common will be discussed here [11].

Falling Ball Technique

The viscosity of a liquid is measured through its resistance to the motion of a sphere in free fall under the force of gravity. The Stokes equation is used to calculate the force exerted on a sphere of radius **r** and density $\rho_s$, moving at a velocity **v** through a liquid of density $\rho$:

$$\eta_0 = \frac{2}{9}\frac{(\rho_s - \rho)}{v} r^2 g \qquad (17)$$

Sometimes a counterbalance is attached to the sphere to slow down or reverse its motion. The viscosity range covered is $10^1$-$10^7$ Poise (1 Poise = 1 dPas ).

Rotation Viscometer [12]

A small cylinder is rotated concentrically within a larger cylindrical crucible containing the liquid. The torque **M** transmitted to the crucible of radius, **R**, when the inner cylinder of radius **r** and length **ℓ** is rotated at an angular velocity of **ω**, leads to the viscosity through the equation:

$$\eta_0 = \frac{1}{4\pi\ell}\left(\frac{1}{r^2} - \frac{1}{R^2}\right)\frac{M}{\omega} \tag{18}$$

The viscosity range covered by this approach is $10^1$-$10^4$ Poise. This accurate technique has been used as a standard method to develop and certify many viscosity standards available from NIST (where, in most measurements, the outer crucible is rotated and the torque on the inner cylinder is measured to extend the range to $10^5$ Poise). For higher viscosities, one can measure the time, **Δt**, taken by the inner cylinder to go between two angles of deflection $\theta_1$ and $\theta_2$, but here, an instrument constant, **C'**, must be calibrated:

$$\eta_0 = C'\frac{\Delta t}{\ln\left(\frac{\theta_1}{\theta_2}\right)} \tag{19}$$

The viscosity range covered is $10^{4.5}$-$10^{7.5}$ Poise. A third method developed by Napolitano [12] extends the range of the same instrument to $10^{5.5}$-$10^{10}$ P by fixing the outer crucible and twisting the inner cylinder through a small angle while measuring the driving torque.

Parallel-Plate Viscometer [14]

A cylinder of glass (volume **V**) can be deformed between two parallel plates and the force **F** measured to reduce its height, **h**, at a rate **dh/dt**. Two cases must be considered in calculating the viscosity, either there is no slip at the plate-cylinder interface, or there is complete slip. Intermediate cases are covered by Sakoske [15]. The range achievable is $10^6$-$10^9$ P.

$$\begin{aligned} \textit{No slip:} \quad F &= -\eta_0\frac{dh}{dt}\left[\frac{3V}{h^2} + \frac{3V^2}{2\pi h^5}\right] \\ \textit{Full slip:} \quad F &= -\eta_0\frac{dh}{dt}\left[\frac{3V}{h^2}\right] \end{aligned} \tag{20}$$

The no-slip condition can be favored with metal plates. Full slip is more difficult to achieve, but can be approximated with graphite inserts between sample and plates.

Fiber-Elongation Viscometer [16]

Here a fiber is suspended between two supports in a furnace. The elongation rate **dℓ/dt** is related to the applied stress, **F/A**, where **A** is the time-dependent cross-sectional area of the fiber.

$$\eta_0 = \frac{F/A(t)}{3\frac{d\ell/dt}{\ell}} \tag{21}$$

This method is quite useful in that the narrow diameter of the fiber allows it to reach the furnace temperature rapidly and the viscosity may be measured soon after the load is applied. However, because of the large surface area to volume ratio of the sample, one must take care to insure that the surface does not change composition by either adsorption of furnace vapors or evaporation of glass components, since this will drastically alter the measurement [17]. Simmons et al. have used this technique to study phase separation in soda-borosilicates [18] and non-linear viscous flow in various glasses [19]. The range of viscosities achievable is quite large, spannning $10^9$-$10^{16}$ P. Problems of fiber hot length and changing fiber diameter during a measurement are discussed in [19].

Beam-Bending Viscometer [20,21]

In the high viscosity range, $10^{11}$-$10^{15}$ P, a bar of glass may be deformed by a force applied transversally to its midpoint. The equation for viscosity depends on **Mg**, the applied force, $\mathbf{I_c}$, the moment of inertia of the bar, **A** the cross-secional area of the bar and **L** the distance between supports:

$$\eta_0 = \frac{gL^3}{2.4I_cV}[M + \frac{AL\rho}{1.6}] \tag{22}$$

Miscellaneous Viscometers

The penetration viscometer measures the penetration of a sphere under a weight to cover the range $10^9$-$10^{13}$ P. The torsion viscometer measures the torsional deflection of a hollow glass tube under a torque in the range $10^{12}$-$10^{15}$ P. The cone and plate rheometer shears a thin disk of glass between a cone and a flat plate to cover the viscosity range $10^1$-$10^8$ P [22,23].

**Dynamic Viscosity**

For melts, ultrasonic relaxation spectroscopy may be used to obtain the longitudinal and shear moduli and compliances over the frequency range 0.5-100MHz [24,25] by measuring the velocity and absorption of ultrasonic waves propagating in the liquid. Light autocorrelation spectroscopy [26] can be used to study fluctuations in the Rayleigh scattered light from a melt to obtain the Fourier transform of the atomic diffusion coefficient and the relaxation of density fluctuations over time scales ranging from microseconds to hundreds of seconds [26].

Measurements in solid glasses include acoustic resonance and damped resonance (internal friction). Acoustic resonance is measured by suspending a bar of the solid between supports coupled to an acoustic driver and a detector. When the frequency of the forced oscillation corresponds to a bar resonance, the width (Q) and the frequency of the resonance yield the modulus and the loss tangent for the material. Bars of different size must be used to obtain measurements at different frequencies. The range however is very limited. For a rectangular bar the following equations apply [27]:

$$\nu_0 = \frac{3.56}{\sqrt{12}}\sqrt{\frac{E}{\rho}\frac{b}{L^2}} \qquad (23)$$
$$2\pi Q^{-1} = \frac{2\pi}{\sqrt{3}}\frac{\Delta\nu}{\nu_0} = 2\pi\tan\delta_e$$

where **E** is Young's modulus, $\rho$ the density, **L** and **b** the length and thickness of the bar, $\upsilon_0$ and $\Delta\upsilon$ are the resonance frequency and the half-width of the resonance.

Internal friction is measured using a torsional pendulum [28], in which a glass rod is initially twisted and allowed to oscillate torsionally. One measures the incremental decrease in amplitude between successive oscillations to obtain the loss tangent.

In both techniques using solid samples, the frequency range is varied by changing either the cross-section or length of the resonant sample. This drastically limits the range of frequencies achievable by these approaches. Therefore, most reported measurements are conducted at only a few frequencies and are conducted over a broad temperature range. The frequency constraint severely limits the ability of investigators to infer a distribution of relaxation times, especially if the distribution is temperature dependent. In fact, distributions of relaxation times are only accessible to these techniques when the distribution is temperature independent. This is termed thermo-rheological simplicity (TRS). Many investigators, using these techniques near the glass transition temperature, have assumed TRS without proof in order to analyze the behavior of these materials. As the reader will see below, many models have been developed which depend on TRS behavior.

## MODELS OF VISCOSITY, VISCOELASTICITY AND STRUCTURAL RELAXATION

The viscosity of a liquid, if not interrupted by a liquid-crystal phase transition, will vary with temperature over about 20 orders of magnitude. The viscosity, as seen above, is the product of the high-frequency shear modulus and the *average* relaxation time. Because of the averaging effect, the behavior of the viscosity with temperature is not a direct measure of changes in atomistic processes. The average relaxation time is susceptible to several factors, which include the simple slowing down of diffusion processes with lowered temperature, and more complex processes such as collective motions which probe the surrounding structure, motions sensitive to variations in local structure and non-exponential stress relaxation. These can affect the distribution of relaxation times or the nature of the relaxation process itself, making simple spring-dashpot models inappropriate. Therefore, in modeling the viscous behavior of a liquid, it is necessary to examine effects which (1) shift the relaxation processes to longer times with lowered temperature and diffusion coefficient, (2) alter the width of the distribution of these relaxation processes and (3) alter the nature of the relaxation processes thus either introducing new contributions to the distributions of relaxation times, or modifying totally the form of the integrated distribution. An example of (3) would be variations in the structure which alter the relaxation process from an essentially exponential one at high temperatures to a non-exponential process at lower temperatures. All the effects mentioned above will modify the *average* relaxation time and consequently the viscosity. Therefore, measurements of the viscosity alone cannot distinguish between atomistic causes of variations in the average relaxation time.

Models, in order to account for the large increase in viscosity with decreasing temperature, generally invoke an underlying thermodynamic or a kinetic singularity in the flow process. Scherer [7] describes the great variety of model approaches and divides them into rheological theories which address the temperature and pressure dependence of the viscosity, kinetic theories which treat the form of relaxation functions, and relaxation theories which follow the behavior of the distribution of relaxation times. Since most model development or comparisons are based essentially on viscosity measurements as a function of temperature rather than on the relaxation time spectrum, too much latitude is present in the tests. Scherer's paper presents over 14 different models, each of which has some limited value in the description of viscous flow. The problem is that the 'static' viscosity is not sufficiently sensitive to the atomistic processes of viscous flow. Below, we will examine the most notable of these models and identify the missing data which might elucidate the microscopic processes determining the mechanical response of liquids to applied stress or strain.

For the discussions to follow, let us define a relaxation function, $\phi(t)$ for an isothermal measurement at **T** after a jump from another temperature:

$$\phi(t) = \frac{f(T,t) - f_0(T,\infty)}{f(T,0^+) - f_0(T,\infty)} \tag{24}$$

where **f(T,t)** is the temperature and time-dependent measurement of the property of interest (stress, strain, specific volume, enthaly, etc.); $\mathbf{f_0(T,\infty)}$ is the equilibrium value at infinite time and the time $\mathbf{0^+}$ represents the small increment of time necessary to allow the elastic response to pass. Three equations will be used to discuss the form of the relaxation function:
(1) The Debye or exponential relaxation:

$$\phi(t) = e^{-t/\tau_0} \tag{25}$$

(2) The Distribution of Exponential Relaxation Times (DERT) with weighting factors $\mathbf{w_i}$:

$$\phi(t) = \sum_{i=1}^{\infty} w_i e^{-t/\tau_i} \tag{26}$$

(3) The stretched exponential or KWW function [29] whose exponent varies between 1 (exponential relaxation) and almost zero (very broad distributions):

$$\phi(t) = e^{-(t/\tau_0)^\beta} \tag{27}$$

Several experimental observations seem to indicate the following bounds on theories of viscosity: (1) structural relaxation is non-linear, and (2) structural relaxation and viscoelasticity appear non-exponential, (DERT) or (KWW), near the glass transition temperature. Additional conditions which are not as universally observed, yet are followed by a large majority of glasses include thermorheological simplicity at low temperatures (near $T_g$) and exponential (Debye) stress relaxation at very high temperatures.

**Rheological Theories**

Below, we will examine the principal empirical viscosity theories and discuss their physical bases and limitations. All theories produce 'static' viscosity equations which will be compared to the general trends in viscosity behavior, such as high-temperature Arrhenian behavior in which the temperature dependence of the viscosity can be expressed as **exp(E/kT)**, intermediate temperature non-Arrhenian behavior with an increase in viscosity greater than predicted by Arrhenian behavior, and thermo-rheological simplicity near the glass transition temperature. This kind of pattern is not followed by all glasses, but it is so widespread among oxide, fluoride and organic glasses, that it provides a good comparison for viscosity theories [1,30].

Eyring [31] calculates a viscosity expression from the probability, under an applied stress, for motion between two equilibrium positions separated by an energy barrier, **ΔG**:

$$\eta(T) = \frac{\sigma_0 \, e^{\frac{\Delta G}{kT}}}{2\,\nu\,\sinh(\sigma_0 V_0 / 2kT)} \tag{28}$$

where $\sigma_0$ is the difference in stress between the two sites; $\nu$ is the jump attempt frequency and $V_0$ is the volume occupied by a molecule. In the limit of small differences in stress, the equation reduces to the Arrhenian form:

$$\eta(T) \approx \frac{kT}{\nu V_0} e^{\Delta G/kT} \approx \eta_0 \, e^{\Delta G/kT} \tag{29}$$

This equation was later modified [32] to account for structural randomness:

$$\eta(T) = \frac{kT}{\nu V_0} \frac{e^{(\Delta G/kT - A^2C^2/2)} \, sech[B(1 + AC^2)]}{[(1 + AC^2)^2 + C^2]} \tag{30}$$

where **A** represents the degree of coupling between the distribution of activation energies and structural variations (**A** = **E** $\mathbf{a_0}$/**kT** in which **E** comes from the distribution of activation energies written as $\Delta\mathbf{G}(\mathbf{a}) = \Delta\mathbf{G}_0 + \mathbf{E}(\mathbf{a}-\mathbf{a}_0)$ and **a** is the jump distance); **B** is a dimensionless stress factor and **C** is the width of the distribution of jump distances.

The Eyring Equation only seems to apply to the high temperature behavior of glasses. However, Goldstein [33] indicates that in the region where Arrhenian behavior is observed, the rate of structural relaxation is as high as stress relaxation therefore the concept of an activated state is inappropriate since the structure breaks down in the time it takes to make a stress relieving jump. This question will return later in the examination of mode-coupling models. Goldstein [33] and Scherer [7] indicate that the Arrhenian behavior may simply be a result of the probability for diffusive motion being directly related to the energy distribution of the atoms which follows Boltzmann statistics.

The Vogel-Fulcher-Tamman Equation [34] has been the most successful in fitting a wide range of viscosities with only 3 adjustable parameters. It is purely empirical and provides for the deviation from Arrhenian behavior with decreasing temperature:

$$\eta(T) = \eta_0 \exp[\frac{A}{T-T_0}] \qquad \textbf{(31)}$$

While the equation diverges at $\mathbf{T_0}$, this is generally below the glass transition temperature and is never reached. However, the VFT Equation fails well before reaching the glass transition temperature, as it overestimates the viscosity at low temperatures where Arrhenian behavior reappears.

Investigators who simulate glass-forming liquids in MD studies have learned the great value of local free volume in facilitating mass transport. But, long before simulations of dynamics of liquids became popular, Cohen and Turnbull [35] proposed a concept of viscosity in which stress relaxation was related to the availability of local free volume for structural redistribution. The model was an outgrowth of work by several investigators [36]. It postulates the existence of free volume as the difference between the local volume around each particle less that occupied by the particle. A distribution of free volume is associated with each particle or structural unit at all temperatures. Molecular transport only occurs when the free volume redistributes itself locally to reach the molar volume of the flow unit, **v***. No free energy is required for the free volume redistribution, and the viscosity can be calculated from the probability of volume redistribution:

$$\eta(v_f) = \frac{kT}{3\pi a_0^2 g u} e^{\gamma v*/v_f} \tag{32}$$

$\mathbf{a_0}$, **g** and **u** are a characteristic jump distance approximately the cage diameter, a geometrical factor, and the particle velocity estimated from kinetic theory, respectively. $\gamma$ is a recovery factor between ½ and **1**. The free volume is obtained from the thermal expansion of the amorphous phase above the glass transition temperature with $\Delta\alpha$ the difference in coefficient of thermal expansion between the liquid and glass, $\beta$ the compressibility, and the second term added to account for **ΔP** variations in pressure:

$$\begin{aligned} v_f &= 0 \qquad \textit{for } T < T_g \\ v_f &= v_m \Delta\alpha(T-T_0) - v_p \beta \Delta P \qquad \textit{for } T > T_g \end{aligned} \tag{33}$$

As the temperature is lowered, the $(\mathbf{T}-\mathbf{T_0})$ term causes curvature above the Arrhenian behavior. However, if $\mathbf{T_0} = \mathbf{T_g}$ then the curvature is too large and $\mathbf{T_0} = \mathbf{T_g} - \Delta\mathbf{T}$ must be used with $\Delta\mathbf{T} \sim$ **20-50°C**. This model has been shown [37] to give a good fit of the viscosity behavior of glasses up to about $\mathbf{10^4}$ **P**. However, near $\mathbf{T_g}$, the modeled viscosity is too large and exhibits too high a curvature. The model has further problems such as heating at constant volume which removes much of the change in viscosity, and ignores configurational rearrangements.

Some of the limitations of the Free Volume model were addressed by several investigators. Macedo and Litovitz [37] argue that while the Free Volume Model calculates the probabilty for making the hole, it neglects the activation energy required for making the diffusive jump. Thus, they develop a Hybrid Equation which combines the Free Volume term with the Eyring term. This approach repaired one of the problems of the Free Volume model in that it exhibits Arrhenian viscosity for flow under constant free volume (isostructural) conditions. However, the model still predicts too much curvature at low temperatures. Cohen and Grest [38] address the question of percolation of critical free volume regions through the structure. The Cohen and Grest modifications will be discussed after the configurational entropy model since they combine both approaches.

The Configurational Entropy Model was developed by Adam, Gibbs and DiMarzio [39,40]. This approach postulates that rearrangements of the structure are only possible when a region has a minimum size able to include at least 2 structural configurational states. Transition probabilities are thus cluster size dependent and the viscosity is a function of the configurational entropy, $\mathbf{S_c}$ (e.g., difference between the liquid and the crystal entropy). Following the concepts of the Kauzmann Paradox [41] which extrapolates the configurational entropy to a vanishing point at $\mathbf{T_2}$, a sub-$T_g$ phase transition temperature, the viscosity diverges at the same singularity:

$$\eta(T) = \eta_0 \, e^{\Delta\mu S^*/(kTS_c)} = \eta_0 \, e^{b/(\Delta C_p \ln(T/T_2))} \propto A \, e^{B/(T-T_2)} \qquad (34)$$

**B** is proportional to the height of the energy barrier, $\Delta\mu$, for cooperative rearrangement; **S*** is the entropy of the smallest cooperatively rearranging subsystem, **S* = k ln2**, and $\mathbf{T_2}$ is the 2nd order phase transition temperature $\mathbf{T_2 \approx T_g - 50°C}$. The model reduces to the Fulcher Equation readily, except that $\mathbf{T_2}$ is fixed by the extrapolation of the configurational entropy measurement. As far as fitting the temperature dependence of the viscosity, this approach has the same limitation as the Fulcher Equation with viscosity increasing too strongly with decreasing temperature. However, it yields the correct isostructural viscosity dependence at low temperatures. Furthermore, the model better [42] exhibits the effect of pressure on $\mathbf{T_g}$ [43,44]. Conceptually, the model resolves the Kauzmann paradox by showing that the structural relaxation process diverges at $\mathbf{T_2}$, kinetically arresting a phase transition so that the $\mathbf{S_c = 0}$ point is never reached.

One of the puzzling issues presented by the Configurational Entropy Model and the Kauzmann Paradox is the question of the underlying phase transition, because, if this transition is second order, it is well known [45] that supercritical fluctuations are present well above the transition temperature. These could be detected, their temperature-dependence analyzed by well characterized scaling laws, and the transition temperature calculated relatively accurately. Such an experiment has not yet been conducted to the knowledge of this writer.

Many investigators promote the concept that there is a strong relationship between the temperature dependence of viscosity and configurational entropy, especially near the glass transition temperature. The form proposed by Adam and Gibbs, while providing a good first approximation may require refining or modification. Sillinger [46], for example, proposes flow by structural rearrangements involving the movement of low, free volume or densely packed regions. His expression uses the configurational entropy to the 2/3 power. Fredrickson and Brawer allow internal rearrangements of the configurations, even if they are close packed, as long as a sufficient number of neighbors are not [47]. More on this later.

**Structural Relaxation**

Let us first make a distinction between structural relaxation and viscoelasticity. Brawer [5] gives an example of non-linear structural relaxation. Consider the graphs of Fig. 3, representing a time-temperature dependent property **f(T,t)** of the viscous liquid. If the ambient temperature is changed from $\mathbf{T_1}$ to $\mathbf{T_2}$, then the property undergoes an instantaneous elastic response changing from the initial equilibrium value of $\mathbf{f_0(T_1,\infty)}$ to $\mathbf{f(T_2,0^+)}$, which is followed by a relaxational response to the equilibrium value at the new temperature $\mathbf{f(T_2,\infty)}$. The relaxation function is written as Eq. 24 above. The relaxation function can then be examined for 3 different cases of temperature change: (a) an incremental change from $\mathbf{T_{2+\epsilon}}$ to $\mathbf{T_2}$, (b) a change from $\mathbf{T_1}$ to $\mathbf{T_2}$, and (c) an incremental change from $\mathbf{T_{1+\epsilon}}$ to $\mathbf{T_1}$. If $\mathbf{T_2 < T_1}$, then the behavior of the relaxation function is as shown in Fig. 3. Curve (c) has the fastest decay because the relaxation process is occuring at a higher temperature. Curve (a) has the slowest decay because its relaxation process is occuring at a lower temperature. The interesting response is shown in curve (b). Here, while the temperature for the relaxation process is the same as that for curve (a), namely $\mathbf{T_2}$, the decay process is intermediate between the two other cases and in fact does not equal the behavior of curve (a). Physically, we can easily understand this. The configurational structure of the liquid does not change instantaneously, therefore, intially, it still has the more open structure of the $\mathbf{T_1}$ state, but, as its structure relaxes to a less open configuration, more representative of the $\mathbf{T_2}$ state, its relaxation process slows down and approaches the decay rate of curve (a). Since the relaxation function is dependent on the magnitude of the temperature jump,

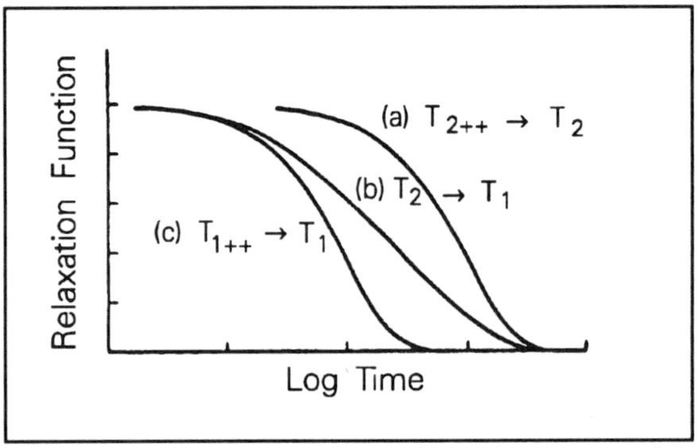


**Figure 3** - Structural relaxation [5]. Curves (a) and (c) correspond to a small temperature change.

then it is a *non-linear* relaxation process. A similar description of non-linear structural relaxation is presented by Scherer [7] for the density behavior of a soda-lime silica glass [48] in which the rate of relaxation is dependent on the direction and magnitude of the temperature jump. The process can appear to be linear only if the temperature jump is sufficiently small that the structural change is relatively undetectable [49].

Non-linear structural relaxation is a direct result of structure-dependent changes in the properties. Thus, structural rearrangements following a small drop in temperature occur more readily than those following a small jump in temperature to the same final value because the more open starting structure of the former facilitates mass transport, while the smaller specific volume of the starting structure in the latter resists structural rearrangements. Thus, non-linearity in structural relaxation is a consequence of the glassy state and does not require non-exponential relaxation.

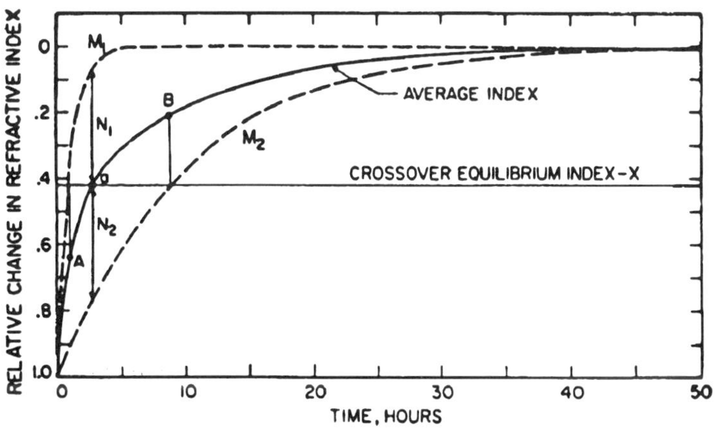


**Figure 4a** - Time-dependent refractive index illustrating the fast ($M_1$) and slow ($M_2$) relaxation times after a decrease in temperature [52].

A direct study of the history-dependence of structural relaxation processes in molten glasses is the set of classic cross-over experiments conducted by Macedo and Napolitano [50,51]. Studying borosilicate crown [50], germania [51], and boron trioxide [52] glasses, they accurately measured the change in refractive index of very small samples with temperature of heat-treatment. For long heat treatment periods, they established the equilibrium relationships $\mathbf{f_0(T,\infty)}$ for the index as a function of temperature. The cross-over test consisted of taking samples heat-treated at one temperature and changing their ambient temperature. Figure 4 shows such a study in boron trioxide. Samples were heat treated at 310.5°C untilequilibrium, then they were transfered to a furnace at a temperature of 225.5°C while the refractive index was monitored as it increased due to the drop in temperature. When the refractive index reached the equilibrium value of the intermediate temperature of 270.2°C, the samples were transferred to an isothermal treatment at that temperature and the refractive index measured as a function of time. The index was seen to drop initially and then asymptotically rise to

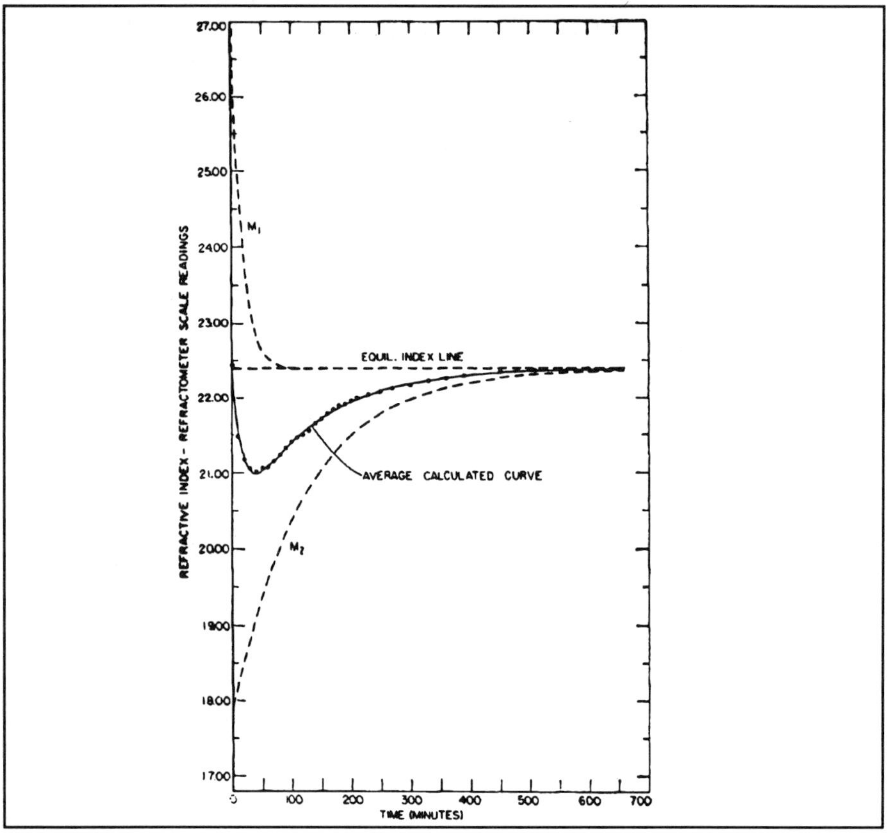

**Figure 4b** - Cross-over behavior for $B_2O_3$ at 270.2C [52].

the equilibrium value for the isothermal treatment temperature. Macedo and Napolitano showed that this unusual cross-over behavior could be well explained by assuming that the structural relaxation processes in the glass occurred at a variety of rates, some fast, and others slow. Approximating the behavior with a fast and a slow exponential relaxation, they were able to show that as the temperature was first lowered below the cross-over temperature, some configurations relaxed quickly to their new equilibrium states, while others relaxed slowly. The index of refraction which averages over the density of all configurations, gave some mean value averaging the two kinds of local densities. When the sample was transferred to a cross-over isothermal treatment, the mean structure exhibited the equilibrium index for that temperature, but the local structure was not at equilibrium. Thus, during the isothermal treatment, the configurations which relax rapidly ($\mathbf{M_1}$) were at the high index density, so they relaxed rapidly back to an intermediate index density, while the slow relaxation configurations ($\mathbf{M_2}$) which were at densities still below the equilibrium continued to increase slowly. The sum of both processes exhibited an initial decrease in index due to the readjustment of the fast processes, followed by the asymptotic increase driven by the slow processes.

As shown in the figure (*average calculated curve*), the process is remarkably well fitted by assuming two relaxation processes. (A continuous distribution of relaxation times could have been equally appropriate.) Other investigators claim that no significance should be associated with sums of exponential relaxations since non-exponential processes can also be used. In fact, the stretched exponential equation of KWW has been often proposed to describe these processes.

What is shown here is that macroscopic measurements of equilibrium behavior do not reveal the complex relaxation processes which control them. Viscosity models based on the 'static' viscosity behavior will inevitably be empirical.

At this point, it is instructive to look at the Tool-Narayanaswamy (TN) model [53] of structural relaxation and its various modifications. An excellent review of the current status of TN is found in Moynihan [54]. Structural relaxation as viewed by the Narayanaswamy school consists of the time-dependent change in macroscopic properties following a jump in thermodynamic variables. The change is associated with the slow rearrangement of the structure of the liquid to reach a new equilibrium. The approach assumes that the same normalized relaxation function will represent the structure regardless of the magnitude of the temperature jump and the beginning and ending temperatures. In an example where the temperature changes abruptly from $T_1$ to $T_2$, the property spans the range between its value for the two temperature equilibria as depicted in Equ. (24). If one ascribes the fictive temperature, $T_f$ to the intermediate value of the relaxation function, then its normalization requires that we define the function similarly to Equ. (24):As $T_f$ varies from $T_1$ to $T_2$, the equilibrium state corresponds to $T_f = T$.

$$\Phi(t) = (T_f - T_2) / (T_1 - T_2) \tag{35}$$

The TN model sums a distribution of relaxation times to form the relaxation function:

$$\Phi(t) = \sum_i g_i \exp[-\int_0^t dt'/\tau_i] \tag{36}$$

with $g_i$ as the temperature-independent weights of the relaxation times $\tau_i$. Away from equilibrium, the relaxation times simply depend on the actual temperature T, and the relaxational state of the system, $T_f$, which assures non-linear character. Narayanaswamy writes the relaxation times through the product of two Arrhenius terms dependent on T and $T_f$:

$$\tau_i = A_i \exp[\frac{x \Delta H^*}{RT} + \frac{(1-x)\Delta H^*}{RT_f}] \tag{37}$$

The parameter x is experimentally adjustable and has been found to require a temperature dependence. The stretched exponential has also served to represent the relaxation time function [Equ. (27)]. Recently, Scherer [55,56] and Hodge [57] based the expression for the relaxation times on the Adam-Gibbs model [Equ. (34)]:

$$\tau_i = A_i \exp[\frac{\Delta H}{RT}\ln(T_f/T_2)] \approx A_i \exp[\frac{\Delta H}{RT(1-T_2/T_f)}] \tag{38}$$

where $T_2$ is the phase transition temperature or the Fulcher $T_0$ term. Nagel and Dixon [58] show that the Fulcher form can be derived from the KWW form using a selected temperature dependence for $\beta$.

Rekhson [59] explains that the TN expression is devised to propose an equivalence between any high temperature schedule and a sum of steps in which the material is slowly cooled to $T_f$. This allows the principle of superposition to hold. Moynihan shows that this approach reduces to replacing the characteristic property of the glass by its fictive temperature:

$$T_f(t) = T_0 + \Delta T_1[1 - \Phi(t-t_1)] + \Delta T_2[1 - \Phi(t-t_2)] + ..... \tag{39}$$

Moynihan suggests that the TN expression using the stretched exponential relaxation time can be generalized by assuming infinitely small discrete steps and a temporal variation in relaxation time $\tau_0(t)$:

$$\Phi(t) = \exp[-\int(dt'/\tau_0)^\beta \tag{40}$$

But using this approach to fit specific heat data from Birge and Nagel [60], he finds $\beta_{Nagel} > \beta_{exp}$, and forcing $\beta$ to the experimental value does not fit the specific heat data. Large **x** values lead to more pronounced disagreement.

Rekhson [59] resolves the dilemna in an elegant self-consistent approach which promises to lead to a profound understanding of non-linearity in structural relaxation. He notes that all experimental tests of the relaxation function require a temperature dependent stretched exponential, $\beta(T)$ and writes it as:

$$\beta(T) = b(1 - T_0/T) \tag{41}$$

Following Rendell and Ngai [61], he calculates the relaxation time:

$$\tau(T) = [\beta\, \omega_0^{(1-\beta)} \tau_0 e^{\Delta H/kT}]^{1/\beta} \tag{42}$$

in which the term $\omega_0 = 10^{10}$. When $\beta = 1$, the above function reduces to a simple Arrhenian expression of single relaxation: $\tau(T) = \tau_0 \exp[\Delta H/kT]$. This leads to an inverse attempt frequency $\tau_0 = 1.4 \times 10^{-14}$ sec. At lower temperatures, the Arrhenian behavior is broken and the value of $\beta$ decreases. Using $T_0$ as the Fulcher constant, and choosing the parameter, **b**, such that $\beta = 1$ at 1100°C where the distribution of relaxation times has been measured to be narrow [24], and $\beta = 0.57$ at 530°C where the distribution is broad, he finds that the value of the term: $(\omega_0^{1-\beta})^{1/\beta}$ is **1** when $\beta = 1$, and $10^{10}$ when $\beta = 0.5$. This term therefore represents the large deviation from Arrhenian behavior in $\tau(T)$ measured at low temperatures.

This approach fits well the viscosity behavior of NBS-710 glass over its entire temperature range.

Turning to structural relaxation, Rekhson replaces the stretched exponential form by the distribution of relaxation times [Equ. (36)] using the expression $\ln \tau_i = \ln \tau_0 + \Delta H_i/(T-T_2)$, and examines the activation energy values, $\Delta H_i$. He finds that as the function relaxes, the short relaxation time processes (small $\tau_i$) occur first with very low activation energies, followed by the slower relaxation processes with higher activation energies. At low temperatures, the slow relaxation processes become frozen, and only the 'faster'processes relax, therefore, the experiments exhibit a drastically reduced activation energy in the process. This leads to the marked decrease in $\beta$ at low temperatures.

In a sense, this could be visualized as the behavior of a compound dashpot, made from soft and hard sections which respectively respond quickly and slowly. When a stress is applied, the first extensions are from the fast sections followed by the increasingly slower sections. The amount of response to any applied stress is dependent on the applied stress, as the strain partitions disproportionally between the soft and hard sections. Therefore, the extension or compression of the spring is non-linear. At low temperatures, if we assume that the slow parts of the spring become frozen, only the softer, more compliant sections will respond, exhibiting a reduced spring constant (or activation energy).

**Viscoelasticity**

Theorists, have in general, exploited two means of obtaining low temperature non-exponential viscoelastic behavior:
(a) requiring that the relaxation processes are cooperative (e.g., that stress or structural relaxation involves rearrangements of large groups of particles),
(b) postulating the existence of structural heterogeneities which can lead to distributions of free volumes, densities, compliant defects, etc.

Measurements of viscoelasticity were conducted by Tauke [24] and Simmons [62] through isothermal ultrasonic propagation studies in molten glasses (boron trioxide and soda-borosilicates). They measured the frequency dispersion of the ultrasonic velocity and absorption. A time transform of the dispersion in the frequency-dependent modulus yields a distribution of relaxation times representing a sum of exponential relaxation processes. While the transform is not unique, and could have used a non-exponential relaxation

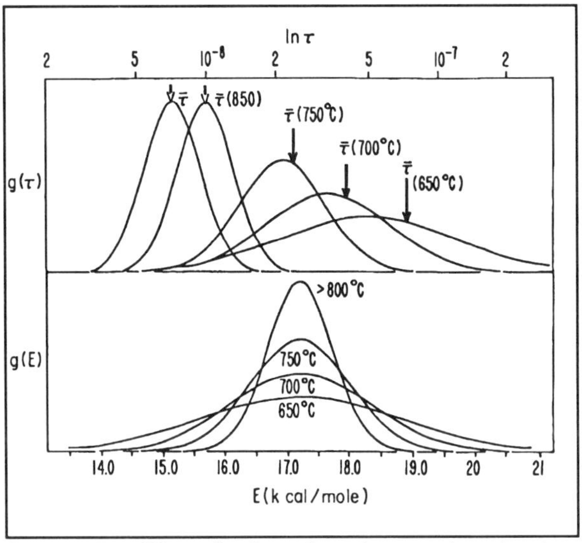


**Figure 5a** - Plot of relaxation time and activation energy spectra for $B_2O_3$ [24].

process as well, both studies showed similar trends, which consisted of narrow distributions of relaxation times at high temperatures, broadening as the temperature was lowered, then becoming broad and temperature-independent (thermo-rheological simplicity) near the glass transition temperature (see Fig. 5 for $B_2O_3$ and soda-borosilicate glass)[63,64]. At high temperatures and viscosities below several hundred poise, the distribution of relaxation times was single indicating that only 1 relaxation time was required to fit the time-transformed dispersion of the frequency-dependent modulus. Consequently, the average and most probable relaxation times, as shown in Fig. 5, were the same, and the viscosity behavior was Arrhenian (Log viscosity linear in inverse temperature) since it is proportional to the *average* relaxation time. As the temperature was lowered, the distribution of relaxation times broadened asymmetrically. (The authors surmised the distribution to be logarithmic in time which allows an activation energy formalism to be used in which the distribution of activation energies is symmetric in energy.) The asymmetric broadening of the distribution of relaxation times clearly shows a growing difference between the *average* and the *most probable* relaxation times and a departure of viscosity from Arrhenian behavior at values above $10^3$ P in $B_2O_3$. Between $10^3$ and $10^8$ P, the viscosity is non-Arrhenian. Then, at low temperatures, and viscosities above $10^8$ P in boron trioxide, it returns to Arrhenian behavior [30] implying that the broadening of the distribution saturates and becomes independent of temperature down to the glass transition. While the average and most probable relaxation times remain different, the temperature dependence of the average relaxation times seems to return to Arrhenian behavior. Since the viscosity again exhibits Arrhenian behavior, the viscoelastic behavior of the melt becomes thermo-rheologically

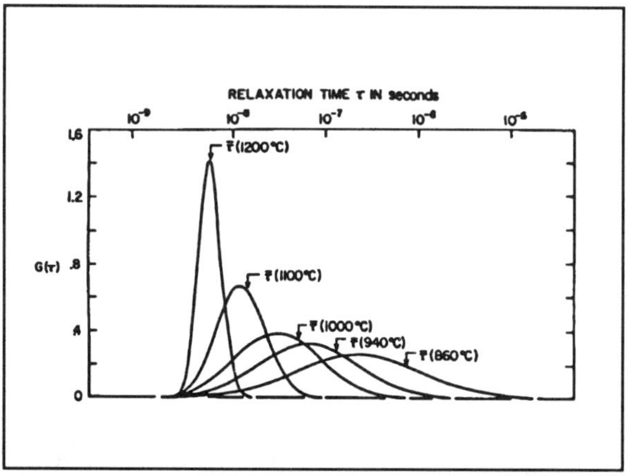


**Figure 5b** - Relaxation time distribution for borosilicate glass [65].

simple. This low temperature Arrhenian viscosity behavior is critical to understanding viscoelasticity and structural relaxation in glasses since (1) it satisfiesthermo-rheological simplicity observed in structural relaxation studies [6]; (2) it is consistent with a broad distribution of relaxation times or non-exponential relaxation near the glass transition, and (3) it is the most bothersome factor to the many empirical viscosity models which predict a continuously diverging viscosity as the glass transition is approached.

These results lead us to questions regarding universality of the effect, the source of broadening distributions of relaxation times, and the relevance of the interpretation of sums of exponential relaxation processes in terms of atomistic physical processes.

Local variations in activation energy can result from a variety of processes, including density fluctuations, composition fluctuations or any other change in local order. Simmons and Macedo [64,65] examined the behavior of a soda-borosilicate glass known to have supercritical fluctuations in composition. Using well established supercritical phase separation theory, they were able to calculate expected composition fluctuation sizes. Since supercritical fluctuations vary in size with temperature, they were able to postulate temperature-dependent structural and compositional heterogeneities. The fluctuation times were also calculated and were found to be longer than the shear relaxation times [66], therefore, the composition fluctuations appeared static to the particles involved in relaxing the shear stress. The investigators assumed that since the frequency dispersion of the modulus could be fitted by a single relaxation time at high temperature, the use of sums of exponential relaxation processes would also be appropriate at lower temperatures. Using a single adjustable parameter, they developed a model which related the broadening of the distribution of relaxation times to the

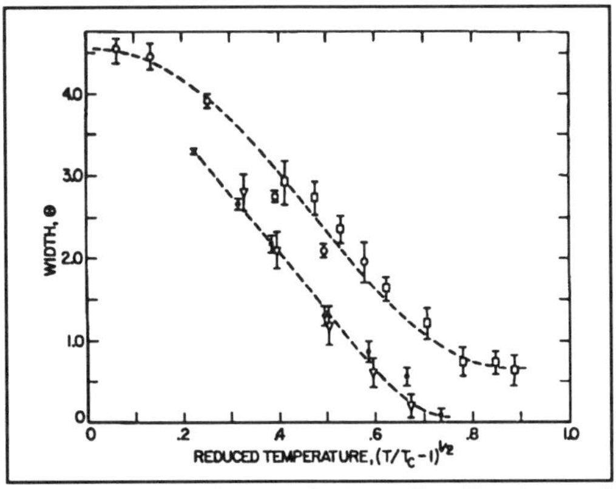


**Figure 6** - Width of the distribution of relaxation times for borosilicate glasses as a function of reduced super-critical temperature.

growth of structural heterogeneities developed by the supercritical composition fluctuations (see Fig. 6). The model simply assumed that the activation energy for stress relaxation of a particle in a liquid was dependent on the composition of its local environment. Thus regions of different composition would respond with different activation energies and relaxation times. This is an old model which has often gone in and out of fashion. Since the fluctuation sizes are very small at temper-atures well above the critical point, this approach easily explains the observation of a single distribution of relaxation times as local variations are too small and easily averaged in each particle's activation neighborhood. At lower temperatures, the fluctuation sizes grow thus creating a broadening distribution of relaxation times as local differences appear in each particle's neighborhood. The major conceptual contribution of Simmons and Macedo was the assumption that the regions which affect relaxation of each particle are finite, postulating some interaction range, $\mathbf{r}_0$, in Fig. 7. Thus as the structural heterogeneities grow beyond this range, further changes in size are relatively ineffective in changing the make up of

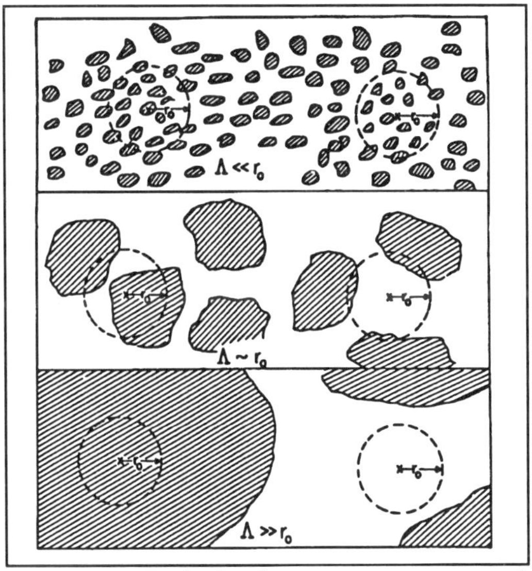


**Figure 7** - Typical microstructure expected to produce the 3 regions of behavior in the distributions of relaxation times.

the local neighborhoods and the associated distribution of relaxation times, consequently leading to a temperature independent region where the structural heterogenieties are greater than the interaction range (Fig. 7).

The Simmons-Macedo Environmental Relaxation Model (ERM) Equation therefore expresses the viscosity in terms of the shift in the average relaxation time due to the broadening distribution of relaxation times. The degree of broadening was obtained by fitting the ultrasonic relaxation data, which also gave the position of the most probable relaxation time, curiously found to follow an Arrhenian behavior throughout the temperature range of the measurements. The ERM Equation thus restricts the temperature dependence of the viscosity to the Arrhenian term in activation energy and the structure or fluctuation size term **Λ(T)**:

$$\eta(T) = \eta_0 \, e^{\frac{E_0}{kT}} \exp\left\{\frac{\Delta E}{kT}\left[\frac{1}{1+[r_0/\Lambda(T)]^2}\right]^{3/2}\right\} \tag{43}$$

with $\mathbf{r_0}$ as the adjustable interaction range parameter. At high temperatures, $\Lambda << \mathbf{r_0}$, and the melt exhibits single exponential behavior as the viscosity exhibits Arrhenian behavior:

$$\eta(T) = \eta_0 \, e^{\frac{E_0}{kT}} \tag{44}$$

At low temperatures, $\Lambda >> \mathbf{r_0}$, the melt exhibits thermo-rheological simplicity and the viscosity again exhibits Arrhenian behavior:

$$\eta(T) = \eta_0 \, e^{\frac{E_0 + \Delta E}{kT}} \tag{45}$$

Non-Arrhenian behavior and temperature dependence of the distribution of relaxation times are observed in the intermediate region where $\Lambda \approx \mathbf{r_0}$. The interaction range (e.g., for cooperative relaxation), $\mathbf{r_0}$, was calculated from the calculated sizes of the supercritical fluctuations, and was found to be 41-50 Å.

The Arrhenius/non-Arrhenius/Arrhenius (A/NA/A) behavior of viscosity with decreasing temperature has been observed in many inorganic and organic glasses as well as molten salts. Angell uses this concept to classify glasses according to degree of curvature [67]. *Strong* glasses follow Arrhenian behavior over the majority of their melt forming region while *fragile* glasses exhibit a very pronounced curvature (Fig. 8). Strong liquids exhibit a very small change in heat capacity due to a pronounced stability in short and medium range order. The distributions of relaxation times are very narrow, nearly single, and the average relaxation time is nearly always Arrhenian. Fragile liquids have structures which become more degenerate with increasing temperature, a large heat capacity at the glass transition, and a complex

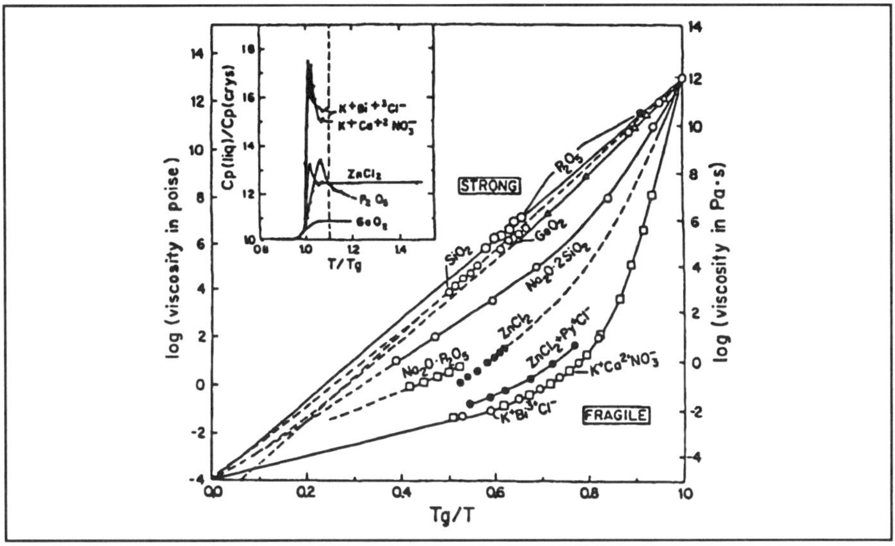


**Figure 8** - Classification of viscosity behavior [1].

relaxation spectrum. The behavior is typically A/NA/A, and the relaxation processes follow non-exponential form, often according to KWW. The image offered of glass-formation by this structural analysis can be analyzed in terms of thermodynamic and structural properties. The model is consistent with the configurational entropy model and the viscosity follows the Fulcher form except near $T_g$. The higher configurational specific heat corresponds to a greater curvature. Structurally, the strong liquids have well defined coordination shells, well-defined short and medium range order, small thermal degeneracy associated with bond breakage, and flow processes defined by defect migration mechanisms. Fragile liquids have high configurational degeneracy, and less well defined coordination shells. Generally the radius ratios exhibited by fragile liquids are much larger than those of the strong liquids.

This approach is also consistent with the ERM concepts of Simmons and Macedo, in which the non-Arrhenian curvature in viscosity arises from the formation of regions with different activation energies. Formation of regions of different local density, composition or order will yield a high configurational specific heat, while glasses with well defined short and medium range will not exhibit significant variations in local activation energy. The major difference, however, is that, in the ERM, distributions in activation energies affect the inverse-temperature dependent exponent, while variations in configurational specific heat only change the pre-exponent, and in the Angell model the curvature in the viscosity results directly from the large configurational specific heat.

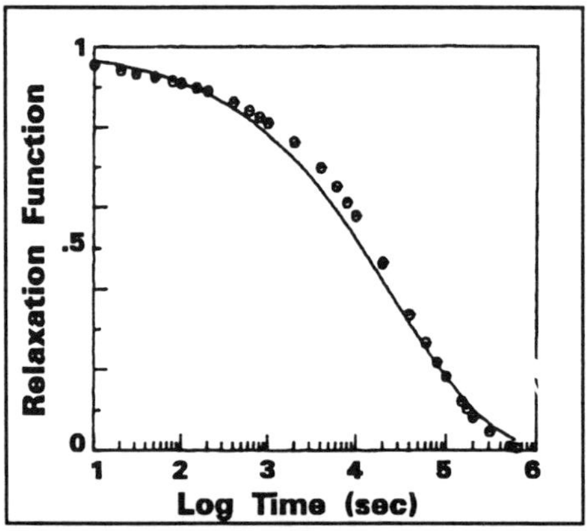


**Figure 9** - Stress relaxation data of Kurkjian [69] fitted with the KWW function [6].

The next question relates to the universality of the process. Rekhson has shown that any relaxation process can be represented by discrete exponential relaxation processes [6,68]. Figure 9 shows a fit of stress relaxation data from Kurkjian [69] by a KWW stretched exponential. Figure 10 shows strain relaxation data of Rekhson et al. [70] fitted by a sum of weighted discrete relaxations. A comparison by Rekhson to a KWW fit shows that the sum of relaxation times gives a better fit. The range of relaxation times used can cover up to 5 orders of magnitude. It is clear that the addition of more adjustable parameters would indeed improve fits of the relaxation function. However, the need for numerous parameters does not automatically eliminate

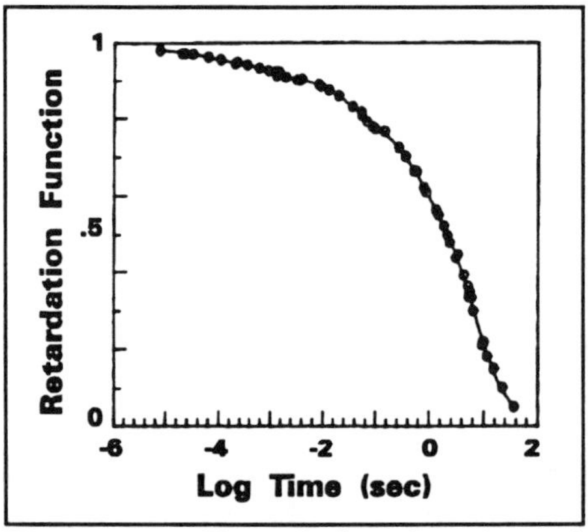


**Figure 10** - Retardation function of Rekhson et al. [70] fitted by a sum of discrete relaxations [6].

the value of summing exponential processes. Therefore, the question of universality or even physical relevance cannot be answered without more extensive measurements of the dispersion in the frequency-dependent modulus.

The question of universality has been addressed in a greater context by Ngai and co-workers [71]. They find that the stretched exponential behavior occurs in a variety of properties, not limited to viscosity. Thus according to their principle of First Universality, the fractional exponent behavior appears to hold over a broad range of phenomena including mechanical relaxation, dielectric relaxation, ionic conduction, luminescence, electronic hopping, spin-lattice interactions, magnetization and many others. The operational equation is similar to the KWW form:

$$\Phi(t) = \Phi_0 \exp[-(t/\tau_p)^{1-n}] \qquad \textit{where } 0 \le n \le 1 \tag{46}$$

The Second Universality addresses the relaxation time $\tau_p$. The relaxation mode interacts with the environment so that the microscopic-scale relaxation time $\tau_0$, which can be exponential, is stretched by the interaction with the environment to become the macroscopically observed relaxation time, $\tau_p$.

$$\tau_p = [(1-n)\,\omega_c^{\,n}\,\tau_0]^{1/(1-n)} \tag{47}$$

If the material exhibits a single relaxation time, this is due to weak coupling to the environment. The concept is similar to an interesting view previously discussed by Litovitz and Montrose [72]. That is, if during the stress relaxation process, the structure rearranges itself, then, the relaxation process will behave non-linearly, in the same way as structural relaxation observed by changing the temperature. Thus, at short times, the structure appears static, showing weak coupling, but it appears fluid at long times, showing enhanced coupling. Viewed from another aspect, during the early part of the relaxation process, the structure will appear static, but it will flow during the latter part of the stress relaxation, effectively short-circuiting the local stress. The result, however, is the opposite of a stretched relaxation process. This effect will occur only when the stress and the structure relaxation processes are of the same order of magitude, possibly, in the region of non-Arrhenian behavior.

Brawer developed a relaxational model [73] in which a region of the liquid must reach a transitional state before relaxation can occur. Thus the liquid is divided into regions with a temperature dependent concentration of low density and high density regions. These density fluctuations produce a distribution of relaxation times which arises from the variation in the state of excitation of the different regions. Since density fluctuations are temperature dependent, the relaxation process will be history-dependent. Conceptually, this model is quite similar to that proposed by Tweer et al. [74] based on the Environmental Relaxation Model, in which various glasses are fitted to the ERM Equation on the pretext that the broadening distribution of relaxation times results from temperature-dependent density fluctuations. However, Tweer presumes a low-temperature 2nd order phase transition below $T_g$ to make the fluctuations grow as the temperature is lowered, while Brawer postulates that a transition is not necessary.

**Mode-Coupling Models**

Much attention has been directed recently toward a set of viscosity models based on hydrodynamic considerations arising from the interaction between particle velocities and density fluctuations [75-79]. The models are based on the concept that decay of density fluctuations is controlled by the longitudinal viscosity while the viscosity is itself dependent on a feedback from the correlation of density fluctuations. Thus, with increasing viscosity, density fluctuations are more correlated, and as their correlation increases, the viscosity grows. On the atomic level, this can be interpreted by considering the result of the motion of a particle through a liquid. Collisions of the particle with the lattice are transfered to other regions of the lattice and accumulate due to collective motions of the other particles. The resulting disturbances eventually return to affect the moving particles, As the coefficient coupling this feedback increases, the particle motion is greatly hindered by the response of the lattice and the viscosity increases. As particles try to diffuse from their cages, the cooperative motion of all the particles in the system causes the cages to become more effective traps. Thus, at high temperatures, where density fluctuations have very limited ranges, the coupling coefficient is very weak, and the viscosity is low. As the temperature is lowered, the range of density fluctuations and the associated coupling coefficient increases, raising the viscosity. The feedback nature of the hydrodynamic considerations is best seen in the representation of the resulting damped non-linear oscillator equation for the density fluctuations [75]:

$$\ddot{\Phi}(t) + \gamma\,\dot{\Phi}(t) + \Omega_0^{\,2}\,\Phi(t) = -4\lambda\,\Omega_0^{\,2}\int_0^t d\tau\,\Phi^2(\tau)\,\dot{\Phi}(t-\tau) \qquad (48)$$

where $\mathbf{\Phi(t)}$ is the density correlation function for a specific Fourier component; $\gamma$ is the damping factor or viscosity and $\lambda$ is the coupling constant which determines the strength of the feedback loop. Note that the right hand term couples the past values of the correlation function and its derivative with its present value.

The solution of this equation yields a viscosity which increases with $\lambda$. For $\lambda < 1$ the density fluctuations decay to zero with long tails. For $\lambda \geq 1$ the density fluctuations do not decay to zero at long times; instead they decay to some finite value which increases from ½ at $\lambda_c$ to 1 at infinite coupling. For $\lambda$ increasing but less than 1, the viscosity becomes frequency-dependent, e.g., viscoelasticity develops.

This approach predicts that a temperature will be reached when the coupling will be sufficiently large ($\lambda = \lambda_c$) that no molecular diffusion will be possible. Thus a kinetic singularity, unrelated to thermodynamic transitions will develop, and the viscosity diverges as $(1-\lambda/\lambda_c)^{-\mu}$, with $\mu > 1$. While this point could be the glass transition temperature, it occurs at too low a viscosity (around 10 P). Angell recently pointed out [80] that the singularity in the viscosity occurs at about the point where viscous flow begins to be dominated by activated atomic jumps out of local structural cages. These jumps could short-circuit the diffusion process and allow the material to continue to flow. Other modifications to the mode-coupling models [78,79] have shown that as the system becomes sluggish, higher order approximations of the solution to the feedback problem remove the singularity. The viscosity of the liquid continues to increase with density, but never diverges.

These studies have great value, in that they predict the effect of the dynamics of the local environment on the viscosity. At temperatures above the predicted sigularity, the rate of relaxation of the local cages will control the viscosity as predicted by these models. Below the predicted singularity, it can be assumed that structural relaxation is slower than stress relaxation and that the local cage structure will be static during the period of the activated jump. In that temperature range, the mechanism for diffusion of particles switches from movement through a relaxing structure to activated jumps through a static structure.

**Atomistic Structure Models**

Cohen and Grest [38] have modified the free volume concepts to consider local variations in free volume. They assume that above $T_g$ the viscous liquid can be divided into solid-like ($v < v_c$) and liquid-like ($v > v_c$)

cells. Atoms in fixed cages in solid-like cells can only vibrate. Exchange of free volume is only possible between liquid-like cells with sufficient liquid neighbors. Flow requires that the local free volume be greater than the molecular volume, and when clusters of liquid-like cells can accumulate their free volume to exceed this minimum, they form a liquid cluster. Within a liquid cluster, there exists a communal configurational entropy due to the accessibility of many configurations. There then exists a critical concentration of liquid clusters to produce an infinitely connected volume (percolation channel). The percolation probability can be calculated from the cluster distribution function. The model recalculates the free volume term in the original free volume theory:

$$\bar{v}_f = \frac{k}{2\zeta_0}\left\{T - T_0 + [(T-T_0)^2 + 4v_a\zeta_0 T/k]^{1/2}\right\} \tag{49}$$

The model predicts a first order phase transition when the percolation probability vanishes and the configurational entropy goes to zero. The viscosity diverges as in the Configurational Entropy Model, at the entropy catastrophe. While the system falls out of equilibrium at $\mathbf{T_g}$, many liquid-like cells remain with frozen voids. These regions can then contribute to the low temperature specific heat and provide tunneling centers for sub-$T_g$ transitions.

Fredrickson, Brawer and Andersen [47,80,81] developed similar concepts in a Kinetic or Facilitated Ising Model. Here, cooperative behavior in multiparticle ensembles is treated by representing the liquid by local regions with either low or high density. Since there are only 2 states, Fredrickson and Brawer assign a spin condition to them: spin up for low density and spin down for high density. This is done only to facilitate book-keeping as there are no spin states associated with the glass. Relaxation of a spin state (e.g. changes in local density) is facilitated by the condition of the surrounding matrix. Two of the contiguous neighboring cells or regions must have a low density (spin up) in order to allow relaxation of the central cell. As the temperature is lowered, the probability of the high density state (spin down) increases, thus relaxation is impeded and its cooperative character becomes more evident as the likelihood of finding sufficient low density neighbors is drastically reduced (e.g., if a cell is surrounded by spin down nearest neighbors, it must wait until at least two of its neighbors flip to the other spin state before it can undergo a relaxational process). Monte Carlo calculations were conducted on a square planar lattice with periodic boundary conditions, to find a high temperature Arrhenian relaxation behavior, and a low temperature KWW behavior with a $\beta$ value in the stretched exponent decreasing from about 0.55 to 0.25 with decreasing temperature, in fair agreement with that obtained in laboratory measurements on glasses. The relaxation time follows the Configurational Entropy Model with the Adam Gibbs relationship (Frederickson finds that in 3-dimensions, the Adam-Gibbs Equation is only followed when the number of low density neighbors necessary for structural relaxation of the central cell is 3 rather than 2) and the viscosity exhibits a temperature dependence developed by Taborek [$\eta = (T-T_0)^{-2}$] at low viscosities and the Fulcher form near $\mathbf{T_g}$, which occurs at $\mathbf{T < T_0}$ from the Taborek term. The Monte Carlo study shows that dynamical correlations are established quickly among regions of different spins, but they require a long decay time. The approach also models sub-$T_g$ relaxations, since as in the Cohen-Grest Model, there are frozen-in regions with low density (spin up). The 2-spin model shows no dynamic or thermodynamic singularity. However, it fails near $\mathbf{T_g}$ by its inability to return to Arrhenian behavior. Consequently, the model is believed to describe well the onset of cooperative behavior, but it fails at high viscosities.

The concept of a relationship between the A/NA behavior of the viscosity, the variation in the distribution of relaxation times and local composition and/or density fluctuations will recur time after time in viscosity theories. The approach proposed by Simmons and Macedo provides a reasonable explanation of the return to Arrhenian behavior at low temperatures in phase-separated systems. Central to this result is the need to limit the range of cooperative behavior, thus when the size of fluctuations is very large, most relaxing regions are fully embedded within the fluctuations and are no longer sensitive to further changes in fluctuation size. This produces the necessary *broad* yet *time-independent* distribution of relaxation times. Scherer [7] agrees that the model essentially yields the temperature dependent behavior of the viscosity, but questions the size of the relaxing region (50 Å in radius with approximately $10^4$-$10^5$ atoms) as being too large for density fluctuations well above $T_g$. However, this is precisely the core of the argument. Well above $T_g$, the fluctuations are much smaller than the relaxing region. Only near $T_g$ are the fluctuation sizes greater than the relaxing region. Furthermore, the calculation of the size of relaxing regions results from a theoretical analysis of the size of supercritical fluctuations and from measurements of the microstructure in phase separated glasses as discussed below, therefore the size scale is fairly well supported.

**Viscous Flow During Phase Separation**

Phase separation processes in glasses are generally very slow, making possible measurements of viscous flow during the formation and growth of the two-phase structure. This prospect has interested numerous investigators and a strong dependence of viscosity on structure has been reported [18,83-88]. The viscosity is sufficiently sensitive to the local structure that its time-dependent change has been used to identify the ocurrence of phase separation in glasses which did not exhibit detectable light scattering [89]. Pyrex® and several high viscosity glasses were first found to exhibit phase separation by measurements of the time dependence of the viscosity.

According to critical point theory [64], the viscosity of a critical mixture is expected to approach a singularity as the temperature is lowered to the 2nd order phase transition. This is due to the infinite growth of composition fluctuations. The 2nd order phase transition is only attainable in glasses at the maximum of the miscibility gap, where the coexistence and spinodal curves meet. Measurements by Simmons et al. [90] above the critical point of the soda-borosilicate system indeed showed an anomalous increase in viscosity, however, the singularity predicted by critical-point theories was not observed. This absence of a singularity was subsequently explained in terms of the limited range of relaxation processes and the resulting insensitivity to large-scale variations in either density or composition postulated in the Simmons and Macedo ERM model. Similar behavior was also observed in the lead borate glass system [91].

Below the coexistence temperature, phase separation theory predicts that the glass will decompose into separate phases according to either of two processes: nucleation and growth or spinodal decomposition. The latter occurs below the critical point of the system and consists of a rapid, spontaneous growth of composition fluctuations which increase in intensity until they reach the coexisting compositions, followed by a diffusion-limited coarsening process.

Both processes, in general, will lead to a change in viscosity and the analysis of the behavior of viscous flow processes is highly dependent upon the experimental conditions. Simmons and co-workers [18] have investigated the effect of phase separation in several soda borosilicate glasses with a high likelihood for separating by spinodal decomposition. In this case, the viscosity will change during the initial spinodal process as the phase composition is changed, as well as during the subsequent coarsening process as the microstructure grows in size. Studies of the soda-borosilicate glass system, particularly the composition at the top of the immiscibility dome, showed that under isothermal conditions phase composition equilibrium is achieved within a few minutes in a temperature range 150-200°C below the immiscibility temperature. Thus, it was possible to isolate the effect of microstructure growth for study by analyzing viscosity changes after the initial few minutes. This is shown in Fig. 11 for a slower glass (Pyrex®) in which the spinodal decomposition time is calculated to be approximately 11 minutes. It is well known that a change of composition is also associated with coarsening, however, in the soda-borosilicate system, this effect is relatively small. Fiber elongation measurements shown in Fig. 12 exhibited increases in viscosity of 4-5 orders of magnitude with isothermal heat-treatments. Calculations showed that the spinodal decomposition induced changes in phase composition were essentially complete within a few minutes in this glass. Loading the fibers at the beginning of the experiment or after a variety of soak times had no effect on the measured viscosity, clearly indicating that the viscosity change is associated with the growth of microstructure as confirmed by electron micrographs. The question of composition change vs microstructure-size change has received the attention of a variety of investigators [93]. In a recent comment, Simmons and Simmons [92] point out that the role of composition change is determined by the experimental conditions and the question is easily resolved by considering the thermal history of the samples. Their analysis [92] shows that the fiber data of Simmons et al. [18] was taken during the period of microstructure growth. Conflicting conclusions reached by an analysis of Takamori and Tomozawa [93] which seemed to indicate a change in properties (chemical durability and viscosity) due to composition shift, originate from data taken on samples whose thermal history was not precisely controlled prior to the experiments. In their analysis, the samples had received prior heat treatments to induce phase separation, consequently, subsequent heat treatments exhibited a slow change in composition as the individual phases readjusted to new equilibrium compositions. This problem was carefully avoided in the viscosity measurements of Simmons et al. [18].

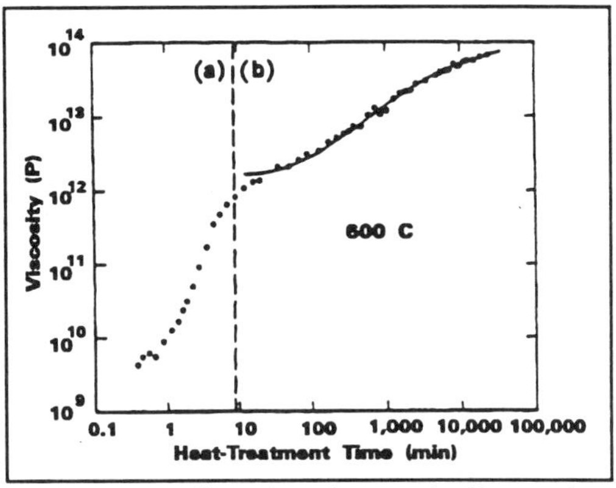


**Figure 11** - Viscosity change in Pyrex due to phase separation. Period (a) is dominated by spinodal decomposition. Period (b) is dominated by coarsening[92].

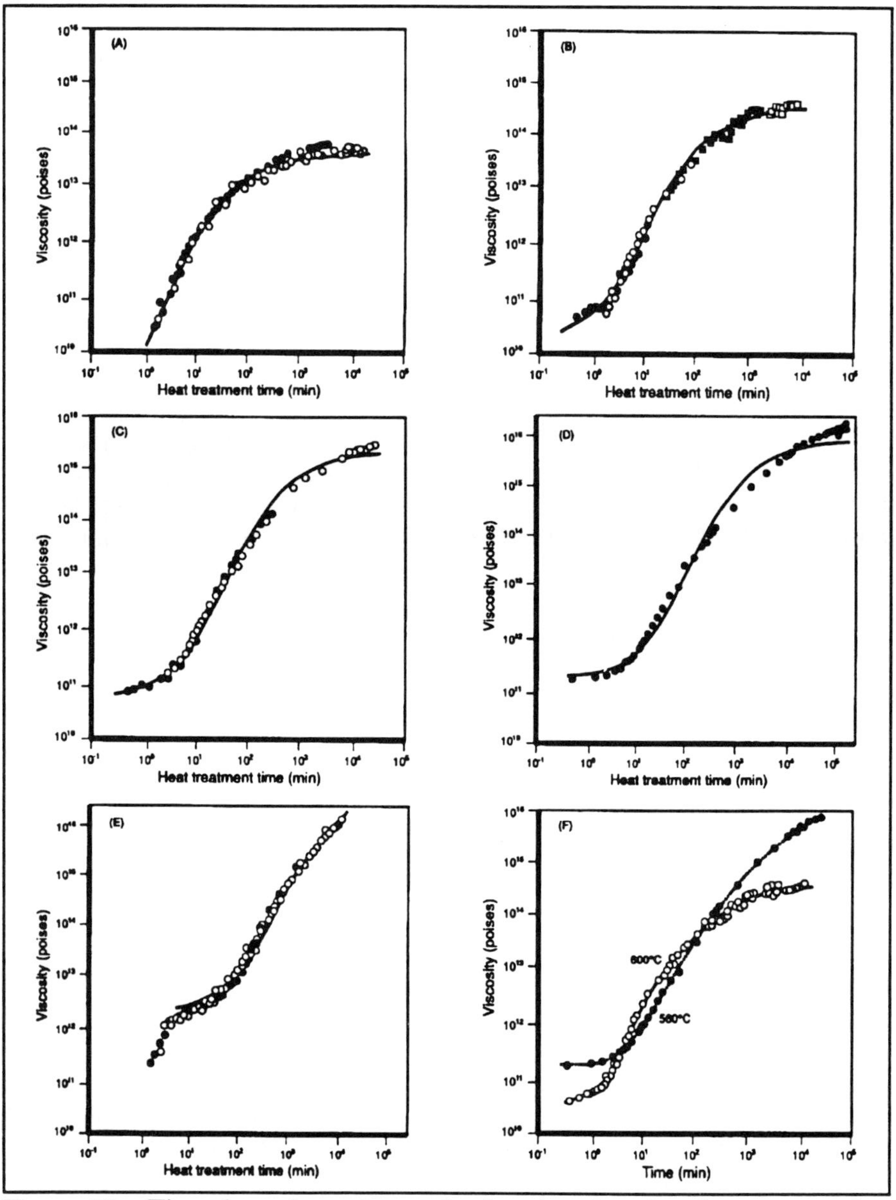


**Figure 12** - Time-dependent change in the viscosity of a phase separating soda-borosilicate glass compared to the predicted behavior of the ERM model [86] at various temperatures.

The Simmons and Macedo ERM model which had been developed to study the dependence of the viscosity on stable composition fluctuations above the critical point was found to be equally applicable to the change in viscosity due to coarsening. The analysis of the data essentially shows that 2 interconnected phases grow in time during heat treatment, with one phase, nearly pure silica, and the other rich in alkali and boron trioxide. The two phases differ greatly in their individual viscosities. The rate of coarsening was found to depend directly on the viscosity of the fluid phase, while the measured viscosity of the two-phase material was a combination of the effect of microstructures with different activation energies as summed by the ERM model. The viscosity of the two-phase material approaches asymptotically that of the less fluid phase. The rate of isothermal viscosity change is related to the rate of growth of the microstructure and the ERM model is used to relate the measured viscosity to the microstructure size by replacing the temperature-dependent size parameter **Λ(T)** in Equ. 43 with a time-dependent size parameter **Λ(t)** obtained from actual microstructure size measurements. The interaction range, $\mathbf{r_0}$, was the same as that derived for the supercritical fluctuation studies. The ERM in this case simplifies to the following:

$$\ln\eta(T,t) = \ln\eta(T,0) + \frac{\Delta E}{kT}\left\{\frac{1}{1+[r_0/\Lambda(t)]^2}\right\}^{3/2} \qquad \ln\eta(T,0) = \ln\eta_0 + \frac{E_0}{kT} \tag{50}$$

where the viscosity of the phase separated glass with very small microstructure is **η(T,0)**; the viscosity of the fluid phase which controls the rate of microstructure growth is **η(T,0) exp(-ΔE/kT)** and the viscosity of the viscous phase which controls the asymptotic viscosity behavior is **η(T,∞) = η(T,0) exp(+ΔE/kT)**. For the soda-borosilicate glass the constants are: $\mathbf{\ln\eta_0 = -50.43}$, $\mathbf{E_0 = 128}$**kcal/mol** and $\mathbf{E_0 + \Delta E = 146}$ **kcal/mol.** **Λ(t)** is measured by TEM and $\mathbf{r_0}$ is **41-56 Å** from the supercritical measurements. The fit of the data by this equation is shown as the solid line in Fig. 12 for all the temperatures measured which include (A) 620C, (B) 600C, (C) 580C, (D) 560C, (E) 540C and (F) a comparison of 600C with 560C. Considering the small number of adjustable parameters and the viscosity range covered, this is indeed an excellent fit.

## NON-NEWTONIAN VISCOUS FLOW AT HIGH SHEAR RATES

So far, our discussions of viscosity have been conducted with the assumption that the only strain rate dependence present was due to viscoelasticity. This effect would appear as an increase in resistance to deformation with increasing strain rate due to the contribution from unrelaxed delayed elasticity and at even higher strain rates, due to elastic response itself. Divers who jump from great heights understand this effect well, as it provides the slap when hitting the water at high speed. Mackenzie has reported elastic fracture of melts upon impact from a high speed bullet. Gob shearing during bottle forming is a viscoelastic fracture of the melt.

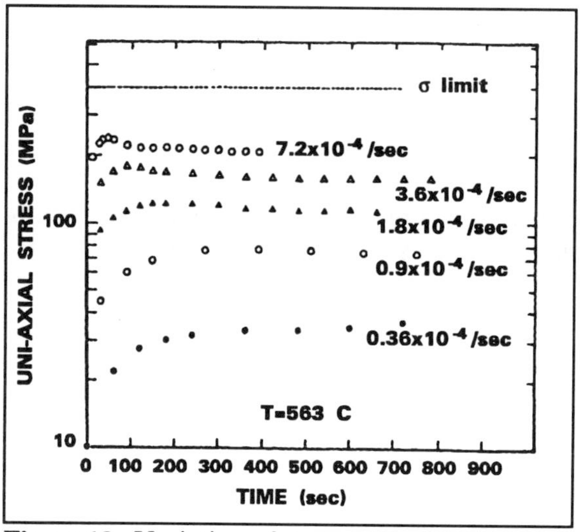


**Figure 13** - Variation of stress with time after the application of constant strain rate. Note the stress overshoot and the non-linear increase with strain rate.

Measurements by a few investigators showed that a decrease in viscosity could be observed at high strain rates [94] or high stresses [95]. Subsequent studies by Simmons and co-workers [19] discovered that a shear thinning process characterized the viscosity of soda-lime silica glasses at high strain rates. The study was conducted at constant strain rate and allowed a detailed analysis of the phenomenon. The structural aspects of the effect were analyzed on the basis of two experimental results: (1) the appearance of stress overshoot before the onset of shear thinning, in which, at applied constant strain rate, the stress increases rapidly to a maximum then decreases to a steady-state value (See Fig. 13), and (2) measurements of similar shear thinning behavior in a simple Lennard Jones fluid simulated by molecular dynamics, in which the liquid structure is observed to form in stratified layers in the direction of the shear (Fig.14).

These results indicate that the shear thinning process is not a result of viscoelasticity, but rather an effect of structural relaxation, by which the stress or strain relaxation of the melt is short-circuited by the relaxation of the melt structure to better facilitate flow. Thus, as the stress increases after the

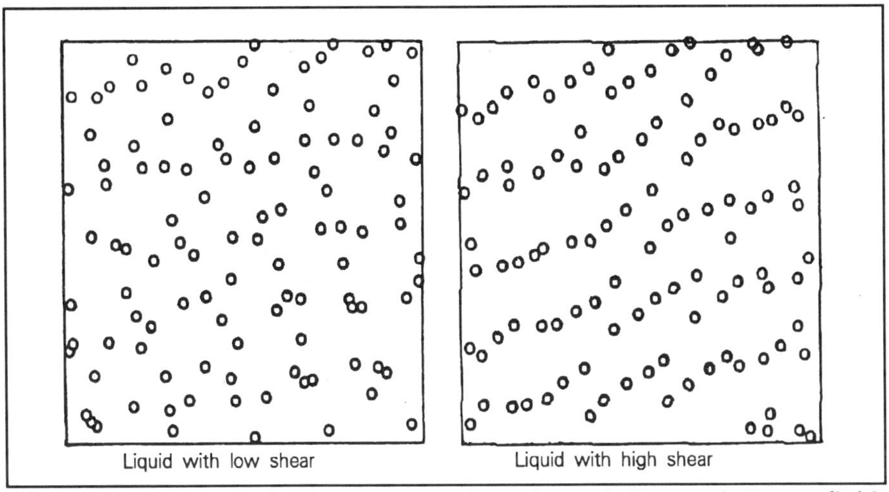


**Figure 14** - MD derived structure of a sheared Lennard Jones fluid, showing alignment of particles at high shear rates.

application of a deformation at a high strain rate, the structure rearranges itself to a more compliant form and the stress relaxation time is reduced, along with the viscosity, exhibiting the observed stress overshoot at constant applied strain rate.

The constant strain rate measurements also showed that after relaxation of the structure, the stress reached a reproducible, steady state value. This indicated that the structure did not fully break down and that it rearranged to a kinetically stable form which was in quasi-equilibrium at the applied strain rate. A plot of the steady state (or long-time) stress vs strain rate showed a saturation of the stress to a maximum

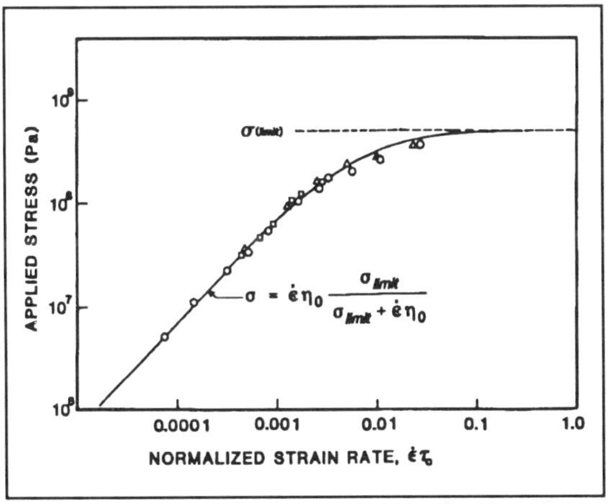


**Figure 15** - Steady-state stress during shear thinning showing the asymptotic approach to a limiting value.

value, $\sigma_{limit}$, as the strain rate increased (Fig 15). This result has the curious implication that if a stress is applied above the $\sigma_{limit}$ value, there is no steady state strain rate which can support it, therefore, Simmons and co-workers interpreted it as the *theoretical shear strength* of the melt. The value of this strength, $\sigma_{limit}$ is an adjustable parameter in the fitting of the steady state shear viscosity as a function of normalized strain rate, $\dot{\epsilon}\tau$ in which $\dot{\epsilon}$ is the strain rate and $\tau$ is the shear stress relaxation time:

$$\eta(\dot{\epsilon}) = \eta_0 \frac{1}{1 + \dot{\epsilon}\tau G_\infty/\sigma_{limit}} \tag{51}$$

This simple function appears to hold universally for different glasses (Fig. 16 shows 3 different compositions, and new ones are constantly being added to the list, including E-glass and others) and over a broad range of temperatures, with the single adjustable parameter $\sigma_{limit}$ [96].

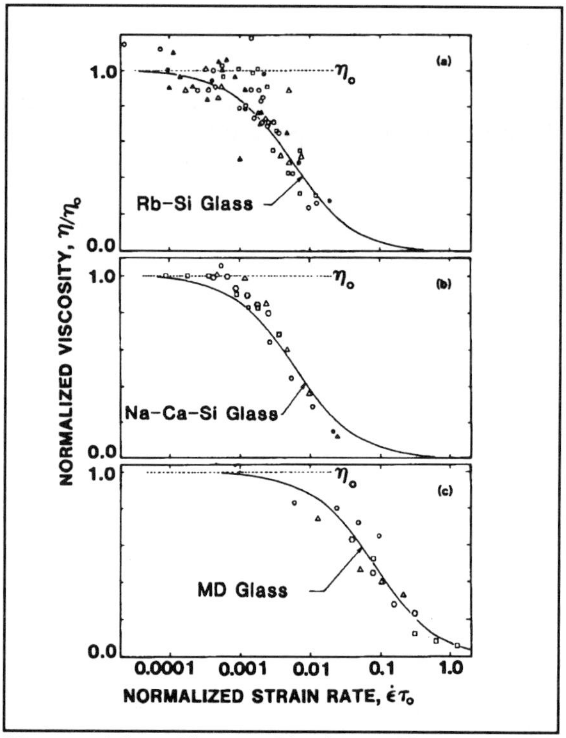


**Figure 16** - Shear thinning effect on the viscosity of 2 real glasses and an MD glass [19].

Figure 17 shows the reduced viscosity, $\eta(\dot{\epsilon})/\eta_0$ as a function of reduced strain rate over a temperature range of 538-810°C and $10^6$-$10^{14}$ P in viscosity. The *theoretical shear strength* calculated from the parameter $\sigma_{limit}$ is temperature-dependent as shown in Fig. 18 where the temperature abscissa has been replaced by the corresponding Newtonian viscosity. Simmons and Simmons [97] review these findings in terms of forming characteristics, and show that the non-Newtonian behavior also exhibits a complex viscosity change with phase separation. The reader is referred to that paper for an analysis of the combined effects of phase separation and high strain rate.

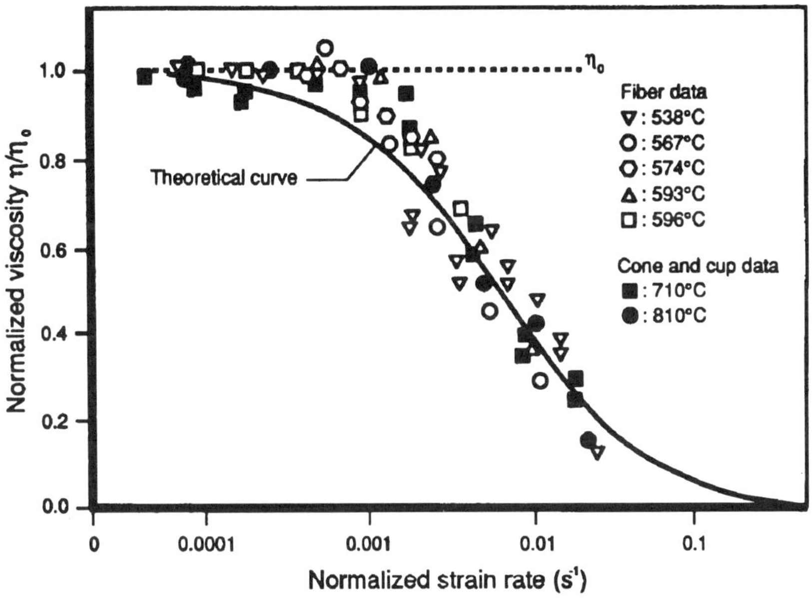


**Figure 17** - Measured steady-state viscosity, reduced by the Newtonian value potted against strain rate, combining results from measurements over the temperature range 538-810 C.

## CONCLUSIONS

In conclusion, we find that a variety of theories have been proposed to describe the large temperature dependence of viscosity from glass-forming liquids. These could be relatively consistent with each other if we assume that several processes can occur in these liquids, but that the controlling process is the one which relaxes applied stress fastest in each temperature range, thus

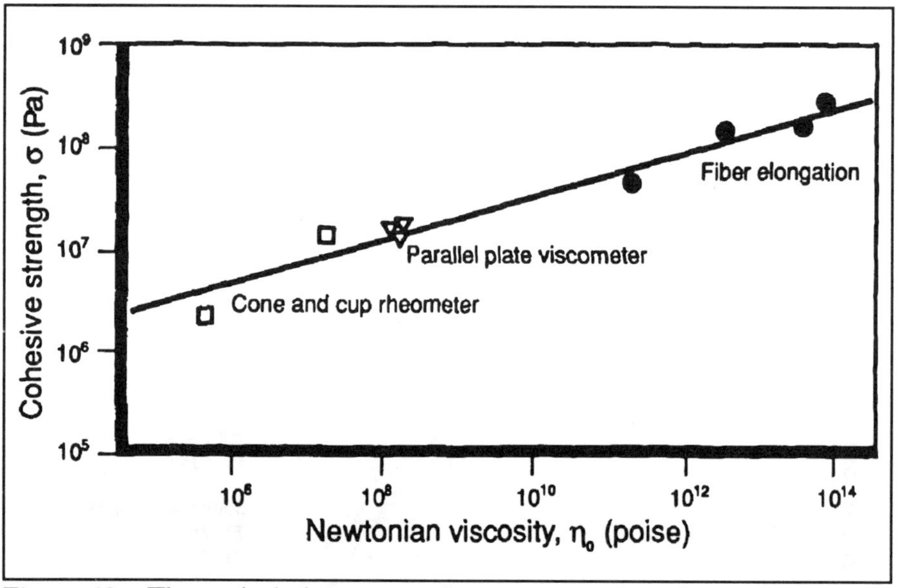


**Figure 18** - Theoretical shear strength parameter plotted as a function of the Newtonian viscosity corresponding to the measurement temperature.

short-circuiting the others. Thus, mode-coupling models dominate at high temperatures since flow in a low density liquid is controlled by hydrodynamic considerations. As the temperature is decreased and the density increased, hydrodynamic processes slow down sufficiently that molecular activation processes are quicker. Activation energies and free volume mechanisms dominate as the viscosity enters the non-Arrhenian region as structural heterogeneities broaden the activation energy distribution. However, the probability for stress relief by activation of all molecules decreases drastically, as the free volume is decreased with reduced temperature. Cooperative flow processes with lower free volume requirements (Cohen and Grest) take over until the free volume percolation limit, followed by configurational rearrangements between structures differering little in free volume (Fredrickson and Brawer) as the temperature is lowered through the glass transition region. As structural heterogeneities grow with decreasing temperature, the viscosity is controlled by a growing, then saturating distribution of relaxation times (Simmons and Macedo, ERM model). Thus, separate mechanisms could control the viscosity as the density is increased from the very fluid liquid to the solid.

## SUGGESTED READING

G. W. Scherer, *Relaxation in Glass and Composites*, J. Wiley (N. Y., 1986).

S. A. Brawer, *Relaxation in Viscous Liquids and Glasses*, Am. Ceram. Soc. (Columbus, OH, 1985).

S.M. Rekhson, "Viscoelasticity of Glass," in *Glass Science and Technology* **3**, N. Kreidl and Uhlmann eds., Academic Press, (1986).

R. M. Christensen, *Theory of Viscoelasticity; An Introduction*, Acad. Press (NY, 1971).

J. R. Van Wazer, J. W. Lyons, K. Y. Kim and R. E. Colwell, *Viscosity and Flow Measurements*, Wiley Interscience (New York, NY, 1963) p 74.

J. Zarzycki, "Glasses and the Vitreous State," Cambridge University Press, Cambridge, UK (1991).

## REFERENCES

1. C. A. Angell, "Relaxation in Liquids, Polymers, and Plastic Crystals - Strong/Fragile Patterns and Problems," *J. Non-Cryst. Solids* **131-33**, 13-31 (1991).
2. S. N. Crichton, R. Mossadegh, C. T. Moynihan, P. K. Gupta and M. G. Drexhage, "Viscosity and Crystallization of $ZrF_4$-Based Glasses," *Mat. Sci. Forum* **20**, 435-41 (1987).
3. J. S. Sanghera, C. -J. Chu and J. D. Mackenzie, *Mat. Sci. Forum* **20**, 685-91 (1987).
4. G. W. Scherer, *Relaxation in Glass and Composites*, J. Wiley (New York, 1986).
5. S. A. Brawer, *Relaxation in Viscous Liquids and Glasses*, Am. Ceram. Soc. (Columbus, OH, 1985).
6. S.M. Rekhson, "Viscoelasticity of Glass," in *Glass Science and Technology* **3**, N. Kreidl and Uhlmann eds., Academic Press, (1986).
7. G.W. Scherer, "Theories of Relaxation," *J. Non-Cryst. Solids* **123**, 75-89 (1990).
8. J. B. Wachtman, "Elastic Deformation of Ceramics and Other Refractory Materials," in *Mechanical and Thermal Properties of Ceramics*, J. B. Wachtman, ed., Nat. Bureau of Standards, Spec. Pub. 303, 139-168 (1969).
9. R. M. Christensen, *Theory of Viscoelasticity; An Introduction*, Acad. Press (NY, 1971).
10. B. Gross, *Mathematical Structure of the Theory of Viscoelasticity*, Hermann (Paris, 1953).

11. J. R. Van Wazer, J. W. Lyons, K. Y. Kim and R. E. Colwell, *Viscosity and Flow Measurements*, Wiley Interscience (New York, NY, 1963) p 74.
12. A. Napolitano, P. B. Macedo and E. G. Hawkins, "A Wide-Range (up to $10^{10}$ P) Rotating Cylinder Viscometer, *J. Res. Nat. Bureau Standards* **69a**, 449-455 (1965).
13. A. Napolitano, P. B. Macedo and E. G. Hawkins, "Viscosity of Boron Trioxide," *J. Am. Ceram. Soc.* **48**, 613-616 (1965).
14. E. H. Fontana, "A versatile Parallel Plate Viscometer for Glass Viscosity Measurements to 100°C," *Bul. Am. Ceram. Soc.* **49**, 594 (1970).
15. G. E. Sakoske and M. J. Leitman, "Analytical Solutions for Pressing a Glass Cylinder in a Parallel Pate Viscometer," *Bul. Am. Ceram. Soc.* **70**, 551 (1991).
16. A. Napolitano and E. G. Hawkins, "Viscosity of a Stnadard Soda-Lime-Silica Glass," *J. Res. Nat. Bureau of Standards* **68A**, 439-448 (1964)
17. W. C. Lacourse, Alfred University, unpublished data.
18. J. H. Simmons, S. A. Mills and A. Napolitano, "Viscous Flow In Glass During Phase Separation," *J. Am. Ceram. Soc.* **57**, 109-117 (1974).
19. J. H. Simmons, R. K. Mohr and C. J. Montrose, "Non-Newtonian Viscous Flow in Glass," *J. Appl. Phys.* **53**, 4075-4080 (1982).
20. H. E. Hagy, "Experimental Evaluation of Beam-Bending Method of Determining Glass Viscosities in the Range $10^8$ to $10^{15}$ Poises," *J. Am. Ceram. Soc.* **46**, 93 (1963).
21. A. Napolitano, J. H. Simmons, D. H. Blackburn and R. E. Chidester, "Analysis of Low Temperature Viscosity Data for Three NBS Standard Glasses," *J. Res. Nat. Bureau Standards* **78A**, 323-329 (1974).
22. I. Braun, *Bull. Res. Council, Israel* **1**, 126 (1951).
23. J. J. Mills, "Summary Report on the Workability of Glass," *IITRI Project No. G8021*, November 6, 1973.
24. J. Tauke, T. A. Litovitz and P. B. Macedo, *J. Am. Ceram. Soc.* **51**, 158-163 (1968).
25. J. H. Simmons and P. B. Macedo, "High-Temperature Shear Ultrasonic Interferometer Using Sensitive Phase-Lock Detection System," *J. Acoust. Soc. of Amer.* **43**, 1295-1301 (1968).
26a. C. Demoulin, C. J. Montrose and N. Ostrowsky, "Structural Relaxation by Digital-Correlation Spectroscopy," Phys. Rev. **A9**, 1740-42 (1974).
26b. C. C. Lai, P. B. Macedo and C. J. Montrose, "Light Scattering Measurements of Structural Relaxation of Glass by Digital Correlation Spectroscopy," J. Am Ceram. Soc. **58**, 120-123 (1975).
27. S. Spinner and W. E. Teft, "A Method for Determining Mechanical Resonance Frequencies and for Calculating Elastic Moduli From These Frequencies," *Am. Soc. for Testing and Mat.* **61**, 1221-1238 (1961).

28. D. Day, *J. Non-Cryst. Solids* **21**, 343 (1976).
29. G. Williams and D. C. Watts, *Trans. Faraday Soc.* **66**, 80 (1970); G. Williams, D. C. Watts, S. B. Dev and A. M. North, *Trans. Faraday Soc.* **67**, 1323 (1971).
30. P. B. Macedo and A. Napolitano, "Inadequacies of Viscosity Theories for $B_2O_3$," *J. Chem. Phys.* **49**, 1887-1895 (1968).
31. H. Eyring, *J. Chem. Phys.* **4**, 283 (1936); also W. T. Laughlin and D. R. Uhlamnn. *Bul. Am. Ceram. Soc.* **47**, 402 (1968).
32. J. H. Li and D. R. Uhlmann, "The Effect of Structural Randomness on the Absolute Rate Theroy of Viscous Flow," *Phys. Chem. Glasses* **11**, 93-98 (1970).
33. M. Goldstein, *J. Chem. Phys.* **51**, 3728 (1969).
34. M. L. Williams, R. E. Landel and J. D. Ferry, *J. Am. Chem. Soc.* **77**, 3701 (1955); also G. Fucher, *J. Am. Ceram. Soc.* **8**, 339 (1925).
35. M. H. Cohen and D. Turnbull, "Molecular Transport in Liquids and Glasses," *J. Chem. Phys.* **31**, 1164-1169 (1959); D. Turnbull and M. H. Cohen, "On the Free-Volume Model of the Liquid-Glass Transition," *J. Chem. Phys.* **52**, 3038-3041 (1970).
36. A. K. Doolittle, *J. Appl. Phys.* **22**, 1471 (1951); also T. G. Fox and P. J. Flory, *L. Appl. Phys.* **21**, 581 (1950).
37. P. B. Macedo and T. A. Litovitz, "On the Relative Roles of Free Volume and Activation Energy in the Viscosity of Liquids," *J. Chem. Phys.* **42**, 245-256 (1965).
38. M. H. Cohen and G. S. Grest, *Phy. Rev.* **B20**, 1077 (1979); G. S. Grest and M. H. Cohen, "Liquids, Glasses and The Glass Transition: A Free Volume Approach," *Adv. in Chem. Phys.* **48**, 455-525 (1981).
39. J. H. Gibbs and E. A. DiMarzio, "Nature of the Glass Transition and the Glassy State," *J. Chem. Phys.* **28**, 373-383 (1958).
40. G. Adam and J. H. Gibbs, "On the Temperature Dependence of Cooperative RElaxation Properties in Glass-Forming Liquids," *J. Chem. Phys.* **43**, 139-146 (1965).
41. W. Kauzmann, *Chem. Rev.* **43**, 219 (1948).
42. P. K. Gupta, *J. Am. Ceram. Soc.* **70**, C152 (1987).
43. M. Goldstein, *J. Chem. Phys.* **39**, 3369 (1963).
44. C. A. Angell and W. Sichina, *Ann. NY Acad. Sci.* **279**, 53 (1976).
45. L. D. Landau and E. M. Lifshitz, *Statistical Physics*, Addison-Wesley (Reading, MA 1958).
46. F. H. Stillinger, *J. Chem. Phys.* **89**, 6461 (1988).
47. G. H. Fredrickson and S. A. Brawer, "Monte Carlo Investigation of a Kinetic Ising Model of the Glass Transition," *J. Chem. Phys.* **84**, 3351

(1986).
48. M. Hara and S. Suetoshi, *Res. Lab. Rep. Asahi Glass Co.* **5**, 126 (1955).
49. S. M. Rekhson and O. V. Mazurin, *J. Am. Ceram. Soc.* **57**, 327 (1974).
50. P. B. Macedo and A. Napolitano, "Effects of a Distribution of Volume Relaxation Times in the Annealing of BSC Glass," *J. Res. Nat. Bureau Standards* **71A**, 231-238 (1967).
51. A. Napolitano and P. B. Macedo, "Spectrum of Relaxation Times in $GeO_2$ Glass," *J. Res. Nat. Bureau of Standards* **72A**, 425-433 (1968).
52. L. Boesch, A. Napolitano and P. B. Macedo, "Spectrum of Relaxation Times in $B_2O_3$," *J. Am. Ceram. Soc.* **53**, 148-153 (1970).
53. O. S. Narayanaswamy, J. Am. Ceram. Soc. **54**, 491 (1971).
54. C. T. Moynihan, S. N. Crichton and S. M. Opalka, "Linear and Non-linear Structural Reaxation," J. Non-Cryst. Solids **131-33**, 420-434 (1991).
55. G. W. Scherer, J. Am. Ceram. Soc. **67**, 504 (1984).
56. G. W. Scherer, J. Am. Ceram. Soc. **69**, 374 (1986).
57. I. M. Hodge, Macromolecules **19**, 936 (1986); also I. M. Hodge, Macromolecules **20**, 2897 (1987).
58 S. R. Nagel and J. Dixon, J. Chem. Phys. **90**, 3885-86 (1989).
59. S. M. Rekhson, "Treatment of Nonlinearity in Structural Relaxation in Glass," VII Internatina Conference on the Physics of Non-Crystaline Solids, Cambridge, UK, August 4-9, 1991.
60. N. O. Birge and S. R. Nagel, Phys. Rev. Lett. **54**, 2674 (1985).
61. R. W. Rendell, K. L. Ngai, G. R. Fong and J. J. Aklonis, Macromolecules **20**, 1070 (1987); also R. W. Rendell, K. L. Ngai and D. J. Plazek, "Volume-Dependent Rate Processes Predicted bu the Coupling Model," J. Non-Cryst. Solids **131-33**, 442-450 (1991).
62. J. H. Simmons and P. B. Macedo, "Viscous Relaxation Above the Liquid-Liquid Phase Transition of Some Oxide Mixtures," *J. Chem. Phys.* **53**, 2914-2922 (1970).
63. P. B. Macedo, "Viscoelasticity of Glasses," in *Mechanical and Thermal Properties of Ceramics*, J. B. Wachtman, ed., Nat. Bureau of Standards, Spec. Pub. 303, 169-188 (1969).
64. J. H. Simmons and P. B. Macedo, "Analysis of Viscous Relaxation in Critical Oxide Mixtures," *J. Chem. Phys.* **54**, 1325-1331 (1971).
65. J. H. Simmons and P. B. Macedo, "Effect of Environment on Viscous Flow in Inorganic Oxide Glasses," *J. Res. Nat. Bureau Standards* **75A**, 175-184 (1971).
66. J. H. Simmons, P. B. Macedo, A. Napolitano and W. H. Haller, "Investigation of Liquid-Liquid Phase Transitions in Oxide Melts by Viscosity Measurements," *Disc. Faraday Soc.* **50**, 155-169 (1970).
67. C. A. Angell, "Structural Instability and Relaxation in Liquids and Glassy

Phases Near the Fragile Liquid Limit," *J. Non-Cryst. Solids* **102**, 205-221 (1988).
68. M. S. Rekhson and S. M. Rekhson, *J. Am. Ceram. Soc.* (1986).
69. C. R. Kurkjian, *Phys. Chem. Glasses* **4**, 128-136 (1963).
70. S. M. Rekhson, N. O Gonchukova and M. A. Chernousov, *Proc. Int. Congr. Glass., Prague* **1**, 329-339 (1977).
71. K. L. Ngai, "Temperature Dependence of the Stretched Exponent in Structural Relaxation of Fragile Gass-Forming Molecular Liquids," *J. Non-Cryst. Solids* **131-33**, 80-83 (1991).
72. C. J. Montrose and T. A. Litovitz, "Structural-Relaxation Dynamics in Liquids," *J. Acoust. Soc. Am.* **47**, 1250-1257 (1970); T. A. Litovitz and C. J. Montrose, "Interaction-Induced Light Scattering in Liquids:Molecular Motions and Structural Relaxation," in *Intermolecular Spectroscopy and Dynamical Properties of Dense Systems*, Soc. Italiana di Fisica (Bologna, 1980) pp 307-324.
73. S. A. Brawer, "Theory of Relaxation in Viscous Liquids and Glasses," *J. Chem. Phys.* **81**, 954 (1984).
74. H. Tweer, J. H. Simmons and P. B. Macedo, "Application of the Environmental Relaxation Model to the Temperature Dependence of the Viscosity," *J. Chem. Phys.* **54**, 1952-1959 (1971).
75. E. Leutheusser, "Dynamical Model of the Liquid-Glass Transition," *Phys. Rev.* **A29**, (1984).
76. U. Bengtzellius and A. Sjolander, *Ann. NY Acad. Sci.* **484**, 229 (1986).
77. T. R. Kirkpatrick and P. G. Wolynes, *Phys. Rev.* **A35**, 3072 (1987).
78. S. P. Das and G. F. Mazenko, *Phys. Rev.* **A34**, 2265 (1986).
79. W. Gotze and L. Sjogren, *Z. Phys.* **B65**, 415 (1987).
80. C. A. Angell, Arizona State University (private discussion).
81. G. H. Fredrickson, *Ann. NY Acad. Sciences* **484**, 185 (1986).
82. G. H. Fredrickson, *Ann. Rev. Phys. Chem.* **39**, 149 (1988).
83. J. H. Li and D. R. Uhlmann, "The Flow of Glass at High Stress Levels - II The Effect of Phase Separation on Viscosity," *J. Non-Cryst. Solids* **3**, 205-224 (1970).
84. O. V. Mazurin, V. P. Kluyev and G. P. Roskova, "Influence of Heat-Treatment on the Viscosity of Some Phase Separated Glasses, *Phys. Chem. Glasses* **11**, 192-195 (1970).
85. P. Bernheim and A. C. D. Chaklader, *J. Non-Cryst. Solids* **4**, 328 (1971).
86. J. H. Simmons, S. A. Mills and A. Napolitano, "Viscous Flow in Glass During Phase Separation," *J. Am. Ceram. Soc.* **57**, 109-117 (1974).
87. R. Mahoney, G. R. Srinivasan, P. B. Macedo, A. Napolitano and J. H. Simmons, "Effect of Subcritical Microstructure on the Viscosity of a

Sodium Borosilicate Glass, *Phys. Chem. Glasses* **15**, 24-31 (1974).
88. O. V. Mazurin and G. P. Roskova, "Properties of Phase Separated Glasses," in *Phase Separation in Glass*, O. V. Mazurin and E. A. Porai-Koshits, North-Holland (Amsterdam, Netherlands, 1984) pp 243-290.
89. W. Haller, J. H. Simmons and A. Napolitano, "Viscosity-Drift Technique for Determination of Critical Liquid-Liquid Immiscibility Temperature of Glasses," *J. Am. Ceram. Soc.* **54**, 299-302 (1971).
90. J. H. Simmons, A. Napolitano and P. B. Macedo, "Supercritical Viscosity Anomaly in Oxide Mixtures," *J. Chem. Phys.* **53**, 1165-1170 (1970).
91. J. H. Simmons, unpublished data.
92. J. H. Simmons and C. J. Simmons, "Reply to Comment on Nonlinear Viscous Flow in Glass Forming," *Bul. Am. Ceram. Soc.* **69**, 1508-10 (1990).
93. T. Takamori, and M. Tomozawa, "Viscosity and Microstructure of Phase Separated Borosilicate Glasses," *J. Am. Cream. Soc.* **62**, 373-77 (1979).
94. J. J. Mills, *Glass Technol.* **14**, 101 (1973).
95. J. H. Li and D. R. Uhlmann, "The Flow of Gass at High Stress Levels - I. Non-Newtonian Behavior of Homogeneous 0.08 $Rb_2O$-0.92 $SiO_2$ Glasses," *J. Non-Cryst. Solids* **3**, 127-147 (1970).
96. J. H. Simmons, R. Ochoa, K. D. Simmons and J. J. Mills, "Non-Newtonian Viscous Flow in Soda-Lime-Silica Glass at Forming and Annealing Temperatures," *J. Non-Cryst. Solids* **105**, 313-322 (1988).
97. J. H. Simmons and C. J. Simmons, "Non-Linear Viscous Flow in Glass Forming," *Bul. Am. Ceram. Soc.* **68**, 1949-1955 (1989).

# CHAPTER 12

## MEASUREMENT OF ELECTRICAL CONDUCTIVITY OF GLASSES

H. Jain
Department of Materials Science and Engineering
Lehigh University
Bethlehem, PA 18015

### INTRODUCTION

There are numerous applications where an understanding of the dynamic behavior of ions in glasses is important. For example, the long range ion movement (as in diffusion) determines the stability of many elelctronic devices. It may determine also the usefulness of a glass as an electrical insulator, a fast ion conductor for solid state batteries, a waste form to immobilize radionuclides, etc. The determination of electrical conductivity offers a convenient method to characterize such movement of ions in ionically conducting glasses.

Both electrical conductivity and electrical relaxation are the manifestations of the movement of charge in a material when it is subjected to an external electric field. Basically the phenomenon of electrical conductivity does not implicate time (or frequency) as a variable, whereas electrical relaxation specifically describes the time dependent electrical response of the material. In common glasses, which are frequently ionic conductors, this distinction is important because electrical conductivity is determined by the long range motion of ions but the relaxation may include their localized motion as well. In this chapter we will primarily focus on electrical conductivity.

In literature we frequently associate the terms dc conductivity and ac conductivity with very different physical meanings. Some authors distinguish dc from ac depending on whether a dc or an ac electric field generator is used in the experiment. Others use dc conductivity to describe the asymptotic electrical response for long times or low frequencies regardless of the experimental procedure. One can usually understand the meaning of these terms from context, but to avoid confusion we suggest the use of dc or ac depending on the experimental technique (frequency), and refer true electrical conductivity (or simply electrical conductivity) to the one which is independent of the frequency of the field or the duration of observation. Then, electrical conductivity would represent the long range motion of charged species and can be correlated with their diffusion. With this definition, as we will see, sometimes it is difficult to experimentally determine electrical conductivity.

The objective of this chapter is to describe the experimental technique and the method of data analysis to determine the electrical conductivity of glasses. We will particularly focus on the experimental problems associated with high resistivity measurements since the majority of glasses are insulators under ambient conditions. Application of an electric field in some ceramic materials, by changing their electrochemical potential, can change their composition. However, most of the common glasses are reasonably stable under small electric fields and we will assume the composition to remain fixed during the experiment. It is suggested that one should use the smallest possible electric field permitted by the sensitivity of one's equipment and verify that Ohm's law is obeyed.

## TECHNIQUE

The basic principles of electrical conductivity measurements are simple: either apply an electric field from a signal generator (ac or dc) and measure its response in the form of current in the sample or measure the conductivity directly by treating the sample as the unknown arm of a Wheatstone type bridge. In either case, the detection system needs to be connected to the specimen through electrical wires and electrodes. In practice the determination of conductivity of glasses becomes complex when the measurement itself interferes with the intrinsic response of the material. Therefore, it is important to ensure that the observation represents the properties of the sample and not any other phenomenon.

There are three independent variables of interest in electrical conductivity measurements: (a) temperature, (b) atmosphere or partial pressure of one of the constituents of the sample and (c) frequency of the electrical signal. Thus, the measurement of electrical conductivity requires a sample of appropriate dimensions, the electrodes, a source of electrical signal, and instruments to monitor the system response and to measure independent variables. These are now described below in detail.

## Sample

The common practice in electrical conductivity measurements of glasses is to prepare either a rectangular (or cylindrical) bar or a disc shaped sample. Clearly, the sample should be free from any macroscopic defects such as bubbles, voids, cracks etc. The choice between bar and disc shapes depends on the electrode configuration to be used. Such shapes of uniform cross section, although not necessary, are helpful in applying the electrodes and determining the geometrical factor accurately. Conductivity of samples of arbitrary shape but constant thickness can be determined using a method due to van der Pauw [1]. The sample dimensions should be large enough that it has sufficient strength for handling and its thickness and electrode area can be measured with reasonable accuracy. On the other hand the size should be small enough so that the temperature across the sample is uniform. A temperature gradient in the sample would not only cause variations in conductivity but also generate thermoelectric voltage which should be avoided to keep the interpretation of electrical measurements simple. In case this is not possible, the current measurement should be repeated keeping the same temperature gradient but reversing the polarity of the external field and the average of the two values of current should be used in analyzing the data.

In comparison to the homogeneity of temperature, the homogeneity of structure and composition of glass is likely to have a greater influence on the electrical properties but is difficult to achieve and more frequently ignored. This is perhaps the major reason why so frequently the electrical conductivity data of different investigators do not agree, e.g. electrical conductivity values of 30 mol% $Na_2O$:70 mol% $SiO_2$ glass have been reported in the literature to vary almost by a factor of 3. Since it is difficult to give an exact description of glass structure one should always describe the glass preparation method including the thermal history in detail. Before initiating electrical measurements internal stresses in the sample should be relieved by annealing near the glass transition temperature and cooling slowly, unless the purpose of the experiment is to study the effect of stresses. Note that a glass having stresses introduced during quenching is likely to have higher conductivity than an annealed specimen of the same composition [2]. The effect of internal stresses often shows up as differing values of conductivity during heating and cooling. Of course other structural changes such as those due to phase separation or devitrification during heating may also show similar behavior. In that case it is advisable to first heat the specimen to the highest temperature of interest and then conduct electrical measurements during cooling.

Glasses are frequently prepared by melting appropriate amounts of salts which have very different melting points and vapor pressures above the melt. As a result the final composition of the glass depends on how long and at what temperature the melt was maintained. Recently it is reported that the temperature at which a melt is equilibrated before quenching can significantly affect the electrical conductivity [3]. Then, it is inadequate to specify a glass by its batch composition, particularly if the volatile component (typically alkali) is also the one which has the strongest influence on conductivity. It is recommended that besides describing the melting procedure chemical analysis of the samples should be presented. Lack of gross chemical inhomogeneity in the sample (such as due to incomplete mixing in the melt) should be demonstrated at least by visual and optical microscope examination, preferably by electron microprobe. Microscopic compositional heterogeneities (e.g., due to phase separation) are parts of the glass structure which should be characterized, particularly if they depend on the preparation procedure and are expected to have a strong effect on the electrical properties.

## Electrodes

Ideally, electrodes and connecting leads should make ohmic contacts with the specimen, have negligible resistance and be chemically inert to both the specimen and the atmosphere. Also, the leads to electrode connection should have mechanical strength sufficient for normal handling. Generally, the biggest problem remains to find electrodes with ohmic contacts that will provide negligible potential drop at the electrode/glass interface. For glasses, which are often ionic conductors, an ideal electrode would be a good conductor of both electrons and the mobile ions in the glass. That is, there would be no energy barrier for charge transfer at electrode/glass interface and the electrode would behave reversibly. For example, molten sodium or sodium amalgum may be used as electrode material on sodium ion conducting glasses. Unfortunately, sodium is highly reactive and can be impractical to use in certain electrode configurations. For some other glasses the mobile ionic species may not be known or appropriate electrode material may not be available. Therefore, instead of searching for a reversible electrode an electronically conducting inert metal electrode is often used. The behavior of the junction between electronically conducting electrode and ion conducting glass can be very complex [4], but one can eliminate the interference from electrode polarization at the electrode/glass interface using ac impedance analysis as described later in this chapter.

Non-ohmic contacts can result even when both electrode and glass are electronic conductors because of the difference in work functions of the two. Often this can be avoided if the metal selected for the electrode has a work function lower than that of an n-type glass specimen. A high density of surface energy states created by a rough surface, or the use of an electrode material which will make an n-type sample more n-type or a p-type sample more p-type will also reduce non-ohmicity of the contacts [5]. Blumenthal and Seitz [5] have listed various metals along with their characteristics, which may be used as elctrodes on ceramics and glasses. Unless cost becomes a major issue, such as in commercial applications, a noble metal is recommended for electrode use on glasses because of its chemical inertness, ability to withstand high temperatures and ease of application. Up to moderately high temperatures ($\sim$ <800°C) Ag, Au or Pt may be used but for higher temperatures only Pt is suitable. Silver being a relatively small ion may diffuse into the glass during measurements at high temperature and thus introduce error in the conductivity value. It is also the least inert metal among the three noble metals and, therefore, should be used with caution on reactive glass compositions. Platinum is fairly inert, but it is a good catalyst and can promote undesirable reactions between the sample and atmosphere. Above 1000°C platinum would also deteriorate in highly reducing conditions but such conditions are not frequently encountered in glass research. In addition to noble metals aluminum electrodes which have been extensively used in the electronic industry can be suitable for low temperature applications. Colloidal graphite electrodes may be used under non-oxidizing conditions but their connection with lead wires can be a problem.

All the metals mentioned above can be applied on the sample by sputtering or vacuum deposition, but one should be careful with the vacuum deposition of platinum which may form an alloy with the heating element. Electrodes of desired geometries can be deposited simply by using a mask. The deposited electrode films should be sufficiently thick and adhering to provide electrical continuity at all temperatures. It is advisable to verify this continuity also after the conductivity measurement to confirm that the film had not detached from the sample, or broken into discontinuous islands due to thermal expansion mismatch and/or surface tension effects.

Vacuum deposition or sputtering produces electrode films of uniform thickness and there is little chance of incorporating impurities, but they require special equipment. Therefore, commercially available paints or pastes are often applied with a brush. Paints may be used to make the electrodes but pastes are more appropriate for connecting thin lead wires to the electrodes. For example, Engelhard Industries* make silver paints No. 16 and 13 which do not require high temperature curing (∿100°C) and can be used after initial drying. Gold paints (No. 4813 or 5364) and platinum paints and pastes (e.g. 05X ink or A4649 paste) usually require curing at a temperature (>520°C) which may be too high for certain glasses.

Temperature

One of the largest sources of error and uncertainty in electrical conductivity data arises from temperature measurements. The importance of accurate temperature measurement cannot be over emphasized considering that conductivity is a strong, usually exponential, function of temperature. To minimize this uncertainty the sample should be placed in the homogeneous temperature zone of the furnace; the temperature probe, usually a thermocouple, should be placed close to the sample and the procedure and precautions should be followed as described in reference [6]. An electrically heated tube furnace with proportional temperature controller can provide a convenient method of heating a sample. The homogeneous temperature region would be generally larger in a horizontally than a vertically placed furnace. However, the latter configuration may be used if dictated by the convenience of sample handling.

The choice of thermocouple depends on the atmosphere around the sample and temperature range of interest. The copper-constantan (Type T) thermocouple is suitable in -200 to 370°C range in oxidizing, reducing, inert or moist atmosphere; the upper temperature limit is primarily set by the onset of oxidation of copper. The iron-constantan (Type J) thermocouple may be used from 0 to 760°C in a non-corrosive environment. Above 540°C sulfur and oxygen in the atmosphere start attacking the iron which may start rusting if moisture is present and can be embrittled below the ice point. The type E chromal-constantan thermocouple develops the highest emf/°C and has a wider useful temperature range of -200 to 900°C in oxidizing and inert atmosphere but it is not suitable in reducing environments or vacuum at higher temperatures. The chromal-alumel (Type K) thermocouple finds widest application in oxidizing or inert atmosphere ranging from -250 to 1260°C. However it should be avoided in reducing or sulfurous atmosphere and in vacuum for

*Specialty Metals Division, Carteret, NJ 07008.

long times at high temperatures. For a still higher temperature range, which is not often needed in glass research, a platinum vs. platinum-rhodium (Type R, S or B) thermocouple is needed. No matter which thermocouple is selected it should be frequently calibrated by using one of the standard methods because an error as high as ±2°C or ±2% can be within manufacturer's specifications [6].

Environment

As compared to temperature and composition, pressure is likely to have a smaller effect on conductivity of glasses and is frequently, if not always, kept constant. Therefore, the conductivity apparatus described here is suitable for measurements at approximately atmospheric pressure and the reader is referred elsewhere for high pressure apparatus designs [7].

The stoichiometry of many crystalline ceramics can be easily controlled at high temperatures by controlling the partial pressure of one of the components [8]. A similar variation of composition or oxidation state of constituent ions is also possible in glasses when environment is changed. Then it is necessary to have a well defined environment around the specimen. However, the composition of most glasses below the glass transition temperature is reasonably stable and the control of environment may not be as critical. In that case the sample may be placed in an inert atmosphere such as of helium which also provides good temperature control due to its high thermal conductivity.

The composition of a less stable oxide, nitride, sulfide or halide glass can be defined by equilibrating it with the environment of known partial pressure of oxygen, nitrogen, sulfur or halogen (X), respectively. Considerable information is available on how to control oxygen partial pressure ($PO_2$) over a very wide range ($10^{-20}$ to 1 atmosphere) using a mixture of $CO/CO_2$, $H_2/H_2O$, $O_2/Ar$ or pure oxygen [9]. In comparison, the methods of controlling $P_{N2}$, $P_{S2}$ or $P_{X2}$ are much less developed but, in principle, it can be done using pure components or $H_2/NH_3$, $H_2/H_2S$ and $H_2/HX$ equilibria, respectively [10].

Measurement Cell

There is no standard design for conductivity measurement cells and, therefore, each laboratory has developed its own design depending on the material, electrode configuration, environment and temperature of interest. In general, a cell should provide good temperature and environment control, and the ability to change specimens with little skill. For example, a schematic design of one cell is described in Figure 1 for measurements under controlled environment at and above room temperature. It uses specimens with sputtered, painted or vacuum deposited electrodes which are connected to main platinum lead wires with the help of gold or silver paste and thin (∿5 mil) platinum or gold wires. The use of intermediary wires eliminates the need of spring loaded connections between the sample and lead wires. This cell uses a vertical configuration which is a somewhat simpler design than that of horizontal configurations (not shown) but the latter is likely to have greater temperature stability.

For measurements below room temperature a commercially available cryostat (such as from Janis Research Company, Wilmington, MA) may be used. In this case, condensation of moisture often becomes a source of error which should be avoided by filling the chamber with a moisture free inert gas. To minimize electrical noise and stray capacitances the use of shielded cables and lead wires is recommended in general. Inside the high temperature furnace this is achieved by coating the outer surface of the alumina (or quartz) tube and tubings with platinum paint and connecting them to ground. The noise around the specimen is reduced by covering it with a platinum coated and grounded alumina (or quartz) dome as shown in Figure 1.

Electrical Measurements

The two most important considerations in selecting the method and instruments necessary for measuring electrical conductivity of a glass are its magnitude and frequency of the signal. When studying high resistance samples ($R \gtrsim 100\ M\Omega$) it is important to ensure that, within desired experimental error, there are no short circuiting paths and the current flows only through the specimen. Common short circuiting paths are the surface of the specimen, sample holder and ambient gas. Obviously, the input impedance of any voltage measuring instrument should be high enough (at least two orders of magnitude higher than that of the sample) not to draw any significant current.

To minimize surface conduction the sample and electrode geometry should be selected such that the surface conduction path is much greater than the volume conduction (the thickness) path. In the simplest case, called the two-probe method, two same size electrodes may be deposited in the centers of large parallel surfaces of a thin specimen. The electrical resistance can be determined using a bridge (described later) or from the ratio of potential drop across and current through the sample. However,

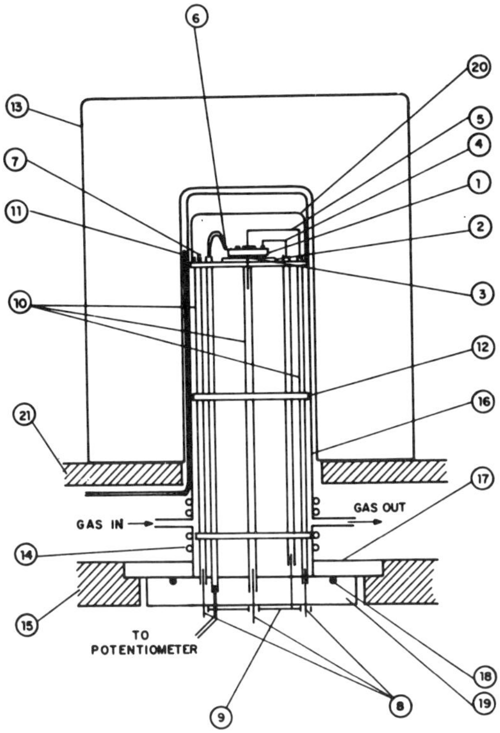


Figure 1. Schematic diagram of a vertical set-up for electrical measurements above the ambient temperature. (1) sample, (2) sapphire disc, (3) bottom lead wire*, (4) ground lead wire*, (5) top lead wire*, (6) thermocouple* (sample temperature measurement), (7) auxiliary lead wire*, (8) co-axial connectors; central leads, (9) co-axial connectors; outer part-connected to bridge ground, (10) platinum coatings on the outer surface alumina tubes, (11) thermocouple; to furnace temperature controller, (12) alumina disc spacer: also one above and one below, (13) furnace, (14) copper tubing carrying cold water, (15) sample chamber support, (16) quartz chamber, (17) brass plate (vacuum sealed to quartz chamber), (18) O-ring, (19) teflon base, (20) quartz dome (21) furnace support.

*Wires inside the tubes are not shown.

the problem of short circuiting via surface is eliminated more effectively using the three-probe method as shown in Figure 2. In this method a guard ring electrode is placed around the central electrode on the top face (Figure 2a) and any current between the guard ring and bottom electrode is excluded from the measurement. The presence of a guard ring also reduces the fringe effects which make a non-uniform field between the central and bottom electrodes. The electrical conductivity is then given by

$$\sigma = \frac{\text{sample thickness, } d}{\text{effective electrode area, } A_e} \times \frac{1}{\text{resistance, } R} \tag{1}$$

The effective electrode area $A_e = \pi r^2$ where

$$r = r_1 + \frac{g}{2} - \frac{2d}{\pi} \ln \cosh (\pi g/4d), \tag{2}$$

with $r_1$ as central electrode radius, g the gap width and d as sample thickness [11]. The last term in the above expression is a small correction term and is often neglected [5]. To avoid short circuiting via sample holder the cell should be designed such that the measurement leads are not connected through any other component in the high temperature zone. Also, the surfaces of insulating components (alumina, teflon etc.) should be kept free of carbon, grease or other impurities which are likely to introduce leakage paths. Leakage current through such extraneous paths can be estimated by determining the resistance of the specimen holder under the expected experimental conditions but without any sample.

When investigating glasses of moderate resistivity, usually surface conduction is not a source of error, instead barrier layers at the electrode/glass interface and non-ohmic contacts may become a major problem. These can be eliminated by using the four-probe method as shown schematically in Figure 3. In this method, a voltage drop in the specimen is measured away from the electrode/glass interface by connecting a high impedance voltage measuring system to the two central leads. If possible two thin platinum wires should be embedded in the specimen, which would act as voltage probes. In low resistivity samples the voltage drop between the probes would be small; this can be increased by increasing the current. However, care should be taken not to cause any Joule heating of the sample by high currents.

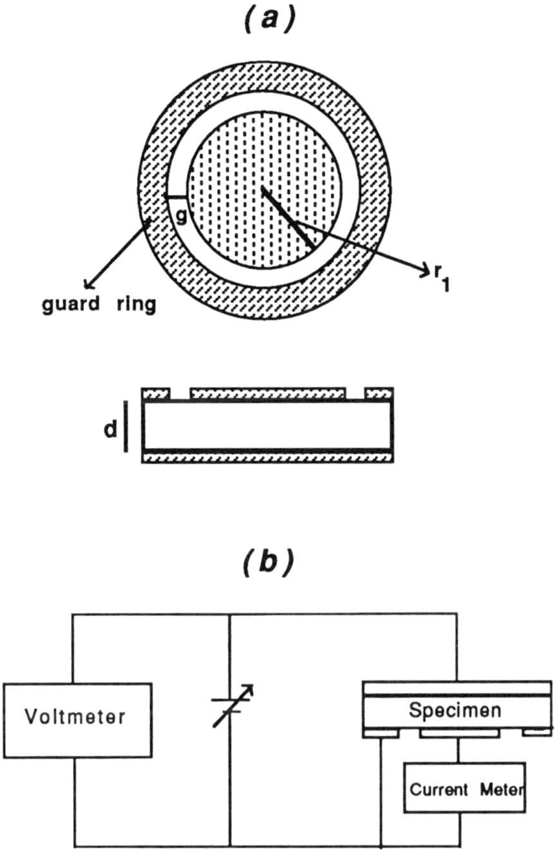


Figure 2. Three electrode configuration for electrical measurements. (a) specimen with electrodes and (b) circuit connections.

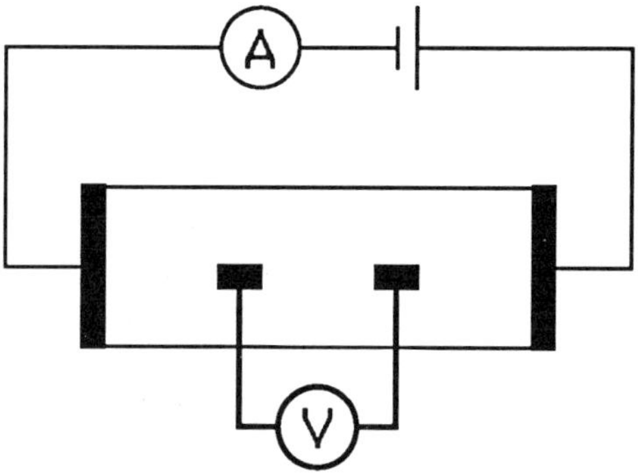


Figure 3. Four electrode configuration for conductivity measurements.

Frequency Dependent Measurements

A need for measuring conductance as a function of frequency arises from two reasons:

(a) Although the dc conductivity reflects the steady state flow of current, its magnitude is often modified by the presence of electrode polarization or contact resistance. The calculated dc conductivity then may not represent long range diffusion of charge carriers in the specimen. For this reason it may be difficult to interpret some of the old conductivity data obtained using dc or one fixed frequency ac signal. It is convenient to describe the electrode polarization on both electrodes (which are identical) as a parallel combination of resistance $R_e$ and capacitance $C_e$. The glass specimen is represented as a similar combination of $R_s$ and $C_s$ so that the sample-electrode assembly can be represented as shown in Figure 4. Because the barrier layer at the electrode/specimen interface is much thinner than the specimen thickness, $C_e >> C_s$ and the time constant (=RC) for electrode phenomena $(\tau_e) >> \tau_s$, the time constant for the sample dielectric. Thus, electrode phenomena dominate at the larger $\tau_e$ or at smaller frequencies. At high frequencies, $C_e$ is short circuited and one only measures the response of the glass specimen. In order to identify the frequency region where electrode phenomena would not interfere, one must measure conductance and capacitance in a wide frequency domain.

(b) Frequency dependent electrical measurements are also necessary if one is interested in the time dependent movement of charges. Such time dependent movements of charge lead to conductivity and dielectric relaxations phenomena.

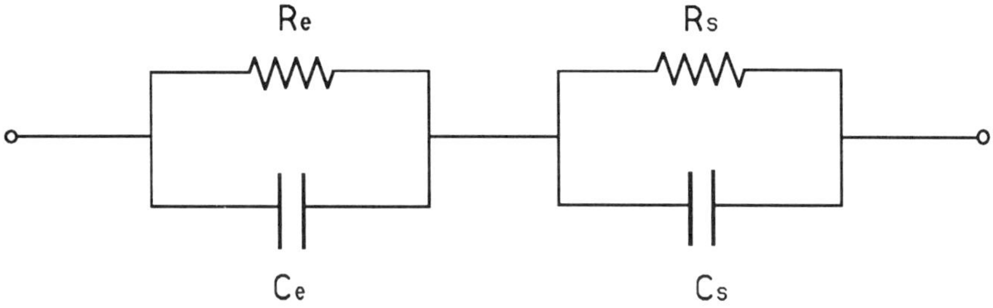


Figure 4. Two parallel R-C elements representing electrodes and specimen.

In principle, the two-, three- or four-probe methods described above can be adapted to conduct ac measurements. An important consideration with ac measurements, particularly at high frequencies, is that the leads should be shielded to minimize inductive coupling and stray capacitances. For very low frequencies (<10Hz) it is convenient to apply a constant potential across the specimen and then observe the time dependence of charging current (which is proportional to sample conductivity) with a recorder or oscilloscope (Figure 2). This measurement in time domain can give the value of ac conductivity as a function of frequency in one experiment. However, one is often interested in also knowing the sample capacitance as a function of frequency. Ac bridge such as a Harris bridge or a Berberian-Cole type bridge, the construction of which has been described in reference [12] and [13], respectively, is particularly suitable for low frequency measurments. The Harris bridge can be used at frequencies as low as $10^{-3}$ Hz and Martin [14] has used the Berberian-Cole bridge on glasses down to $5 \times 10^{-2}$ Hz. Since the construction of a bridge requires extensive time and efforts, these days tendency is to use one of the commercially available bridges or LCR meters which can be also controlled by a computer.

Most of the early ac conductance and capacitance measurements on glass were conducted below 1MHz partly because of experimental difficulties and partly because of lack of interest in the physics of glasses at higher frequencies. The operation of traditional bridges in the MHz and higher frequency region becomes difficult as the stray capacitance and admittance start dominating. These days electrical measurements at MHz and higher frequencies have become easier with the availability of several computer controlled network analyzers. Nevertheless, in the present overview we limit ourselves to the details of intermediate frequency ($10-10^6$ Hz) measurements which can be best related to electrical conductivity.

The impedance (conductance, G, and capacitance, C) measuring instruments in the intermediate frequency range are divided into bridges or meters. Modern ac bridges are based on Wheatstone bridge principle where various impedances are adjusted so that no current flows through the detector and a null point is obtained. This fact allows the measurement of very high impedances with high accuracy but balancing the bridge is time consuming. On the other hand, meters measure voltage and current that are calibrated for unknown impedances using Ohm's law. The meters may not be very accurate with high impedance samples, but they are easier and faster to use than the bridges. Thus, the choice between a bridge and a meter depends on whether the glass sample has very high resistance (or very low dielectric loss), in which case a capacitance bridge is more suitable.

The accuracy and range of measurements of a classical capacitance bridge can be greatly enhanced by replacing two arms of the basic Wheatstone bridge by windings on a transformer core as shown in Figure 5. The ac signal from the generator (GEN) when applied to primary windings, $N_p$, induces voltages, $V_s$ and $V_x$, in secondary windings whose ratio is $N_s/N_x$ where $N_s$ and $N_x$ are number of turns in the two windings, respectively. When the bridge is balanced and no current flows in the detector (DET),

$$G_x + i\omega C_x = \frac{N_x}{N_s} (G_s + i\omega C_s) \qquad (3)$$

where $\omega$ ($=2\pi f$) is the angular frequency of ac signal.

Thus the values of conductance ($G_x$) and capacitance ($C_x$) of the sample can be determined in terms of standards $G_s$ and $C_s$. Alternatively, one can also measure loss tangent or dissipation factor $D_x$ ($=G/\omega C$) and $C_x$ as a function of frequency. It is important to note from Figure 5 that the bridge circuit treats

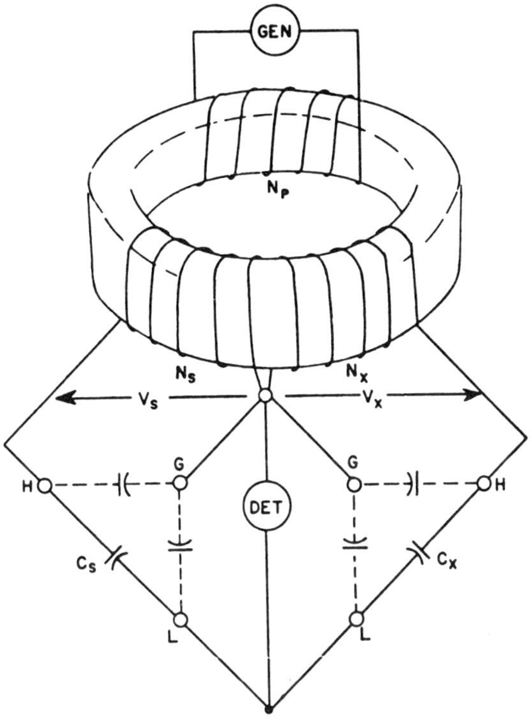


Figure 5. Wheatstone type capacitance bridge with transformer ratio arms [15].

the sample as a parallel combination of a resistor and a capacitor. The ratio $N_x/N_s$ can be varied very accurately and has little dependence on temperature, atmosphere and aging.

The bridge can be used in both two-probe and three-probe configurations. In the three-probe method any shunting between the guard electrode (or ground) G, and the high (H) or low (L) terminals of the sample (Figure 5), only affects the sensitivity of the detector but not the balance conditions if the transformer core has negligible resistance and leakage inductances.

Examples of commercially available transformer ratio bridges are Quadtech* (formerly General Radio) model 1615 and 1616 (20Hz-100kHz). It is advisable to use a frequency counter with these bridges to monitor the signal frequency. Typically, after a sample has reached equilibrium conditions, an ac signal of predetermined frequency is applied from the oscillator (generator). The sample conductance and capacitance are measured by varying $N_s$, $N_x$ and standard $G_s$ and $C_s$ until minimum deflection is recorded in the detector. The same procedure of balancing the bridge is repeated at a series of frequencies covering the available range. Clearly this is a time consuming procedure and attempts have been made to automate the bridge balancing procedure [16]. At present, the automated bridges of similar capabilities are not available commercially.

For glass samples of resistance less than 10MΩ the use of capacitance bridge may not be necessary and an LCR meter would be a more convenient and faster alternative. Several meters are commercially available [e.g. Hewlett Packard** model 4274A (100Hz-100kHz), Quadtech* Digibridge 1689 (12Hz to 100kHz), Solartron Schlumberger*** 1260 ($10^{-5}$Hz-32MHz), Wayne Kerr**** 6425 (20Hz to 300kHz), etc.] which can be interfaced with a microcomputer and collect data at many more frequencies in a very short time. The meter can be set to treat a sample equivalent to either a series or a parallel combination of L, C and R elements. The choice between the series and parallel arrangement would depend on which describes the conduction through the glass/electrode assembly more accurately, but either set of parameters can be easily transformed into the other set, as shown in Table 1, for negligible inductive effects. Here subscript s and p refer to series and parallel circuits, respectively. The digital recording of the data makes the subsequent analysis very convenient and one can automate most of the experiments except for the specimen preparation steps.

---

*Bolton, MA 01740
**Woburn, MA 01801
***Santa Clara, CA 95052
****Burlington, MA 01803

Table 1. Interrelations Between Parallel and Series Circuit Elements

$$C_p = \frac{C_s}{1 + (\omega R_s C_s)^2} \quad , \quad G_p = \frac{(\omega R_s C_s)^2}{R_s \, [1 + (\omega R_s C_s)^2]}$$

$$C_s = \frac{C_p \, (\omega^2 C_p^2 + G_p^2)}{\omega^2 C_p^2} \quad , \quad R_s = \frac{G_p}{G_p^2 + \omega^2 C_p^2}$$

$$\text{Dissipation factor, } D = \frac{G_p}{\omega C_p} = \omega R_s C_s$$

## DATA AND ITS ANALYSIS

When an LCR meter or capacitance bridge is used, one obtains conductance (G) and capacitance (C) of the sample directly. Data in terms of other electrical properties (D, $R_s$, Q or $C_s$) can be easily converted to G and C using the relations given in Table 1. If only charging or discharging current through the specimen is measured as a function of voltage, using Ohm's law one obtains conductance alone as a function of time (frequency). In this case if no time dependence of current is observed one might assume that the electrode polarization is absent and the conductance represents true conductivity. For most of the common glass samples with noble metal electrodes this is not likely to be true and information on capacitance will be needed for further analysis.

Typical capacitance and conductance data are shown in Figures 6 and 7, respectively, for a borate glass at various temperatures. Note that the strong frequency and temperature dependence of capacitance becomes weaker at higher frequencies. From the trend in Figure 6 it is clear that at a given temperature the capacitance would become frequency independent at sufficiently high frequencies. At such high frequencies the electrode capacitance ($C_e$ in Figure 4) is short circuited and one only measures the volume capacitance of the glass which is proportional to its dielectric constant. In this frequency region the conductance is also expected to be unaffected by the

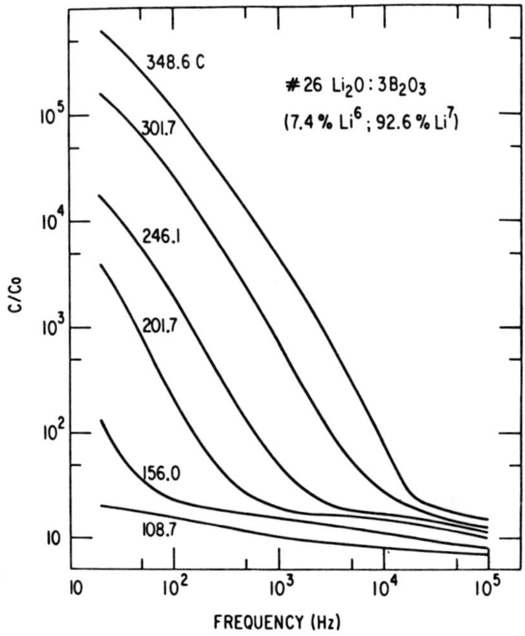


Figure 6. Normalized capacitance vs. frequency for a lithium triborate glass at six temperatures.

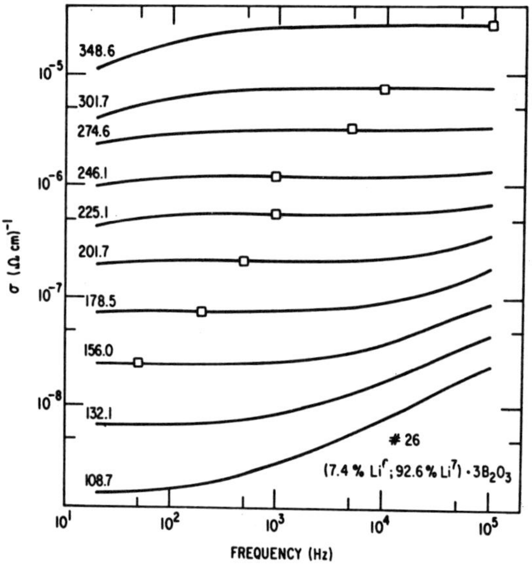


Figure 7. Electrical conductivity vs. frequency for a lithium triborate glass (same as in Fig. 6) at various temperatures [17].

electrode phenomenon. Clearly, the capacitance data at lower frequencies represents a combination of the electrode and glass capacitances (not necessarily as shown by the equivalent circuit in Figure 4).

The most obvious observation from Figure 7 is that conductance is a strong function of frequency which itself depends on temperature. The frequency dependence of conductance in Figure 7 can be divided into three regions. At high temperatures, $G(\omega)$ is frequency independent over most of the range measured but decreases with frequency at the lower end. Since the decrease in conductance correlates well with the strong increase in capacitance at low frequencies, the most likely reason for both observations is the presence of electrode phenomenon. Then the frequency independent conductance at high frequencies would be a property of the glass specimen alone. When temperature decreases the level of conductance decreases, as expected from ionic conduction. At the same time frequency dispersion of conductance, which represents electrode phenomenon, moves to lower frequencies and finally becomes unobservable below 200°C. On the other hand a new frequency dispersion, showing increase in conductance with frequency, appears at high frequencies. As the temperature decreases further this high frequency dispersion extends to lower and lower frequencies and remains the only source of conductance at room temperature. In general, the three regions of frequency dispersion of $G(\omega)$ can be observed at any temperature if $G(\omega)$ is measured over a sufficiently wide frequency range. In practice the frequency range is limited and one may observe only one or two regions at a given temperature.

Occasionally one might observe at the high end of the frequency region that $G(\omega)$ starts decreasing as frequency increases. Such anomalous behavior is likely to be observed with very high resistance materials due to some spurious effects from surface contamination, stray capacitance or inductance, etc. In such cases it is advisable to clean the insulating paths and properly shield the lead wires.

It is now clear from the complex behavior of $G(\omega)$ in Figure 7 that one can make a serious error in interpreting the conductance of a sample which is measured at a fixed frequency. Much of the older conductivity data in literature were obtained using dc method or ac method at one fixed frequency and therefore, we should be careful with their quantitative interpretation. From the viewpoint of electrical conductivity one would like to know the value of $G(\omega)$ which can be related to the long range motion of charge carriers. Considering that the low frequency is dominated by electrode phenomenon and true electrical conductivity should not depend on frequency, it is clear that the plateau in Figure 7 represents the electrical conductivity of the specimen. Unfortunately, in some cases this plateau can be too small to get a reliable value of conductivity. Therefore, one makes use of complex impedance analysis based on the equivalent circuit of Figure 4, as described below, to separate volume conductivity from electrode effects.

The real and imaginary parts of complex impedance $Z^*(\equiv Z' - iZ'')$ are defined as

$$Z' = \frac{G}{G^2 + \omega^2 C^2} \tag{4}$$

$$Z'' = \frac{\omega C}{G^2 + \omega^2 C^2} \tag{5}$$

A plot of Z" vs. Z' with $\omega$ as an implicit parameter, should show two semicircles for the equivalent circuit of Figure 4. In practice where the frequency range is limited one may observe only one or parts of the arcs of semicircles as shown in Figure 8. The intersection of the high frequency arc with the real axis (shown by the arrows) gives the true resistance of the sample. The squares in Figure 7 represent conductance obtained from this analysis. To confirm that this indeed is the property of the specimen alone, one can change the sample thickness keeping the electrode area constant and repeat the experiment. The electrode phenomenon would depend on the electrode area, and conductivity on the sample volume. If the resistance obtained by the above method represents a property of the glass, its value in two experiments should change in the ratio of thickness change. This is clearly demonstrated in Figure 8(b) where decrease in thickness by a factor of 1.75 decreased the resistance by 1.76.

A close examination of Figure 8(a) shows that, instead of forming a semicircle, the data form an arc of a semicircle with center below the real axis. Therefore, the electrical circuit of Figure 4 is not an exact representation of the glass-electrode assembly. However, when the depression of center in the sample is not too large and the change in thickness experiment confirms the resistance to be a volume property, the above described complex impedance analysis offers the best method to obtain true conductivity as shown for many glasses including borates [18,19], phosphates [20], germanates [21], and

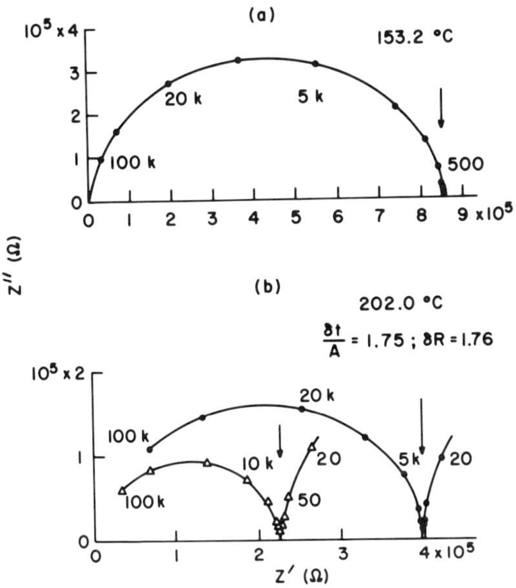


Figure 8. Complex impedance plots (Z" vs. Z') for (a) 0.17 NaF-1.03 $Na_2O$-3 $B_2O_3$ and (b) lithium triborate glasses [18].

silicates [22]. Unfortunately, one can also find examples where the arc for volume conduction is skewed on the low frequency side or does not have a sharp minimum which can be identified with true resistance [23,24]. In such cases one should first try to improve the electrode preparation procedure ensuring that the glass surface has not changed between polishing and metal deposition such as by moisture attack. If this fails, one may try changing the electrode metal. No improvement in the shape of impedance arcs after these attempts may mean that the glass sample cannot be described as a parallel combination of an ideal capacitor and an ideal resistor. Perhaps the conduction process is inherently frequency dependent so that there is no frequency independent region in the $G(\omega)$ vs. $\omega$ plot. Surprisingly, this is found in some examples of a presumably ideal material, viz. quartz crystal as shown in Figure 9. In that case the meaning of electrical conductivity breaks down as defined in the beginning of this chapter. At present there is no consensus on how to answer this problem.

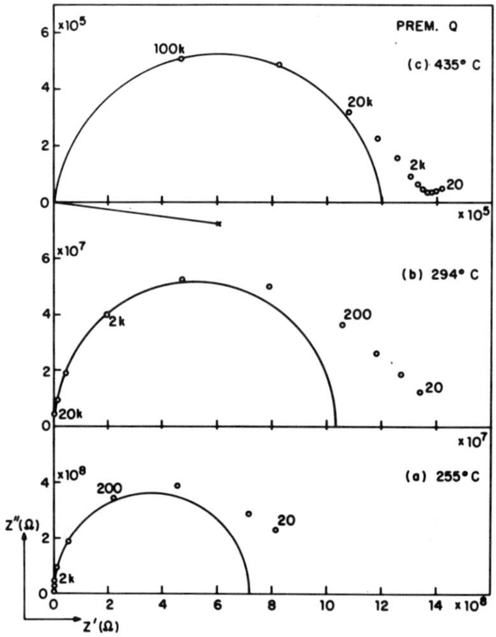


Figure 9. Complex impedance plots for a Z-cut Premium Q quartz crystal at three different temperatures [25].

Once the true conductivity values have been obtained for a glass at various temperatures, it is convenient and customary to present this information in terms of activation enthalpy (H) and a pre-exponential factor (A) by fitting the experimental data to Arrhenius equation [26]:

$$\sigma = \frac{A}{T} \exp - \frac{H}{kT} \qquad (6)$$

where k and T are Boltzmann constant and absolute temperature, respectively. A typical set of data usually plotted as $\ln\sigma T$ vs. $1/T$, (see Figure 10) shows the appropriateness of equation 6 for several orders of magnitude change in conductivity. However, a careful analysis would often show a downward deviation of data (by a few percent) from equation (6). This deviation can be reduced, if not eliminated, by excluding T before the exponential in equation (6). Nonetheless, this author feels that equation (6), which is based on our understanding of diffusion in crystals, is more appropriate to use. For further discussion on this topic the reader is referred to [27].

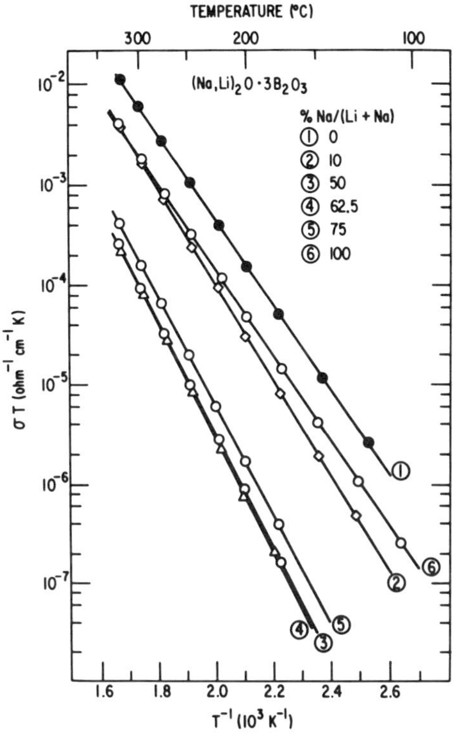


Figure 10. Arrhenius plots for six lithium-sodium triborate glasses [18].

Equation (6) is convenient to present the conductivity of a sample at various temperatures in terms of only two parameters, viz. A and H. With the knowledge of A and H, it is trivial to extrapolate conductivity beyond the experimental temperature range. However, one should be aware of the fact that in general the values of A and H obtained from the measurements below the glass transition temperature ($T_g$) will not be applicable for temperature above $T_g$.

The overwhelming presence of the continuous increase in $G(\omega)$ with $\omega$ at low temperatures and high frequencies (Figure 7) represents conductivity relaxation. It is a common observation not only in glasses but also in crystalline oxides, polymers, semiconductors, etc. [28,29]. At present, a simple relation between electrical relaxation and dc conductivity is not established and, therefore, we will not discuss this subject here. The interested reader is referred to recent literature [30] on the current status of our understanding of electrical relaxation.

## INTERPRETATION OF THE RESULTS

Having obtained the numerical values of A and H, we are obviously curious to know their physical significance. To satisfy this curiosity, we depend on the theory of diffusion developed for crystals. Note that ionic conducitivity under discussion is related to ion diffusion via the Nernst-Einstein relation. Then the pre-exponential factor, A, contains the attempt frequency for ion jump, square of the jump distance, an ion concentration term, an entropy term, and a geometrical factor [26]. The exact expression for A depends on the further details of the diffusion process [31]. Usually the various terms in A are not known for glasses and, therefore, much less attention is directed towards understanding A in terms of the glass structure than for understanding H. Fortunately, H, being in the exponent of equation (6), is much more important than A for determining the behavior of conductivity. This is illustrated by the data shown in Figure 11. Here, we note that $\sigma$ for a lithium-sodium trisilicate glass at 200°C shows a strong minimum when lithium is gradually replaced by sodium due to the well-known mixed alkali effect. The maximum in H is consistent with such a variation of $\sigma$, but the maximum in A is not consistent with the observed minimum in $\sigma$. Essentially, the variation of A is overwhelmed by the variation of H.

In a general sense, H in equation (6) represents activation enthalpy, that is, the height of the enthalpy barrier that an ion must overcome to complete each conductive jump. A higher value of H means an energetically more difficult jump and therefore $\sigma$ should decrease (assuming no change in A). Thus, the peak in H in Figure 11 indicates that the conductive jump requires greater energy in the mixed alkali glasses than in the single alkali glasses. This inference has prompted the investigators to seek an explanation of the mixed alkali effect in the energetics of ion jump.

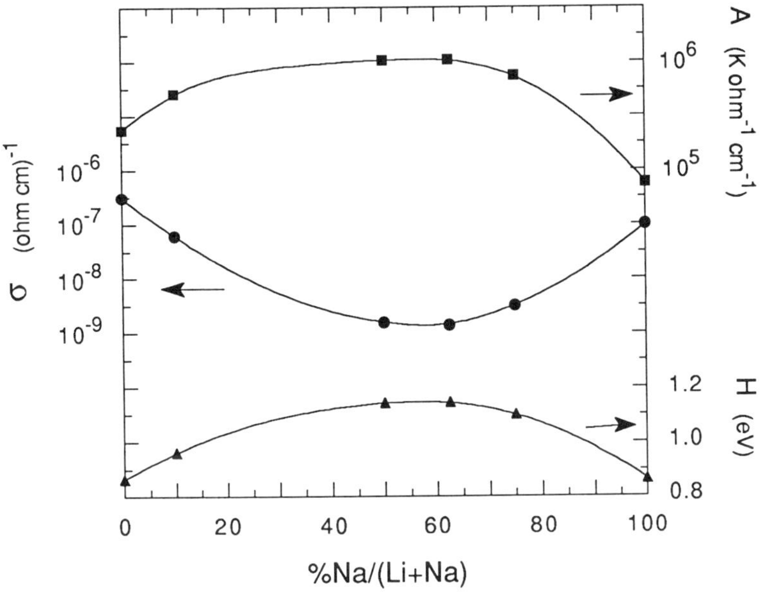


Figure 11. Variation of electrical conductivity (σ) at 200°C, the pre-exponential factor (A) and activation enthalpy (H) with alkali composition of a lithium-sodium triborate glass series. Note that the peak in H predicts a minimum in σ, but the peak in A predicts just the opposite [18].

It is often tempting to relate the variation of H to the structural changes resulting from composition variation in a glass series. However, usually we do not know the precise meaning of what consititutes H. In the Anderson and Stuart picture [32], H is made of primarily two terms: (a) the energy required to break the bond between a cation and its charge compensating center (e.g., a non-bridging oxygen), and (b) the energy required to dilate the network for the jumping of the cation. In an alkali silicate glass, H decreases with increasing alkali concentration. This has been explained qualitatively in terms of the breakdown of glass network with the addition of non-bridging oxygens. However, we do not know how much the two energy terms change separately with increasing alkali concentration. That is, we do not know the precise role of the structural changes. Clearly additional experiments must be conducted to determine the relative contributions of the two energy terms. It is possible that in some glasses the first term is negligible, i.e., all the ions are free to jump at any given time. In that case, one speaks of the glass as a strong electrolyte. On the other hand, if some energy must be supplied for a cation to first become free to move, only a fraction of the ions will be mobile at a given instant. Then, one considers glass as a weak electrolyte.

Finally, we should realize that electrical conductivity is not an ion specific property. That is, it does not indicate what are the conducting species. If there are several conducting species, partial conductivity due to each specie must be determined with the techniques not described here. In the example of Figure 11, the conductivity of mixed alkali compositions is due to both Li and Na, but we only know the total conductivity. It is shown by the tracer diffusion experiments that the relative contribution to total conductivity from an alkali decreases as its relative concentration decreases in any mixed alkali glass series (see e.g.[22]). Thus if we focus on the partial conductivity due to any of the two ions, we find a monotonic change rather than the $\sigma$-minimum of Figure 11. The activation enthalpy for each partial conductivity also varies monotonically. This example illustrates an important shortcoming of the electrical conductivity for describing ion movement in a glass with more than one charge carrier. Nevertheless, due to the convenience of its measurement, it is perhaps the most popular method for characterizing ion dynamics in all kinds of glasses.

## SUMMARY

In this chapter, we have described the methods of measuring the electrical conductivity of common glasses. It is pointed out that the major sources of uncertainty are likely to be the incomplete characterization of the glass itself and the sample temperature. Three and four probe electrode configurations are superior to two probe configurations and between the two the former is preferred for specimens likely to have significant surface conduction. Unless reversible electrodes are used, it is important to measure conductance and capacitance as a function of frequency and temperature. Then complex impedance analysis should be used to obtain volume conductivity. At low temperatures and high frequencies, these measurements would show rapidly increasing conductance with frequency, which represents conductivity relaxation.

ACKNOWLEDGEMENT

The author is grateful to the U.S. Department of Energy for supporting this work under Grant No. DE-FG02-90ER45419.

REFERENCES

1. L. J. van der Pauw, "A Method of Measuring Specific Resistivity and Hall Effect of Discs of Arbitrary Shapes", Phillips Res. Rept. 13, 1-9 (1958).
2. J. T. Littleton and W. L. Wetmore, "The Electrical Conductivity of Glass in the Annealing Zone as a Function of Time and Temperature", J. Am. Ceram. Soc. 19, 243-245 (1936).
3a. R. H. Magruder, D. L. Kinser, R. A. Weeks and J. M. Jackson, "Effect of Melt Temperature on the dc Conductivity of $GeO_2$ Glasses", J. Appl. Phys. 57, 345 (1985).
b. R. A. Weeks and D. L. Kinser, "Effects of Liquid Equilibration Temperature on the Properties of Germania Glass; A Review", Diffusion and Defect Data 53/54, 117-126 (1987).
4. M. Kleitz, P. Fabry and E. Schouler, "Descriptions of Junctions Between Electronic and Solid Ionic Conductors", in Electrode Processes in Solid State Ionics, M. Kleitz and J. Dupuy, eds., Reidel Publishing Co., Dordrecht-Holland (1976), p. 1-17.
5. R. N. Blumenthal and M. A. Seitz, "Experimental Techniques" in Electrical Conductivity in Ceramics and Glass, N. M. Tallan, ed., Marcel Dekker, New York (1974), p. 35-178.
6. Manual on the Use of Thermocouples in Temperature Measurement, ASTM STP 470B, American Society for Testing and Materials (1981).
7. N. E. Hill, W. E. Vaughan, A. H. Price and M. Davies, Dielectric Properties and Molecular Behavior, van Nostrand Reinhold Co. (1969), p. 172.
8. Nonstoichiometric Oxides, O. T. Sorensen, ed., Academic Press (1981).
9. Page 76 of reference 5.
10. O. Kubaschewski, E. L. Evans and C. B. Alcock, Metallurgical Thermochemistry, Pergamon Press, New York (1967).
11. W. G. Amey and F. Hamburger, "A Method of Evaluating the Surface and Volume Resistance Characteristics of Solid Dielectric Materials", Proc. ASTM 49, 1079-1091 (1949).
12. W. P. Harris, Nat. Acad. Sci.-Nat. Res. Counc., Pub. No. 1484, 72 (1967).

13. J. P. Berberian and R. H. Cole, "Low Frequency Bridge for Guarded Three-Terminal and Four-Terminal Measurements of Admittance", Rev. Sci. Instrum. 40, 811-817 (1969).
14. S. W. Martin: Fast Ion Conduction in Glass, Ph.D. Thesis, Purdue University (1986).
15. Instruction Manual GR1616 Precision Capacitance Bridge, General Radio Company (now Quadtech Inc.), 1971.
16. Andeen-Hagerling, Inc., Chagrin Falls, OH 44022.
17. H. L. Downing, N. L. Peterson and H. Jain, "Mixed Isotope Electrical Conductivity in Lithium Borate Glasses", J. Non-cryst. Solids 50, 203-213 (1982).
18. H. Jain, H. L. Downing and N. L. Peterson, "The Mixed Alkali Effect in Lithium-Sodium Borate Glasses", J. Non-cryst. Solids 64, 335-349 (1984).
19. H. L. Tuller, D. P. Button and D. R. Uhlmann, "Fast Ion Transport in Oxide Glasses", J. Non-cryst. Solids 40, 93-118 (1980).
20. S. W. Martin and C. A. Angell, "Dc and ac Conductivity in Wide Composition Range $Li_2O$-$P_2O_5$ Glasses", J. Non-cryst. Solids 83, 185-207 (1986).
21. J. N. Mundy and G. L. Jin, "Ionic Transport in Sodium Aluminogermante Glasses", Sol. St. Ionics 21, 305-325 (1986).
22. H. Jain, N. L. Peterson and H. L. Downing, "Tracer Diffusion and Electrical Conductivity in Sodium-Cesium Silicate Glasses", J. Non-cryst. Solids 55, 283-300 (1983).
23. R. Syed, D. L. Gavin, C. T. Moynihan and A. V. Lesikar, "Shape of Complex Resistivity Arcs for Ionically Conducting Glasses", J. Am. Ceram. Soc. 64, C118-C120 (1981).
24. M. Tomozawa and M. Yoshiyagawa, "A.c. Electric Characteristics of Mixed Alkali Glasses", Glasstech. Ber. 56, 939-944 (1983).
25. H. Jain and A. S. Nowick, "Electrical Conductivity of Synthetic and Natural Quartz Crystals", J. Appl. Phys. 53, 477-484 (1982).
26. J. M. Stevels, "The Electrical Properties of Glass", Handbuch der Physik, Vol. 20 (Springer-Verlag, Berlin, 1957), p. 350-391.
27. H. Jain, "Temperature Dependence of Electrical Conductivity of Glasses", J. Non-cryst. Solids 66, 517-522 (1984).
28. K. L. Ngai, "Universality of Low Frequency Fluctuation, Dissipation, and Relaxation Properties of Condensed Matter I & II", Comments Solid St. Physics 9, 127-155 (1979).
29. A. K. Jonscher, Dielectric Relaxation in Solids, Chelsea Dielectrics Press, London (1983).

30. H. Jain, "Ion Movement Relaxation in Inorganic Glasses-Salient Features", J. Non-cryst. Solids 131-133, 961-968 (1991). Also see theoretical papers in Section 2.8 of the same volume.
31. A. S. Nowick, W-K. Lee and H. Jain, "Survey and Interpretation of Pre-exponential of Conductivity", Solid St. Ionics 28-30, 89-94 (1988).
32. O. L. Anderson and D. A. Stuart, "Calculation of Activation Energy of Ionic Conductivity in Silica Glasses by Classical Methods", J. Am. Ceram. Soc. 37, 573-580 (1954).

# CHAPTER 13

# MAGNETIC PROPERTIES OF COMMON INSULATOR GLASSES

Jeffrey T. Kohli, NYS College of Ceramics, Alfred University
Presently with Corning Inc., Corning, NY

## INTRODUCTION

The study of magnetic properties of materials began at least 2,000 years ago when the Greeks observed that stones containing magnetite attracted pieces of iron. The magnetic properties of materials are now a vast subject of interest for materials scientists, engineers, chemists, and physicists. New materials are constantly being developed and each new material has a different set of magnetic properties. The recent interest in magnetic recording media, amorphous metals, various types of sensor materials, and new superconducting ceramics have all led to a heightened interest in magnetic properties and magnetic measurements. New magnetic materials have had a profound effect on our society. Perhaps most importantly, they have increased our ability to store and disseminate information. Magnetic tapes have been used for storage of information in high tech computer systems and for such commonplace applications as musical recordings. Today, advanced magnetic materials are being developed for use as optical recording media in which photonic signals from lasers are used to "write" information on a substrate via light stimulated magnetization. Moreover, magneto-optical devices such as the laser isolator are used in high powered laser systems or in fiber-optic systems where optical signals are used to transmit information through thin strands of glass. Many people have even envisioned contactless conveyor systems as well as transportation systems which would require unique magnetic materials. It will be fascinating to see how magnetic devices will affect our society in the future.

## BASIC MAGNETIC THEORY

Although ordinary glass and ceramic materials are not thought of as magnetic materials, all materials have magnetic properties. The magnetic characteristics of materials are largely determined by the electrons associated with the atoms or ions in the material. If we consider that the electrons of an atom orbit the nucleus in quasi-confined orbits, these electrons have a current associated with their motion. The currents created by these electrons give rise to magnetic moments. Moreover, electrons have an inherent spin so that each electron also has a spin magnetic moment.

When materials are placed in a magnetic field (**H**), the magnetic moments tend to align parallel with the applied field. The quantity **H** is also known as the magnetic intensity. The magnetic dipole moment per unit volume is the magnetization (**M**). The unitless proportionality constant between magnetization and the magnetic intensity is the magnetic susceptibility ($\chi$). We therefore have the relation

$$\mathbf{M} = \chi \mathbf{H} \tag{1}$$

where the magnetic susceptibility is a constant for most materials.

In light of the various responses of materials to a magnetic field, it is useful to introduce the concepts of magnetic induction and permeability. The magnetic induction ($\mathbf{B}$) is related to the magnetization and magnetic field by the relation

$$\mathbf{B} = \mu_0(\mathbf{H} + \mathbf{M}) \qquad 2a)$$

or

$$\mathbf{B} = \mu \mathbf{H} \qquad 2b)$$

where

$$\mu = (1 + \chi_m)\mu_0$$

The quantity $\mu$ is the magnetic permeability of the material and $\mu_0$ is the permeability of free space.

The permeability of a material allows us to define further the magnetic properties of materials. Michael Faraday divided materials into three categories: 1) paramagnetic materials, 2) diamagnetic materials, and 3) ferromagnetic materials. The first two categories encompass materials whose magnetization varies directly with the strength of the applied field. In the case of ferromagnetism, on the other hand, the magnetization does not vary linearly with the applied field, and the permeability of ferromagnetic materials is not only a function of **H** but also depends upon the previous state of the material's magnetization. Faraday's three categories of materials differ in the following manner

| | |
|---|---|
| paramagnetic materials | $\mu \cong 1$, but $\mu > 1$ |
| diamagnetic materials | $\mu \cong 1$, but $\mu < 1$ |
| ferromagnetic materials | $\mu$ is very high and dependent upon **H**, except in very large magnetic fields |

Subsequent to Faraday's classification of magnetic materials, further studies led to the discovery of two other general categories of substances, specifically: 4) ferrimagnetic materials, and 5) antiferromagnetic materials.

We can further differentiate para- and diamagnetic materials from ferromagnetics by the permanence of the orientation of magnetic dipoles created by an applied field. When either a paramagnetic or diamagnetic material is withdrawn from a magnetic field, the dipole moments induced in the material spontaneously return to a state of random orientation. In ferromagnetic materials, on the other hand, dipoles remain oriented to some degree after the material is withdrawn from a field. The hysteresis loop exhibited by ferromagnetic materials, when induction is plotted as a function of applied magnetic field, illustrates the retention of domain alignment after the removal of the field.

When the magnetization of a ferromagnetic material is plotted against increasing temperature, the magnetization decreases and eventually reaches zero at a characteristic temperature known as the Curie temperature, where the material undergoes a phase change. This phase change is associated with the disappearance of long range magnetic order.

Materials which exhibit ferromagnetism are paramagnetic above their Curie temperature. Paramagnetism is characterized by the fact that the reciprocal of the magnetic susceptibility varies linearly with temperature. This dependence of the susceptibility with temperature is known as the Curie-Weiss law, where the susceptibility follows the expression

$$\chi = C/(T-\theta_p) \qquad 3)$$

In the above expression, $\theta_p$ (the paramagnetic Curie temperature) and C (the Curie constant) are constants which are related to the fundamental nature of the atoms or ions in a material. The temperature (T) and $\theta_p$ are expressed in Kelvin. For ferromagnetic materials, the Curie temperature and the paramagnetic Curie temperature are rarely exactly the same but are nevertheless very similar in most cases. In many paramagnetic materials which do not exhibit ferromagnetism at any temperature, the paramagnetic Curie temperature is equal to 0 K. In such cases, the Curie-Weiss law reduces to $\chi = C/T$.

Since the type of glasses considered in this text generally exhibit only paramagnetic and diamagnetic behavior, except at temperatures approaching 0 K in some cases, it is suggested that the reader refer to general physics and magnetic materials texts for more information about the other categories of materials. This chapter will further investigate the causes of para- and diamagnetism and the techniques which are used to study these two types of magnetism. We will briefly mention more elaborate characterization and structural techniques which utilize the inherent magnetic traits of glasses in the final section of this chapter.

## PARAMAGNETISM AND DIAMAGNETISM

As mentioned earlier, magnetic properties result from the electrons which belong to the atoms or ions of a material. Electrons have dipole moments as a result of their orbital motion and their spin. An electron orbiting in a loop has a magnetic dipole moment ($\mu$) which is proportional to its angular momentum and is normal to the plane of the orbit. The dipole moment associated with the orbital motion is given by

$$\mu = (-e/2m)\,\mathbf{L} \qquad 4)$$

where e is the charge , m is the mass, and **L** is the angular momentum of the electron. The quantity $\mu/L$ is known as the gyromagnetic ratio. The magnetic dipole associated with an electron's spin, on the other hand, is given by

$$\mu = -g_s(e/2m)\,\mathbf{S} \qquad 5)$$

where $g_s = 2.0023$, and **S** is the spin angular momentum of the electron. In most cases $g_s$ may be taken as 2.00, and therefore the spin gyromagnetic ratio is approximately e/m.

Diamagnetism results from electrons which are spin-paired within the orbitals of atoms. The spin magnetic moments of spin-paired electrons cancel one another. However, electrons orbiting within atoms induce magnetic moments. When a magnetic field is applied, the motion of the electrons is altered and the magnetic moment is changed. Since all atoms and ions, except for the hydrogen atom, have at least two spin-paired electrons, diamagnetism is inherent to all materials. Lenz's law of electromagnetic induction states that currents which are induced by a magnetic field are oriented such that their magnetic fields oppose the original inducing field, i.e. the dipole moments are directed opposite to an applied field. Diamagnetic materials, therefore, are repelled by magnetic fields. As a result of this phenomenon, the magnetization and the susceptibility of diamagnetic materials are negative.

Paramagnetism, on the other hand, is caused by electrons which are not spin-paired. Unpaired electrons contribute appreciable dipole moments to a material whether they are free (conduction electrons) or confined to an atom or ion. In the presence of a magnetic field, these dipole moments align themselves in the direction of the field, giving the material a positive magnetic susceptibility. The magnitude of a spin moment, unlike an orbital moment, is constant and does not change in response to an applied field. If an atom has unpaired electrons, the paramagnetic contribution of the unpaired electrons is usually significantly greater than the diamagnetic contribution. Of course, this does not mean that a material which contains paramagnetic ions is highly paramagnetic. Whether a material is net paramagnetic or diamagnetic simply depends upon the type of atoms or ions in the material and the relative concentrations of paramagnetic and diamagnetic species.

Unlike diamagnetism, paramagnetism varies as a function of temperature. At very low temperatures, the moments of paramagnetic materials align relatively easily with applied fields. At higher temperatures, the moments tend to be randomly aligned due to the thermal motion of the atoms in the material. The work required to rotate a dipole 180° in a magnetic field is given by $2\mu B$, where $\mu$ is the dipole moment. For a typical magnetic moment of 1 Bohr magneton and a field of 1 Tesla, the work required to rotate the dipole 180° is

$$2\mu B = 2(9.27 \times 10^{-24}\,\mathrm{J/T})\,(1\mathrm{T}) = 1.85 \times 10^{-23}\,\mathrm{J} \qquad 6)$$

At room temperature (300K), the thermal energy (kT) is approximately 200 times greater than the difference in potential energy when the dipole is aligned parallel or antiparallel to the applied field,[1] which indicates that, even at room temperature, the majority of the dipoles in a paramagnetic material will be randomly oriented as a result of the thermal energy of the system.

There are two types of both para- and diamagnetism, which are caused by : 1) bound, or 2) free electrons. We will not describe the contributions to the magnetic susceptibility due to free electrons because we are considering materials which are insulators. The classical bound electron contribution to the diamagnetic susceptibility of materials was derived by Langevin[2]

$$\chi_{dia} = \frac{-NZq^2\langle r^2\rangle\mu_0}{6m} \qquad 7)$$

where N is the concentration of atoms or ions, Z is the atomic number of the atom (or number of electrons in the ion), q is the electronic charge, $\langle r^2\rangle$ is the average of the square of an electron's orbital radius (usually taken as the radius of the atom or ion), and $\mu_0$ and m have their usual meanings.

Bound electron paramagnetism is frequently referred to as Van Vleck paramagnetism, but classical theory really follows from developments made by Langevin around 1905. The bound electron paramagnetism is

$$\chi_{para} = \frac{N\,p_{eff}^2\,\mu_B^2}{3kT} \qquad 8)$$

where $p_{eff} = g[J(J+1)]^{1/2}$ when both spin and orbital contributions to the moments of the electrons are considered, $\mu_B$ is the Bohr magneton, k is Boltzman's constant, and T is the temperature in degrees Kelvin. The Bohr magneton is equal to $eh/4\pi m$, where h is Planck's constant. The magnetic moment associated with the spin of a single electron is equal to one Bohr magneton.

## PARAMAGNETIC IONS IN GLASS

It is rather simple to identify those elements or ions which impart paramagnetic behavior to glasses; one need only look at a table of the electronic distributions of the elements. If we ignore the many types of electronic defects which have been identified in glasses, then the transition metals, the rare earths, and many of the actinides account for the paramagnetic behavior of glasses. Although the scarcity and radioactivity of the actinides have precluded extensive research with these elements, there have been several studies of the magnetic susceptibility of glasses containing plutonium,[3] thorium,[4] and uranium.[4-6]

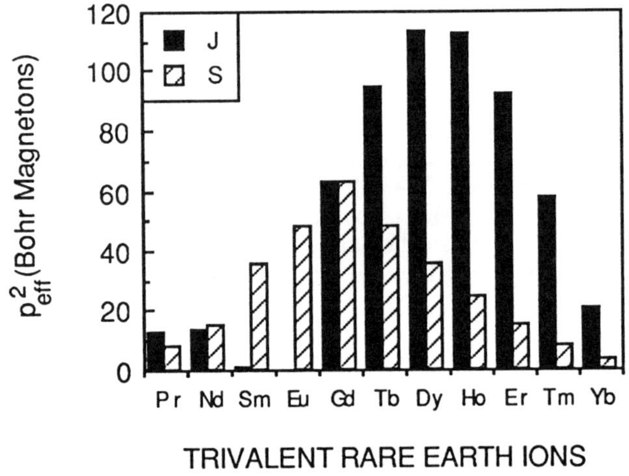


**Figure 1. Calculated $p_{eff}^2$ versus trivalent rare earth ion identity.**

The transition metals have unpaired electrons in their 3d suborbital, while the rare earth elements have unpaired electrons in their 4f suborbital. The unpaired electrons are responsible for the interesting optical, magnetic, and magneto-optical effects that these elements impart to glasses. The 3d suborbital of the transition metals is partially shielded by 4s electrons, but because the 3d electrons are so close to the valence electrons of these elements, the ligand field caused by surrounding ions may significantly alter the orbital motion of the unpaired 3d electrons. The 4f electrons of the rare earths, on the other hand, are well shielded by outer electrons in the 5s, 5p, and 6s orbitals so that the orbital motion of the unpaired 4f electrons remains largely unaffected by ligand fields.

The quantity $p_{eff}^2$ has been plotted against the identity of the trivalent rare earth ions in Figure 1 using the quantum number J (the spin-orbit interaction quantum number), and alternatively S (the spin-only quantum number). If the orbital quantum number L is equal to zero, i.e. the spin-only scenario where $p_{eff} = g[S(S+1)]^{1/2}$, then g=2. The values used for g in the calculation where $p_{eff} = g[J(J+1)]^{1/2}$ were obtained from Weber.[7] In the rare earth ions, both the orbital momentum and the spin of the 4f electrons contribute to the magnetic nature of these ions. Since paramagnetic susceptibility is proportional to $p_{eff}^2$, it is not surprising that the experimental trend in the magnetic susceptibility of rare earth aluminosilicate glasses (see Figure 2) matches the calculated trend of $p_{eff}^2$ when $p_{eff} = g[J(J+1)]^{1/2}$ is used. The same general trend observed in Figure 2 was observed by Van Vleck[8] when the susceptibility of rare earth salts was measured. The data reported in Figure 2 were obtained with a Johnson Matthey JME susceptibility balance at room temperature, i.e. 21 to 23°C. The samples were in powder form. The accuracy of the balance was cross-referenced with water and neodymium oxide powder to within ± 2%, and the reproducibility of the measurements was within ± 0.5 $x10^{-7}$ (SI).

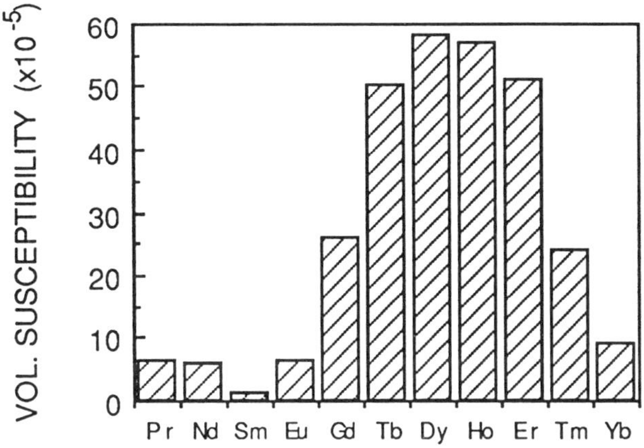


**Figure 2. Volume susceptibility of glasses having the molar composition $20R_2O_x$-$20Al_2O_3$-$60SiO_2$ as a function of the rare earth ion in the glass. R indicates a rare earth ion. Data taken by the author. Susceptibility in SI.**

When considering the behavior of the transition metal ions, on the other hand, the equation for the magnetic susceptibility (eqn. 8) yields calculated numbers which follow experimental susceptibility data more closely if only the spin of the unpaired 3d electrons are considered, i.e. $p_{eff} = g[S(S+1)]^{1/2}$. This behavior results from the loss of orbital momentum caused by the electrostatic interaction between the unpaired 3d electrons in the transition metals and the surrounding ligand field. The loss of orbital momentum occurs because the unpaired 3d electrons are not screened from the ligand field of other ions, whereas the unpaired 4f electrons in the rare earth ions are sufficiently shielded.

Although Curie-Weiss behavior is typical of most inorganic glasses containing paramagnetic ions, several investigators have noted nonlinearities in the reciprocal susceptibility vs. temperature data at low temperatures in some materials. The nonlinearities typically arise below 50 K. In general, this effect can be attributed to the onset of spin-glass ordering or a spin-freezing transition. When this behavior is observed, it is usually associated with the presence of clustered magnetic ions either in amorphous or crystalline microregions within the material. Within these regions, strong interactions between magnetic ions may occur causing ferromagnetic or antiferromagnetic behavior, depending upon the magnetic ions in the material, their proximity, and their interactions. In one study,[9] however, where the investigators studied materials containing plutonium, the departure from Curie-

Weiss behavior below 50 K was attributed to the local site symmetry surrounding the plutonium ions.

The study of materials which do not follow the Curie-Weiss law is relatively complex and is frequently augmented with Mössbauer spectroscopy when the glasses contain iron. A considerable amount of work on these materials has been reported in the last 10-15 years, yielding information on spero-, aspero-, and sperimagnetic spin configurations in amorphous insulating materials. These materials are frequently produced by extremely rapid quenching from the melt and therefore cannot be considered traditional glasses, even though they may be amorphous. More information on these studies can be gleaned from references 10 and 11.

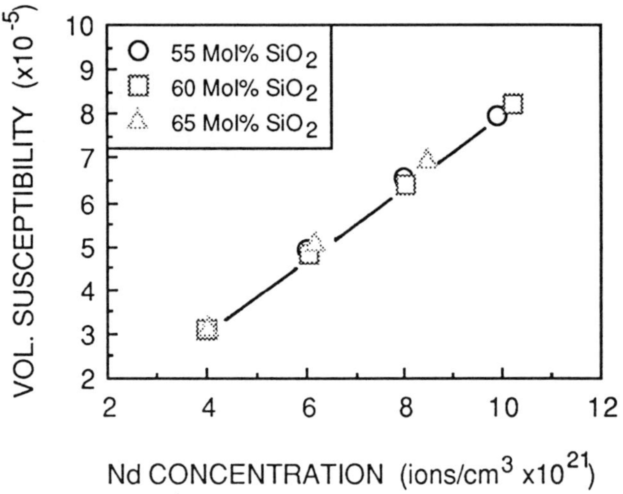


**Figure 3. Room temperature volume susceptibility of neodymium aluminosilicate glasses versus neodymium ion concentration. Susceptibility in SI.**

In simple oxide glass compositions containing paramagnetic ions, the magnetic susceptibility of a series of glasses varies linearly with the concentration of the paramagnetic ions. An example of this behavior is provided by a series of neodymium aluminosilicate glasses prepared by the author (see Figure 3). Notice that the susceptibility increases with the concentration of the rare earth ion, irrespective of the concentration of either alumina or silica in the glass.

Several mixed rare earth aluminosilicate glasses have also been made by the author. Since the different types of rare earth ions do not interact significantly in these compositions, the susceptibility of the mixed rare earth glasses varies linearly with the concentrations of the rare earth ions (see Figure 4) . The data in Figure 4 are plotted as a function of the concentration of one of the two rare earth ions for simplicity. The data shown in Figures 3 and 4 were also obtained at room temperature from powdered samples using the Johnson Matthey susceptibility balance.

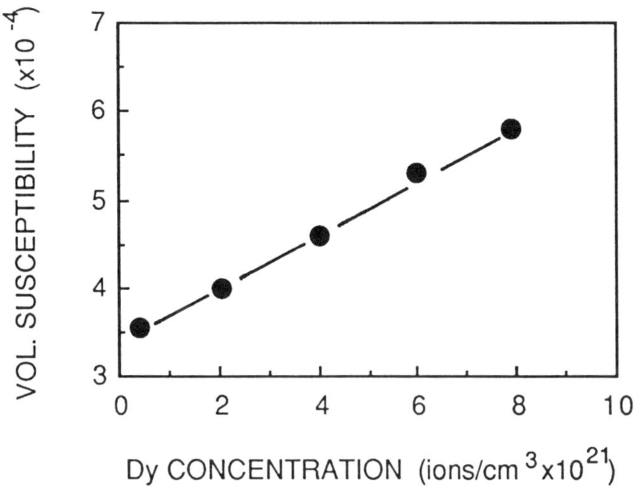


**Figure 4.** **Room temperature volume susceptibility of dysprosium and gadolinium aluminosilicate glasses (having the overall molar composition $20R_2O_3$-$20Al_2O_3$-$60SiO_2$) versus dysprosium ion concentration. R indicates a rare earth ion. Susceptibility in SI.**

## MEASUREMENT OF MAGNETIC SUSCEPTIBILITY

### Usefulness

The measurement of the magnetic susceptibility of glasses can be useful in many respects. Determination of the different valence states of transition metal, rare earth, and actinide elements is extremely important to manufacturers of glasses used for optical and magneto-optical applications, as well as scientists who are interested in the structure of glasses. As an example, measurement of the magnetic susceptibility of glasses containing uranium ions can be used to distinguish between the presence of $U^{+4}$ and $U^{+6}$, the former ion having a magnetic moment of 3.5 $\mu_B$ and the latter having a moment equal to zero. Although the valence state of paramagnetic ions in glass can often be determined by optical spectroscopy, inexpensive commercial susceptibility balances are now available, and measurement of the susceptibility of glasses does not require time-consuming grinding and polishing work which is necessary for optical studies.

Another practical use of susceptibility measurements of glasses is to check for the presence of impurities such as iron. Very small concentrations (at ppm levels) of the transition metal ions may be easily detected with a proper susceptibility balance. To test for ferromagnetic contamination in paramagnetic solids, it is necessary to measure the susceptibility of the material as a function of magnetic field strength. The magnetization of a purely paramagnetic material varies linearly with field, so the magnetic susceptibility is independent of the strength of the magnetic field (see equation 1) for these materials. If there are ferromagnetic contaminants in a paramagnetic material, the susceptibility will not be completely independent of the magnetic field.

Susceptibility measurements might also be used to follow the formation of some metal colloids in glasses which have been treated in hydrogen. If a magnetic ion, such as iron (ferrous or ferric), is reduced by exposure of the glass to hydrogen so that iron colloids are formed, the magnetic susceptibility of the glass should change because the number of unpaired electrons in the orbits of iron is increased as a result of reduction. Theoretically however, the measurement of magnetic susceptibility would not be useful for observing the formation of silver colloids in oxide glasses. Silver is present as a monovalent cation in glass. Monovalent silver does not have a magnetic moment because the 3d suborbital is filled with ten electrons. When the silver ion is reduced and metal colloids form, the atom still has a filled 3d suborbital, yielding no theoretical change in the magnetic moment.

Magnetic susceptibility measurements are also used in superconductivity studies. Although most superconducting materials are crystalline or polycrystalline, there have been significant recent efforts to crystallize superconducting phases from glass fibers. The critical current density, resistivity, and critical temperature of superconducting materials are important parameters measured by researchers studying superconductivity. However, because superconducting materials have different magnetic properties within different temperature ranges, it is important to understand the magnetic susceptibility of such materials.

## Traditional Techniques

The number of methods which have been developed for the measurement of magnetic susceptibility is considerable. Some methods have been developed exclusively for measurements of either solids, liquids, or gases. There are only two well recognized traditional methods used to measure magnetic susceptibility in solids, however. Many of the other methods are only variations of these two techniques. The Faraday method was developed in the 1850's by Michael Faraday, while the other, the Gouy method,[12] was developed shortly thereafter by L. G. Gouy in the 1880's. Because measuring the magnetic susceptibility of a substance is relatively straightforward, it is of no coincidence that these two methods are quite similar. The major difference between the two methods is that the Gouy method uses a uniform magnetic field, while the Faraday method employs a non-uniform magnetic field.

### *Gouy Method*

The Gouy method, which is perhaps the most common technique for measuring the susceptibility of solids, involves suspending a cylindrical sample between the poles of a magnet such that one end of the specimen is in a region of high magnetic field strength, while the opposite end experiences a very small field effect (see Figure 5). In this case the specimen experiences a force (f) proportional to the field. The force, integrated over the region of maximum field to the region of minimum field, may be written:

$$f = \chi_v v H(dH/ds) \qquad 9)$$

where $\chi_v$ is the magnetic susceptibility per unit volume, v is the volume of the specimen, H is the strength of the magnetic field, and s is the distance along the major axis of the cylindrical specimen, i.e. dH/ds is the gradient of the field in the s direction. Integration of equation 9 then yields the resultant force

$$f = 1/2\, \chi_v H^2 A \qquad 10)$$

where A is the cross-sectional area of the cylindrical specimen. The Gouy method cannot be used for ferromagnetic samples because the entire sample is not saturated by the magnetic field.

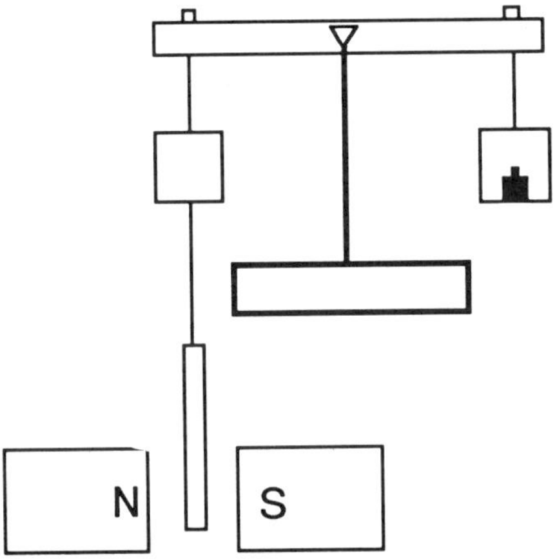


**Figure 5. Schematic of a Gouy magnetic susceptibility balance.**

The Gouy method may be used with powdered specimens. The powder technique is probably the most suitable method for determining the susceptibility of glasses. Since glasses are isotropic, there is no directionality of the susceptibility within a sample as there may be in a single crystal, or a grain-oriented ceramic material. For this reason, it is convenient to grind a small amount of an amorphous inorganic material into a fine powder and carefully pack the powder in a glass capillary tube of uniform cross-section. Once the capillary tube is packed with the powder, it may be suspended within the magnetic field of a Gouy balance (see Figure 5). Of course, the capillary tube itself has a magnetic susceptibility, but this may be accounted for by measuring the susceptibility of the empty tube and subtracting its contribution from the measured susceptibility of the sample in the capillary tube. This method is much simpler than attempting to fashion a perfectly cylindrical specimen from a piece of glass.

As one can imagine, inaccuracies may result from measuring the magnetic susceptibility of a solid material by grinding the material into a powder. The most obvious potential problem arises from non-uniform packing of the sample in the capillary tube. Moreover, the size and shape of the particles will greatly affect the packing density of the particles in the tube. One may minimize the errors caused by packing inconsistencies by making several measurements in which the powder is packed into the tube, the susceptibility is measured, and then the powder emptied after each measurement. The average of the measured susceptibilities may then be reported. Contamination of the sample from the grinding medium may also become a serious consideration. This problem can be minimized by using a wear-resistant grinding medium that does not contain significant quantities of transition metals which may affect the measured value. In most cases, a particle packing percentage of greater than 75% (i.e. less than 25% porosity) cannot be attained without extraordinary milling procedures. This means that one should not try to measure the volume occupied by the powder in the capillary tube because a large fraction of this volume is occupied by pores. Rather it is wisest to simply measure the weight of the powder in the tube, and hence measure the mass susceptibility of the powder. The volume susceptibility of the material is simply the product of the mass susceptibility and the density (measured from a bulk piece of the glass).

Although the powder method may not be the most accurate method to determine the magnetic susceptibility of inorganic glasses, this technique may yield very satisfactory results if one is reasonably meticulous. The prime objective in many studies, is to determine the change in the susceptibility of a series of materials with variations in composition. In such studies, accuracy is not as important as the ability to determine distinct trends as a function of material composition. As long as the powders are ground using the same method, the packing is relatively consistent from sample to sample. If the experimentalist is also careful to avoid cross-contamination from different samples by carefully cleaning the capillary tube after each sample is measured, one may obtain very satisfactory results. In common inorganic glasses, one may expect to observe a linear change in magnetic susceptibility with changes in material composition, as predicted by equations 7 and 8. Large deviations from linearity indicate improper experimental procedure. One may obtain accuracy as high as ± 1% with sound experimental procedures.

Other researchers have devised simple variations of the Gouy method. An example is the susceptibility balance designed by D.F. Evans for Johnson Matthey Inc. This balance is similar to the traditional Gouy balance, however, instead of measuring the force which a magnet exerts on a sample, the Evans balance measures the force exerted by a stationary sample on two suspended permanent magnets positioned on opposite sides of the sample. The deflection of the suspended arm, which holds the magnets, is related to the magnetic susceptibility of the specimen. This balance is also designed for convenient measurement of powdered specimens.

*Faraday Method*

The Faraday method is very similar to the Gouy method in that a sample of material is suspended between the poles of a magnet and the force acting on the specimen is measured. The magnetic field in the Faraday method, however, is produced by inclining the magnetic poles toward one another, or alternatively by shaping the poles so that a non-homogeneous field is produced (see Figure 6). The sample is placed between the poles of the magnet such that the strength of the field changes with distance along the axis of symmetry (s). The force (f) along the s axis is given by

$$f = m\chi_m H(dH/ds) \qquad 11)$$

where m is the mass of the sample, and $\chi_m$ is the mass susceptibility of the sample.

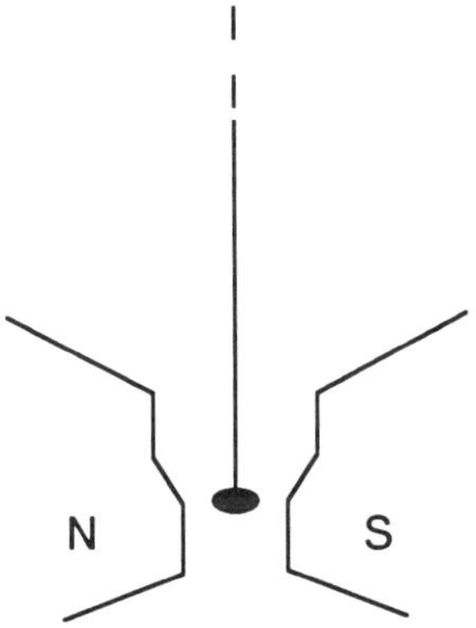


**Figure 6. Schematic of the poles in a Faraday balance.**

An advantage of the Faraday method is that only a small sample is required for the measurement. The Faraday method is convenient and may be very sensitive if a suitable balance is used. This technique may be used to measure the susceptibility of ferromagnetic materials because the entire sample is placed within the magnetic field.

*The Sucksmith Ring Method*

The Sucksmith ring was developed by W. Sucksmith[13] in the 1920's. This device incorporates some of the most desirable features of both the Gouy and the Faraday methods. A simplified version of the Sucksmith ring is shown in Figure 7.

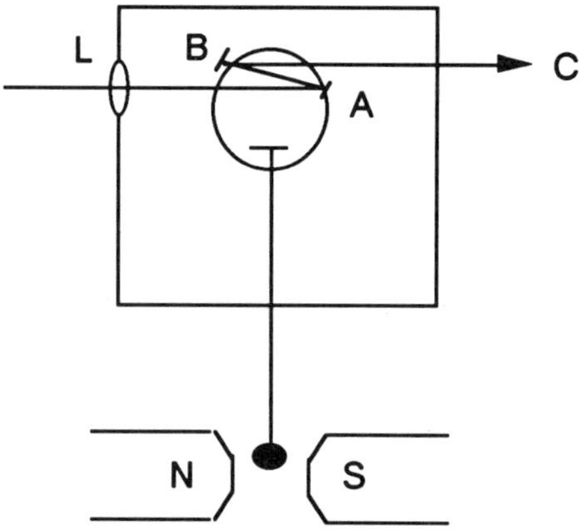


**Figure 7. Schematic of the Sucksmith balance.**

Unlike the Gouy and Faraday methods, Sucksmith did not use a traditional balance in his design. Instead, he suspended a sample from a phosphor bronze ring which is fixed at its highest point on the circumference. A wire is attached at the lowest point on the circumference of the ring and the sample is fixed at the opposite end of the wire. Two mirrors are located within the ring as shown in Figure 7 so that light entering through lens L focuses onto mirror A and is reflected to mirror B, forming an image of the light filament on a scale at point C. The movement of the light filament is extremely large compared to the movement of the sample in the magnetic field. The diameter and thickness of the ring may be altered depending upon the desired sensitivity of the apparatus. The device is usually enclosed in a box so that the temperature of the apparatus may be closely maintained. The apparatus may also be adapted so that measurements may be made *in vacuo*, or at elevated or depressed temperatures.

*Curie-Cheveneau Method*

The Curie-Cheveneau method was developed by Cheveneau[14] in 1910. A sample is placed within a cylindrical tube. The tube is then suspended between the poles of a magnet from a wire attached to a torsion arm. The movement of the torsion arm is gauged by either a pointer or the combination of a lamp, a mirror, and a scale. This device is convenient because the sample is automatically positioned in the region of maximum repulsion or attraction, depending upon whether the material is diamagnetic or paramagnetic. The advantage of this technique is that measurements may be made quickly and with small samples. Unfortunately, the sensitivity of the apparatus is not very high in most cases, and it is difficult to provide a means for controlling temperature.

*Rankine Method*

The Rankine balance was developed by A.O. Rankine[15] in the 1930's. In this balance, the sample is fixed while the magnet is free to move in reaction to the magnetic characteristics of the material being studied. This method requires very small magnetic fields relative to the techniques discussed thus far. The technique has also been used to observe extremely small changes in the magnetic susceptibility of materials, making the technique suitable for measurement of the formation and decay of free radicals.[6]

## Newer Techniques

### *Change in Flux Methods*

Many of the change in flux methods involve either the use of direct-current, alternating-current, or radio-frequency coils. The principle of many of the change in flux techniques is to measure the change in inductance of a solenoid when a sample is placed within or near the coil(s) of the apparatus.

A simple example of a direct-current method is that developed by Barnett.[16] A field of about 75 to 200 Oersteds is produced in this instrument by a long magnetizing coil which is wound around two secondary coils connected in opposition to one another. When a sample is introduced into one of the secondary coils, the sample is quickly displaced to the other secondary coil, changing the mutual inductance of the two coils. The change in the mutual inductance is measured by a galvanometer, the deflection of which is related to the volume susceptibility of the sample.

The alternating-current techniques which have been developed generally use an inductance bridge. Since these techniques involve complex circuitry in most cases, it is suggested that the reader see references 17 and 18 for detailed information. When the inductance bridge techniques were first developed, the technique was tedious due to the need for several adjustments to balance the real and imaginary parts. These techniques have been improved, however, and are frequently used today to obtain fast, sensitive, and continuous monitoring of the magnetic susceptibility of materials.

Although radio-frequency methods were first developed in England[19] and France[20] about 1950, these methods are infrequently employed today. When a sample is placed in the tuning coil of a high-frequency oscillator, the frequency of oscillation is altered by the presence of the sample. The change in frequency is measured by beating the sample oscillator against that of a reference (oscillating at a known frequency). The primary disadvantage[21] of this technique is that the temperature control of the oscillators must be extremely rigid (within ± 0.001 K). For this reason, it is very difficult to take measurements at temperatures far from room temperature.

Nuclear magnetic resonance (NMR) has also been used to determine the magnetic susceptibility of materials.[21] Like most nuclear magnetic resonance techniques, the information is obtained by chemical-shift, i.e. the value of the susceptibility is obtained in reference to the resonance spectrum of a standard material. Although NMR techniques traditionally have been used to investigate the magnetic properties of nuclei when a material is placed in a radio-frequency field, one may study the absorption of energy by nuclei in different magnetic environments. Thus, one may obtain information regarding magnetic susceptibility of a material by placing standard nuclei within a sample and measuring the frequency at which resonance occurs. The beauty of the NMR technique is that powders may be studied with no apparent adverse effects arising from the packing of the powder in a rotating tube. Moreover the NMR techniques are very sensitive relative to other techniques.

Another variation of the change in flux method for measuring magnetic susceptibility is the vibrating-sample magnetometer (VSM). The VSM technique involves placing a small sample in a magnetic field that is modulated by an alternating-current (whose frequency is typically about 60 to 100 Hz). The electronic coils are frequently positioned with their axes perpendicular to the magnetizing field, thereby minimizing the noise arising from field fluctuations. Typically, a permanent magnet is vibrated at a remote location on the same mechanical arm as the sample. The permanent magnet vibrates within a pair of reference coils, while the sample vibrates within a separate set of coils so that the signal created by the sample can be referenced to that from the permanent magnet. Vibrating-sample magnetometers are most often used for determining magnetization. Since magnetic susceptibility is simply a constant defined by the ratio of the magnetization (M) and the magnetic field (H), the magnetic susceptibility can easily be obtained from a magnetometer. VSM devices are relatively simple in design and can be easily modified to allow for temperature control of the sample.

Although many other techniques have been employed by various researchers (see references 22 and 23 for example), it should be evident to the reader that the measurement of magnetic susceptibility in solid materials is straightforward in principle. The development of so many different techniques has stemmed from the diverse conditions under which researchers have made magnetic susceptibility measurements (Kapitza and Webster[24] even developed a device which measures the susceptibility of a material under stress).

## REPORTING EXPERIMENTAL RESULTS

As shown in equations 7 and 8, the diamagnetic susceptibility is theoretically independent of temperature for most materials, while the paramagnetic susceptibility is temperature dependent. For this reason, it is important to report the temperature at which the susceptibility of materials is measured. Furthermore, if one wants to determine the effect of temperature on magnetic susceptibility, a temperature control device may be added to the apparatus. In such cases, it is important, obviously, to ensure that the specimen has reached thermal equilibrium.

As with all experiments, if the data are to be published in a journal, it is wise to report as much as possible about the conditions under which magnetic susceptibility was measured. If a commercial susceptibility balance is used, it is suggested that the investigator report the manufacturer's name as well as the model number. If an apparatus was used that was designed by the investigator, it is advisable to describe its design in detail. No matter what type of device is used, it is also advisable to report experimental errors and compare experimental values obtained from standard materials with values reported in the literature. When an investigator is thorough in these respects, a reader will place a higher degree of confidence in the data.

The use of various systems of units has been a nuisance to researchers in all fields of science and engineering, but in the field of magnetics the dilemma is almost overwhelming to the novice. The problem is somewhat more pronounced by the fact that magnetic susceptibility is commonly reported per unit volume of material, per unit mass of material, on a molar basis, using Bohr magnetons, or electrostatic units. Moreover, traditionally, magnetics researchers have used cgs units and have been reluctant to change. Although there is a strong movement in the U.S. to convert from all other systems of units to Systeme International (SI), the conversion to SI will be tedious in the field of magnetics. Another problem associated with magnetic units is the distinction between induction in free space and that which is actually present within a medium. In many books, the free space induction is also referred to as the field in SI, and measured in Teslas. Magnetic force, which is also referred to as the field, is measured in SI units of Amperes/meter. Strictly speaking, magnetic susceptibility has no units because it is simply a proportionality constant between magnetization and magnetic intensity. Conventionally, the volume susceptibility of a material is regarded as a dimensionless quantity while the mass susceptibility is considered to have the dimensions of reciprocal density. When susceptibility is reported as dimensionless, the researcher should indicate whether the values are in cgs or SI. Values reported in SI and cgs differ by a factor of $4\pi$. An interesting discourse on the problems associated with magnetic units is given in the introduction of Crangle's The Magnetic Properties of Solids.[25]

## OTHER EXPERIMENTAL TECHNIQUES USED IN GLASS SCIENCE RELATED TO MAGNETIC PROPERTIES

There are three other common techniques based on magnetic behavior used currently in the evaluation of glass structures. These techniques are 1) nuclear magnetic resonance (NMR), 2) electron spin resonance (ESR), and 3) Mössbauer spectroscopy. These techniques are described briefly in the ensuing paragraphs. For further information, see the various chapters in this text regarding these techniques.

Nuclear magnetic resonance was discovered in 1945 by two independent research groups. In 1952 Bloch and Purcell were awarded the Noble Prize for their work. At its inception, NMR was used only to determine the magnetic moments of nuclei, but it has recently become useful for studying the structural nature of glasses. NMR has been used extensively to study the coordination of boron and aluminum ions in glass, and to determine the local structure around silicon ions in silicate glasses (see references 26-29 for example).

ESR has been used for the study of electronic defects in glasses for the last 45 years. ESR is also often referred to as electron paramagnetic resonance (EPR) because the technique involves resonant absorption of microwaves by unpaired electrons. ESR may be used to study radiation-induced defects, paramagnetic ions, so-called structural defects inherent to some glass compositions, strain-induced defects, and the chemical annealing of defects by treatment in hydrogen. Griscom[30] has recently written a thorough review paper on the applications of ESR in glasses.

The Mössbauer effect was first reported by Rudolf Mössbauer in 1958. The Mössbauer effect involves the resonant scattering of gamma rays that are absorbed and emitted by the nuclei of solids without recoil energy loss. Mössbauer's work demonstrated that the theoretical limit of resolution, or the linewidth which is determined by the decay time, could be achieved using certain isotopes. This resolution is $10^{-10}$ to $10^{-15}$E, where E is the energy of the gamma ray.[31] Although the Mössbauer effect has been observed with more than forty of the elements in the periodic table, the technique has only been used extensively to study glasses containing iron. Since the linewidth of gamma rays are so narrow, Mössbauer experiments must be carried out with sources and absorbers having the same nuclear transition. When iron is studied, the 14.4 keV transition between the ground and first excited state of $^{57}Fe$ requires a gamma source containing $^{57}Co$. The technique can be used to determine valency, ligand coordination, and the degree of covalency in some cases.[10]

## REFERENCES


1. P.A. Tipler, Physics, 2nd Edition, pp. 805-806, Worth Publishers, Inc., New York, NY, 1982.
2. P. Langevin, J. de Physique, **4** 678 (1905); Annales de Chemie et Physique, **5** 70 (1905).
3. A.T. Aldred, "Magnetic Susceptibility of Sodium Disilicate Glasses Containing $PuO_2$," J. Non-Cryst. Solids, **40** 347-352 (1980).
4. W. Ostertag, "Magnetic Susceptibility of Uranium(IV) Glasses," J. Am. Ceram. Soc., **52** [1] 61-62 (1969).
5. C. Dupas, "Magnetic Properties of Fluoride Glasses," Matls. Sci. Forum, **6** 731-738 (1985).
6. P.W. Selwood, Magnetochemistry, 2nd Edition, pp. 182-188, Interscience Publishers, Inc. New York, NY, 1956.
7. M.J. Weber, "Faraday Rotator Materials," Report for the U.S. Government (DOE) and University of California, Lawrence Livermore Laboratory, p. 6, 1982.
8. J.H. Van Vleck, Theory of Electric and Magnetic Susceptibilities, p. 244, Oxford University Press, London, 1932.
9. G. Raphael and R. Lallement, "Susceptibilite Magnetique De $PuO_2$," Solid St. Commun., **6** 383-385 (1968).
10. M. Eibschutz and M.E. Lines, "Magnetic Insulator Glasses," Hyperfine Interactions, **27** 47-58 (1986).
11. M.E. Lines, "Computer Modelling of Amorphous Magnetic Materials," J. Magn. Magn. Matls., **36** 1-25 (1983).
12. L.G. Gouy, Compt. rend., **109** 935 (1889).
13. W. Sucksmith, "An Apparatus for the Measurement of Magnetic Susceptibility," Phil. Mag., **8** 158 (1929).
14. C. Cheveneau, Phil. Mag., **20** 357 (1910).
15. A.O. Rankine, "Demonstration of Paramagnetism and Diamagnetism in Fields of Low Intensity," Proc. Phys. Soc. (London), **46** [1] 391 (1934).
16. S.J. Barnett, "Method of Determining Magnetic Moments and for Measuring Susceptibilities and Permeabilities," J.Appl. Phys. Quart., **23** 975 (1952).
17. C.M. Brodbeck, R.R. Bukrey, and J.T. Hoeksema, "Integrated Circuit AC Mutual Inductance Bridge for Magnetic Susceptibility Measurements," Rev. Sci. Instrum., **49** [9] 1279-1281 (1978).
18. M. Kumano and Y. Ikegami, "Semiautomatic Mutual Inductance Bridge for Magnetic Susceptibility Measurements," **50** [7] 921-922 (1979).

19. H.G. Effemy, D.F. Parsons, and J.O'M Bockris, "Investigation of the Heterodyne Beat Method of Measurement of Magnetic Susceptibility," J. Sci. Instr., **32** 99-102 (1955).
20. A. Pacault, Rev. Sci. Acad. Sci., Paris, **84** 169 (1946); **86**, 38 (1948); Bull. Soc. Chim. France, **D371** (1949).
21. L.N. Mulay, "Techniques for Measuring Magnetic Susceptibility," pp. 431-553 in Physical Methods of Chemistry, Part IV, Determination of Mass, Transport, and Electrical-Magnetic Properties. Edited by A. Weissberger and B.W. Rossiter. Wiley-Interscience, John Wiley & Sons, Inc., New York, NY, 1972.
22. J.R. Hendrickson and J. Philbrook, "Resonant Torsional Apparatus for Contactless Measurements of Electrical Conductivity and Magnetic Susceptibility of Solids," Rev. Sci. Instrum., **50** [7] 849-855 (1979).
23. S.E. Nave and P.G. Huray, "Micromagnetic Susceptometer," Rev. Sci. Instrum. **51** [5] 591-596 (1980).
24. P. Kapitza and W.L. Webster, Proc. Roy. Soc. A, **132** 442 (1931).
25. J. Crangle, The Structures and Properties of Solids 6, The Magnetic Properties of Solids, pp. 156-164, Edward Arnold (Publishers) Limited, London, 1977.
26. G.L. Turner, R.J. Kirkpatrick, S.H. Risbud, E. Oldfield, "Multinuclear Magic Angle Sample-Spinning Nuclear Magnetic Resonance Spectroscopic Studies of Crystalline and Amorphous Ceramic Materials," Am. Ceram. Soc. Bull., **66** [4] 656-663 (1987).
27. E. Lippmaa, M. Magi, A. Samoson, G. Engelhardt, and A.R. Grimmer, "Structural Studies of Silicates by Solid-State High-Resolution $^{29}$Si NMR," J. Am. Chem. Soc., **102** 4889-4893 (1980).
28. N. Janes and E. Oldfield, "Prediction of Silicon-29 Nuclear Magnetic Resonance Chemical Shifts Using a Group Electronegativity Approach: Applications to Silicate and Aluminosilicate Structures," J. Am. Chem. Soc., **107** 6769-6775 (1985).
29. G. Englehardt, N. Nofz, K. Forkel, F.G. Wihsmann, M. Magi, A. Samoson, and E. Lippmaa, Phys. Chem. Glasses, **26** [5] 157-165 (1985).
30. D.L. Griscom, "Electron Spin Resonance in Glasses," J. Non-Cryst. Solids **40** 211-272 (1980).
31. G.K. Wertheim, A. Hausmann, and W.Sander, The Electronic Structure of Point Defects as Determined by Mössbauer Spectroscopy and by Spin Resonance, Vol. 4 pp. 1-11 in the series Defects in Crystalline Solids. Edited by S. Amelinckx, R. Gevers, and J. Nihoul. North-Holland Publishing Company, Amsterdam, 1971.

## OTHER SUGGESTED READINGS

L.F. Bates, Modern Magnetism, 4th Edition, Cambridge University Press, London, 1963.

M.M. Schieber, Experimental Magnetochemistry - Nonmetallic Magnetic Materials, North-Holland Publishing Company, Amsterdam, 1967.

# CHAPTER 14

## QUANTITATIVE FRACTURE SURFACE ANALYSIS OF GLASS MATERIALS

J.J. Mecholsky, Jr.
Department of Materials Science and Engineering
University of Florida
Gainesville, FL 32611

## INTRODUCTION

The intention of this chapter is to identify the sources of failure in glass and prescribe the analysis techniques available. Most of the chapter will be devoted to analyzing fracture surfaces and deducing the origin and state of stress at failure.

The origins of failure can be classified as occurring from three sources: Mechanical, (Chemical and Thermal) Processing, and Intrinsic. These sources will be discussed by way of introducing the basis and usefulness of fractographic analytical techniques. That is, most defects can be analyzed as equivalent semi-circular cracks and are amenable to fracture mechanics techniques.

The section on the Analysis of Fracture Origins will present the applicable principles of fracture mechanics, fracture surface analysis (concepts or mirror, mist and hackle) and the combination of the two analyses. By combining the observation and analysis of the crack size and shape at the failure origin with the surrounding fracture topography known as mirror, mist, hackle and crack branching, we can obtain additional important information about the fracture process. We not only can identify the source and stress at failure, but we can determine the presence of residual stress and stable crack growth. This section will be devoted to describing the techniques by which we can quantitatively define the fracture process. Thus, we will discuss non-planar cracks, residual stress and stable crack growth.

## SOURCE OF DEFECTS

Most origins of fracture in brittle materials like glass are from sharp cracks.[1] Even when the source is an inclusion of foreign material, the fracture generally originates from a crack at the boundary between the bulk material and the inclusion.[2] The shape and size of crack are important to the fracture process, so it is necessary to understand and to investigate their sources. The types of sources can be categorized by their origination: extrinsic and intrinsic. Extrinsic origins include mechanical (handling, intentional indentation, finishing operation, etc.) and chemical and thermal processing (sputtering, etching, quenching, etc.). Intrinsic origins include defects in the structure inherent in the specific glass manufacturing technique.

### Extrinsic

#### *Mechanical*

Almost all of the mechanically induced cracks can be idealized as a semi-elliptical, sharp crack of depth, a, and half-width, 2b. Examples of mechanically induced flaws or cracks are shown in Figures 1, 2 and 3. Figure 1 shows a crack in a chemically tempered solar mirror which was introduced by a sharp instrument (most likely a screwdriver). The sharp instrument broke through the compressive layer to produce a semi-elliptical flaw in the tensile region of the mirror. This flaw then grew slowly in a moist environment to cause a delayed failure in the mirror. Figure 2 shows two types of cracks that can be produced from machining operations. The direction of grinding relative to the final stressing direction produces different sized and shaped cracks and, thus, different strengths.[3] The cracks perpendicular to the direction of grinding are smaller and have an aspect ratio (a/b) of 1.6, and the cracks along the direction of grinding are larger and have an aspect ratio of 0.5. Polishing is a less severe random grinding operation and will produce all shapes of flaws.[4] Generally, the aspect ratio of the flaws that cause failure approaches the worst case in grinding, i.e., 0.5.

Since machining can result in complicated flaw shapes, we can intentionally introduce controlled flaws which simulate damage from machining by indenting with a Vicker's or Knoop hardness diamond.[5] Figure 3 shows the result of indenting with a Vicker's hardness diamond and then subsequent fracture of a glass bar. Because of the work of Lawn, Marshall and others,[6-8] this has become a reproducible, predictable and very useful technique for determination of fracture mechanics parameters. This technique usually results

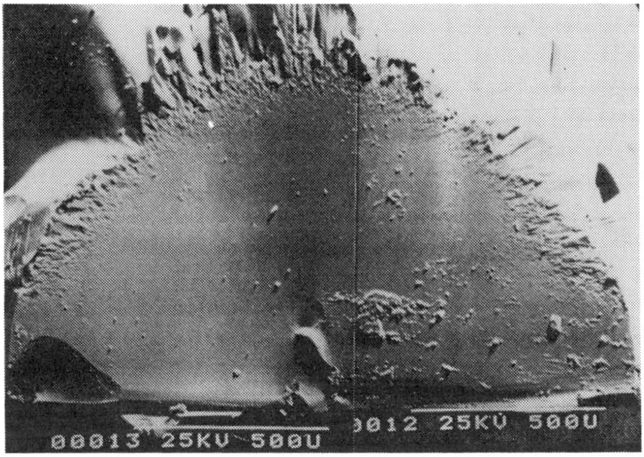


Fig. 1. Delayed failure in an ion-exchanged solar glass mirror. The final critical crack is semi-elliptical in nature and starts within the tensile region of the mirror, i.e., just inside the compressive layer on the surface.

in a semi-elliptical crack with an aspect ratio of ~0.8.[9] An excellent description of this technique with its versatile applications is presented elsewhere.[10]

Scribes and scratches leave defects in the glass very similar to machining damage.[11] In fact, most mechanically introduced cracks have the same general character and make these relatively easy to analyze. However, the exact relationship between the strength of glass and the shape of the fracture initiating crack depends on the particular local and far field stress state and, in most cases, is described very accurately with the use of fracture mechanics.[12] This will be discussed in a later section.

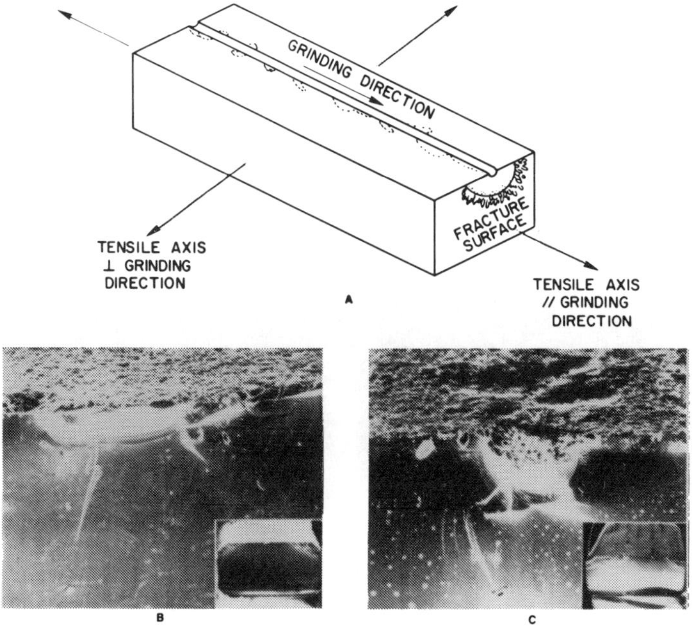


Fig. 2. Cracks produced from grinding are not symmetric but related to the direction of grinding relative to the stressed direction. The schematic in A shows that the cracks in the grinding direction are slightly elongated and result in lower strength when the applied stress (perpendicular to the cracks) activates these cracks.

***Chemical and Thermal Processing***

Although processing defects are not generally associated with glass as much as with polycrystalline ceramics, they are still a source of failure especially with optical and infrared fibers.[13] Figure 4 shows failure from a bubble produced during the drawing of an optical fiber from a preform with a large bubble in it.[13] Pores or bubbles can be produced as a result of manufacturing flat glass or bottles or as a result of thermal processing such as

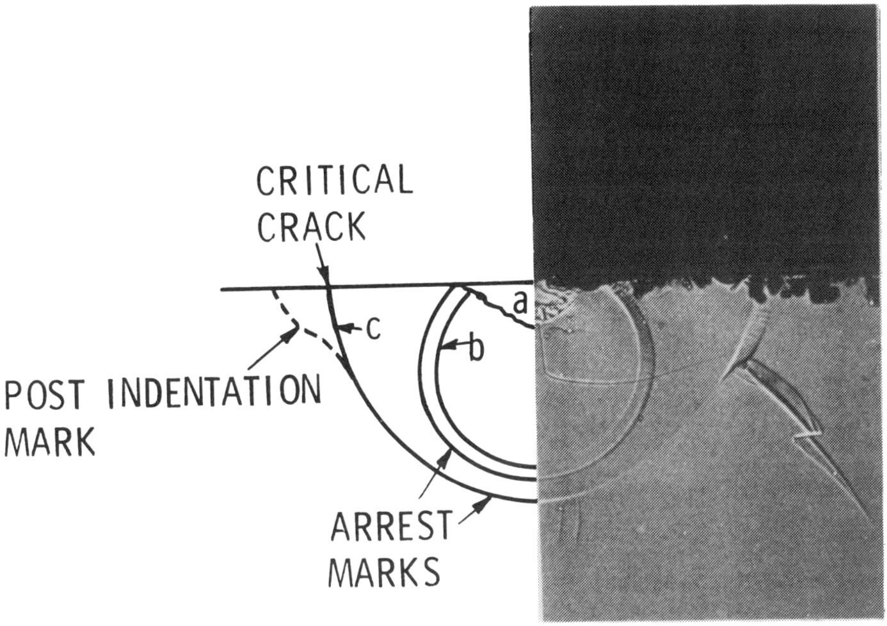


Fig. 3. Fracture surface of glass failed at a crack introduced by indentation. The stages of crack propagation during the indentation process can be observed by the demarcations surrounding the indentation site. The final (critical) crack size is semi-elliptical.

(incorrect) flame polishing. Generally these sources of failure are not as severe as sharp cracks.[14]

Stones (inclusions), cords (chemical gradients) and seeds (porous regions) have been a source of failure in manufacturing processes of glasses for a long time[14] and can still be a source of failure, but because of improved inspection techniques these are not as common as they once were. With the increasing interest in glass matrix composite materials, there will be cracks from thermal expansion or elastic mismatch stresses, incorrect bonding at reinforcement-matrix interfaces, and other processing defects common to polycrystalline ceramics like porous regions and local density gradients.[15]

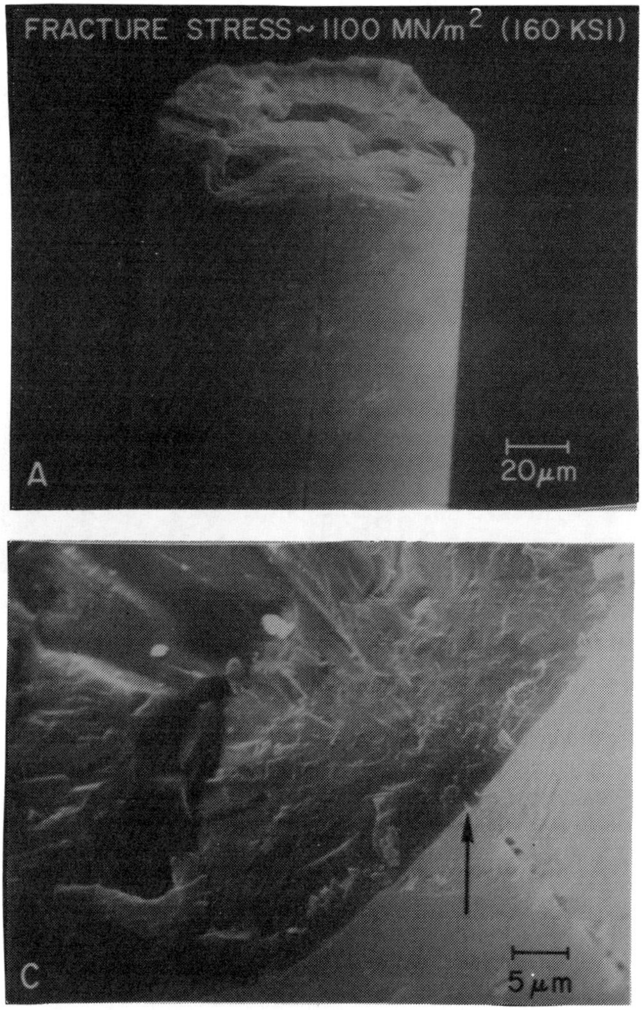


Fig. 4. A silicate fiber failed from a processing defect. This fiber developed a defect due to a bubble in the preform being extended out during the fiber drawing process. Notice the succession of bubbles along the fiber surface. These defects tend to be less severe than sharp semi-elliptical cracks.

If surface corrosion occurs, either due to surface crystallization, phase separation or chemical attack due to the environment, then the nature of defects found or formed on the surface may be altered. Surface corrosion can produce pits which act as stress concentrators and lead to defects which will act as the source of failure. If many cracks already exist on the surface and only limited surface corrosion occurs, then the effect may be to alter the severity of the existing cracks. Depending on the exact nature of the corrosion, this may be either deleterious or beneficial.

## Intrinsic

Since glass is in a metastable thermodynamic state, there are many compositions that are prone to crystallization.[16] In most cases, the crystals which are derived from the glass have a larger volume and coefficient of thermal expansion than the parent glass. These differences generally lead to strain mismatches which can lead to cracks at the glass-crystal interface.[14] Of course, this phenomenon can be advantageous if controlled, as for commercial glass ceramics.[17] However, uncontrolled crystallization has been a problem in glass manufacture since the beginning of commercialization of glasses. Isolated crystals in the bulk which are at or near the surface can act as sources of failure due to the difference in volume and thermal expansion between the crystal and glass. The most recent examples of this isolated crystallization has occurred with the production of fluoride bulk glasses and fibers. In addition, surface crystallization can also occur in some glasses. Again, when this phenomenon is controlled, strengthening of glasses can result.[17] However, when surface crystallization occurs in an uncontrolled (and unplanned) manner, then these crystallites can either act as the source of failure or aid the failure process due to residual tensile stresses associated with the crystallization.

Glass-in-glass phase separation is a phenomenon that occurs due to thermodynamic and kinetic parameters which permit composition differences to exist on a microscopic scale.[16] Typically phase separation exists on the tens to hundreds of Angstroms scale. As such, usually these regions do not act as the source of failure themselves because there are defects that are larger from which the glass has to choose for the worst defect. Occasionally, there can be large isolated regions of compositional differences, but these are generally due to improper mixing before the melting of glass and are called "cords." However, the phenomenon of phase separation can lead to changes in the mechanical behavior of glass. It is disputed in the literature whether phase separation raises or lowers the strength. The two phases generally will have considerably different chemical compositions and thus different properties. This fact can lead to different local behavior such as selected leaching (corrosion). In fact, this process is used to produce Vycor glass by controlling the phase separation. However, if the phase separation is unintentional or unavoidable, then its existence can affect the properties of any existing cracks. In general, the most likely effect is to blunt the crack and make it less severe. However, if the phase separation results in an isolated phase, then dissolution of the phase may occur in the proper environment and will adversely affect the mechanical behavior.

Glass structure has inherent defects, e.g., non-bridging oxygen in a silicate structure, that result from the formation process. These sites are considered to have higher energy states and thus are potential sites for failure. When the glass structure is strained due to far field stresses (external applied loads), the local structure responds by resisting the motion through the bonding energy between atoms. It is generally considered that a defect site will be more active in the fracture process than an equilibrium site. Broken bonds existing due to the formation process can act as stress concentrations that will add to the stress from external loading and can lead to failure.[18] Certainly if enough chemical bonds are broken, the whole glass network can be considered adversely affected and less resistant to crack propagation. With the advent of optical and infrared fibers, we now have to be concerned with defects of the size of atomic units.[19] This may force us to find new ways to analyze fracture.

Another source of failure may be from strained bond sites (as in the stress corrosion phenomenon).[20] The bonds between Si-O-Si in a silicate glass may not be broken but may be strained at lower angles than the equilibrium angle and, thus, create a site for crack growth enhancement.[21]

Finally, it can be hypothesized that impurities at interstitial or substitutional sites will affect the local bonding and adversely affect crack initiation and propagation. In nearly perfect glasses these can become defect sites for fracture.

Although most of the sources of failure mentioned in this section are rarely seen because a much larger and more severe defect already exists in the material, as we get better at manufacturing nearly pristine glasses, we will discover new failure sites. Of course, this is a difficult area to investigate because of the size of the defects and until we discover suitable means with which to identify these sites, the mystery of some failures will remain so.

## ANALYSIS OF FRACTURE ORIGINS

### Fracture Mechanics

In an attempt to quantify our understanding of fracture of brittle materials, several investigators identified an approach to the strength of materials based on the size of the defects. In a true merging of mathematics and reality, Irwin[22] built on the work of Griffith,[23] Orowan[24] and Inglis[25] and developed what is now known as Linear Elastic Fracture Mechanics (LEFM). The purpose of this paper is not to derive the useful equations needed to study fracture but rather to apply these to familiar and unique situations found in the fracture of glass. For semi-elliptical cracks of depth, a, and half-width, 2b, which are small relative to the thickness, the (Mode I) stress intensity, $K_I$, is given as:

$$K_I = \frac{\sqrt{1.2\pi}}{\phi}\sigma_a c^{1/2} \tag{1}$$

where c = a or b whichever is smaller, $\sigma_a$ is the applied stress, and $\phi$ is an elliptical integral of the second kind which varies between $\phi = 1$ for a slit crack (a/b = 0) to $\phi = 1.57$ for a/b = 1.0. The condition at failure is that $K_I > K_{Ic}$ at which point the initial crack propagates spontaneously. $K_{Ic}$ is the critical stress intensity factor or fracture toughness. Thus, $K_I$ is a measure of the intensification of the external loading at a crack (or defect) such that the stresses (and strains) approach the theoretical limit. $K_{Ic}$ is a measure of the glass structure's resistance to this loading, i.e., a macroscopic measurement of bonding energies. In most cases, to a good approximation $K_{Ic}$ can be considered a materials constant for glasses. The strength, $\sigma_f$, of a glass is related to the defect or crack size c,

$$\sigma_f = \frac{\phi K_{Ic}}{\sqrt{1.2\pi}\sqrt{c}} \tag{2}$$

It has been shown[3] that for most glass fractures, the failure initiating crack can be modeled as an equivalent semicircular crack of radius c where $c = \sqrt{ab}$. This modifies Equation (2) with $\phi = 1.57$ to:

$$\sigma_f = \frac{K_{Ic}}{1.24\sqrt{c}} \tag{3}$$

An equivalent result can be obtained from an energy balance approach, i.e., Griffith[23] states that in an infinite body containing a through thickness crack of length 2c, elastic fracture will occur at a critical value of the strain energy release rate per unit thickness, $G_I$, given by:

$$G_c = 2\gamma_f \tag{4}$$

where $2\gamma_f$ is the elastic work to fracture, $\gamma_f$ being the fracture surface energy per unit area, and $G_c$ is the critical strain energy release rate or fracture toughness. Thus, Equations (2) and (4) can be related[26,27] through the relationship of $K_{Ic}$ and $G_c$:

$$G_c = A' \frac{K_{Ic}^2}{E} \tag{5}$$

where E is Young's modulus and $A' = 1$ in plane stress and $1-\nu^2$ in plane strain; $\nu$ is Poisson's ratio. Thus, Equation (3) becomes:

$$\sigma_f = \frac{\sqrt{2E\gamma_f}}{1.24\sqrt{A'}\sqrt{c}} \tag{6}$$

Equations (3) and (6) are very useful for analyzing fractures of glass. The values for $K_{Ic}$, $G_c$ ($2\gamma_f$) and E for many glasses are given in Table I. With these constants and a knowledge of the size and shape of the fracture initiating crack, we can determine the strength of glass containing cracks and any additional stresses surrounding that crack.

**Table I**

**Fracture Properties for Representative Glasses**

| | $M_1$ | $M_2$ | $M_3$ | E GPa | $G_c$ $J/m^2$ | $K_{Ic}$ MPa $m^{1/2}$ |
|---|---|---|---|---|---|---|
| 1. Silica [28]* | 2.2 | 2.4 | 2.7 | 72 | 7.4 | 0.73 |
| 2. Aluminosilicate [28] | 2.1 | 2.4 | --- | 89 | 7.4 | 0.81 |
| 3. Borosilicate [28] | 1.9 | 2.1 | --- | 64 | 8.0 | 0.71 |
| 4. Soda Lime [28] | 1.8 | 2.0 | 2.3 | 73 | 7.0 | 0.72 |
| 5. Lead Silicate [28] | 1.6 | 1.8 | --- | 52 | 7.6 | 0.63 |
| 6. Leached High Silica [28] | 0.9 | 1.2 | --- | 18 | 5.0 | 0.30 |
| 7. Glassy C [28] | 1.2 | 1.7 | --- | 25 | 17.0 | 0.65 |
| 8. $As_2S_3$ [28] | 0.6 | 0.7 | --- | 17 | 4.2 | 0.30 |
| 9. $Ge_{33}As_{12}Se_{55}$ [28] | 0.6 | 0.7 | --- | 22 | 2.0 | 0.21 |
| 10. ZBLA [29] | 0.8 | --- | --- | 55 | 1.7 | 0.31 |
| 11. Lead Borate [30] ($30PbO\text{-}70B_2O_3$) | 1.7 | --- | --- | 19 | 3.4 | 0.80 |
| 12. Zinc Silicate [31,32] | 2.1 | 2.6 | 3.2 | 82 | 9.9 | 0.90 |
| 13. $3BaO \bullet 5SiO_2$ [31] | 1.3 | 1.5 | --- | 64 | 3.9 | 0.50 |
| 14. $Li_2O \bullet 2SiO_2$ [31] | 2.2 | 2.6 | --- | 52 | 9.4 | 0.70 |

*[ ] cf. references

## Fracture Surface Analysis

The stresses at the crack tip can be related to the far field (applied) stresses through equations like Equation (2). At fracture, the stress intensity, $K_I$, is greater than the fracture toughness, $K_{Ic}$. The propagating crack forms demarcations on the fracture surface which are precursors to crack branching. Although there is still much debate[33] on the exact mechanism and driving force for the formation of these regions, they have proven to be useful in analyzing failure and of relating local effects to far field effects. These regions have come to be known as the mirror, mist and hackle regions. Examples of these regions are shown in Figure 5. More details of these regions are discussed elsewhere.[34,35]

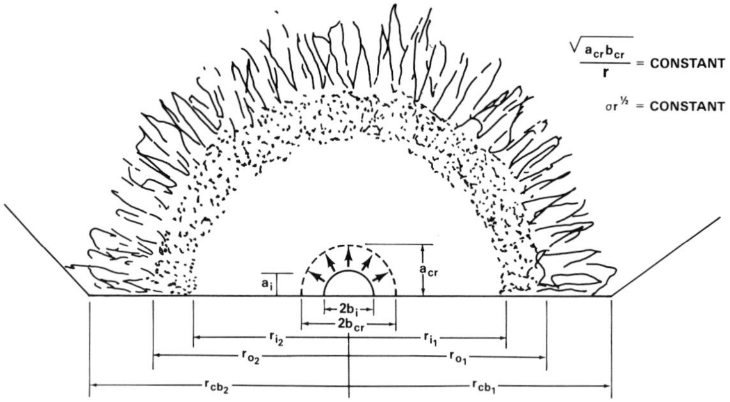


Fig. 5. Schematic of a fracture surface of glass. The arrows indicate that some stable crack growth can occur up to a critical point when rapid propagation commences (dashed lines). Mirror-mist ($r_1$), mist-hackle ($r_2$) and crack branching ($r_3$) boundaries are indicated. The primes indicate that the fracture may not be symmetric.

We observe that from the fracture origin* the crack propagates in a relatively smooth plane (the mirror region) to the boundary, $r_1$, and progressively gets rougher by deviating slightly out of plane in a region that resembles mist (the mist region), between $r_1$ and $r_2$, finally getting very rough with the crack deviating locally from the main plane of fracture (the hackle region), between $r_2$ and $r_3$, and finally branching into two or more cracks at $r_3$. Beyond branching, the process can repeat itself on each branch of the propagating crack.[36] It has been experimentally observed that the distance from the fracture origin to the boundary between the mirror and mist region, $r_1$, and from the origin to the boundary of the mist-hackle region, $r_2$, and from the origin to crack branching, $r_3$, are all related to the applied (far field) stress at fracture, $\sigma_f$, in an empirical equation:

$$\sigma_f \, r_j^{1/2} = constant = M_j \tag{7}$$

where $r_j = r_1$, $r_2$ or $r_3$ corresponding to different constants $M_j = M_1$, $M_2$ or $M_3$ for each.

Most of the strength measurements from which the fracture mirror data are obtained have been performed using bend tests. One of the past errors made in fracture mirror analysis has been shown to result from measurement of the mirror into the thickness of the specimen and, therefore, into a varying stress field. Mecholsky et al.[28] pointed out that if all the fracture mirror data are measured along the tensile surface, that is, under constant stress, then good agreement is obtained between the various investigators. While Abdel-Latif et al.[37] have suggested that the mirror constant is actually a function of specimen size for flexural test measurements, there are little experimental data other than their work to corroborate this hypothesis; the large deflections associated with small specimens-large spans, which can explain the differences in their mirror constant values for different conditions, have not been addressed in Reference 37.

Although it is not the purpose of this paper to discuss the formation of crack branching, I will summarize the present state of knowledge. The main ideas presented are derived from two references: Ravi-Chandar and Knauss,[38] and Rossmanith.[39] In the mirror zone the crack propagates at relatively low

* The difference between the initial crack ($a_i$, $2b_i$) and the final, critical crack ($a_{cr}$, $2b_{cr}$) is discussed in a later section.

stress intensity values and maintains a relatively smooth and continuous crack front.

In the mist zone, the crack propagates at higher stress intensity levels; defects ahead of the main crack front are nucleated at or prior to the arrival of the main crack. Thus, the "mist" is a result of a number of small cracks propagating along with the apparent crack front. In the hackle region, the stress intensity is even higher. Several cracks ahead of the ensemble crack front initiate and grow simultaneously. The cracks in this zone are similar to but larger than those in the mist region. Branching is a result of the interaction of these cracks with one another. Branching is a discontinuous process with a number of micro-branches parallel to the main crack front.

The distances between the mist and hackle regions and between the hackle and macroscopic branching distances are related to the stress levels of failure[35] which, in turn, are related to the original size of the fracture initiating crack. The width and smoothness of the mist region correlates with the inverse square of the fracture stress, higher strengths yielding narrower and rougher mist regions. Ball et al.[35] support the idea that the formation of mist and, thus, branching may occur at particular values of strain energy release rates (energy) or stress intensities (cf. Equation 5).

In addition to the observations of the mirror, mist, hackle and crack branching regions on the fracture surface, there have been observations of regions of multiple mist.[36] Multiple mist regions are the recurrence of a second, or a second and third, consecutive mist region at increasing radii from the fracture origin. Abdel-Latif et al.[36] showed that these regions follow relations like Equation (7) but with larger constants. They suggested that an energy criterion that incorporates dynamic effects is more appropriate than a velocity or stress intensity criterion for describing the formation of mist, hackle and crack branching. An alternate explanation of these observations is that these multiple regions occur when the stress level has decreased either due to crack branching (where there are at least two cracks) or local perturbations, like cracks or bubbles, and the stress intensity has correspondingly decreased to the mist formation level. However, as Abdel-Latif et al.[36] point out, additional studies are needed to resolve this point.

Probably the most useful measurement is the distance to crack branching because it is relatively easy to measure and much literature has addressed this quantity in other materials.[38] It is generally thought that the distance to crack branching is not influenced by far field stresses and boundary conditions (and thus Equation (7) is valid), but that the angle of crack branching varies with the boundary conditions and far field stresses.[39]

Even though the distance to crack branching may be the most useful quantity next to the crack size, the other boundaries should not be neglected; if the specimen is not large enough, we will not observe crack branching.

It has been suggested[40,41] that there exists relationships between these radii, which reflect the action of far field stresses, to the fracture origin which reflects the action of local stresses. This relationship can be expressed in the convenient form:[41]

$$K_j = Y_j \sigma_f r_j^{1/2} \tag{8}$$

where $Y_j$ is a crack border correction factor obtained for cracks of a specified size and shape and for specified loading conditions, e.g., tension or flexure, and $j = 0,1,2,3$; $K_0 = K_{Ic}$ and $r_o$ is the critical flaw size (in mode I loading). The $j = 0$ case is equivalent to the generally accepted fracture mechanics equation for a semi-elliptical surface crack [Equation (3)]. The cases of $j = 1,2,3$ correspond to the formation of the mirror-mist, mist-hackle and crack branching boundaries [Equation (7)] respectively. Thus, one set of equations [Equation (8)] can describe all of these fracture demarcations.

This approach* has appeal because it can be used in failure analysis and provides an analytical approach to fractography. Recognize that $\sigma_f r_j^{1/2}$ for $j = 1,2,3$ are the mirror constants, $M_j$. Equation (8) shows that there is a relationship between the stress intensity and the mirror constant. Several researchers[40,41,44,45] have recognized the importance of the relationship between the mirror constants and stress intensity and, in particular, the critical stress intensity factor $K_{Ic}$. In fact, the use of Equation (8) suggests that mirror boundary formation occurs at constant $K_I$ values.[35,38,40,41]

For many glasses the $M_j/K_{Ic}$ ratio is constant as expected from Equation (8). The importance of this ratio has not been elucidated. It has been shown[40] that $M_2/K_{Ic}$ is approximately constant for many glasses, glass ceramics, single and polycrystalline materials. However, it is important to analyze the

* Shetty et al.[42] questioned whether this approach represented realistic observations; subsequent research[43] shows that it is correct.

relationship between the stress intensity at crack branching to the stress intensity at the initiation of fracture, i.e., $K_{Bj}/K_{Ic}$. Several researchers[38-42] suggest the importance of the relationship between all $M_j$'s and $K_{Ic}$ and that $M_3$ is directly related to $K_{Bj}$.[28] The fact that there is a relationship between these two quantities means that the initiation of crack propagation can be related to the propagation of the crack. Even though kinetic effects are present during propagation, they evidently are accounted for in the measurements of the radius at crack branching.

Taking the ratio of Equation (8) with j = 0 and Equation (8) with j = 3, one obtains a relation between the fracture toughness and the stress intensity at crack branching. This is related through the mirror to flaw size ratio which can be measured.

$$\frac{K_{Bj}}{K_{Ic}} = \frac{Y_3}{Y_0}\sqrt{\frac{r_3}{c}} \tag{9a}$$

or upon substituting Equation (7) for Equation (8) with j = 3:

$$\frac{M_3}{K_{Ic}} = .42\sqrt{\frac{r_3}{c}} \tag{9b}$$

where $c = r_0$ and $K_{B3} = K_3$ have been substituted into Equation (8). $Y_0$ and $Y_3$ are geometric constants for the crack and branching radius, respectively. Figure 6 shows the graph for Equation (9a). It has also been observed that the stress intensity near crack branching (at the mist-hackle boundary), $K_2$, is related to the elastic modulus.[46] This means that information of the localized fracture process is being transmitted to the fracture surface and can be related to bond strength of the atoms. Notice that if we assume that the crack shape does not change during propagation,* then $Y_0 = Y_3$ and

* In a tensile stress field, this is reasonable. In a flexural stress field this will not be valid, but then the stress is also changing; thus, we assume that in a flexure test the measurements are made along the tensile surface. All that really has to be true is that there is a constant relation between the two geometries.

$$\frac{K_3}{K_{Ic}} = \sqrt{\frac{r_3}{c}} \tag{10a}$$

or, generalizing for all the boundaries,

$$\frac{r_j}{c} = \left(\frac{K_j}{K_{Ic}}\right)^2 \tag{10b}$$

where j = 1,2 and 3.

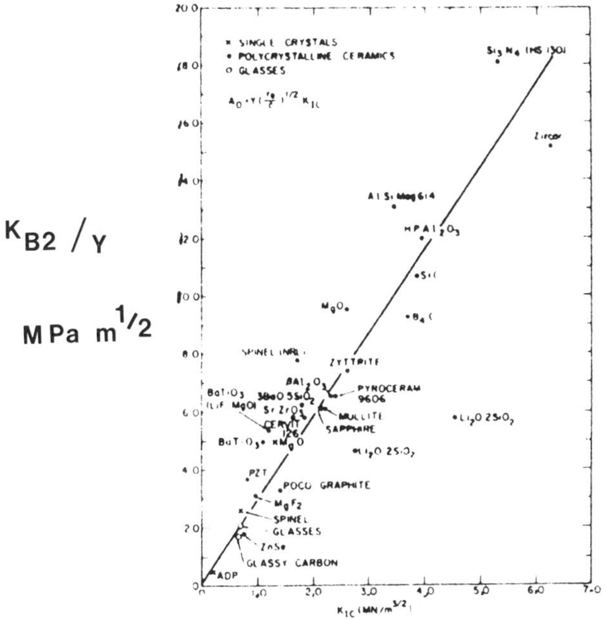


Fig. 6. Crack branching constant, $K_{B2}$ as a function of the critical stress intensity factor, $K_{Ic}$. Equation (9a) is tested for many materials. Notice that the trend is for $K_B/K_{Ic}$ to be a constant. This particular graph is shown for $K_j = K_2$, i.e., the mist-hackle boundary, but is also valid for $K_1$ and $K_3$.

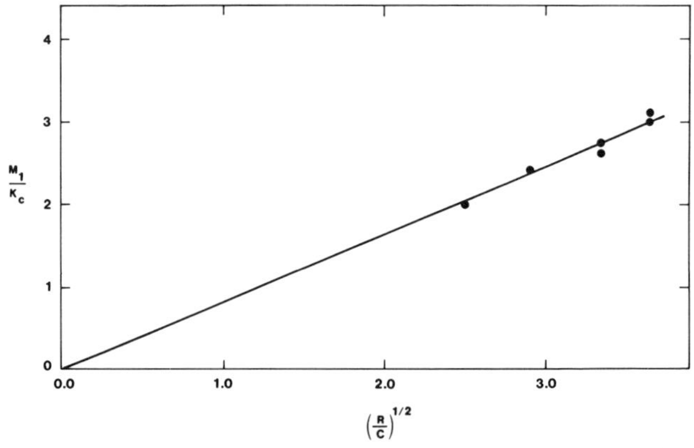


Fig. 7. $M_1/K_c$ as a function of $(r_1/c)^{1/2}$. Experimental verification of Equation (9b). Data from Reference 30 are replotted to test relations of the form of Equation (9b). We see that a straight line through the origin results. This is interpreted to mean that there is a relationship between initiation and propagation of the fracture causing crack.

Shinkai and Ishikawa[30] have shown that the mirror to flaw size ratio and mirror constants vary with the composition of lead borate glasses. However, if we plot their data (air, room temperature) for $M_1/K_c$* as a function of $(r_1/c)^{1/2}$ in Figure 7, we see that the slope is constant.** For this relationship to hold implies that the stress at initiation of rapid propagation and at the formation of mirror, mist, hackle and crack branching is constant, that the geometry of the crack doesn't change, at least in the point of measurement, or is insignificant, and that $M_j/K_{Ic}$ is approximately constant for many materials. The question may legitimately be asked how $M_j/K_{Ic}$ can be constant for many materials and yet not be constant for several (borate) glasses? This paradox can be resolved if we consider that there may be two relationships involved. $M_j$ is related to $K_{Ic}$ through $r_j/c$ [Equation (9b)]. If $r_j/c$ is constant then $M_j/K_{Ic}$ will be constant. For particular glass compositions, $M_j/K_{Ic}$, which varies as $r_j/c$, may be different for each material, but not change very much from material to material.[28] The $M_j/K_{Ic}$ relationship was intended to show a trend for many brittle materials and not imply that $M_j/K_{Ic}$ is one constant for all materials.

* $K_c \sim K_{Ic}$

** Similar relationships hold for $M_2$ and $M_3$.

## Application of Fracture Surface Analysis to Observed Origins

In applying Equations (8) and (9), we often need to modify the equations due to the presence of mixed mode failure, residual stress or slow crack growth due to stress corrosion processes. In fact, we can use our knowledge of the relationship between the size of the initial crack and the fracture mirror boundaries, especially the crack branching radius, to provide additional information for studying the fracture process. We shall first discuss the case of out of plane (mixed mode) or distorted flaws, then proceed to analyze flaws with residual stress, and finally describe the process of stable crack growth.

### *Nonplanar Flaws*

While observation of the fracture initiating flaw is important, flaws in many cases are nonplanar, too small, have unclear boundaries or have been eliminated because of chip removal so that one cannot determine the critical flaw size. One of the powers of fractography is in using the fracture surface features that form outside the flaw to describe fracture behavior. The mirror-mist and mist-hackle boundaries are easiest to discern in glasses (and fine grain dense ceramics). The crack branching radius is only discernible in glasses and ceramics when the sample is large enough relative to the fracture stress. Typically, either the initial flaw at the fracture origin or fracture surface features beyond this region are found in glasses. These features can be used to calculate the fracture stress [Equation (7)] and fracture toughness [Equation (9b)] of the material.[46]

Rice[15] showed that the mirror-to-flaw size ratio increases as the angle of the flaw to the tensile axis increases in mixed-mode loading (Figure 8). This means that different parts of a nonplanar flaw or a flaw not normal to the stress can act essentially independently. Thus, asymmetric mirrors can form because the stress intensity on one portion of the crack is higher than on another portion. It has been demonstrated by Freiman et al.,[47] in the case of the formation of mirrors which are asymmetric about the flaw on the tensile axis, that the larger radius is the correct measurement. They show that the mirror constant obtained from this radius agrees well with other measurements. The reason for the use of the larger radius is that it corresponds to the smallest stress and, therefore, the stress at which the crack first started to propagate. Thus, even though we may not know the actual geometry or size of the fracture initiating crack, we can obtain an effective size and shape by using the crack branching distance and Equation (9b). The calculated value, "c", would then be an equivalent semicircular crack amenable to established analyses.

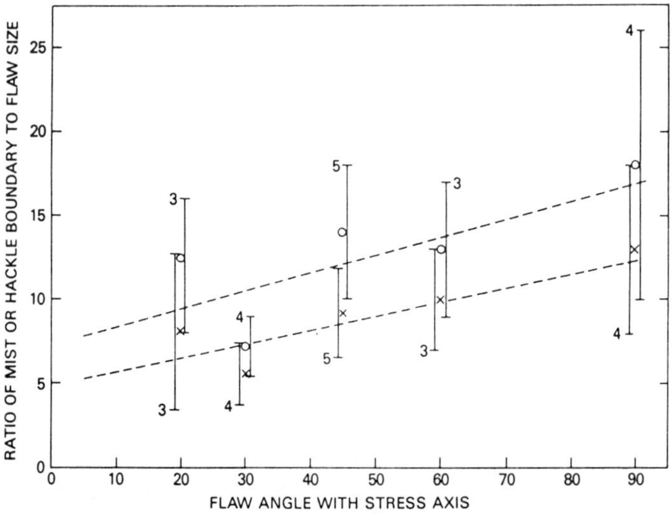


Fig. 8. Mirror-to-flaw size ratios for mixed-mode failure. Rice (Reference 15) plotted data from Reference 47 to show that there is a systematic increase in the $r_j/c$ ratio with increasing angle of the flaw relative to the stress axis.

### *Residual Stress*

An implicit assumption in establishing these relations [Equations (8) and (9)] for the mirror pattern is that the material is free of extraneous stresses or that the stresses are known across the prospective crack plane. This assumption does not always hold in practice. For instance, significant residual compressive stresses may be generated in glasses and ceramics cooled rapidly from high temperatures. However, provided the residual compression, $\sigma_R$, is reasonably uniform over the area of the mirror, Equation (1) remains valid if $\sigma_a$ is replaced by an "effective failure stress" $\sigma_{a'} = \sigma_a - \sigma_R$.[48,49] Indeed, fracture-mirror measurements have been used to determine the actual levels of residual stress in surface-strengthened glass rods.[49]

Residual stresses due to thermal or chemical treatments influence the formation of fracture mirrors in these cases. An attractive approach to study the effect of residual stress is by plotting $\sigma$ versus $r^{-1/2}$ (rather than the log-log plot). This approach yields the value of residual stress contributing to failure at the stress axis intercept. This type of analysis has been applied to glasses having various thermal treatments.[33,50,51] Conway and Mecholsky[52] have suggested that the fracture mechanics approach can also be applied to describe the effect of residual compressive stresses on mirror, mist and hackle formation. The result of this analysis is shown in Figure 9 where the effect of residual compressive stress on the shape of the mirror boundary is shown in the upper figure. Thus, the mirror-to-flaw-size ratio for samples with global residual compressive stress, e.g., from tempering or ion exchange, will be the same on the tensile surface, but will change on the interior to result in the shape shown in the upper figure of Figure 9. The equation that governs this behavior[47] is

$$Y\,(\sigma_R + \sigma_a)\, r^{1/2} = K_B \tag{11}$$

The typical flaw may experience other components of residual stress, especially those associated with the processes actually responsible for the generation of the flaw itself,[5] which may be far from homogeneous. Such is the case with the deformation-induced surface microcracks produced in sharp-point contact.[6] The inhomogeneity of the residual field can then significantly alter the mechanics of failure.[7] Since the strength of a great many brittle solids is controlled by contact flaws, it is appropriate to investigate the influence of this additional factor on the mirror/flaw-size relations.

The mirror-to-flaw size ratio for cracks which are stress free is:

$$\frac{r_j}{c^*} = \left(\frac{K_j}{K_c}\right)^2 \tag{10b}$$

The mirror-to-flaw size ratio for cracks with local residual stress is:

$$\frac{r_j}{c^*} = \left(\frac{4K_j}{3K_c}\right)^2 \tag{12}$$

Fractographic measurements from Reference 53 are summarized in Table II. The mirror constants $M_j$ [Equation (7)] are relatively independent of flaw history, as predicted by theory. The mirror/flaw size ratios, $r_j/c^*$, are highly sensitive to flaw history. Taking ratios appropriate to residual-stress-free, equilibrium crack growth (i.e., annealed specimens, oil test environment) as baseline quantities for normalizing the data for flaws subject to nonzero residual

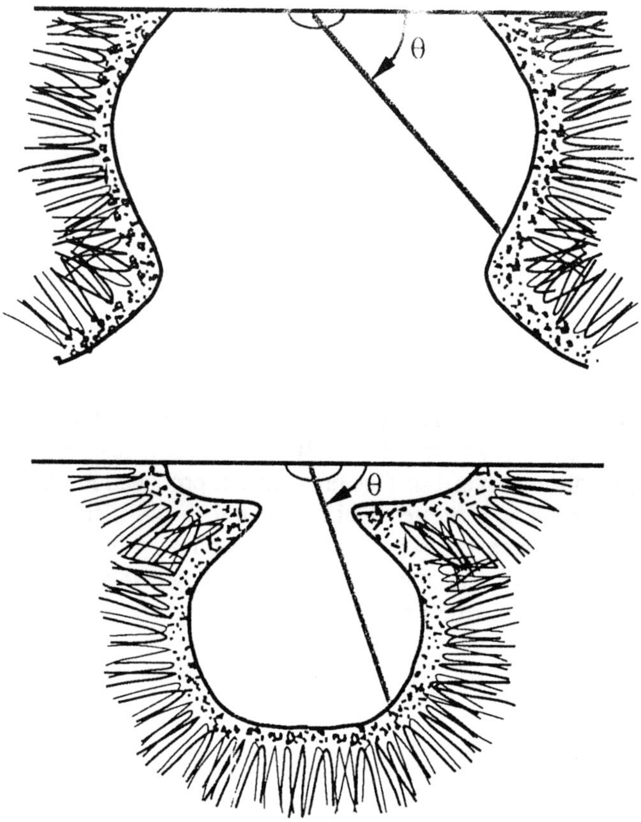


Fig. 9. Schematic illustrating the effect of residual stress on the shape of the fracture mirror boundary. The upper figure shows the shape of a typical fracture surface failed in bending without residual stress. The lower figure shows the shape of the fracture mirror boundary for a glass fractured in bending with global residual compressive stress on the surface for the same size crack as in the upper figure.

stress (i.e., as-indented specimens), factors close to those theoretically predicted, viz. 1.78 for equilibrium growth (oil environment) are found. This finding suggests that grinding flaws are also subject to residual-stress effects and, in some sense, may be regarded as a special case of deformation-induced indentation cracking.[54]

**Table II**

**$r_j/c$ for Indentation Cracks in Soda-Lime-Silica Glass**

| Boundary | Annealed/Oil* | As-Indented Oil* (with residual stress) | As-Indented** / Annealed |
|---|---|---|---|
| $r_1$ | 6 | 11 | 1.80 |
| $r_2$ | 7 | 13 | 1.76 |
| $r_3$ | 9 | 17 | 1.78 |

* A factor of 1.2 was multiplied to the numbers in Reference 53, as suggested, to account for the ellipticity of the crack.

** Ratios calculated before rounding values in columns.

It is apparent that residual contact stresses are not a major factor in the mechanics of mirror formation, but can exert a strong influence on the instability conditions which take the flaw to failure. Thus, the use of mirrorconstants in the evaluation of initial flaw sizes requires that attention be given to specimen history.

*Stable Crack Growth*

A schematic drawing of a typical fracture surface on which slow crack growth has occurred illustrates the important features observed on the fracture surface during a stress corrosion process, e.g., during delayed failure (Figure 5). The initial flaw (solid line, Figure 5) produced by machining, handling, or indentations is observed on the fracture surface. The line of demarcation is usually easily seen because the stress state during testing is almost never the same as that which produced the flaw. The initial flaw grows subcritically under load to a size at which the stress intensity reaches a critical value, $K_{Ic}$. At this point, catastrophic failure occurs - the processes leading to mist, hackle and crack branching follow. These processes occur at specific values of $K_1$, in concurrence with Equation (8), independent of the amount of subcritical crack growth. Since the transition from stable to unstable crack growth in air is continuous and smooth, there is no observable critical flaw boundary. The critical flaw size, $c^*$, can be calculated, however, from the fracture mechanics expression[12,33] [Equation (2) with $c = c^*$]:

$$c^* = \frac{\sigma^2 K_{Ic}^2}{1.2\pi\sigma_a^{*2}} \tag{2b}$$

where $\sigma_a^*$ is the applied stress (that is, $\sigma_f = \sigma_a^*$) and $\phi$ is a geometric constant.

The fracture surface for glass that failed after approximately 200 h under a stress of 60 MPa is shown in Figure 10. The mirror ($r_1$)-to-flaw size ratio of 80:1 compares with 11:1* observed for glasses tested rapidly to failure.[3,28] This ratio difference between delayed and rapid failure can be explained by the subcritical crack growth of the initial flaw to the critical flaw size. A summary of data[55] obtained for soda-lime-silica glass bars shows that the mirror constants are unaffected by the testing rate; however, the ratios of the mirror radii to the initial flaw size are variable and are a function of the time to failure; that is, $f(t_f)$. Because the mirror-to-critical-flaw-size ratios do remain the same as in rapid failure, fracture mechanics expressions based on subcritical crack growth can be used to determine the time under load.

A good representation of slow crack growth in brittle materials is given by:[56]

$$V = V_0 \frac{K_1^n}{K_0^n} = AK_1^n \tag{13}$$

where V is the velocity of the crack, $V_0$, A, and n are experimentally determined crack growth parameters, and $K_0$ is a normalizing factor usually taken as 1 MPa/m$^{1/2}$. Since V is dc/dt, the failure time, $t_f$, can be determined through integration

* Notice that this value of 11:1 implies that there is local residual tensile stress present around the crack in machined surfaces before slow crack growth.

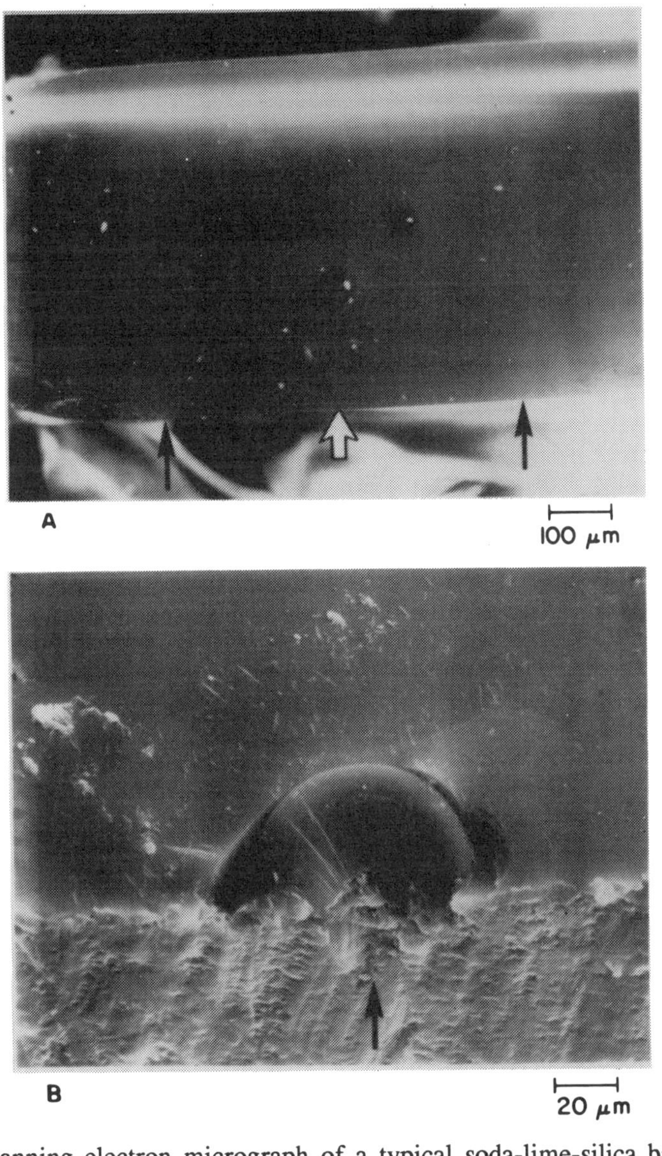


Fig. 10. Scanning electron micrograph of a typical soda-lime-silica bar fractured after time under load. (A) Fracture origin (white arrows) and the beginnings of the fracture mirror boundaries (black arrows) are shown. (B) Pre-existing flaw at higher magnification. The ratio of inner mirror radius to initial flaw size ($r_j/\sqrt{a_j b_j}$) is 80, indicating a relatively long time under load.

$$t_f = \int_{c_i}^{c^*} \frac{dc}{V} \tag{14}$$

where $c_i$ is the size of the initial flaw and $c^*$ is the size of the critical flaw. The result of this operation, using Equations (14) and (1), is given in Equation (15).[54]

$$t_f = \frac{2}{A(n-2)} \left[ \frac{(1.2\pi)^{1/2}}{\phi} \sigma_a c_i^{1/2} \right]^{-n} c_i \tag{15}$$

This failure time can be expressed as a function of the mirror dimension by substituting Equation (7) into Equation (15). The following expression is obtained:

$$t_f = \frac{2}{A(n-2)} \left[ \frac{(1.2\pi)^{1/2}}{\phi} M_j \left( \frac{c_i}{r_j} \right)^{1/2} \right]^{-n} c_i \tag{16}$$

Equations (15) and (16) relate the time to failure under constant stress to the initial flaw size and geometry [Equation (15)] and the fracture mirror size [Equation (16)], respectively. The data in Reference 50 are plotted in Figure 11 based on the form suggested by Equation (15). The predicted curves were obtained by inserting values of n and A into Equation (15).[54] The two curves shown in Figure 11 correspond to geometric factors of $\phi$ for a semicircular flaw (1.57) and for a flaw with an aspect ratio of 0.8 (1.42). Although the theoretical curves pass through the approximate center of the data, there is extensive scatter. A least-squares fit to the data had a correlation coefficient of 0.55.

Equation (15) is based on the assumption that the initial flaw lies in a plane perpendicular to the maximum applied stress, $\sigma_a$, which has been shown to be not necessarily the case. Most flaws grow from an initial elliptical shape to a semicircular mirror; the shape change with time is not accounted for in the derivation of Equation (15).

Previous studies suggest that the fracture mirror is a measure of the stress at fracture even when the flaw is inclined at an angle to the plane of fracture.[3,47] The time-to-failure data were, therefore, plotted according to Equation (16) (Figure 12) using all the data for which mirror-mist boundaries could be determined (22 points). The theoretical curves were obtained by inserting the values of fracture mirror constants for soda-lime glass in Equation (16) along with the values of n and A. The dashed line is the least-square fit to the data and has a correlation coefficient of 0.72. The critical flaw sizes calculated from Equation (1) are also shown. For every $t_f/c_i$ point, there is a corresponding value of $t_f/c_{crit}$* for the same mirror radius. The best-fit line to the $t_f/c_{crit}$ data point correspond to an $r_1/c_{crit}$ ratio of $11.6 \pm 1.2$,[4] which is in good agreement with that obtained with rapid failure.[3]

The significantly reduced scatter in the lifetime predictions obtained from the fracture-mirror rather than the flaw size data is due to the fact that, at the point at which catastrophic failure occurs, the crack is wholly contained within the fracture plane. The geometry at the initial flaw can be quite complicated and vary from specimen to specimen, whereas the geometry at the mirror boundary is usually the same for each specimen. Thus, the assumption that $\phi$ is 1.57 (for a semicircular shape) is probably true for Equation (16) but not for Equation (15). In addition, because of the shape of the flaw and its angle to the applied stress, $\sigma_a$, the stress at which crack propagation begins probably varies from the applied stress, depending on the angle of the flaw to $\sigma_a$. The mirror, on the other hand, is wholly contained within the fracture plane and the stress at formation is the applied stress. The formation of the mirror evidently compensates for any irregularities at the initial flaw.

The $K_1$ at which catastrophic failure occurs does not depend on the previous crack growth history, as evidenced by the agreement in values of $r_j/c^*$ for large variations in $t_f$. This evidence means that knowledge of $r_j$ gives information regarding $c^*$ that, combined with the observed $c_i$, allows the extent of the subcritical crack growth to be determined. Obviously, this analysis can only be performed after the specimen is broken and, thus, cannot be used for time-to-failure predictions. However, for in-service failures, when no other information is given, measuring the mirror and the initial flaw size provides a reasonable way of estimating the total time that the structure was subjected to load. This estimate is only good to a couple of orders of magnitude in time, but it is as good as any predictive capability that exists now.[56] An implied assumption is that the stress on the flaw was constant for this period.

* $C_{crit} \equiv c^*$.

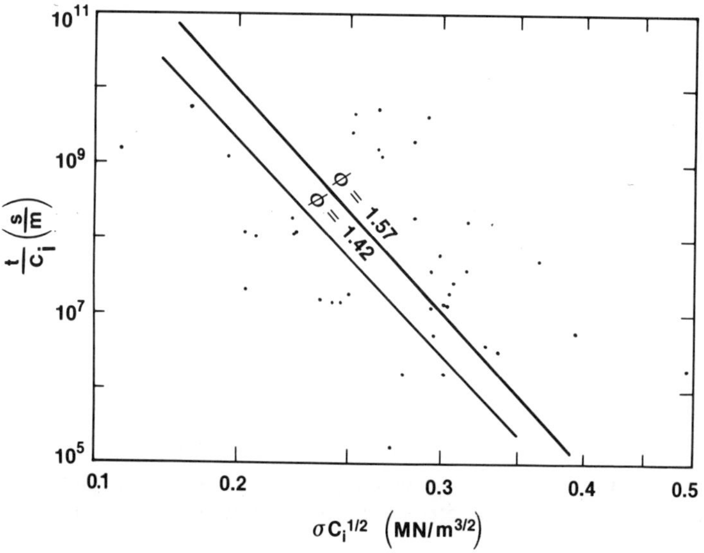


Fig. 11. Time-to-failure (i.e., $t/c_i$) as a function of applied stress, $\sigma$, and initial flaw size, $C_i = \sqrt{a_j b_j}$. The solid lines represent the calculated time-to-failure [Equation (15)] for a semicircular flaw ($\phi = 1.57$) and for a flaw with $a/b = 0.8$ ($\phi = 1.42$).

The foregoing paragraphs described slow crack growth in air under constant stress. When water or other chemically active liquids are present, we see additional markings on the fracture surface. Several of these features are shown in Figure 13. We see the arrest mark (controlled flaw) which is the trace of the arrested crack. Upon introduction of new stress this crack starts again at a slightly different angle. The cavitation scarp (Figure 13) is a demarcation left on the fracture surface and is a result of cavitation in water as the crack rapidly grows from a low velocity ($10^{-4}$ m/s) to a relatively high velocity ($10^{-1}$ m/s). This process has been described in detail in a paper by Michalske.[57] Before the point of transition between slow and fast fracture (at the scarp), we see evidence of transition hackle.[57] Wallner lines,[58] which are the loci in intersections of the fracture front with stress wave reflections off the surface, provide a means of identifying the direction of fracture.

Many fracture features can be used to determine the direction of propagation. For example, in Figure 13 we know from the direction of crack

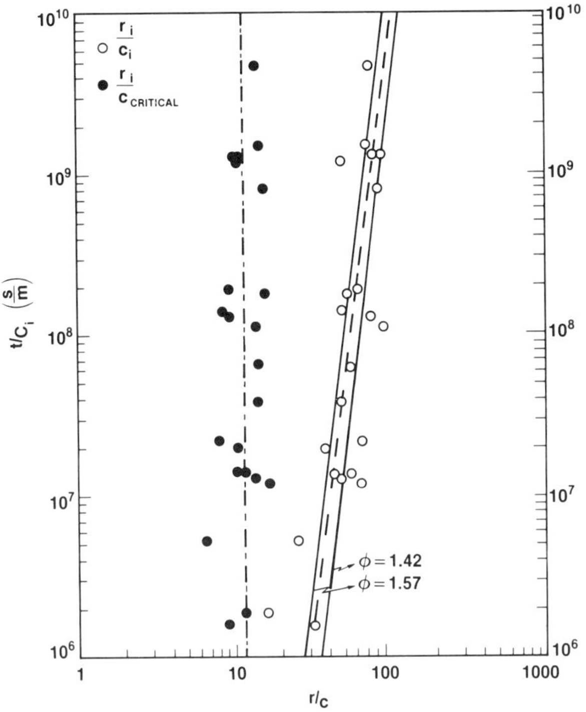


Fig. 12. Ratio of inner mirror radius to flaw size as a function of time under load (i.e., $t/c_i$). $C_i$ is $\sqrt{a_j b_j}$ and $C_{critical}$ is determined from Equation (3). The solid lines represent Equation (16) for a semicircular flaw ($\phi$ = 1.57) and for a flaw with a/b = 0.8 ($\phi$ = 1.42). The dashed line is a least-squares fit to the data with a correlation coefficient of 0.73. The two-dash line represents the $r_1/c$ value of 11 from Table II and is drawn parallel to the ordinate as an aid to the eye.

propagation (arrows) that the crack did not proceed symmetrically but rather started at one end of the crack and curves around to the other. Hence, the radius of the scarp at the point of initiation of rapid fracture marks the boundary

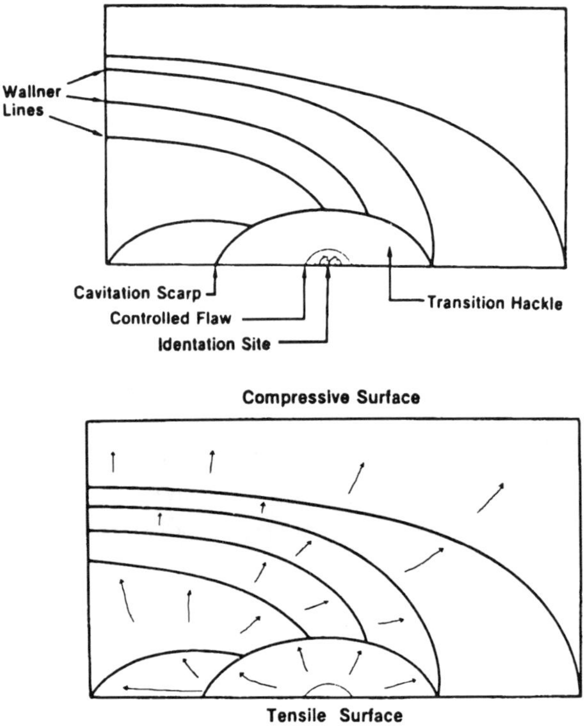


Fig. 13. Schematic and fracture surfce of a glass failed in water. The demarcations are indicated on the schematic and are related to the fracture surface. The direction of crack propagation is shown by the arrows. Rapid propagation began on one side (where c is indicated) of the crack at the surface, and not symmetrically.

of the "critical crack".* This helps us in analyzing cracks because now we can identify the radius of the crack from the edge of the initial crack to the point of rapid propagation on the boundary of the scarp. In Table III, we calculate the critical stress intensity factor, $K_{Ic}$, for zinc silicate and fluoride glasses from the scarp boundary on the fracture surfaces of bars failed in delayed failure in water. It was assumed that the critical crack size was from the edge of the initial crack to the scarp boundary. The agreement between these calculations and the measured $K_{Ic}$ in air for fast fracture ($10^{-3}$ - $10^{-1}$ m/s) is good. The

* Critical crack size, c, is defined as the radius of the crack from the initiation of propagation to the point of transition from slow ($\sim 10^{4}$ m/s) to fast ($\sim 10^{1}$ m/s) fracture (i.e., the scarp boundary).

**Table III**

**Fracture Toughness, $K_{Ic}$, (MPa $m^{1/2}$)**
**Calculated from Scarp Boundary**

| | Calculated | | Measured (Technique) |
|---|---|---|---|
| | Initial Crack* (Eq. 1) | Scarp Boundary (Eq. 3) | |
| Zinc Silicate Glass | 0.50 ± 0.07 | 0.90 ± 0.10 | 0.90 ± 0.10 (DCB/Short Bar) |
| Fluoride Glass | 0.15 ± 0.05 | 0.25 ± 0.03 | 0.23 ± 0.07 (Indent) |

* These values are the initial stress intensities and may be low because they do not include any effect of local residual stress.

agreement in $K_{Ic}$ for delayed failure in water and for fast fracture implies that a point on the scarp boundary measured from the initial crack can be considered the "critical" crack size (for fast fracture).

In failure analysis, the presence of a scarp would indicate the presence of water. By collectively looking at the arrest mark, scarp, Wallner lines and surrounding topography, we can identify the critical crack size with some assurance. The relationship of the initial flaw dimensions to those of the mirror, mist, hackle and crack branching boundaries also permits an estimate of the time under load.[2,57]

## SUMMARY AND CONCLUSIONS

Fracture in glass occurs when preexisting cracks propagate due to excessive applied tensile stresses. These cracks can be induced by mechanical means such as from grinding or polishing damage, by processing, or by intrinsic defects, i.e., natural bonding imperfections in the glass structure. Since for most cases these cracks can be adequately represented by a semicircular flaw, the fracture process can be quantitatively analyzed using fracture mechanics and fracture surface analysis. Most evidence shows that the propagation of the crack is determined by varying levels of stress intensity or energy.[41,59] Because of these relationships (Equation 8), much information is contained on the fracture surface. We can determine the fracture origin, the fracture stress and the existence or not of external influences like residual stresses and stress corrosion processes. The relationship between the fracture initiating crack and the surrounding topography implies that there is a fundamental relationship between the beginning of crack propagation and crack branching. The fracture mirror-to-crack size ratio provides information concerning residual stress and crack growth history.[33,53]

The fracture mirror ($r_j$) to (initial) flaw size ($c_i$) ratio (e.g., $r_1/c_i$) for residual stress-free cracks in soda-lime-silica glass is ~6:1 and for cracks containing local residual tensile stress is ~11:1. If slow crack growth occurs, then $r_1/c_i$ is a function of the time under load and generally $r_j/c_i >> 11$ and can reach values of 100 or more.

Even though much information has been determined since the first observations of fracture mirror boundaries, there is much left to be determined and understood. With the advent of optical and infrared fibers, we now have the means for studying and possibly observing fracture initiating defects at very high stresses, i.e., the fracture stress $\sigma_f$ can approach the theoretical strength, $\sigma_{th}$ ($\sigma_f > 1/2\ \sigma_{th}$). Another technique for studying crack origination is by using subthreshold indentation cracks on glass as suggested by Dabbs and Lawn.[60] Combining fracture surface analysis with subthreshold cracking could provide important information as to the fracture process. We can preserve the many sections that result from high strength breaks by testing silicone coated fibers. This coating tends to hold the many pieces together and thus provides us with observations on the fracture surfaces of very high strength fractures.

While observations of fracture surfaces have provided considerable insight into the fracture process and have been used extensively in fracture forensics, much more information can be obtained by diligent pursuit of new fracture features and relationships.[61] It is extremely useful to study the size, shape and stress intensity of fracture initiating cracks and to relate them to the surrounding topography for the special loading conditions mentioned in this paper. Fortunately, several excellent reviews of the principles illustrated in this chapter exist in the literature.[14,15,33,45,59,61]

# APPENDIX

## Sample Calculations Using The Principles of Fracture Surface Analysis

1. Location of the Failure Origin

The fracture origin is found by following the mist and hackle markings which radiate out from the source of failure. The origin is approximately centrally located between these markings. The approximately semi-elliptical crack is not necessarily symmetric between the mirror-mist boundaries on either side due to several factors: (1) sometimes the origin of failure can occur from one side of the crack, usually at the surface; (2) the formation of the mist and hackle markings can be influenced by boundary conditions such as other (smaller) cracks present on the surface and such as the end of the specimen. In these cases, Wallner lines can be used to locate the fracture origin. In the case of slow crack growth, this can be seen in Figure 13. [Note: Randall[26] showed that even irregularly shaped cracks can be approximated by semi-elliptical cracks; using the formulation of Equation 3 further reduces the complication by describing an equivalent semicircular crack.]

2. Measurement of the Size of the Fracture Initiating Crack

The measurement of the equivalent semicircular crack size is accomplished by measuring the depth and width of an approximately semicircular crack. For example, the crack in Figure 3 has a depth 50 $\mu$m and a width of 100 $\mu$m. In this case the crack happens to be semicircular. However, if it were not, we can obtain the equivalent semicircular crack by multiplying the depth by the half-width and taking the square root of the result, i.e., $[50\ \mu m\ 50\ \mu m]^{1/2} = 50\ \mu m$.

The strength of the glass in this case can be obtained using Equation 8 with $j = 0$ which corresponds to $r_0 = c$, where c is the equivalent semicircular crack size.

$$\sigma_f = \frac{K_{IC}}{Y_0\ c^{1/2}}$$

$Y_0$ is determined to be either 1.24 for a stress-free crack or 1.65 if there is local residual stress such as with indentation induced stresses. In this case, the crack was introduced using indentations and so $Y_0 = 1.65$. This particular sample is soda-lime-silica glass and the critical stress intensity factor, $K_{Ic}$, is 0.72 MPa $m^{1/2}$ obtained from Table I. The stress is calculated using Equation 3 or 8 to be 61.7 MPa.

3. Delayed Failure

As an example of using the principles of fracture surface analysis, we can calculate the stress at failure and critical crack size for the glass in Figure 10 which has experienced delayed failure. The equivalent semicircular crack size can be calculated by measuring the depth, a, and the width, 2b, in Figure 10b. These values are found to be a = 40 $\mu$m and b = 44 $\mu$m. Thus, c = 42 $\mu$m. The mirror-mist boundary, $r_1$, is measured to be 3360 $\mu$m. This last measurement provides the stress at failure from Equation 7 using the data in Table I ($M_1$ = 1.8 $MPa^{1/2}$), i.e., $\sigma$ = 31.1 MPa (4.5 ksi):

$$\sigma_f = \frac{M_1}{r_1^{1/2}} = \frac{1.8\ MPa\ m^{1/2}}{(3360\ x\ 10^{-6}\ m)^{1/2}} = 31.1\ MPa$$

Since the critical stress intensity factor for soda-lime-silica glass is $K_{Ic}$ = 0.72 $MPa^{1/2}$ (Table I), then we can calculate the expected stress at failure from Equation 3 using the crack size measured.

$$\sigma_f = \frac{K_{Ic}}{1.24\ \sqrt{c}} = \frac{0.72\ MPa\ m^{1/2}}{1.24\ (42\ x\ 10^{-6}\ m)^{1/2}} = 89.6\ MPa$$

This value is much larger than that obtained from the "mirror" measurement. Our conclusion is that the measured crack size is the initial crack and not the final one. The final (critical) crack size is obtained using Equation 3 with the values for $K_{Ic}$ and the actual stress at failure of 31.1 MPa. Thus, $c_{cr}$ = 346 $\mu$m.

## REFERENCES


1. R.W. Rice, J.J. Mecholsky, S.W. Freiman, and S.M. Morey, "Failure Causing Defects in Ceramics: What NDE Should Find," NRL Memorandum Report 4075, Naval Research Laboratory, Washington, DC, 20375 (October 1979).
2. J.J. Mecholsky, S.W. Freiman, and S.M. Morey, "Fracture Surface Analysis of Optical Fibers," in Fiber Optics, Advances in Research and Development, B. Bendow and S.S. Mitra, eds., Plenum Press, NY (1979).
3. J.J. Mecholsky, S.W. Freiman, and R.W. Rice, "Effect of Grinding on Flaw Geometry and Fracture of Glass," J. Am. Ceramic Soc. 60, 3-4 (1977).
4. J.J. Mecholsky, S.W. Freiman, and R.W. Rice, "Effect of Surface Finish on Strength and Fracture of Glass," Proc. XIth International Congress on Glass, Prague, Czech. (1977).
5. B.R. Lawn and T.R. Wilshaw, Fracture of Brittle Solids, Cambridge University Press, London, England (1975).
6. D.B. Marshall and B.R. Lawn, "Surface Flaws in Glass," in Strength of Glass, C.R. Kurkjian, ed., Plenum Press, NY (1985).
7. J.T. Hagan and M.V. Swain, "The Origin of Median and Lateral Cracks at Plastic Indents in Brittle Materials," J. Physics D: Applied Phys. 11:2091 (1978).
8. D.B. Marshall, A.G. Evans, B.T. Khuri-Yakub, J.W. Tien, and G.S. Kinu, "The Nature of Machining Damage in Brittle Materials," Proc. Roy. Soc. Lond. A385:461 (1983).
9. J.J. Mecholsky, unpublished data.
10. B.R. Lawn and D.B. Marshall, "Indentation Fractography," in Fractography of Glasses, R.C. Bradt and R.E. Tressler, eds., Plenum Press (1992).
11. M.V. Swain, Proc. Roy. Soc. Lond. A366, p. 575 (1979).
12. D. Broek, Elementary Engineering Fracture Mechanics, Sijthoff and Noordhoff, The Netherlands (1978).

13. J.J. Mecholsky, "Fracture Analysis of Optical Fibers," Proc. Fiber Optics in the Nuclear Environment, Defense Nuclear Agency, Adelphi, MD (1980).
14. V.D. Frechette, "Failure Analysis of Brittle Materials," in Advances in Ceramics, 28, Am. Ceramic Soc., 1990.
15. R.W. Rice, "Ceramic Fracture Features, Observations, Mechanisms and Uses," in Fractography of Ceramic and Metal Failures, ASTM STP 827, J.J. Mecholsky, Jr., and S.R. Powell, Jr., eds., American Society for Testing and Materials, Philadelphia, PA (1984).
16. P.W. McMillan, Glass Ceramics, Academic Press, Inc., New York, NY, 1964.
17. G.H. Beall, "Structure Properties and Applications of Glass Ceramics," in Adv. in Nucl. and Cryst. in Glasses, Am. Ceram. Soc., L.L. Hench and S.W. Freiman, eds., 251 (1971).
18. E.R. Fuller, Jr., and R.M. Thomson, "Lattice Theories of Fracture," in Fracture Mechanics of Ceramics, 4, Bradt, Hasselman and Lange, eds., Plenum Press, NY (1978).
19. F.V.D. Marcello, A.C. Hart, Jr., J.C. Williams, and C.R. Kurkjian, "High Strength Furnace-Drawn Optical Fibers," in Fiber Optics, pp. 125-135, op. cit. (1979).
20. T.A. Michalske and S.W. Freiman, "A Molecular Mechanism for Stress Corrosion in Vitreous Silica," J. Am. Ceramic Soc. 66[4] pp. 284-288 (1983).
21. T.A. Michalske and B.C. Bunker, "A Fracture Model Based on Strained Silicate Structures," accepted for publication by J. Appl. Physics (1985).
22. G.R. Irwin, "Relation of Stresses Near a Crack to the Crack Extension Force," 9th Intl. Cong. Appl. Mech., Brussels, 1957.
23. A.A. Griffith, "The Phenomena of Rupture and Flow in Solids," Phil. Trans. Roy. Soc., A221:163 (1921).
24. E. Orowan, "Energy Criteria of Fracture," Weld, J. Res. Suppl. 20, 1573 (1955).
25. C.E. Inglis, "Stresses in a Plate Due to the Presence of Cracks and Sharp Corners," Trans. Inst. Naval Architects, 55, pp. 219-241 (1913).
26. P.N. Randall, "Plain Strain Crack Toughness Testing of High Strength Metallic Materials," in ASTM STP 410, pp. 88-126 (1967).
27. A. de S. Jayatilaka, Fracture of Engineering Brittle Materials, Applied Science Publishers, London (1979).
28. J.J. Mecholsky, R.W. Rice, and S.W. Freiman, "Prediction of Fracture Energy and Flaw Size in Glasses from Mirror Size Measurements," J. Amer. Ceramic Soc. 57, 10 (1974).
29. J.J. Mecholsky, unpublished data.

30. N. Shinkai and H. Ishikawa, "Crack Branching of Binary PbO-$B_2O_3$ Glasses," J. NonCryst. Solid, 52(1-3) pp. 385-394 (1982).
31. J.J. Mecholsky, "Fracture Mechanics Analysis of Glass Ceramics," Adv. in Ceramics 4, Nucl. and Cryst. in Glasses, Am. Ceram. Soc., 261 (1982).
32. J.T. Healey and J.J. Mecholsky, "Application of Backscattered Electron Images to the Study of Fracture in Glass Ceramics," Scanning 4, 62-68 (1981).
33. J.J. Mecholsky, S.W. Freiman, and R.W. Rice, "Fractographic Analysis of Ceramics," in Fractography in Failure Analysis, ASTM STP 645, B.M. Strauss and W.H. Cullen, Jr., eds., American Society for Testing and Materials, pp. 363-379 (1978).
34. N. Shinkai, "The Fracture and Fractography of Flat Glasses," in Fractography of Glasses, op. cit.
35. M.J. Ball, D.J. Landini, and R.C. Bradt, "Fracture Mist Region in a Soda-Lime-Silica Float Glass," ASTM STP 827, op. cit. pp. 110-120.
36. A.I.A. Abdel-Latif, R.C. Bradt, and R.E. Tressler, "Multiple-Mist Regions On Glass Fracture Surfaces," in Fractography and Material Science, ASTM STP 733, L.N. Gilbertson and R.D. Zipp, eds., ASTM 1981, pp. 259-27.
37. A.I.A. Abdel-Latif, R.C. Bradt, and R.E. Tressler, "Dynamics of Fracture Mirror Boundary Formation in Glass," Intl. J. of Fracture 13, 3, pp. 349-359 (1977).
38. K. Ravi-Chandar and W.G. Knauss, "An Experimental Investigation into Dynamic Fracture: II. Microstructural Aspects," Intl. J. of Fracture 26, pp. 65-80, (1984).
39. H.P. Rossmanith, "Crack Branching in Brittle Materials: Part I. Analytical Aspects," University of Maryland Research Report, University of Maryland, College Park, MD. Photomechanics Lab (1980).
40. J.J. Mecholsky, R.W. Rice, and S.W. Freiman, "Fracture Surface Analysis of Ceramics," J. Mat. Sci. 11(1976).
41. H.P. Kirchner and J.W. Kirchner, "Fracture Mechanics of Fracture Mirrors," J. Am. Ceramic Soc. 62, 3-4, pp. 198-202 (1979).
42. D.K. Shetty, G.K. Bansal, A.R. Rosenfield, and W.H. Duckworth, "Criterion for Fracture Mirror Boundary Formation in Ceramics," J. Am. Ceramic Soc. 63, 1, pp. 106-108 (1980).
43. H.P. Kirchner and J.C. Conway, "Comparison of the Stress Intensity and Johnson and Holoway Criteria for Crack Branching in Rectangular Bars," J. Am. Ceram. Soc. 70 [8] 565-69 (1987)
44. D. Bahat, G. Leonard, and A. Rabinovitch, Intl. J. of Fracture 18[1], pp. 29-38 (1982).

45. J.J. Mecholsky and S.W. Freiman, "Determination of Fracture Mechanics Parameters through Fractographic Analysis of Ceramics," ASTM STP 678, S.W. Freiman, ed., American Society for Testing and Materials, pp. 136-150 (1979).
46. S.W. Freiman, J.J. Mecholsky, and P.F. Becher, "Fractography: A Quantitative Measure of the Fracture Process," Ceramic Transactions V. 17 in Fractography of Glasses and Ceramics, Frechette and Varner, eds., The American Ceramic Soc. (1991).
47. S.W. Freiman, A.C. Gonzalez, and J.J. Mecholsky, "Mixed Mode Fracture in Soda-Lime-Glass," J. Am. Ceramic Soc. 62, pp. 3-4 (1979).
48. E.B. Shand, J. Am. Ceramic Soc. v. 37, 12, pp. 559-572 (1954).
49. J.J. Mecholsky and M.G. Drexhage, "Comparison of Optical and Fractographic Measurement of Residual Stress in Compressively Strengthened Glass Rods," J. Am. Ceramic Soc. 63, 7-8 (1980).
50. H.P. Kirchner and R.M. Gruver, "Fracture Mechanics of Ceramics," (1) op. cit. (1973).
51. E.G. Shand, "Breaking Stresses of Glass Determined from Fracture Surfaces," The Glass Industry (April, 1967).
52. J.C. Conway, Jr. and J.J. Mecholsky, Jr., "Use of Crack Branching Data for Measuring Near-Surface Residual Stresses in Tempered Glass," J. Am. Ceramic Soc. 72[9] 1584-87 (1989).
53. D.B. Marshall, B.R. Lawn, and J.J. Mecholsky, "Effect of Residual Contact Stresses on Mirror/Flaw Size Relations," J. Am. Ceramic Soc. 63, pp. 7-8 (1980).
54. M.V. Swain, pp. 257-72 in Fracture Mechanics of Ceramics, 3, op. cit. (1978).
55. J.J. Mecholsky, A.C. Gonzalez, and S.W. Freiman, "Fractographic Analysis of Delayed Failure in Soda Lime Glass," J. Am. Ceramic Soc. 62, pp. 11-12 (1979).
56. S.M. Weiderhorn and J.E. Ritter, in Fracture Mechanics Applied to Brittle Materials, ASTM STP 678, op. cit.
57. T.A. Michalske, "Fractography of Slow Fracture," ASTM STP 827, op. cit.
58. H. Wallner, Zeitschrift for Physik, 114, pp. 368-378, 1939; Ceramic Abstracts 19, 6, p. 137, 1940.
59. R.W. Rice, "Perspective on Fractography" in Advances in Ceramics, 22, 3-56, The American Ceramic Soc. 1988.
60. T.P. Dabbs and B.R. Lawn, "Fatigue of High Strength Soda-Lime Glass: A Constant Stressing Rate Study Using Subthreshold Flaws," Phys. Chem. Glasses, 23, [93] (1982).
61. J.J. Mecholsky, "Quantitative Fractography: An Assessment," in Fractography of Glasses and Ceramics, Ceramic Trans. 17, op. cit.

# INDEX